PLANT ANALYSIS:
Comprehensive Methods and Protocols

PLANT ANALYSIS:
Comprehensive Methods and Protocols

B.K. Garg

Published by:

Scientific Publishers (India)
5 A, New Pali Road, P.O. Box 91
Jodhpur 342 001 (India)

E-mail: info@scientificpub.com
Website: www.scientificpub.com

Branch Office
Scientific Publishers (India)
4806/24, Ansari Road, Daryaganj
New Delhi - 110 002 (India)

Print, 2023

ISBN: 978-81-7233-751-3

Typeset by Rajesh Ojha
Printed in India

PREFACE

At the outset I wish to express my thanks to Shri Pawan Kumar M/s Scientific Publishers (India), Jodhpur for his inspiration and encouragement to write this book on laboratory techniques and analytical methods for plant analysis. His constant cooperation and generous help in arranging typing of this voluminous work in a short span of time made it possible to bring out the present publication in a record time of three months.

Analytical methods and protocols for diverse estimations in plants are a vital part of successful experimentation and execution of research projects in the field of Plant Physiology, Biochemistry, Molecular Biology and related plant sciences in Agriculture. There are numerous methods and techniques for analysis of plant nutrients, leaf metabolites, enzyme activities and other physiological and biochemical measurements. However, these are available in widely scattered sources like different journals, books, manuals and other forms. It is rather difficult for a student and researcher to lay his hands on all methods and techniques which demands lot of time and effort. The purpose of this book has been to compile all possible plant measurements at one place in a systematic manner for the optimum use of research workers and students.

The present book "Plant Analysis: Comprehensive Methods and Protocols" consists of 23 chapters covering different aspects of plant analytical techniques in a simple and easy manner. In chapter one guidelines for experimentation and plant analysis have been outlined. Chapter two deals with various methods of digestion of plant samples (including dry and wet digestion) for nutrient analysis. Chapter three discusses methods for determination of all major nutrient elements (N, P, K, Ca, Mg, S, Na and Micronutrients). Determination of leaf pigments (Chlorophylls and carotenoids) and chlorophyll fluorescence technique have been described in chapter four.

Chapter five deals with the estimation of organic molecules in plants, such as amino acids, proteins, nucleic acids, and other metabolites related to fatty acid metabolism. In chapter six, the concept of pH and acids and bases in plants has been introduced alongwith composition of all important buffers. Protocols for assay of various enzymes of nitrogen metabolism, ammonia assimilation, reactive oxygen species (Chapter seven), photosynthetic enzymes and metabolites (Chapter nine) and other important plant enzymes (Chapter eight) have been elucidated alongwith relevant reaction equations. Similarly estimations of enzymes of infected plants and parasites (Chapter ten) and disease resistance related estimators have been described in a simplified manner in Chapter eleven.

The book also caters to the need of researchers in the field of plant water relations and stress physiology as measurement of growth indices and leaf area (Chapter twelve), stress indicators indices (Chapter fourteen), plant water status (Chapter fifteen) and aerenchyma (Chapter sixteen) have been thoroughly described. Measurement of photosynthetic and respiration rates in the field (using IRGA), transpiration (poromotus) and canopy temperature (infrared tele thermometer) have been vividly explained in chapter thirteen, which will be very useful for plant physiologists and Biochemists in their research work. Special topics like Electrophoresis (Chapter seventeen), Molecular Biology Techniques (Chapter eighteen), Determination of polyamines and phenolics by TLC/HPLC (Chapter nineteen) and Histochemistry of enzymes and metabolites (Chapter twenty) have also been dealt comprehensively. The process of symbiotic nitrogen fixation in legumes is very important for acquistation of atmospheric nitrogen in such plants. Various methods of N_2-fixation measurement and related estimations have been covered in chapter twenty two. Likewise, all important methods of root studies have been systematically presented in chapter twenty one. Assay and bioassays for plant hormones which are essential for plant growth and development have been given due place in the book (Chapter twenty three).

Thus, the present book is a complete account of all important techniques and methods for students and researchers in plant sciences and agriculture. I hope this compilation will be of immense use in various research laboratories devoted to Plant Physiology, Biochemistry and Molecular Biology Research. Suggestions and valuable comments are most welcome for further improvement of the present book.

Lastly, I wish to put on records my sincere thanks and regards to Dr. S.P. Vyas (Ex Principal Scientist) and Dr. Uday Burman (Principal Scientist, Plant Physiology) for their kind help and cooperation in compilation of this work. I am extremely grateful to my wife for her patience and constant encouragement during the writing of this manual.

B.K. Garg

CONTENTS

CONTENTS

CHAPTER - 1

GENERAL GUIDELINES FOR EXPERIMENTATION AND PLANT ANALYSIS

INTRODUCTION

To make laboratory work more meaningful it is advisable to follow certain basic guidelines. Before starting an experiment read carefully the protocol, the assigned objective and the hypothesis involved. Then prepare the plan of work before actually starting the experiment. Do not be a "filter paper scientist". Record all primary operations and data directly in your bound notebook. Data recorded in discs of filter paper on loose sheets of paper are likely to get lost. Data should be recorded in tabular form in well defined format. All the data obtained should be recorded in your report no matter how unsatisfactory the data may appear to be. Do not feel perturbed if data, negative to your expectation, are obtained. Negative data are more meaningful and pave the way for exploration of protocol and testing of other components. It also provides opportunity to indicate the loopholes in your experimentation.

Graphical presentation of data should be done in order to quicker understanding of the experimental results. Finally, a concise report should be presented stating the methodology, results obtained and a brief discussion about the findings. At the end of day, your working bench should be as clean as it was in the beginning of the experimental operations.

CONTROL OF ACCURACY

Accuracy may be defined as the degree of conformity to the truth. A good researcher/student needs to obtain very accurate measurements at the working bench. He is also required to present an accurate account of the experiment. Before actual experimental work a precise plan of work is mandatory. In addition, the protocol for experiment should precisely be followed for accurate results. Problems faced during work, arising due to the variations in plant materials, researchers, working conditions, etc., should clearly be stated in the report so that those who read it should have prior information about the difficulties which they can encounter during work.

MEASUREMENTS AND ERRORS

Laboratory work always involves certain measurements which are liable to errors. The potential source of these errors should always be appreciated. Errors may be random, or caused due to carelessness or inaccurate instruments or impure chemicals. Errors can also arise from careless experimental work such as using apparatus wrongly, preparing and measuring solution inaccurately. Even modern instruments can produce strange results if the operational instructions are not followed correctly or the optimum conditions for the instrument is not met. Such errors can be rectified and with good practice can be eliminated.

The limits of instruments or tools being used must be defined in determining the accuracy of the final results. It is unreasonable to mention a value of 5.453 for any measurement if the least measurable count in the instrument is 0.03. These errors are known and can be taken into account.

Finally, there are random errors which are individually unpredictable. A person conducting a number of measurements under identical conditions obtains a slightly different results each time. This random error can be minimised by taking a large number of measurements and calculating the average value. For many purposes especially in biochemical measurements, duplicate estimations are sufficient, provided there is good agreement between them. The degree of agreement between replicate measurements is termed as the precision. Precision does not mean accuracy, since the measurement may be highly precise but inaccurate due to faulty instrument or technique.

An additional factor to be considered when working with living material is the biological variations. For biochemical measurements, any single value cannot be considered as correct but it has a range of so called normal values. For example if an animal or plant material is free from stress and healthy then any biochemical value may fall within the normal range. Say protein or carbohydrate content or blood serum constituent. The extent of normal range depends upon constituent being measured. Human serum chloride has a relatively narrow range of values (95-107 m equivalent/l) while fasting blood glucose varies more widely from 60-100 mg/100 ml.

STANDARDS AND BLANKS

To obtain an accurate value from an estimation, errors must be reduced to a minimum. This can be done by careful working and the use of standard solutions. Standard solutions of the substance to be measured should be included with any test even when a calibrated instrument and standard reagents are used. This provides a useful check on the accuracy of a method since the techniques of handling instrument and precision in observing protocol may vary from person to person .

Ideally, the standard solution should be treated in an identical manner to the test solution under estimation. A standard curve can then be constructed showing the theoretical concentrations and measured values. The standard curve should encompass the range of values to be measured for the test solution. If a standard curve has previously been constructed, only two standards may be included while measuring the test solutions.

Blank solutions should always be included in any measurements. The same volume of distilled water or solvent used replaces the substance to be estimated and the blank is then treated in the same way as the test and standard. Any value obtained for the blank need to be adjusted from the values for the test and standard in the final calculation, since the blank value is due to the reagents used or any other contamination. Blanks in certain cases are used for setting the instrument zero, thus are automatically accounted for in the measurements. Several blanks or controls need to be used when measurements are done on enzymes.

PRACTICAL BOOK OR DIARY

All laboratory exercises and operations must be recorded in a stiff covered note book especially a foolscap size. Loose sheets and papers easily get lost and vitiate the experiments. Writing on loose sheets and filter paper is to be strongly discouraged. Write your name, course and address on the note book so that it is traceable in the event of a loss of the note book.

WRITING OF AN EXPERIMENT

All experiments should have a title and be put on the top of the page alongwith the date. The title should be short and self explanatory so that any one reading the experiment should easily understand the theme of the experiment. There are various ways to present the experimental details but a guideline is being presented here for suggestion (Chapman and Pratt, 1961; Chopra and Kanwar, 1976).

(A) Theory or Principle

The principle of a particular piece of equipment being used for the measurement be outlined briefly. If a chemical reaction is being carried out the reactions involved should be recorded as far as possible.

(B) Materials or Apparatus Used

A brief account of the materials used alongwith reagents and apparatus must be mentioned.

(C) Experimental Methodology

This exactly describes the way you perform the experiment. This should be written in past tense and not in the first person. You can note the protocol from a book or Journal and conduct your experiment, but when writing in your note book the method should be in the past tense and second person. An experiment in a book, for example, may give an instruction as :

"Add 0.2 ml of reagent to 5 ml of test solution, heat at 100°C for 5 minutes. Cool and measure absorbance at 620 mm against reagent blank".

The above instruction may be followed while working but in your practical book it should be written as :

"0.2 ml of reagent was added to 5 ml of test solution and heated at 100°C for 5 minutes. On cooling, the absorbance was measured at 620 nm against reagent blank".

(D) Results

The results should describe what you observed during performing your experiment. Each and every details of the observations must be recorded here. Insufficient information if recorded may create problems later on in the interpretation of the results.

Sometimes you may observe slightly indifferent than the one described in the protocol. This should be recorded honestly, and subsequently the reasons for the variation may be searched. Sometimes unexpected observations during experiment may prove quite useful when repeated in future. The detailed experimental conditions need to

3. Moisture Factor

Weighing of perfectly oven-dried samples is, however cumbersome (involves continuous oven drying and use of desiccator, and is still prone to error) as plant material may absorb moisture during the weighing process, particularly if relative humidity is high in the laboratory.

To get around this difficulty, use of the Moisture Factor is suggested instead. The Moisture Factor for each batch of samples can be calculated, by oven-drying only a few sub-samples from the lot (e.g., 5 from a batch of 100-200 samples).

$$\text{Moisture Factor} = \frac{\textit{Weight of air dry sample}\,(g)}{\textit{Weight of oven dry sample}\,(g)}$$

Thereafter, air dry samples are weighed, considering the Moisture Factor. For example, if Moisture Factor = 1.09, then weight of oven dry and air dry samples will be as follows :

Oven dry weight (g)	Air dry weight (g)
0.25	0.27
0.50	0.55
1.00	1.09
2.00	2.18

The Moisture Factor approach is also used for weighing soil samples, and expressing the analytical results on oven dry soil weight basis.

Quality Control and Standardization Procedures

Quality control is an essential part of good laboratory practice. During routine analyses, errors may gradually appear due to contamination, changes in reagent quality, environmental differences, operator error and instrument calibration or failure. *Maximum reproductibility* and *adequate accuracy of results are the important objectives.* Repeated measurement of an air-dried soil sample should provide consistent results when analyzed over time for most routine chemical procedures. The deviation of an observed value from its absolute "true" value results from either *systematic or random errors.*

Once identified, systematic errors are more easily corrected than those which occur at random. Three precautions are essential for laboratory quality control and should be routinely included among the test samples. These precautions involve the use of *blanks, repeats,* and *internal references,* as elaborated below.

Whenever a new procedure is introduced to the laboratory, its accuracy should be evaluated and compared to the test already in use. Both methods should be compared for a homogeneous test sample using ten-fold replicates, with the standard deviation calculated for each set. This provides a *measure of precision.* Then a known amount of reagent should be added to the homogeneous test sample, the procedures repeated, and the *mean* and *standard deviation* calculated. The agreement between the increase in the values obtained to the known increase in test sample concentration provides a *test of accuracy.* For procedures in which the test material is known to interact with the added reagent, as with phosphorus-sorption soils, this test can be conducted by reagent solutions (Plummer, 1971).

1. Blanks

Blanks are reaction vessels that are subjected to identical procedures as the sample in a given batch which have no added test material. Blanks allow correction for any background contamination introduced from reagents, filter papers or other systemic sources of error. Provided the blank values are consistent, the mean value can be subtracted from the sample value. When blanks yield large values, this suggests excess extraneous contamination; in such cases, the entire batch analysis be repeated.

2. Repeats

At least 1 in 10 samples selected from the test materials and placed at random within the batch should be analyzed in *duplicate.* The choice of 1 in 10 is a suggested compromise between the ideal of analyzing all samples in duplicate, considering the time, effort and expense of doing so. Obviously, the analytical results for given pairs of duplicate repeats should closely resemble one another, in general, repeat values should fall within ±2.5 – 5.0% of their mean, depending on the analysis in question; any greater discrepancy must be investigated. If repeat values are not consistent, the entire batch should be re-analyzed.

3. Internal References

Internal reference samples are necessary for each type of material and analysis practiced within the laboratory. The internal sample should not be the same as the homogeneous material routinely used in the testing new methods and analytical technique. A sample obtained from a large, well-mixed and homogeneous composite bulk sample should be included in each batch analyzed. Variation from the mean as calculated over previous batches may be indicated as an error.

Analytical results for the internal reference may be plotted on a quality control chart to monitor the performance of the analyses over time. Corrective action could be taken if a single value exceeds the ±3 standard deviation limits or if two successive values exceed the ±2 standard deviations. Periodically, the critical limits could be re-assessed by re-calculation of the overall standard deviation of the internal reference sample as more data are accumulated.

4. Standardization of Methods

Results can only be validity compared to one another when these have been obtained using standardized methods. Collaboration between laboratories can be improved by exchanging reference materials and then comparing their results. Such materials are referred as ``External References". An example of such standardization is the exchange network of ISRIC (International Soil Reference and Information Center) in Wageningen, The Netherlands, operating an international soil and plant analytical exchange programs.

REFERENCES

Bell,, R.W. 1997. Diagnosis and purification of boron deficiency for plant production. *Plant and Soil.* 193: 149-168.

Chapman, H.D. and Pratt, P.F. 1961. *Methods of Analysis for Soils, Plants and Water.* Univ. California, Barkley, CA, USA.

Chopra, S.L. and Kanwar, J.S. 1976. *Analytical Agricultural Chemistry.* Kalyani Publishers, New Delhi, 514 p.

Gupta, P.K. 1999. *Soil, Plant, Water and Fertilizer Analysis.* Agro-Botanica Publishers.

Katyal, J.C. and Sharma, B.D. 1980. A new technique of plant analysis to resolve iron chlorosis. *Plant and Soil.* 55: 105-108.

Munson, R.D. and Nelson, W.L. 1990. Principles and Practices in Plant Analysis. pp. 223-248 In: R.L. Weiterman (Ed.). *Soil Testing and Plant Analysis Soil Sci. Soc.* Am. Madison WI, USA.

Plummer, D.T. 1971. *An Introduction to Practical Biochemistry.* Tata McGraw-Hill Publishing Company Ltd., Bombay-New Delhi. 369 p.

CHAPTER - 2

DIGESTION OF PLANT SAMPLES FOR NUTRIENT ANALYSIS

INTRODUCTION

Digestion of plant samples is the pre-requisite step to determine total nutrients present in plants. The precision and interpretation of plant analysis results depends on how accurately the plant samples are digested, after their proper collection and processing (cleaning, drying and grinding) (Kensworthly, 1964; Tandon, 1993; Jones *et al.*, 1991). In plants, the nutrients exist in various organic combinations such as carbohydrates, proteins, fats, etc. and digestion of plant material releases them in mineral forms. The weight of plant sample for digestion varies from 0.2 to 2.0 g. depending on (i) the type of plant species, (ii) stage of plant growth and (iii) the plant parts. However, the usual weight of plant sample taken for digestion for total elemental analysis is 0.5 g. Digestion of plant samples can be done by (a) dry ashing or (b) wet oxidation. Dry ashing is done by igniting the plant material in a muffle furnace at 550-600°C followed by the extraction of ignited material in dilute acid (HCl or HNO_3). Whereas, wet oxidation involves digestion of plant material in a mixture of two or three concentrated oxidizing acids ($HClO_4$, H_2SO_4 and HNO_3) at a specified high temperature. The details of techniques and procedures for dry and wet oxidation of plant samples adopted for soil and plant analysis are available (Chapman and Prat, 1961; Piper, 1966; Jackson, 1973; Chopra and Kanwar, 1976; Greweling, 1976; Tandon, 1993; Gupta, 1999). However, in this chapter only main methods of plant digestion for analysis of macro-, micronutrients and toxic elements are discussed.

METHODS OF DIGESTION

There are two methods for digestion of plant samples (i) dry ashing and (ii) wet oxidation. Each of these methods have merits and demerits. The choice of method depends on the availability of equipments, reagents and number of samples to be analyzed. Earlier, dry ashing was used commonly but presently it is restricted to the laboratories which are not so well equipped. Wet digestion is more common method in well equipped modern laboratories, with provision of exhausting the acid fumes (fume hoods) produced during the digestion.

1. DRY ASHING OF PLANT SAMPLES

The dry ashing technique is simple and free from the problem of acid fumes. It is used for the analysis of K, Na, Ca, Mg, Cu, Zn, Fe, Mn, Mo and B (Issac and Johnson, 1985). It is further sub-divided as (a) simple dry ashing and (b) dry ashing with basic compound such as magnesium nitrate. The nutrient elements such as Ca, Mg, K, Na, and most of the trace elements, which do not volatilize at high ignition temperature (around 550°C) can be digested by simple dry ashing. Whereas, others like N, P and S, which volatilize at high ignition temperature should not be digested by dry ashing and they should be wet digested. However for S, dry ashing can be done with magnesium nitrate, to avoid the volatilization losses (Greweling, 1976).

Advantages

It is simple, free from problem of fumes of boiling acids and is preferred method for B and Mo estimation.

Disadvantages

It is slow and can not be used for N, P and S estimation because they get volatilized at high ignition temperature (550°C) required for ashing.

(a) Simple Dry Ashing Method *(For elements except S and P)*

Principle

The plant samples for the determination of nutrient elements (K, Na, Ca, Mg, Cu, Zn, Fe, Mn, Mo and B), which do not volatilize at high temperature can be digested by dry ashing in furnace using suitable silica, porcelain or platinum crucible and extracting them in dilute HNO_3.

Apparatus

Muffle furnace, silica/porcelain/platinum crucible of 50 ml, capacity, electric hot plant, 50 ml volumetric flask, 1*l* volumetric flask, 1*l* beaker, wash bottle, etc.

Reagents

(i) **Concentrated Hydrochloric Acid (HCl) :** Transfer 100 ml AR grade concentrated HCl in 250 ml beaker.

(ii) **5 M Nitric acid (HNO_3) :** Transfer 310 ml of concentrated HNO_3 (16 N) in one litre volumetric flask and make the volume with distilled water and store in reagent bottle.

(iii) **2M Nitric acid (HNO_3) :** Transfer 125 ml concentrated HNO_3 (16 N) in one litre volumetric flask and make the volume with distilled water and store in reagent bottle.

Procedure

(i) Take 0.5 g well dried and ground plant sample in a crucible and place the crucible inside the muffle furnace.

(ii) Raise the temperature of muffle furnace slowly and then fix at 550°C for 4 to 6 hours, so that the entire plant material turns to white ash i.e., no black ash is left.

(iii) The crucible is then removed from the furnace and cooled. The contents of crucible are then dissolved in 5 ml of 2 M HNO_3 and transferred to 50 ml volumetric flask.

(iv) Makeup the volume with distilled water and allow it to stand for some time for the silica to settle down. Then filter and store this aliquot for subsequent elemental analysis.

(v) If large amount of visible black ash remains in the crucible, to remove it add 2-3 ml HNO_3 to cooled ash and evaporate the sample to dryness on an electric hot plate set at 100-120°C. Then place it in the muffle furnace and heat it for about 1 hour at 500°C.

(vi) Remove the crucible from the muffle furnace and after cooling moisten the ash with few drops of distilled water and then add 2 ml of concentrated HCl. Evaporate it to dryness on a hot plate (100-120°C) and continue to heat it for another 1/2 to 1 hour. Then follow the rest of the procedure as explained under step-iii above to finally obtain 50 ml aliquot required for analysis.

(b) Dry Ashing of Plant Samples with Magnesium Nitrate (For Sulphur)

Elements like sulphur, phosphorus and chlorine are volatile elements (at high ignition temperature of 550°C) hence, cannot be determined correctly by dry ashing as done for K, Na, Ca, Mg, Cu, Zn, Fe, Mn, B and Mo. To prevent the volatilization losses of S during dry ashing process, some basic compounds such as magnesium nitrate, sodium carbonate and sodium peroxide are used with the plant samples. The method of dry ashing for S estimation is described below.

Principle

Dry ashing of plant samples for S estimation can be done by using magnesium nitrate (Mg $(NO_3)_2$), which prevents the losses of S by volatilization at high ignition temperature (550°C) required for ashing in the muffle furnace. The organic form of plant S is converted into sulphate form on ashing, which is extracted in HCl acidified warm water.

Apparatus

Muffle furnace, procelain crucible of 50 ml capacity, electric hot plate, 50 ml volumetric flask, 1.0 volumetric flask, 250 ml and 1.0 beakers, wash bottle, etc.

Reagents

(i) **Magnesium Nitrate Solution** : Weight and transfer 955 g of Mg $(NO_3)_2$. $6H_2O$ to 1 litre beaker and then add 350 ml of distilled water and one g of MgO on it. Boil the solution and filter through Whatman filter paper No. 42 and make the final volume to one litre with distilled water.

(ii) **Concentrated Hydrochloride Acid (HCl)**: Transfer 100 ml AR grade concentrated HCl in 250 ml beaker.

Procedure

(i) Take 0.5 g of finely ground plant sample in a 50 ml procelain crucible and add 5 ml of Mg $(NO_3)_2$ solution to it. Now swirl the contents gently. Heat the crucible on hot plate at about 180°C till dryness of the contents. Then transfer the hot crucible to the furnace.

(ii) Slowly raise the temperature of the furnace to 500°C and ignite the contents of the crucible until entire carbon gets oxidized. Remove the crucible from the furnace and allow to cool.

(iii) Add 5 ml concentrated HCl to the crucible, evaporate to dryness on hot plate at 110-120°C and bake it for one hour to dehydrate silica.

(iv) Add few drops of distilled water to moisten the ash and then add one ml of concentrated HCl and warm it on hot plate to dissolve the salts.

(v) Transfer the content of the crucible to a 50 ml volumetric flask and make up the volume with warm distilled water. Allow the silica to settle down. If the solution is not clear, add few drops of concentrated HCl till the solution becomes clear. This solution contails total S as well as other elements of plant sample oxidized to mineral form.

(c) Dry Ashing with Sodium Peroxide (for Sulphur)

The plant sample is ignited in a Nickel crucible at high temperature (400-500°C) with an oxidative mixture of Na_2CO_3 and Na_2O_2 in a muffle furnace resulting in the oxidation of organic form of S to sulphate form, which is extracted with acidified (HCl) distilled water.

Apparatus

100 ml Nickel crucible, Ni or Pt rod, electric hot plate, 600 ml beaker, 100 ml volumetric flask, aliquot storage bottles, wash bottle, etc.

Reagents

(i) **Sodium carbonate (Na_2CO_3) :** This is used as anhydrous powder during the digestion. For its convenient use, transfer approximately 20-30 g. of this salt in a 50 ml beaker.

(ii) **Sodium peroxide (Na_2O_2) :** This is also used as anhydrous powder during the digestion.

(iii) **Concentrated Hydrochloric acid (HCl)** : Transfer 100 ml AR grade concentrated HCl in 250 ml beaker.

Procedure

(i) Weigh 0.5 g of finely ground plant sample in a 100 ml Nickel crucible.

(ii) Add 2.5 g anhydrous Na_2CO_3 to the sample and mix thoroughly with the Ni or Pt rod and moisten with 1 ml distilled water.

(iii) To the above mixture, add Na_2O_2 taken in splits of 0.25 g at a time. Mix the material thoroughly after addition of each split and continue addition until mixture becomes nearly dry and granular.

(iv) Place the crucible on electric hot plate and heat carefully, stirring it occasionally, until contents are fused. Care should be taken that the material does not get ignited at this stage, otherwise experiment will fail.

(v) Remove the crucible from hot plate and allow it to cool. Then cover the hardened mass in the crucible with more Na_2O_2 up to depth of about 5 mm.

(vi) Heat crucible first gradually and then raise the temperature of the hot plate until fusion takes place again. Rotate crucible occasionally to bring the total contents in contact with oxidizing material.

(vii) Continue heating for 10 minutes after fusion is complete.

(viii) Cool for some time and then place the warm crucible along with contents into a 600 ml beaker and add about 60 ml distilled water carefully.

(ix) When initial violent action ceases, wash the material out of crucible. Make it slightly acidic with HCl by adding small portions at a time.

(x) Transfer the material into 100 ml volumetric flask. Cool the contents of flask and make the volume. Filter the contents with Whatman No. 42 filter paper.

(xi) The filtered solution contains total S as well as other elements of the plant oxidized to mineral form.

Precautions for Dry Ashing

(i) The plant samples to be dry ashed, must be fully dry and weighed accurately.

(ii) Avoid use of silica crucible (which may cause stickling of nutrient elements like Cu on its walls). To avoid this problem, use porcelain or platinum crucible to avoid this defect.

(iii) The temperature of the furnace should be increased slowly to reach its ignition mark of 550ºC.

(iv) The temperature of the muffle furnace should not be allowed to go beyond 550ºC ± 15ºC.

2. WET DIGESTION OF PLANT SAMPLES

Wet digestion of plant samples employs the oxidation of plant material by using suitable combination of two or three concentrated oxidizing acids ($HClO_4$, HNO_3, and H_2SO_4 taken in specific ratios) (Table 1). The type of acid mixtures to be used for digestion will depend on the type of nutrient elements to be determined in the plant samples. The suitability of oxidizing acid mixtures for different nutrients is given in Table 1.

Table 1. Suitability of oxidising acid mixtures for different nutrients.

S.No.	Acid mixture	Elements to be determined
1.	$HNO_3 + HClO_4$	All nutrients* except N
2.	$H_2SO_4 + H_2O_2$	Only for N
3.	$H_2SO_4 + HClO_4$	All nutrients* except S
4.	$HNO_3 + H_2SO_4 + HClO_4$	All nutrients* except N and S

*All nutrients do not include Cl and Mo in the present case.

Principle

The basic principle of wet digestion is that a known weight of plant material is treated with the specified mixture of two or three concentrated oxidizing acids at boiling temperatures resulting in release of the plant nutrient elements from their original organically bound forms to the mineral forms.

(a) Wet Digestion by Di-acid Mixture of HNO_3 and $HClO_4$ (All nutrients, except N)

This method is most suitable for the analysis of P, K, S, Ca, Mg, Na, Al, Se and all micronutrients except B and Mo in plant material. A known weight of plant sample is oxidized by the di-acid mixture of HNO_3 and $HClO_4$ (4 : 1 Ratio). The digest is diluted with distilled water to obtain an optimum volume of aliquot.

Advantages : It is easier, less time consuming and convenient to analysis large number of samples.

Disadvantages: The boiling acids produce fumes, which is problem in ordinary laboratories.

Apparatus

Analytical balance, electric hot plate, 50 ml conical flask, acid fume hood, tongs, 5 ml tilt measure, 500 ml beakers, dropper, 50 ml volumetric flask, funnels, whatman No. 1 filter paper, wash bottle, aliquot storage bottles, etc.

Reagents

(i) Concentrated Nitric acid (HNO_3, 16 N)

(ii) Concentrated Perchloric acid ($HClO_4$, 11.6 N)

Di-acid mixture: Prepare the di-acid mixture by mixing the concentrated HNO_3 and $HClO_4$ in the ratio of 4 : 1, respectively. To prepare 1 litre of the di-acid mixture at a time, mix 800 ml of concentrated HNO_3 in 200 ml of concentrated $HClO_4$ taken in a 2 litre beaker. Allow the mixture to cool and store it in a reagent bottle.

Procedure

(i) Weigh 0.5 g well dried and powdered plant material in a 50 ml conical flask.

(ii) Add 5 ml diacid mixture and allow it to stand over night or about 2-3 hours for pre-digestion.

(iii) Place the flask on the electric hot plate and start heating for digestion at 180°C - 200°C until dense white fumes of acids are evolved and a clear colorless solution/aliquot is obtained. If the solution is not clear, remove the flask from hot plate cool for some time and add 1-2 ml of the diacid mixture and heat it again to obtain a clear solution.

(iv) Remove the flask from the hot plate when a little amount (0.5 ml) of the digest is still left. The contents of the flask should not be allowed to totally dry out.

(v) Allow the contents of the flask to cool down and then add approximately 20 ml distilled water to dissolve the contents.

(vi) Transfer the entire contents of the digestion flask to a 50 ml volumetric flask and make up the volume with distilled water.

(vii) Filter the aliquot through Whatman No. 1 filter paper and store in a plastic aliquot storage bottle for its subsequent analysis.

(b) Wet Digestion with H_2SO_4 and H_2O_2 (Only for Total N)

In this method the plant sample is charred with concentrated H_2SO_4 and then complete oxidation is accomplished by use of H_2O_2. To estimate the total nitrogen in plants, ground plant sample of known weight can be digested with concentrated H_2SO_4 taken in a Kjedahl flask with excess of 30% H_2O_2 and K_2SO_4 - Catalyst mixture, which, give nascent oxygen and raise the boiling point of the digestion system.

Apparatus

Analytical balance, Kjedahl digestion-set, 250 ml Kjeldahl flask, 10 ml-tilt measure, 50 ml volumetric flask, 500 ml beaker, plastic aliquot storage bottles, wash bottle, etc.

Reagents

(i) Concentrated Sulphuric acid (H_2SO_4)-36 N

(ii) Hydrogen Peroxide (H_2O_2)-30%

(iii) Potassium Sulphate-Catalyst Mixture: This mixture is prepared by mixing 10 g potassium sulphate (K_2SO_4) and 1 g catalyst mixture (20 parts of $CuSO_4$ and 1 part of Se powder).

Procedure

(i) Take 0.5 g dry and grinded plant sample and transfer in 250 ml Kjeldahl flask.

(ii) Add 20 ml concentrated H_2SO_4 slowly and carefully with the help of tilt measure such that whole of the plant material is well covered at the bottom. Allow it to stand over night or at least 2-3 hours for pre-digestion.

(iii) Place the flask in the Kjeldahl digestion set and heat the sample at 150-200°C for about 15-20 minutes. The contents of the flask would turn from black to brown.

(iv) Remove the Kjeldahl flask from the digestion set and cool for 10-15 minutes. Then add 1-1.5 ml of H_2O_2 and place the flask in the digestion set for another 5-10 minutes. Repeat this process 2-3 times or unless the solution becomes clear.

(v) Add 1 g of potassium sulphate-catalyst mixture as listed above in reagents and heat the contents of flask to complete the digestion at high temperature (360°C).

(vi) After completion of digestion, cool and transfer the entire contents of flask to a 50 ml volumetric flask and make up the volume with distilled water.

(vii) Filter the aliquot through Whatman No. 1 filter paper and store in a plastic aliquot storage bottle for its subsequent analysis.

(c) Wet Digestion by Modified Kjeldahl Method (For Total-N including NO_3.N)

In simple Kjeldahl method, only a fraction of plant NO_3 will be reduced to ammonia during digestion. The mature plants have low contents of NO_3. However, if plants samples are taken at an early stage of growth or have received high rates of applied N, there may be substantial quantities of NO_3 present in plant tissue, which would not get reduced to ammonia during digestion. So if the NO_3 is present in substantial quantities, the digestion must be modified (Bremner and Mulvaney, 1982) as explained under.

Principle

Nitro compounds formed by the reaction of salicylic acid with NO_3 in acid medium are reduced to corresponding amino compounds by heating the mixture with sodium thiosulphate and zinc dust. The main product of nitration is 5-nitrosalicylic acid and small amount of 3-nitrosalicylic acid. Drying of plant samples in oven eliminates the interference of water in analysis as nitration of salicylic acid does not take place if more than a trace of water is present.

Apparatus

Analytical balance, Kjeldahl digestion-set, 250 ml Kjeldahl flask, 10 ml-tilt measure, 50 ml volumetric flask, 500 ml beaker, plastic aliquot storage bottles, wash bottle, etc.

Reagents

(i) **Sulphuric salicylic acid:** Dissolve 1g of salicylic acid in 30 ml of concentrated H_2SO_4.

(ii) **Sodium thiosulphate:** 24 mesh dried powder.

(iii) **Potassium Sulphate-Catalyst Mixture:** This mixture is prepared by mixing 10g potassium sulphate (K_2SO_4) and 1 g catalyst mixture which contains 20 parts of $CuSO_4$ and 1 part of Se powder.

Procedure

(i) Weigh 0.5 g dried plant tissue and transfer it to a 100 ml Kjeldahl flask.

(ii) Add 20 ml Sulphuric salicylic acid mixture in the flask and swirled so as to let the dry samples come in contact with the reagents.

(iii) Allow it to stand overnight.

(iv) Add 5 g of sodium thiosulphate powder and heat gently for about 5 minutes avoiding frothing. Now cool the contents and add 10 g of potassium sulphate-catalyst mixture. Again put on Kjeldahl apparatus for digestion at full heat (360°C-400°C) until the solution is clear. The digestion is continued for 1 hour after the solution is cleared.

(v) After completion of digestion, digest is cooled and diluted with distilled water to make the volume of 50 ml.

(d) Wet Digestion by Diacid Mixture of H_2SO_4 and $HClO_4$ (All nutrients except S)

A known weight of plant sample taken in a conical flask is treated with excess of di-acid mixture of H_2SO_4 : $HClO_4$ (9:1) and digested on electric hot plate at 180°C-200°C. The digest is diluted with distilled water to obtain an optimum volume of aliquot.

Apparatus

Analytical balance, electric hot plate, 50 ml conical flask, acid fume hood, tongs, 5 ml tilt measure, 500 ml beakers, dropper, 50 ml, volumetric flask, funnels, Whatman No. 1 filter paper, wash bottle, aliquot storage bottles, etc.

Reagents

(i) Concentrated Sulphric acid (H_2SO_4, 36 N)

(ii) Concentrated Perchloric acid ($HClO_4$, 11.6 N)

Prepare the di-acid mixture by mixing the concentrated H_2SO_4 and $HClO_4$ in the ratio of 9:1, respectively. To prepare 1 litre of the di-acid mixture at a time, mix 900 ml of concentrated H_2SO_4 in 100 ml of concentrated $HClO_4$ taken in a 2 litre beaker. Allow the mixture to col and store in a reagent bottle.

Procedure

(i) Weigh 0.5 g dry and powdered plant material in a 50 ml conical flask.

(ii) Add 5 ml di-acid mixture and allow it to stand over night or for at least 3-4 hours for pre-digestion.

(iii) Place the flask on the electric hot plate and start heating for digestion at 180°C-200°C until dense white fumes of acids are evolved and a clear solution/aliquot is obtained. If the solution is not clear remove the flask from hot plate cool for some time and then add 1-2 ml of the diacid mixture and heat it again to obtain the clear solution.

(iv) Remove the flask from the hot plate when a little amount (0.5 ml) of the digest is still left. The contents of the flask should not be allowed to totally dry out.

(v) Allow the contents of the flask to cool down and then add approximately 20 ml of distilled water to dissolved the contents.

(vi) Transfer the entire contents of the digestion flask to a 50 ml volumetric flask and make up the volume with distilled water.

(vii) Filter the aliquot through Whatman No. 1 filter paper and store it in plastic aliquot storage bottle for its subsequent analysis.

(e) Wet Digestion by Tri-acid Mixture of HNO_3, H_2SO_4 and $HClO_4$ (All nutrients except N and S)

The digestion of plant material with a tri-acid mixture of concentrated Nitric acid (HNO_3), Sulphuric acid (H_2SO_4) and Perchloric acid ($HClO_4$) is specially recommended for the estimation of P and K. A known weight of plant sample can be digested with excess of tri-acid mixture of HNO_3 : H_2SO_4 : $HClO_4$ (10 : 1 : 4) taken in a digestion flask on an electric hot plate at a temperature of 180°C-200°C. The digest is diluted with distilled water to obtain an optimum volume of aliquot.

Apparatus

Conical flasks (100 ml capacity), electric hot plate, volumetric flasks, funnels, beakers, measuring cylinder, aliquot storage bottles, Whatman No. 1 filter paper, etc.

Reagents

(i) Concentrated Nitric acid (HNO_3, 16 N)

(ii) Concentrated Sulphuric acid (H_2SO_4, 36 N)

(iii) Concentrated Perchloric acid ($HClO_4$, 11.6 N)

Prepare the tri-acid mixture by mixing HNO_3, H_2SO_4 and $HClO_4$ acids, in ratio of 10:1:4 (v/v), respectively, in a beaker. To prepare 1.5 litre of the tri-acid mixture at a time we can take and mix 1 litre of HNO_3, 100 ml of H_2SO_4 and 400 ml of $HClO_4$ in a 2 litre beaker. Allow the mixture to cool and then store in a glass reagent bottle.

Procedure

(i) Weight 0.5 g dry and powdered plant material in a 100 ml conical flask.

(ii) Add 10 ml of the tri-acid mixture and allow it to stand over night or about 2-3 hours for predigesting.

(iii) Place the flask on the electric hot plate and start heating for digestion at 180°C-200°C until dense white fumes of acids are evolved and a clear colorless solution/aliquot is obtained.

(iv) Remove the flask from the hot plate when about 1-2 ml of the digest is still left. The contents of the flask should not be allowed to totally dry out.

(v) Allow the contents of the flask to cool down and then add approximately 20 ml of distilled water to dissolve the contents.

(vi) Transfer the entire contents of the digestion of the digestion flask to a 50 ml volumetric flask and make up the volume with distilled water.

(vii) Filter the aliquot through Whatman No. 1 filter paper and store it in a plastic bottle for its subsequent analysis of P and K.

Precautions

(i) Always use well dried and powdered plant samples.

(ii) Select a proper diacid or triacid mixture for digestion of plant material depending on the nutrient element(s) to be estimated in the plant samples.

(iii) Use suitable devices like tilt measure or droppers for adding the concentrated acids or acid mixtures in the digestion flask.

(iv) Avoid the contact of acid fumes to any part of the body. Use proper kits.

(v) To avoid effect of fumes always use digestion chambers provided with fume hoods.

(vi) The digestion flask should be removed from the hot plate, when 1-2 ml of the digest is still left at the bottle to avoid charring.

REFERENCES

Bremner, J.M. and Mulvaney, C.S. 1982. Nitrogen-total. In : *Methods of Soil Analysis Part -2. Chemical and Microbiological Properties* (Ed. A.L. Page *et al.). Agronomy Monograph* No. 9: 595-624. American Society of Agronomy, Madison, WI, USA.

Chapman, H.D. and Prat, P.F. 1961. *Methods of Analysis of Soils, Plants and Water.* University of California, USA.

Chopra, S.L. and Kanwar, J.S. 1976. *Analytical Agricultural Chemistry.* Kalyani Publishers, New Delhi.

Greweling, T. 1976. *Chemical Analysis of Plant Tissues.* Agronomy No. 6. Cornel University Agricultural Experimental Station, Ithaca, NY. *Search Agriculture* 6(8) : 1-35.

Gupta, P.K. 1999. *Soil, Plant, Water and Fertilizer Analysis*. Agro. Botanica, Bikaner.

Issac, R.A. and Johnson, W.C. 1985. Elemental analysis of plant tissue by plasma emission spectroscopy : collaborative study. *Journal Association of Analytical Chemists* 58 : 436-440.

Jackson, M.L. 1973. *Soil Chemical Analysis.* Prentice-Hall of India, New Delhi.

Jones Jr., J.B., Wolf, B. and Mills, H.A. 1991. *Plant Analysis Handbook : A Practical Sampling, Preparation, Analysis and Interpretation Guide.* Micro-Macro Publishing Inc. Athens, Georgia, USA. 39 p.

Kensworthy, A.L. 1964. *Fruit, Nut and Plantation Crops - A Guide for Collection of Foliar Samples for Nutrients Element Analysis.* Department of Horticulture, Michigan State Univ. East Lansing USA, 39 p.

Piper, C.S. 1966. *Soil and Plant Analysis.* Hans Publishers, Bombay.

Tandon, H.L.S. 1993. *Methods of Analysis of Soils, Plants, Water and Fertilizers.* (ed.) Fertilizer Development and Consultation Organization. New Delhi.

CHAPTER -3

DETERMINATION OF PLANT NUTRIENTS

INTRODUCTION

The determination of nutritional need of a crop is an important aspect of nutrient management for the farmer. Complete life cycle normally, of living organism requires a supply of large number of nutrients from outside. This supply is called plant nutrition. Under normal conditions of growth, however, all green plants are cutotrophic and they require from outside the supply of only inorganic substances. A knowledge of nutrient concentration in growing plants can serve as a tool for correcting any deficiencies if carried out early enough to safeguard yield. It can also be used to evaluate the efficiency of a recent application. Analysis of harvested plant parts or mature tissues is like a post mortem. The information it provides can help to plan nutrient application only in subsequent years on that field.

This chapter describes the details of some widely accepted methods for the determination of macro and micro nutrients in plant material.

NITROGEN

One common plant analysis is that of nitrogen (N) by Kjeldahl method (Bremner, 1960). However, wet ashing with H_2SO_4 and H_2O_2 is also used for eliminating the use of selenium in the former method.

1. Kjeldahl Nitrogen

Apparatus

Block-digester.

Distillation unit.

Automatic titrator connected to a pH-meter.

Vortex tube stirrer.

Reagents

1. Catalyst Mixture (K_2SO_4-Se), 100 : 1 w/w ratio
2. Sulfuric Acid (H_2SO_4), concentrated
3. Ethylene Diaminetetraacetic Acid Disodium Salt (EDTA), M.W. = 372.2
4. Sodium Hydroxide Solution (NaOH), 10 N
5. Boric Acid Solution (H_3BO_3), saturated.
6. Sulfuric Acid Solution (H_2SO_4), 0.01 N
7. Standard Stock Solution : 1.2 g NH_4^+ - N per l.

Procedure

A. Digestion

1. Mix and spread finely ground (Cyclone mill) plant sample in a thin layer on a sheet of paper until it looks uniform.
2. Select representative sub-samples of about 1 g by taking at least 10 small portions from all parts of the sample with a spatula, and put them into a plastic vial.
3. Dry the sub-sample at 60°C in an oven (overnight), and then cool in a desiccator.
4. Weigh 0.25 g (grain) or 0.50 g (straw) of dry plant material, and transfer quantitatively into a 100-ml digestion tube.
5. Add a few pumice boiling granules, and add about 3 g catalyst mixture using a calibrated spoon.
6. Add 10 ml concentrated sulfuric acid using a dispenser, and stir with Vortex tube sitrer until mixed well.
7. Place tubes in a block-digester set at 100°C for 20 minutes, and remove the tubes to wash down any material adhering to the neck of the tube with the same concentrated sulfuric acid. Thoroughly agitate the tube contents, and then place the tubes back on the block-digester set at 380°C for 2 hours after clearing.
8. After digestion is complete, remove tubes, cool, and bring to 100 ml volume with DI water.
9. Each batch of samples for digestion should contain at least one reagent blank (no plant), and one chemical standard (weigh 0.1 g EDTA standard digest), and one standard plant sample (internal reference).

B. Distillation

1. Set distillation and titration apparatus and steam out the apparatus for at least 10 minutes.
2. Prior to distillation, shake the digestion tube to thoroughly mix its contents. Then pipette 10 ml aliquot into a 100 ml distillation flask.
3. Carefully add 10 ml N sodium hydroxide solution, and immediately connect the flask to distillation unit and begin distillation.
4. Collect about 35 ml distillate in the collecting dish.
5. Remove distillation flask and connect an empty 100-ml distillation flask to the distillation unit. Drain water from the condenser jacket and steam out apparatus for 90 seconds before connecting the next sample.
6. Titrate the distillate to pH 5.0 with standardized 0.01 N H_2SO_4 using the Auto-Titrator; record titration volume of acid.
7. Each batch of distillations should include a distillation of 10 ml ammonium N standard with 0.2 g MgO and 10 ml DI water with 0.2 g MgO. Recovery of ammonium-N standards should be at least 98%. Recovery of EDTA, corrected for reagent blank, should be at least 97%.

CALCULATIONS

Percentage recovery of Ammonium N standard :

$$\%Recovery = \frac{(V - B) \times 14.01 \times 100}{C \times D}$$

Where : V = Volume of 0.01 N H_2SO_4 titrated for the sample (ml).

B = Distillate blank titration volume (ml)

N = Normality of H_2SO_4 solution.

C = Volume of NH_4-N standard solution (ml)

D = Concentration of NH_4-N standard solution (μg/ml)

14.01 = Atomic weight of N.

$$\%Recovery = \frac{(V - B_1) \times N \times R \times 186.1 \times 100}{W_{t1} \times 1000}$$

Percentage Nitrogen in plant :

$$\%N = \frac{(V - B_1) \times N \times R \times 14.01 \times 100}{W_{t2} \times 1000}$$

Where, R = Ratio between total digest volume and distillation volume.

B_1 = Digested blank titration volume (ml)

W_{t1} = Weight of EDTA (g)

W_{t2} = Weight of dry plant (g)

2. Nitrate-Nitrogen

The routine use of heavy metals as catalysts is not environmentally sound. From this standpoint, treatment of plant material with a mixture of H_2O_2 - H_2SO_4 in the absence of metal catalysts has been proposed as an alternative digestion procedure for replacing Kjeldahl-N determination in soils and plants (McGill and Figueiredo, 1993).

Reagents

A. Sulfuric Acid (H_2SO_4), concentrated

B. Hydrogen Peroxide (H_2O_2), 30%

Procedure

1. Weigh 0.5 g dry plant material into a 100 ml digestion tube.
2. Add 3-4 pumice boiling granules, and then add 5 ml concentrated sulfuric acid, and mix
3. Keep overnight
4. Heat on a block-digest at a moderate temperature i.e., 100-150°C
5. Swirl to restric foaming. If foaming enters the neck of the digestion tube, add 2 ml 30% hydrogen peroxide.
6. Heat the tubes for 30-60 minutes on the block-digester.
7. Cool the tubes, then add 2 ml hydrogen peroxide.
8. Raise the temperature of the block-digester to 280°C.
9. Heat the tubes for 10 minutes at 280°C.
10. Cool, then add 2 ml hydrogen peroxide, and heat for 10 minutes.
11. Repeat 9 and 10 until solution remains clear after 10 minutes of heating.
12. Cool, and make up to the 100 ml volume with DI water.

Measurement

Nitrogen content can be measured in this digest by the Distillation Method. Also, measurement of phosphorus can be done Colorimetrically, after filtering the digest through Whatman No. 1 or 5 filter paper, as described by Murphy and Riley (1962). The results for both N and P are highly correlated with the standard Kjeldahl digestion method.

3. Total Nitrogen

This method is based on digestion of plant materil in a sulfuric-salicylic acid mixture (Buresh *et al.,* 1982).

Reagents

A. Sulfuric Salicylic Acid Mixture (concentrated H_2SO_4 containing 2.5% w/v salicylic acid)

Dissolve 62.5 g reagent grade salicylic acid ($C_7H_6O_3$) in 2.5-l concentrated sulfuric acid.

B. Catalyst Mixture (K_2SO_4-Se), 100:1 w/w ratio.

C. Sodium Thiosulfate ($Na_2S_2O_3$, $5H_2O$), crystal

D. Ethylene Diaminetetraacetic Acid Disodium Salt (EDTA), M.W. = 372.2

Procedure

A. Digestion

1. Mix and spread finely ground plant sample in a thin layer, on a sheet of paper or plastic until the sample looks uniform.
2. Take a representative sub-sample of about 1 g by systematically withdrawing at least 10 small portions from all parts of the sample with a spatula, and put them into a plastic vial.
3. Dry the sub-sample at 60°C in an oven (overnight), and then cool in a desiccator.
4. Weigh 0.25 g (grain) or 0.50 g (straw) dry plant material, and then transfer quantitatively into a dry 250-ml digestion tube.
5. Add 20 ml sulfuric salicylic acid mixture while rotating the tube to wash down any sample adhering to the neck of the tube, and allow to stand 2 hours or longer with occasional swirling.

6. Add 2.5 g sodium thiosulfate through a long-stemmed funnel to the contents of the tube and swirl gently a few times, and allow to stand overnight.
7. Add 4 g catalyst mixture, and 3-4 pumice boiling granules, and place tubes on the block-digester pre-heated to 400°C.
8. Place a small glass funnel in the mouth of the tubes to ensure efficient refluxing of the digestion mixture and prevent loss of H_2SO_4, and proceed with the digestion until the mixture clears.
9. Remove the tubes from the block-digester and allow tem to cool for about 20 minutes. Then wash down any material adhering to the neck of the tube with a minimum quantity of DI water.
10. Thoroughly agitate the tube contents, place tubes back on the block-digester, and digest for 2 hours after clearing. No particulate material should remain in the tube after digestion.
11. After the digestion is finished, allow the digest to cool, and add water slowly shaking until the liquid level is about 2 cm below the graduation mark.
12. Allow tube to cool to room temperature, and add DI water to bring the volume to the 250 ml mark.
13. Each batch of samples for digestion should contain at least one reagent blank (no plant), and one chemical standard (weight 0.1 g EDTA standard digest), and one standard plant sample (internal reference).

B. Distillation

1. Set distillation and titration apparatus and steam out the apparatus for at least 10 minutes.
2. Prior to distillation, shake the digestion tube to thoroughly mix its contents, and pipette an aliquot in a 300-ml distillation flask.
3. Carefully add 7 ml or 15 ml 10 N sodium hydroxide solution for 25 ml or 50 ml aliquot, respectively, and immediately connect flask to distillation unit and begin distillation.
4. Collect about 35 ml distillate in the collecting dish.
5. Remove distillation flask and connect an empty 100 ml distillation flask to the distillation unit. Drain water from the condenser jacket and steam out apparatus for 90 seconds before connecting the next sample.

6. The distillate is then titrated to pH 5.0 with standardized 0.01 N H_2SO_4 using the Auto-Titrator; record titration volume of acid.
7. Each batch of distillations should contain at least two standards and two blanks (reagent blanks). Recovery of EDTA, corrected for reagent blank, should be at least 97%.

CALCULATIONS

Percentage recovery of EDTA standard:

$$\%Recovery = \frac{(V-B) \times N \times R \times 186.1 \times 100}{W_{t1} \times 1000}$$

Percentage Nitrogen in plant :

$$\%N = \frac{(V-B) \times N \times R \times 14.01 \times 100}{W_{t2} \times 1000}$$

Where : V = Volume of 0.01 N H_2SO_4 titrated for the sample (ml).

B = Digested blank titration volume (ml)

N = Normality of H_2SO_4 solution.

14.01 = Atomic weight of N.

R = Ratio between total digest volume and distillation volume.

W_{t1} = Weight of EDTA (g)

W_{t2} = Weight of dry plant (g)

186.1 = Equivalent weight of the EDTA.

Colorimetric Method for Nitrogen Estimation

A colorimetric procedure can be adopted by using auto analyser system for determination of NH_4 - N in the plant digest (Crooke and Simpson, 1971).

Digestion

50 mg dried and well ground plant material is taken in 50 ml conical flask and 3 ml of 9:1 H_2SO_4 and $HClO_4$ mixture is added. The flask are heated gently over a hot plate for 5-10 minutes and thereafter, the heating temperature is raised when the frothing in mixture ceased. Digestion is continued until the solutions become colourless. The digest is cooled and diluted to 100 ml with distilled water.

Reagents

1. Buffer solution (0.1 M Na_2HPO_4, 5% Na-K tartarate, 5.4% NaOH). Dissolve 26.8 g $Na_2HPO_4 . 7H_2O$ in 600 ml of ammonium free water. Add 50 g Na-K tartarate and 10.8 g 50% (w/w) solution of NaOH. Dilute to 1 litre with distilled water.
2. Na-salicylate Na-nitroprusside solution (15%-0.03%). Dissolve 150 g Na-salicylate and 0.30 g Na nitroprusside in 1 litre of distilled water. Store in a light resistant bottle.
3. Na-hypochlorite solution, Dilute 6 ml of 5.25% Na-hypochloride solution to 100 ml with distilled water. Prepare daily.
4. Nitrogen standard solution : Dissolve 4.715 g dry $(NH_4)_2 SO_4$ in a litre of distilled water to prepare 1000 mg N/l. From this standard, solutions containing 5.0, 10.0, 15.0, 30.0, 40.0 and 50.0 mg N/l are prepared.

Determination of N

1 ml aliquot from plant digest is taken in 25 ml volumetric flask. 5.5 ml of solution (1) is added and stirred thoroughly. Then 4 ml of solution (2) is added and mixed again. Now, 1 ml of solution (3) is added and mixed again thoroughly. The reagent must be added in the stated order to avoid the formation of solid residues. The solution is allowed to stand for 45 minutes at 25°C (or 15 minutes at 37°C) to ensure complete colour development. The absorbance is then read in a spectrophotometer at 650 nm and expressed as mg/g dry wt.

PHOSPHORUS

Total P in plant material can be determined either by wet digestion procedure or by dry ashing procedure. Both methods are satisfactory. However, dry ashing is a simpler, easier, non-hazardous and economical option. Later, P content in the digestsd or dissolved ash aliquots are measured colorimetrically.

Apparatus

Spectrophotometer or colorimeter, 410-nm wavelength.

Block-digester.

Vortex tube-stirrer.

Reagents

A. Ammonium Heptamolybdate-Ammonium Vanadate in Nitric Acid

- Dissolve 22.5 g ammonium heptamolybdate $[NH_4)_6 MO_7O_{24}. 4 H_2O]$ in 400 ml DI water (a).
- Dissolve 1.25 g ammonium metavanadate (NH_4VO_3) in 300 ml hot DI water (b).
- Add (b) to (a) in a 1-l volumetric flask, and let the mixture cool to room temperature.
- Slowly add 250 ml concentrated nitric acid (HNO_3) to the mixture, cool the solution to room temperature, and bring to 1-l volume with DI water.

B. Standard Stock Solution

- Dry about 2.5 g potassium dihydrogen phosphate (KH_2PO_4) in an oven at 105°C for 1 hour cool in desiccator, and store in a tightly stoppered bottle.
- Dissolve 0.2197 g dried potassium dihydrogen phosphate in DI water, and bring to 1-l volume with DI water. This solution contains 50 ppm P (Stock Solution).
- Prepare a series of Standard Solutions from the stock Solution as follows :

Dilute 1, 2, 3, 4, and 5 ml stock Solution to 100-ml final volume by adding DI water. These solutions contain 0.5, 1.0, 1.5, 2.0, and 2.5 ppm, P, respectively.

Procedure

A. Wet-Digestion Procedure

- Digest the plant material (as described for Kjeldahl-N in plants).

- Filter plant digest with Whatman No. 1 filter paper, and collect filtrate in a small bottle.

or as an alternative procedure

Dry-ashing Procedure

- Dry ash the plant material (as described for Macro-and Micronutrients by dry ashing).

- Dissolve the ash in 2 N HCl.

B. Measurement

1. Pipette 10 ml of the digest filtrate or aliquot of the dissolved ash (depending on the procedure used) into a 100-ml volumetric flask, add 10 ml ammonium-vanadomolybdate reagent, and dilute the solution to volume with DI water.

2. Prepare a standard curve as follows :

- Pipette 1, 2, 3, 4 and 5 ml standard solution, and proceed as for the samples.
- Also make a blank with 10 ml ammonium vanadomolybdate reagent, and proceed as for the samples.
- Read the absorbance for the blank, standard, and samples after 30 minutes at 410-nm wavelength.

3. Prepare a calibration curve for standards, plotting absorbance against the respective P concentrations.

4. Read P concentration in the unknown samples from the calibration curve.

CALCULATION

Percentage Total Phosphorus in plant:

$$\%P = ppm\ P\ (from\ calibration\ curve) \times \frac{R}{W_t} \times \frac{100}{1000}$$

Where : R = Ratio between total volume of the digest/aliquot and the digest /aliquot volume used for measurement

W_t = Weight of dry plant (g)

Note : The plant digest by the hydrogen peroxide and sulfuric acid can also be used for phosphorus measurement in plants.

Determination of Phosphorus (P) by Yellow Method

Principle

Vanadate molybdate and orthophosphate react together to give a yellow colour complex in HNO_3 medium (Method of Koenig and Johnson, 1942).

Digestion

50 mg dried and well ground plant material is taken in 50 ml conical flask and 3 ml of 9 : 1 H_2SO_4 and $HClO_4$ mixture is added. The flasks are heated gently over a hot plate for 5-10 minutes and there

after, the heating temperature is raised when the frothing in mixture is ceased. Digestion is continued until the solutions become colourless. The digest is cooled and diluted to 100 ml with distilled water.

Preparation of standard curve

Phosphate standard solution: Dissolve 0.2195 g of analytical grade KH_2PO_4 and dilute to 1 liter. This solution contains 50 μg P/ml (50 ppm).

Transfer 0, 1, 2, 3, 4 and 5 ml of standard solution to 50 ml volumetric flask to get 0, 1, 2, 3, 4 and 5 ppm of P, respectively. Add 10 ml of Vanadomolybdate reagent to each flask. Makeup the volume with deionised water and shake thoroughly. Read the absorbance of solution after 2 minutes or before 2 hours of 725 nm against reagent blank. Once a linear calibration curve is establishment, the slope of the curve is determined and than the concentration of unknown solution can be calculated by using the equation A = mc and expressed as mg/g wt.

Absorbance = slope x concentration or

Concentration = Absorbance divided by slope

Procedure

Reagents

Ferrous ammonium molybdate

Ammonium molybdate (50 g) is dissolved in about 400 ml of 10 N H_2SO_4 under constant stirring to avoid, cacking and the volume is made to 500 ml with distilled water. 10 ml of this solution is diluted to 70 ml with distilled water and 5 g of $FeSO_4$, $7H_2O$ is dissolved in it, making the final volume to 100 ml (Prepare a fresh before use).

3 ml of aliquot is taken in test tube and 2 ml of ferrous ammonium molybdate reagent added. Absorbance is measured after 2 minutes or before 2 hours at 725 nm against reagent blank.

POTASSIUM AND SODIUM

Potassium and sodium in the plant samples are usually determined by digesting the plant in diacid mixture (wet oxidation) Then K and Na in acid digest are determined using Emission Flame Photometric Method.

Emission Flame Photometric Method

Principle

It is based on the principle that atoms of some specific elements absorb energy from the flame and get excited to the higher orbit, Such atoms release energy of a characteristic wavelength, which is specific for those elements and is proportional to the concentration of stoms of those elements in the sample.

To estimate the postassium and sodium,the instrument known as "Flame Photometer" is generally used, which works on the principle of flame photometery. If a solution containing metallic salt (or some other metallic compound) is aspirated into a flame, a vapour that contains atoms of the metal may be formed. Some of these gaseous metal atoms may be raised to an energy level, which is sufficiently high to permit the emission of radiations characteristic of the metal when the atoms drop down to less energetic states or to the ground state. Excited atoms and ions give ``line" radiations at very definite wavelengths. The most prominent lines of sodium and potassium and the radiations most usable for flame analysis in soils and plants works are the doublets at 589.0 and 589.6 µm for Na and 766.0 and 769.0 µm for K.

Equipment and apparatus

Flame photometer, funnels, beakers, pipette, conical flasks, volumetric flasks, and filter papers.

Types of flame photometers

Flame photometers are of two types - single beam and double beam.

(i) Single beam type

Single beam equipment commonly referred to as direct reading type, comprises only one set of optics. Light emitted from the core of the flame just above the inner cone is collected by reflector and focussed by a lens of heat resistant glass through interchangeable optical filters (interference type filter) on to a single photo-detector. Alternatively, light from the burner passes into monochromator and radiation leaving the exit slit is focussed on photo-detector unit.

(ii) Double beam type

In the double beam type instrument, a second light path is provided for the radiation emitted by the internal standard element (e.g. lithium) that is added in a fixed amount to each test solution as well as calibration standard. The wavelengths of internal standard and analytical materials are isolated by special filters in a dual optical system and each beam is focussed on separate photoelectric cell.

Single beam type flame photometers are in common use and their operation is described below :

Operation

Although the ordinary type of manual flame photometer can be operated with great ease, but the modern flame photometers are operated by computer. The actual steps involved may vary with the type of instrument. However, the common steps required to estimate Na and K are given as under :

Reagents

- **Standard K solution** : Dissolve 1.908 g dried potassium chloride (KCl) salt in distilled water in one litre volumetric flask and make up the volume to 1 litre with distilled water. This solution will be the stock solution of 1000 ppm K. From the stock solution, make the working solutions of 5, 10, 15, 20, 25, 30, 35 and 40 ppm K by dilution.
- **Standard Na solution** : Dissolve 2.543 g of dried sodium chloride (NaCl) salt in distilled water in one litre volumetric flask and make up the volume to 1 litre with distilled water. This solution will be the stock solution of 1000 ppm Na. rom the stock solution, make the working solutions of 5, 10, 15, 20, 25, 30, 35 and 40 ppm Na by dilution.

Procedure

A. Calibration of flame photometer

The following steps are used for calibration of flame photometer :

i. Select the desired wavelength or optimal filter

ii. Switch on the instrument.

iii. Switch on the compressor to get air supply.

iv. Adjust the air control to give a pressure of 0.75 kg/cm^2 (10 PSI).

v. Leave the instrument on for about 5 minutes to get it stabilized.

vi. Turn on the gas supply and light the flame.

vii. Adjust the gas control until formation of separate blue cones in the flame.

viii. Set the galvanometer to zero by control knob against blank solution.

ix. Set the galvanometer to 100 with the highest concentration standard solution.

x. Repeat steps no. 8 and 9 till the above setting is obtained consistently.

xi. Proceed with the determination of the standards and samples.

xii. For switching off the instrument, close the gas supply first, run distilled water for about 2 minutes and then switch off the compressor. Turn off the knobs and switches.

Procedure

i. The instrument should be connected to mains through a good voltage stabilizer.

ii. Keep the standards and samples free from any suspended material for preventing the clogging of the capillary. If the sample contains solid suspended particles, filtration should be done before analysis.

iii. Air pressure should be maintained at 0.75 kg/cm^2 (10 PSI).

iv. The gas inlet should be opened after opening the air inlet and closed before shutting of the air supply.

v. Do not feed highly concentrated acid and salt solutions.

vi. Do not leave the flame unattended.

vii. Add few drops of isobutyl alcohol to improve spray ability of the sample solution/extract.

viii. The standard curve should be drawn afresh each time the instrument is used.

ix. Periodically replace the pressure tubing of the fuel gas.

x. Keep the instrument protected from dust particles and acid fumes.

B. Preparation of Standard curve

i. Take the above mentioned reagents i.e. a series of standard solution of Na and K (0, 5, 10, 15, 20, 25, 30, 35 and 40 ppm).

ii. After calibration of the flame photometer, repeat calibration steps vii and ix once again. Feed all the standard solutions one by one starting from the lowest concentration and record reading of each concentration for K/Na.

iii. Draw the standard curve on a simple graph paper taking concentration on X-axis and reading of the flame photometer on Y-axis obtained against each concentration.

C. Measurement of Na or K Concentration in Test Solution

The digested plant and/or grain samples can be tested for estimation of Na and K using flame photometer. Depending upon the concentrations of K and Na in the digests of plant samples, a digest can be used either directly or after dilution for flame photometric determinations.

Take the plant digest (either directly or afert dilution) and feed to the flame photometer and record the reading in the same way as explained in section B above. Let the reading of the test solution be x.

Calculations

Calculate Na or K content in sample, using the standard curve.

Weight of the plant material taken	= 0.5 g
Volume made after digestion	= 50 ml
Dilution factor	= 50/0.5 = 100 times
Flame photometer reading (say)	= x
Na or K concentration (ppm) from standard curve against reading (x)	= Y
Total Na or K (ppm) in plant sample	= Y x 100 dilution factor
Total Na or K (%) in plant sample	= Y x 100/1000 = 0.01 Y

CALCIUM AND MAGNESIUM

The digestion of plant material for Ca and Mg determination, is done with a di-acid mixture (concentrated HNO_3 and $HClO_4$) to bring the total plant Ca and Mg into solution/aliquot. The aliquot is used to determine of Ca and Mg concentrations.

There are two laboratory methods (Jackson, 1967) to determine Ca and Mg in a plant digest/aliquot: (i) Titrimetric method (ii) Emission Spectrophotometry method.

Titrimetric Method

This method is commonly used to determine Ca and Ca + Mg as an alternate method, owing to its speed and simplicity next to emission spectrophotometry (Jackson, 1967). During the titration of Ca or Ca plus Mg with EDTA, there can be the interference of heavy metals (Cu, Co, Fe, Mn, Ni, etc.) if they are present in appreciable quantities. This interference can be removed by using adequate amounts of NaCN and hydroxylamine hydrochloride, which will make stronger complexes with the heavy metals. Besides the pH of the titrating solution is raised to 12 for Ca and 10 for Mg by using NaOH and NH_4OH, respectively. The accuracy of this method is ±3% (Hesse, 1971), which is considered reasonably well for its simplicity, therefore, commonly used for laboratory analyses.

Determination of Calcium

Principle

Calcium in plant digest (solution/aliquot) can be determined by titration of a known volume of the aliquot against the standard solution of EDTA using Calcon as indicator at pH 12 in the presence of NaOH. Adequate concentrations of NaCN and hydroxylamine hydrochloride are also used to remove the interference of heavy metals. The endpoint is the change of colour from pink to blue, when whole of Ca gets complexed with EDTA.

Apparatus

Magnetic stirrer, glass coated iron needles, ceramic china dishes of 100 ml capacity, wash bottles, micro burette, pipettes, beakers, volumetric flasks etc.

Reagents

1. **Standard solution of Ca (0.01 N) :** Weigh 0.5 g dry $CaCO_3$ and transfer it to 400 ml beaker. Add 10 ml of 1 N HCl in it and boil to expel CO_2. Cool it and add about 200 ml distilled water to dissolve the contents. Transfer it to 1 liter volumetric flask and finally make up the volume. Store in a glass reagent bottle.
2. **Standard solution of EDTA (0.01 N) :** Dissolve 2.0 g EDTA-di-sodium salt in sufficient volume of distilled water taken in a 400 ml beaker. Transfer the contents to 1 liter volumetric flask. Add 0.05 g of $MgCl_2.6H_2O$ and make up the volume to 1 liter. The normality of this solution is standardi-

zed by the following procedure (titration with standard Ca solution).

Standardization of 0.01 N EDTA solution: Pipette 5 ml of standard Calcium solution (0.01 N) in a 100 ml China dish. Dilute it to 10 ml with distilled water. Add 10 drops each of NaCN and hydroxylamine hydrochloride solutions. Then add 2.5 ml NaOH solution and 10-15 drops of Calcon indicator. Pink colour will appear. Now in another China dish, prepare a blank solution exactly in the same manner, taking 5 ml of distilled water instead of standard Calcium solution. The blaank solution will be blue in colour, but if not, it will turn blue after the addition of 2-3 drops of EDTA solution. Keep this blue blank along side for comparing the end point. Now take the earlier China dish containing the pink coloured standard calcium solution and place it on the magnetic stirrer plate. Put a glass coated iron needle in it and start stirring. Titrate the sclution with EDTA until the colour of the solution is just changed from pink to blue matching the blank. Note volume of EDTA used. On the basis of calculations, $N_1V_1 = N_2V_2$, necessary dilutions/concentrations are made so that the normality of the EDTA solution is exactly equal to 0.01 N.

3. **Calcon indicator solution (0.5%)** : Dissolve 0.5 g of Calcon reagent (Solo chrome dark blue and Erichrome blue black R) in 100 ml of 95% ethanol taken in 250 ml beaker. Store in a plastic bottole. Prepare fresh every 2 weeks.
4. **Sodium hydroxide solution (4 N or 16%)** : Weigh 160 g pure NaOH in a 600 ml beaker. Dissolve it by adding adequate distilled water and trasnfer it to 1 liter volumetric flask. Make the volume by distilled water and store the solution in a glass reagent bottle.
5. **Sodium cyanide solution (2%)** : Accurately weigh 4 g NaCN in a 400 ml beaker. Add 200 ml distilled water and dissolve the salt. Store the solution in a glass bottle of optimum size. NaCN is a poison, so handle it with care.
6. **Hydroxylamine Hydrochloride solution (5%)** : Weigh 5 g reagent hydroxylamine hydrochloride in a 250 ml beaker. Add 100 ml of distilled water to dissolve it. Store the solution in a plastic bottle. It should be prepared fresh every week.

Procedure

(i) In a 100 ml China dish, add 5 ml stock solution of plant digest (0.5 g plant material in 50 ml aliquot/solution) with pipette and dilute it to 10 ml with distilled water.

(ii) Add 10 drops each of 2% NaCN and hydroxylamine hydrochloride solutions.

(iii) Add 2.5 ml 4 n NaOH solution and then add 10-15 drops of Calcon indicator. Pink to reddish colour will appear.

(iv) In another China dish, prepare a blank solution exactly in the same way, taking 5 ml of distilled water instead of the test solution (aliquot). In this case, blue colour will appear, but if not, add few (2-3) drops of the EDTA solution to get the same. Keep this blue blank along side for comparing the end point of the test solution.

(v) Place the earlier China dish containing the pink coloured test solution on the magnetic stirrer top. Put a glass coated iron needle in it and start stirring.

(vi) Titrate the solution with 0.01 N EDTA until the pink colour just changes to blue matching the blank. Note the volume of the EDTA used, say x ml. Calculate Ca in the aliquot and express the results in appropriate units.

Calculations

Weight of the plant material taken for the digestion = 0.5 g

Final volume of the plant digest/aliquot made = 50 ml

Therefore, dilution factor = 50/0.5 = 100 times

Volume of the aliquot taken for titration = 5 ml

Volume of 0.01 N EDTA used for titration = x ml

Total Ca + Mg (me/100 g plant) =

$$\frac{\textit{ml of EDTA used X Normality of EDTA X dilution factor}}{\textit{ml of aliquot taken for titration}} \times 100$$

$$= \frac{\times ml\ X\ 0.01\,N\ X\ 100\ X\ 100}{5\ ml} = 20\times$$

Or, total Ca in plant sample (%)

$$= \frac{20\times X\ \textit{equivalent weight of Ca}}{1000}$$

$$= \frac{20\times X\,20}{1000} = 0.4\times$$

Interpretation of Results

Depending on the concentration of Ca determined in the plant material, a sample can be rated as sufficient or deficient.

Sufficient = 0.2 – 1.0 g Ca per 100 g plant material

Deficient < 0.2 g Ca per 100 g plant material.

Determination of Magnesium

Determination of magnesium by titration by EDTA is done at pH 10 in the presence of Erichrome Black-T indicator. The end point of titration is more distinct than calcium alone, where, indicator colour change at end point is not so sharp. In titration at pH 10, it is not only Mg which is titrated but both Ca + Mg are titrated together and Mg is determined from the difference of Ca + Mg and Ca.

Principle

Magnesium in a plant digest (solution/aliquot) can be determined by titration of a known volume of the aliquot against the standard solution of EDTA using Erichrome Black-T as indicator at pH 10 in the presence of NH_4Cl-NH_4OH buffer. Adequate concentrations of NaCN and hydroxylamine hydrochloride are also used to remove the interference of the heavy metals. The end point of titration will be the change of colour from purple to blue, when whole of Mg gets complexed with EDTA. If Ca is also present in the aliquot/solution this titration will estimate both Ca and Mg i.e., Ca + Mg.

Apparatus

Magnetic stirrer, glass coated iron needles, ceramic china dishes of 100 ml capacity, which bottles, micro burette, pipettes, beakers, volumetric flasks, etc.

Reagents

1. **Standard solution of Magnesium (0.01 N):** Weigh 0.1216 g unoxidized Mg metal and dissolve it in minimum volume (10-15 ml) of 1 N HCl taken in a 600 ml beaker. Add nearly 400 ml of distilled water and transfer the contents of the beaker to 1 liter volumetric flask. Finally, make the volume to 1 litre. If, the magnesium metal is not available, $MgSO_4.7H_2O$ can be used for making the required standard solution of magnesium by dissolving 1.2325 g of the salt in distilled water and then making the final volume to 1 litre.
2. **Standard solution of EDTA (0.01 N) :** Prepare this solution as described earlier for calcium.
3. **Erichrome Black-T indicator solution** (0.5%) : Dissolve 0.5 g Erichrome Black-T reagent in 100 ml 95% ethanol

taken in 250 ml beaker and store it in a plastic bottle. Prepare fresh solution every 2 weeks.

4. **Ammonium chloride-Ammonium hydroxide buffer:** Take 67.5 g of NH_4Cl in a 1 liter beaker and add 400 ml of distilled water to dissolve the salt. To this add 570 of 88% NH_4OH solution and make volume to 1 liter.
5. **Sodium cyanide solution (2%) :** Prepare this solution as described earlier.
6. **Hydroxylamine hydrochloride solution** (5%) : Prepare this solution as described above for calcium.

Procedure

(i) In a 100 ml China dish, add 5 ml stock solution of the plant digest (0.5 g plant material in 50 ml aliquot/solution) with pipette and dilute it to 10 ml with distilled water.

(ii) Add 10 drops each of 2% NaCN and hydroxylamine hydrochloride solutions.

(iii) Add 10 ml of NH_4OH - NH_4Cl buffer and 10 drops of Erichrome Black-T indicator. Purple colour will appear.

(iv) In another China dish, prepare a blank solution exactly in the same way, taking 5 ml of distilled water instead of the test solution (aliquot). In this case blue colour will appear, but if not, add few (2-3) drops of EDTA solution to get the same. Keep this blue blank along side for comparing the end point of the test solution.

(v) Place the earlier China dish containing the purple coloured test solution on the magnetic stirrer top. Put a glass coated iron needle in it and start stirring.

(vi) Titrate the solution with 0.01 N EDTA until the purple colour changes to blue matching the blank. Titration near the end point should be done very slowly. Note the volume of the EDTA used, say y ml.

Calculations

Weight of the plant material taken for the digestion = 0.5 g

Final volume of the plant digest/aliquot made = 50 ml

Therefore, dilution factor = 50/0.5 = 100 times

Volume of the aliquot taken for titration = 5 ml

Volume of 0.01 N EDTA used for titration = y ml

Total Ca + Mg (me per 100 g plant)

$$= \frac{ml\ of\ EDTA\ used\ X\ Normality\ of\ EDTA\ X\ dilution\ factor\ X\ 100}{ml\ of\ aliquot\ taken\ for\ titration}$$

$$= \frac{y\ ml\ X\ 0.01N\ X\ 100\ X\ 100}{5ml} = 20\ y$$

Total Ca (me per 100 g of plant material) = 20 x

Therefore, total Mg (me per 100 g of plant material)

= (Ca + Mg) – Ca

= 20 y – 20 x

= 20 (y – x)

Or, total Mg in plant sample (%)

$$= \frac{20\ (y-x)\ X\ Equivalent\ weight\ of\ Mg}{1000}$$

$$= \frac{20\ (y-x)\ X\ 12}{1000} = 0.24\ (y-x)$$

(c) Interpretation of Results

Depending on the concentration of Mg determined in the plant material, a sample can be rated as sufficient or deficient :

Sufficient = 0.1 – 0.4 g Mg per 100 g plant material

Deficient < 0.1 g Mg per 100 g plant material

Determination of Ca and Mg by Emission Spectrophotometry

In the flame emission spectrophotometry technique, the quantity of emission (radiation energy) produced by the element to be determined is compared with its known quantity in the solution. As compared to titrimetric and gravimetric method, it is rapid and sensitive method to determine K, Na, Ca, Mg and other ions with accuracy of 1-3% (Jackson, 1967). This method saves lot of time, when large number of soil and plant samples are to be analyzed.

Principle

The emission spectrometry/flame photometry is based on the basic principle that when a fine mist of solution containing various atoms or ions like Na, K, Ca, Mg, etc., is sprayed over a flame with the help of an atomizer, the atoms or ions absorb heat energy and get excited.

They emit energy in the form of characteristic electromagnetic radiations, when they come back to the ground state. In a flame photometer there is a provision to isolate or filter the characteristic radiations of a particlar atom/element through specific optical filters. The energy emitted will be directly proportional to the number of atoms excited which will depend on their concentration in the test solution. The energy so emitted is made to fall on the photocell of the flame photometer, which converts these radiations into electric energy. This energy is amplified and read on the galvanometer of the flame photometer.

Determination of Calcium

Apparatus

Emission spectrophotometer/Flame photometer, volumetric flasks (50 ml, 100 ml, 500 ml, 1000 ml), beakers, pipettes, etc.

Reagents

1. **1000 ppm stock solution of Ca :** Weigh 2.4973 g of $CaCO_3$ and trasnsfer it into a 600 ml beaker. Add to it 10 ml of 1 N HCl and boil to expel CO_2. Allow it to cool and then add about 500 ml distilled water. Transfer the contents in a 1 litre volumetric flask and make the final volume.
2. **0.4 N HCl base electrolyte solution :** It is prepared by the dilution of the concentrated HCl (11.6 N). This is used to prepare standard solutions of Ca from the 1000 ppm stock solution for preparing standard curve.

Procedure

A. Preparation of standard curve for Ca

(a) Preparation of standard solutions of Ca

Prepare 50 ppm solution of Ca by pipetting 25 ml of the 1000 ppm stock solution in a 500 ml volumetric flask and making the final volume with 0.4 N HCl base electrolyte solution. From the 50 ppm Ca solution, prepare a series of standard solutions of 0, 2.5, 5, 10, 15, 20, 30 and 40 ppm by taking 0, 5, 10, 20, 40, 60 and 80 ml of the solution in 100 ml volumetric flasks and making the final volume up to the mark with 0.4 N HCl base electrolyte solution.

(b) Calibration of flame photometer

(i) Switch on the flame photometer and apply Ca filter.

(ii) Feed distilled water to the instrument and set the zero of the galvanometer scale with the set zero knob.

(iii) Now feed the blank (0.4 N HCl base electrolyte) and reset the zero of the galvanometer.

(iv) Then feed the instrument with the highest concentration i.e., 50 ppm Ca solution and adjust the 100 of the scale with the amplifier knob.

(v) Repeat steps III and IV to ensure proper calibration.

(c) Readings of standard solutions

Feed all the standard solutions (0, 2.5, 5, 10, 15, 20, 30, 40 and 50 ppm) one by one starting from the lowest concentration and record the galvanometer readings against each one. Repeat the above step once more for ensuring the consistency of the observations.

(d) Plotting of the curve

(i) Take a simple graph paper and plot concentrations of the standard solutions of Ca on the x-axis.

(ii) On the y-axis of the graph paper, plot the readings of the instrument recorded against each concentration, i.e., 0, 2.5, 5, 10, 15 20, 30, 40 and 50 ppm.

(iii) Draw a curve starting from the origin (zero) and joining all the points up to the highest point (100). This will give the standard curve for comparing the concentrations in the test solutions.

B. Analysis of test solution for Ca

(a) Observations

1. Calibrate the instrument by following all the steps for the calibration of the flame photometer as listed above [A.(b)].
2. Feed the test solution in which Ca concentration is to be determined and note down the reading.
3. Compare the reading of the test solution with the standard curve and note down the concentration against the reading, let it be x ppm.

(b) Calculations

Observed concentration of Ca in test solution = x ppm

Weight of plant material taken for the digestion = 0.5 g

Total volume of the digest/aliquot made = 50 ml

Therefore, dilution factor = 50/0.5 = 100 times

So, total Ca in plant sample

= observed concentration X dilution factor

= x ppm X 100 times

= 100 x ppm.

Or, total Ca (%) in plant sample = 0.01 x

Determination of Magnesium

Apparatus

Emission spectrophotometer/Flame photometer, volumetric flasks (50 ml, 100 ml, 500 ml, 1000 ml). beakers, pipettes, etc.

Reagents

1. **1000 ppm stock solution of Mg :** Dissolve 1.000 g of metallic Mg foil in 10 ml of 1 N HCl and dilute it to 1 litre with distilled water. If metallic Mg foil is not available, $MgSO_4.7H_2O$ can be used. To make 1000 ppm solution of Mg, dissolve 10.1352 g of $MgSO_4.7H_2O$ salt in water and make up the final volume to 1 litre.
2. **0.4 N HCl base electrolyte solution :** It is prepared by diluting concentrated HCl (11.6 N). It is used to prepared standard solutions of Mg from the 1000 ppm stock solution for preparing standard curve.

Procedure

A. Preparation of standard curve for Mg

(a) Preparation of standard solutions of Mg

Prepare 50 ppm solution of Mg by pipetting 25 ml of the 1000 ppm stock solution in a 500 ml volumetric flask and mking the final volume with 0.4 N HCl base electrolyte solution. From the 50 ppm Mg solution, prepare a series of standard solutions of 0, 2.5, 5, 10, 15, 20, 30 and 40 ppm by taking 0, 5, 10, 20, 30, 40, 60 and 80 ml of the solution oin 100 ml volumetric flasks and making the final volume up to the mark with 0.4 N HCl base electrolyte solution.

(b) Calibration of flame photometer

(i) Switch on the flame photometer and apply Mg filter.

(ii) Feed distilled water to the instrument and set the zero of the galvanometer scale with the set zero knobs.

(iii) Now feed the blank (0.4 N HCl base electrolyte) and reset the zero of the galvanometer.

(iv) Then feed the instrument with the highest concentration i.e., 50 ppm Mg solution and adjust the 100 of the scale with the amplifier knob.

(v) Repeat steps 3 and 4 to assure proper calibration.

(c) Readings of standard solutions

1. Feed all the standard solutions (0, 2.5, 5, 10, 15, 20, 30, 40 and 50 ppm) one by one starting from the lowest concentrat-ion and record the galvanometer readings against each one.
2. Repeat the above step once more for ensuring the consistency of the observations.

(d) Plotting of the curve

(i) Take a simple graph paper and plot concentrations of the standard solutions of Mg on the x-axis.

(ii) On the y-axis of the graph paper, plot the readings of the instrument recorded against each concentration.

(iii) Draw a curve starting from the origin (zero) and joining all the points up to the highest point (100). This will give the standard curve to be used for comparing the concentrations in the test solutions.

B. Analysis of test solution

(a) Observations

1. Calibrate the instrument by following all the steps for the calibration of the flame photometer as listed above under section A(b).
2. Feed the test solution in which Mg concentration is to be determined and note down the reading.

3. Compare the reading of the test solution with the standard curve and note down the concentration against the reading, let it by y ppm.

(b) Calculations

Observed concentration of Mg in test solution = y ppm

Weight of plant material taken for the digestion = 0.5 g

Total volume of the digest / aliquot made = 50 ml

Therefore, dilution factor = 50/0.5 = 100 times

So, total Mg in plant sample

= observed concentration X dilution factor

= y ppm X 100 times

= 100 y ppm.

Or, total Mg (%) in plant sample = 0.01 y.

SULPHUR

The determination of total S in plants involves two steps i.e. (a) digestion of plant sample by suitable reagents to convert total S into anionic sulphate form in solution and (b) estimation of sulphate S in the solution. The method for dry ashing and wet digestion of plant samples are described in Chapter 2 of this book.

After conversion of total S in plants to anionic sulphate form in the digest either by dry ashing or by wet digestion, the sulphate S may be reduced using special apparatus (Johnson and Nishita (1952) methylene blue. However, the reduction of sulphate S and its determination is tedious. The sulphate S can be determined by turbidimetric method (Chesnin and Yien, 1950 and Tabatabai and Bremner, 1970), which is simple and rapid.

Turbidimetric Method

This method to determine sulphur in plants is commonoly used in laboratories for its simplicity, speed and reasonable accuracy. It is based on the principle of spectrophotometry.

Principle

In the turbidimetric method, sulphate S in plant is precipitated to barium sulphate by barium chloride crystals. The intensity of the barium sulphate turbidity is measured on spectrophotometer at 420 nm wave length.

Apparatus and Equipment

Volumetric flasks (50 and 100 ml), Pipettes, Spectrophotometer, 30 and 60 mesh sieves.

Reagents

(i) Barium chloride crystals ($BaCl_2\ 2H_2O$); Ground to pass 30 mesh sieve but retained on 60 mesh sieve.

(ii) Gum acacia solution (0.25%)

(iii) Potassium sulphate (K_2SO_4)

Procedure

(i) Take 2 to 10 ml aliquot of plant digest depending upont the S contnet in a 50 ml volumetric flask.

(ii) Add distilled water to the flask to make 15 to 20 ml contents in flask.

(iii) Add 1.0 g of barium chloride (30-60 mesh) crystals in the flask and shake vigorously for 1 to 2 minutes

(iv) Add 1 to 2 ml of 0.25% gum acacia solution in the flask to keep the suspension in the colloidal form. Again shake for 1-2 minutes and make up the volume to 50 ml with distilled water. White turbidicity of barium sulphate is developed, which remains stable for 5-30 min.

(v) The intensity of the turbidity is measured on spectrophotometer at 420 nm as % transmittance. It is converted to µg S ml^{-1} with the help of the standard curve of S, prepared afresh on the same spectrophotometer.

Preparation of standard curve

(i) Prepare a working standard solution of 100 µg S ml^{-1} by dissolving 0.5434 g dried reagent grade K_2SO_4 in distilled water and dilute to one litre.

(ii) Pipette out 0, 1, 2, 5, 8, 10, 15 and 20 ml solution from the working solution of 100 µg S ml^{-1} in a 100 ml volumetric flask. Add distilled water to these flasks to make approximately 50 ml content.

(iii) Add 1.0 g barium chloride (30 to 60 mesh) crystals to the flask and shake for one min.

(iv) Add 2 ml 0.25% gum acacia solution to each flask and again shake for 1-2 min.

(v) Make up the volume to 100 ml by adding distilled water and again shake for 1-2 min. It will give 0, 1, 2, 5, 8, 10, 15 and 20 μg S ml^{-1} solution. The turbidity of barium sulphate will be developed.

(vi) Read the intensity of turbidity of the solution of above flask at 420 nm on spectrophotometer as % transmittance.

(vii) Draw standard curve between concentration of S and % transmittance on a semi log paper and calculate the concentration of S in the aliquot taken in the flask.

Calculation of sulphur in plants

Weight of plant sample taken (W) = 1 g

Total volume of plant digest mode (V) = 100 ml

Volume of aliquot taken from plant digest (V_1) = Say 5 ml

Total volume of aliquot made (V_2) = 50 ml

(for reading % Transdmittance)

$$\text{Dilution factor (DF)} = \frac{V}{W} \times \frac{V_2}{V_1} \times \frac{100}{1} \times \frac{50}{5} = 1000 \; times$$

Let the concentration of S in (V_2) aliquot from standard curve be = Cμg Sml^{-1}

Total S in plant sample = (C X DF) = 1000 Cμg Sg^{-1}

Alternatively, the total S content in the plant sample can be calculated as under :

$$Total\ S\ in\ plant = \frac{\mu g\ Sml^{-1}\ in\ aliquot \times 50\ ml}{Volume\ of\ aliquot\ taken\ (ml)} \times \frac{Total\ volume\ of\ digest\ (ml)}{Wt.\ of\ plant\ sample\ taken\ (g)}$$

$$= \frac{C \times 50}{5} \times \frac{100}{1} = 1000\ C\ \mu g\ Sg^{-1}$$

The S content (%) in plant sample as under :

$$= \frac{1000\ C \times 10^2}{10^6} \times \frac{C}{10} = 0.10C\%\ S$$

Interpretation

The sulphur concentration in plant samples may vary according to species, plant age and its part. Low concentration leads to deficiency symptoms of the S in new leaves, which reduces the growth, development and yield. Sulphur in excess amount is non toxic in nature, but in industrial areas the acid rains may damage the crop. Singh (1999) has reported the critical levels of S in different crops (Table 1).

Table 1. Critical levels of S in select crop plants (45-55 day old).

Crop	S concentration in dry matter (%)		
	Deficient	Moderate	Sufficient
Rice, wheat, maize, millets	<0.20	0.20-0.30	>0.30
Groundnut, mustard. soybean, cowpea, brinjal, French bean, cucumber	<0.25	0.25-0.40	>0.40
Sunflower, linseed	<0.35	0.35-0.55	>0.55
Horse gram, pea, chickpea	<0.45	0.45-0.75	>0.75
Potato, cauliflower, spinach	<0.40	0.40-0.75	>0.75

Precautions

(i) The standard curve for each lot should be prepared afresh and separately.

(ii) While preparing the standard curve for S all the reagents and steps used in methods of plant digest should be followed.

(iii) Avoid contamination from chemicals, water, rubber tubing and stoppers, filter papers and glasswares etc. which may have S as impurities.

(iv) Always use standard analytical chemicals and reagents which are free from sulphur.

MICRO-NUTRIENTS

Micronutrients were first recognized as a limiting factor in crop production in 1920 in Florida, USA. For the normal growth and development, crop plants require 18 elements, of these nine (Zn, Mn, Fe, Cu, B, Mo, Ni, Co and Cl) essential elements, are required in micro amounts, hence called micronutrients. They play important roles in plant metabolism as they form stable complexes with naturally occurring ligands. Although, they are required by the plant in very small quantities byt they play major role in enzyme activiation. The plant samples may be collected, processed and digested for chemical analysis as per Chapter 2 in this book.

Generally, Atomic Absorption Spectrophotometer (AAS) is widely used to determine different micronutrient elements in plant samples.

The use of instrument will depend on the purpose of analysis. If one element is to be estimated, then AAS is the best. But if more than one element is to be estimated in the same sample then Inductively Coupled Plasma Atomic Emission Spectroscopy (ICP-AES) can be used. With these instruments direct reading of the element/nutrient concentration in the digest can be taken and if any dilution of the aliquot is needed, the same is done and the dilution factor is applied.

DETERMINATION OF MICRONUTRIENTS

Among the nine micronutrients (Fe, Cu, Zn, Mn, B, Mo, Cl, Co and Ni), Fe, Cu, Zn, Mn, Co and Ni are heavy metals. These heavy metal elements are suitably estimated on Atomic Absorption Spectrophotometer (AAS), because their atoms do not get excited under ordinary flames, hence, cannot be estimated correctly by Emission Spectrophotometer. Whereas, B and Mo are determined separately with spectrophotometer (Colorimeter). Besides, Cl is usually not tested as it is not deficient in soil or plant. Ni when exceeds the permissible limit it becomes toxic to the plant, hence has been included in the list of toxic elements.

Determination of Zn, Mn, Cu, Fe, Ni and Co

Principle

The six micronutrients (Zn, Mn, Fe, Cu, Ni and Co) can be determined in acid digest of plant samples, with the help of AAS (Atomic Absorption Spectrophotometer). The AAS, is based on the principle that atoms of metallic elements (Zn, Mn, Fe, Cu, Ni, Co), which normally remain in ground state under flame conditons, absorb energy when subjected to radiation of specific wavelength. The absorption of radiation is proportional to the concentration of atoms of that element. The absorption of radiation by the atoms in independent of the wavelength of absorption and temperature of the flame.

Apparatus

Conical flasks, volumetric flasks, hot, plate, pipettes, balance, wash bottle, Whatman No. 42 filter paper, AAS.

Reagents

Zinc Sulphate ($ZnSO_4.7H_2O$), Potassium Per Manganete ($KMnO_4$) Copper Sulphate ($CuSO_4.5H_2O$), Ferrous Ammonium Sulph-ate $(NH_4)_2SO_4$ $FeSO_4$. $6H_2O$, Cobaltus Chloride ($CoCl_2$) and Concent-rated Sulphuric Acid (H_2SO_4) and Oxalic acid ($C_2H_2O_4.2\ H_2O$).

Preparation of standard curves

Preparation of standard solutions of reagents

i. **Zinc Standard Solution**: Dissolve 0.439 g zinc sulphate ($ZnSO_4.7H_2O$) of AR grade in 200 ml of double distilled water in a 500 ml beaker. Add 5 ml of dilute H_2SO_4 (1:5) and make up the volume to 1.0 litre. This is a standard solut-ion of 100 mm Zn. Dilute 10 ml of this standard solution to 100 ml in a volumetric flask to prepare stock solution of 10 ppm. Dilute 1, 2, 4 and 6 ml of stock solution (10 ppm) in a 100 ml volumetric flask to get working standards (0, 0.1, 0.2, 0.4 and 0.6 ppm Zn).

ii. **Manganese Standard Solution** : Dissolve 0.288 g potassium permanganate ($KMnO_4$) AR grade in 300 ml of double distilled water in a 500 ml beaker. Add 20 ml concentrated H_2SO_4, warm to about 60°C and add oxalic acid solution drop by drop to make the solution colourless. Cool and transfer to a 1.0 litre measuring flask and make volume upto the mark. This is 100 ppm stock solution of Mn. Make use of this solution for preparing working standards (0, 0.2, 0.4, 0.8, 1, 2, 5 and 10 ppm).

iii. **Copper Standard Solution** : Dissolve 0.392 g copper sulphate ($CuSO_4.5H_2O$) of AR grade in 300 ml of double distilled water and make up volume to 1.0 litre. This is standard solution of 100 ppm Cu. To prepare working standards, dilute 100 ppm Cu stock solution to obtain 1, 2, 4, 6 and 8 ppm Cu.

iv. **Iron Standard Solution** : Dissolve 0.702 g of AR grade ammonium ferrous sulphate $(NH_4)_2SO_4$ $FeSO_4.6H_2O$ in 300 ml of double distilled water in a 500 ml beaker. Add 5 ml of dilute H_2SO_4 (1 : 5) and transfer to a1 litre measuring flask and make volume up to the mark. This is a standard solution of 100 ppm Fe. Finally prepare working standards containing 1, 2, 4, and 6 ppm Fe by diluting appropriate volumes of 100 ppm Fe.

v. **Cobalt Standard Solution** : Dissolve 0.220 g cobalts chloride ($CoCl_2$) AR grade in about 300 ml double distilled water taken in a 500 ml beaker. Dissolve the salt and transfer to a 1.0 litre volumetric flask and make up the volume upto the mark. This will give 100 ppm solution of Co.

From this solution prepare working standard of 1, 2, 4, 6 and 8 ppm.

vi. **Nickel Standard Solution :** Dissolve 1.0 g of Ni metal in 50 ml of HNO_3 (1 : 1 dilution). After the dissolution of Ni dilute the solution to 1.0 litre with double distilled water. This is a standard solution of 1000 ppm of NI. Make use of this solution for the preparation of working standards (0, 0.2, 0.4, 0.8, 1.0. 2, 4, 8 and 10 ppm).

Plotting of standard curves

Feed the working standard solutions of each nutrient element to AAS and take the readings against their concentrations and then prepare the standard curves for each element by plotting AAS readings against Zn, Mn, Cu, Fe and Co concentrations.

Procedure

(i) First, apply the respective cathode lamp for the particular nutrient being determined and then feed the blank of the working standards to the AAS.

(ii) Feed the aliquots (plant digest containing 0.5 g plant material in 50 ml volume) directly or after a required dilutions if needed.

(iii) Note down the readings (and dilution factor if dilution was done) on the AAS. Let the reading be `M'.

(iv) Compare the readings of each element with their standard curves prepared already, and calculate concentration of each nutrient element in ppm.

Calculations

Weight of plant material taken = 0.5 g

Volume made after acid digestion = 50 ml

Dilution factor = 50/0.5 = 100 times

Concentration of Zn, Mn, Fe, Cu, Ni or Co (ppm) in plant sample = M x 100.

Where M stands for the Zn, Mn, Fe, Cu, Ni or Co concentration in aliquot as read from standard curve against the sample readings.

Precautions

(i) Always grind plant samples in stainless stell grinding machine.

(ii) Use only double distilled water, corning/borosil glass apparatus and AR grade chemicals.

Determination of Boron (B)

Boron in the acid digest of plant sample can be determined by a spectrophotometer using azomethine-H reagent (Berger and Truog, 1939; Sippola and Ervco, 1977).

Equipment and apparatus

Spectrophotometer, a dispenser boron-free glassware, viz., conical flasks, volumetric flasks, pipette.

Reagents

(i) **Concentrated HNO_3 and $HClO_4$**

(ii) **Azomethine-H reagent :** Dissolve 0.45 g of azomethine-H in 100 ml of 1% L-ascrobic acid solution in polypropylene reagent bottle. Fresh reagent should be prepared each week and stored in a refrigerator.

(iii) **Buffer solution :** Dissolve 250 g of ammonium acetate and 15 g of EDTA disodium salt in 400 ml of distilled water and slowly add 125 ml of glacial acetic acid and mix.

Preparation of standard solutions

Dissolve 0.570 g boric acid (H_3BO_3) in distilled water and dilute to one litre to obtain a stock solution of 100 ppm B. Dilute 5 ml of stock solution to 100 ml with distilled water which will give a B concentration of 5 ppm. Pipette 0, 0.25, 0.50, 1.0 ,2.0 and 4.0 ml of 5 ppm solution to a 25 ml volumetric flask to have working standards of 0, 0.05, 0.10, 0.20, 0.40 and 0.80 ppm B, respectively.

Preparation of standard curve

Read the intensity of the color of the solution on a spectrophotometer at 420 nm. Draw a standard curve by plotting the % transmittance transmittance (Y-axis) against B concentrations (X-axis) on a semi log paper.

Procedure

(i) Pipette 1 ml aliquot from the acid digested solution in a 25 ml volumetric flask and add 4 ml of buffer solution and mix well.

(ii) Add 4 ml of azomethine-H reagent and mix the contents.

(iii) The colour is allowed to develop for 1.0 h, and the volume is made up to the mark.

(iv) Intensity of the colour is measured on a spectrophotometer at 420 nm.

(v) Calculate the concentration of B against the reading from the standard curve as under.

Calculations

Weight of the plant material taken = 0.5 g

Volume made after acid digestion = V ml

Volume of aliquot taken = V_1 ml

Final volume made = 25 ml

Transmittance (%) as read from spectrophotometer = T

Concentration of B in the solution (ppm) against the transmittance T = C

First dilution = (V/0.5) times

Second dilution = ($25/V_1$) times

Total dilution = (V/0.5) x ($25/V_1$) times = Z

Content of B (ppm) in plant sample = C x Z

Determination of Molybdenum (Mo)

Principle

Molybdenum in the plant samples can be determined by ashing and fusion of sample. A knonw volume of the filtrate, containing measured amount of plant material ashed and fused with anhydrous Na_2CO_3, is taken and Mo content is determined spectrophotometerically.

Equipment and Apparatus

Spectrophotometer, hot plate, refrigerator, water bath, muffle furnace, separating funnel, glass stoppered flasks, volumetric flasks, pipettes.

Reagents

(i) Anhydrous Na_2CO_3.

(ii) **50% Potassium iodide solution :** Dissolve 50 g potassium iodide in 100 ml of double distilled water (DDW).

(iii) **50% Ascorbic acid solution :** Dissolve 50 g ascorbic acid in 100 ml of DDW.

(iv) **10% Tartaric acid solution :** Dissolve 10 g Tartaric acid in 100 ml of DDW.

(v) **10% Thiourea solution** : Dissolve 10 g thiourea in 100 ml of DDW and filter. Always prepare a fresh just before use.

(vi) **Dithiol solution (Toluene-3, 4-dithiol) :** One g of dithiol (51°C) is added in 100 ml of 10% NaOH solution. The contents are warmed up to 51°C with frequent stirring for 15 minutes on a hot plate. Then 1.8 ml of thioglycolic acid is added and stored in refrigerator.

(vii) Isoamyul acetate solution.

(viii) **Ferrous ammonium sulphate solution :** Dissolve 63 g of ferrous ammonium sulphate in distilled water, add 10 ml concentrated H_2SO_4 and make the volume to 1.0 litre.

(ix) Ethyl alcohol (Ethanol).

Preparation of standard solutions of Molybolate

Dissolve 0.184 g of AR grade ammonium molybdate in 100 ml of 0.1 N NaOH, make it slightly acidic with dilute HCl and make volume to 1.0 litre to obtain a stock solution of 100 ppm. Dilute 1, 2, 3, 4, 5, 6, 8, 10 mls of 100 ppm standard Mo solution to 1000 ml with water in volumetric flask to obtain respective working standards of 0.1, 0.2, 0.3, 0.4, 0.5, 0.6, 0.8, 1.0 ppm of Mo.

Preparation of standard curve

Read the colour intensity of different working standard solutions of Mo on a spectrophotometer at 680 nm by using red filter. Draw a standard curve by plotting the % transmittance (Y-axis) against Mo concentrations (X-axis) on a semi log paper.

Procedure

(i) Transfer 5 g material (1 to 10 g) in a platinum dish.

(ii) Ash it in an electric furnace at 500°C. as per Chapter 2 in this book.

(iii) After ashing, fuse the ash with 2 g of anhydrous Na_2CO_3 at 1000°C in the electric furnace, ensuring complete contact of the entire ash with the flux.

(iv) Cool the dish and drop the cake in a 250 ml beaker by inverting the dish.

(v) If cake does not detach, place the dish in beaker and disintegrate the cake in 100 ml water cotnaining 2% ethanol (v/v).

(vi) Measure the suspension in a cylinder and filter through Whatman No. 42 filter paper in a 250 ml beaker.

(vii) Transfer 20 ml of the filtrate in 250 ml separatoring funnels and add 0.25 ml of the ferrous ammonium sulphate solution and 20 ml of double distilled water and shake the funnel vigorously, up the down.

(viii) Add excess of the KI solution and clear the librerated iodine by adding ascorbic acid drop by drop on continuous stirring/ shaking.

(ix) Add one ml of tartaric acid and 2 ml of the thiourea solution and again shake vigorously.

(x) Add 5 drops of the dithiol solution and allow the mixture to stand for 30 min.

(xi) Add 10 ml of the isoamyl acetate and separate out the contents (green color) in glass cuvettes to read the colour intensity (% T) at 680 nm on a spectrophotometer using red filter.

Calculations

Weight of the plant material taken, = 5 g

Volume made after acid digestion = 100 ml

Volume of aliquot taken for color development = 20 ml

Mo content (ppm) in plant sample = $C \times \frac{100}{20} \times \frac{1}{5} = C$

Where, C stands for Mo concentration obtained from standard curve against the reading (%T).

Interpretation of plant test results for micronutrients

Table 2 shows the deficient and sufficient levels of all micronutrient elements, for interpretation of results of plant analysis.

Table 2. Micronutrients content in mature plant tissue for classification as deficient and sufficient.

	Micronutrien concentration (mg/kg dry matter)	
Micronutrient	Deficient	Sufficient
Zn	<20	25-150
Fe	<50	50-250
Mn	<20	20-500
Cu	<4	5-20
B	<15	20-100
Mo	<0.01	0.5-0.1
Co	<10	10-20
Ni	<0.1	0.1-10

Source : Mortvedt *et al.* (1972), Tandon (1984), Jones *et al.* (1991) and Bhargav and Raghupati (1993).

TOXIC METALS

Recently, public awareness about metals (cadmium, lead, nickel and chromium) contamination in environment has increased. The disposal of sewage/industrial effluents and city wastes on agricultural land is becoming a major problem. Such materials contain various toxic metals (cadmium, lead, nickel and chromium) and pollute the soils. Their accumulation/high concentration in soils, causes toxicity (plants, organisms, animals and humans) and adversely affect the nutrient transformations, microbial population and enzyme activities in plants.

ANALYTICAL METHODS

The plant samples may be collected, processed and digested for chemical analysis as per Chapter 2 of this book. Many classical colorimetric procedures have been described to determine the toxic metals in plant samples. However, presently atomic absorption spectrophotometer (AAS) or Inductively Coupled Plasma. Atomic Emission Spectroscopy (ICP-AES) methods are used. With these instruments, direct reading of the digest can be taken.

Determination

Generally, Atomic Absorption Spectrophotometer is used to determine various toxic metals in plant samples. If more that one elements are to be estimated in plants at one time, ICP-AES technique is used.

Principle

The AAS is based on the principle that atoms of metallic elements (Cd, Ni, Pb and Cr), which normally remain in ground state, absorb energy when subjected to radiation of specific wavelength. The absorption of radiation is proportional to the concentration of atoms of that element. The absorption of radiation by the atoms is independent of the wavelength of absorption and temperature of the flame.

Apparatus

Conical flasks, volumetric flasks, hot plate, pipettes, balance, wash bottle, Whatman No. 42 filter paper, AAS.

Reagents

Ni metal, Concentrated Nitric Acid (HNO_3), Cd metal, concentrated Hydrochloric Acid (HCl), Potassium Dichromate ($K_2Cr_2O_7$) and Pb metal

Preparation of standard solutions

(i) **Nickel Standard Solution :** Dissolve 1.0 g of Ni metal in 50 ml of HNO_3 (1 : 1 dilution). After the dissolution of Ni, dilute the solution to 1.0 litre with double distilled water. This is a standard solution of 1000 ppm of Ni, Use this solution to prepare working standards (0, 0.2, 0.4, 0.8, 1.0, 2, 4, 8 and 10 ppm).

(ii) **Cadmium Standard Solution :** Dissolve 1.0 g of Cd metal in 50 ml of HCl (1 : 1 dilution). Then dilute to 1.0 litre with double distilled water. This is a standard solution of 1000 ppm. Use this solution to prepare working standards (0, 0.05, 0.1, 0.2, 0.4, 0.6, 0.8, 1.0 and 2.0 ppm).

(iii) **Chromium Standard Solution :** Dissolve 2.8282 g of reagent grade potassium dichromate ($K_2Cr_2O_7$) in water, adjust to pH 6 to 7, and dilute to 1.0 litre with double distilled water. This is a standard solution of 1000 ppm. Use this solution to prepare working standards (0, 0.5, 1.0, 2.0, 4, 6, 8 and 10 ppm).

(iv) **Lead Standard Solution :** Dissolve 0.100 g of 99.9% Pb metal in 1:1 nitric acid (HNO_3). Dilute to 1.0 litre to give 100 ppm Pb solution. Finally prepare working standard containing 1, 2, 4, 6, 8 and 10 ppm Pb, by diluting 100 ppm Pb stock solution.

Preparation of standard curves

Feed the working standard solutions of each metal to AAS and record the readings against their concentrations. Prepare the standard curves by plotting AAS readings against Ni, Cd, Cr and Pb concentrations.

Procedure

1. First, apply the respective cathode lamp for the particular nutrient being determined and then feed the blank of the working standards to AAS.
2. Feed the aliquots (plant digest contianing 0.5 g plant material in 50 ml volume) directly or after a required dilutions if needed.
3. Note down the readings (and dilution factor if dilution was done) on the AAS. Let the reading be `M'.
4. Compare the readings of each element with their standard curves prepared already, and calculate concentration of each nutrient element in ppm.

Calculations

Weight of plant material taken = 0.5 g

Volume made after acid digestion = 50 ml

Dilution factor = 50/0.5 = 100 times

Concentration of Cd, Ni, Pb or Cr (ppm) in plant sample = M x 100

Where, M stands for the Cd, Ni, Pb or Cr concentration in aliquot as read from standard curve against the sample readings.

Precautions

(i) Always grind plant samples in stainless steel grinding machine.

(ii) Use only double distilled water, corning/borosil glass apparatus and AR grade chemicals.

REFERENCES

Berger, K.C. and Truog, E. 1939. Boron determination in soils and plants. *Industrial and Engineering Chemistry Analysis* 11 : 540-45.

Bhargava, B.S. and Raghupati, H.B. 1993. Analysis of plant materials for macro and micronutrients. Pp. 49-82. In : *Methods of Analysis of Soils, Plants, Waters and Fertilizers* (Ed. H.L.S. Tandon) FDCO, New Delhi.

Bremner, J.M. 1960. Determination of nitrogen in soil by the Kjeldahal method. *Journal of Agricultural Sciences* 55: 11-33.

Buresh, R.J., Austin, E.R. and Craswell, E.T. 1982. Analytical methods. In: N-15 Research. *Fertilizer Research* 3: 37-62.

Chesnin, L. and Yien, C.H. 1950. `Turbidimetric determination of available sulphate. *Soil Science Society of America Proceedings* 15 : 149-151.

Cro̍oke, W.M. and Simpson, W.E. 1971. Determination of ammonia on Kjeldhal digest of crops by an automated procedure. *Journal of Science, Food and Agriculture* 22: 9-10.

Hesse, P.R. 1971. *A Textbook of Soil Chemical Analysis.* John Murray Publisher, London.

Jackson, M.L. 1967. *Soil Chemical Analysis.* Prentice Hall of India, New Delhi.

Johnson, C.M. and Nishita, H. 1952. Microestimation of sulfur in plant materials, soils and irrigation water. *Analytical Chemist* 24 : 736-742.

Jones, J.B. Jr., Wolf, B. and Mills, H.A. 1991. *Plant Analysis Handbook : A Practical Sampling, Preparation, Analysis, and Interpretation Guide.* Micro-macro Publishing Inc. Athens, Georgia, USA.

Koenig, R.A. and Johnson, C.R. 1942. Colorimetric determination of phosphorus in biological materials. *Industrial and Engineering Chemistry* 16: 233-237.

McGill, D. and Figueiredo 1993. Total nitrogen pp. 201-211. In: M.R. Carter (ed.) *Soil Sampling and Methods of Analysis.* Lewis Publ. Boca Raton, FL, USA.

Mortvedt, J.J., Giordand, P.M. and Lindsay, W.L. 1972. *Micronutrients in Agriculture.* Soil Science Society of America, Inc. Madison, Wisconsin, USA.

Murphy, J. and Riley, J.P. 1962. A modified signle solution method for the determination of phospholi in natural waters. *Analytica Chimica Acta* 27: 31-36.

Singh, M.V. 1999. Sulphur Management of Oilseed and Pulse Crops. Bulletin No. 3. *Indian Institute of Soil Science,* Bhopal, India, pp. 54.

Sippola, J. and Ervco, R. 1977. Determination of boron in soils and plants by azomethine-H method. *Finnish Chemistry Letters* 138: 140.

Tabatabai, M.A. and Bremner, J.M. 1970. Simple turbidimetric method for determination of total sulfur in plant material. *Agronomy Journal* 62 : 805-806.

CHAPTER - 4

LEAF PIGMENTS AND CHLOROPHYLL FLUORESCENCE

INTRODUCTION

In most cases, plant productivity is limited by availability of photosynthates and its utilization. Production of photosynthates depends on photosynthesis, which is regulated by rate of light and dark reactions, availability of CO_2, moisture and chloroplastic pigments etc. All green coloured plant parts contain pigments, which absorb light and play important roles in photosynthesis and photo-damage. These are chlorophyll a, chlorophyll b, xanthophylls and carotenoids.

Estimation of chlorophyll using acetone

Leaf pigments are more soluble in organic solvents. Efficiency of organic solvent is increased, if it is mixed with water. Thus aqueous organic solvents are not only specific in extracting chlorophylls and related pigments, other low molecular weight substances like carbo-hydrates, amino acids and phenols etc are also extracted. Acetone (80%) has been widely used to extract leaf pigments.

Principle

Plant pigments present in the chloroplasts are responsible for light reactions. These have very poor solubility in water. Numerous water-miscible organic solvents are used to extract the chlorophyll from various types of plant materials. Arnon (1949) developed a simple method to extract chlorophyll from the leaf tissue by extracting it with 80% acetone and calculating the amount of chlorophyll using the extinction coefficients at 664 and 647 nm. This method has been extensively used for estimating the chl `a' and chl `b' from various types of plant tissues.

Materials

Leaf tissue, pestle and mortar, measuring cylinder, beaker, pipetttes, acetone, distilled water, centrifuge and spectrophotometer.

Procedure

For chlorophyll extraction, 100-200 mg leaf tissue or leaf disc of known area or complete leaf or leaflets are homogenized using a pestle and mortar with 5 ml of 80% acetone at low temperature. The homogenized sample is centrifuged at 10000 x g for 15 min. The supernatant is collected in a test tube and the pellet is re-extracted with 80% acetone till it is free from any chlorophyll. After centrifugation, the supernatants are pooled together and the volume made to 10 ml with 80% acetone.

Observations

The absorbance (A) is recorded at 664 and 647 nm with 80% acetone as blank. For accurate determination of chlorophyll, the extinction coefficients of Graan and Ort (1984) can be used based on absorbance readings made at 647 and 664 nm. The efficiency of acetone methods has been imporved by using buffered (pH 7.5) acetone (Krishnan *et al.*, 1996).

Chlorophyll a ($\mu g\ ml^{-1}$) = 13.19 A664 – 2.57 A647

Chlorophyll b ($\mu g\ ml^{-1}$) = 22.10 A647 – 5.26 A664

Total Chlorophyll ($\mu g\ ml^{-1}$) = 7.93 A664 + 19.53 A647

Precautions

(i) The extract should not be exposed to light.

(ii) The extract should not be acidic.

(iii) Homogenise the material at low temperature and in diffused light.

Extraction of pigments from leaf tissue without maceration (DMSO method)

Principle

The old methods for extraction of chlorophyll used several water miscible solvents, viz., methanol, ethanol, acetone, pyridine and acetone plus ethyl acetate (Strain and Snec, 1966). Generally, 80 or 90% acetone has been used to extract chlorophyll from algae, gymnosperm and angiosperms (Arnon, 1949; Bruinsma, 1963; Slack *et al.,* 1973). According to Arnon (1949) chlorophyll extraction involves grinding the plant tissue in 80% acetone with subsequent centrifugation to remove solid plant material. Furthermore, the presence of degradation products renders the chlorophyll extract unstable and demands that spectroscopic analysis be carried out immediately after extraction.

Shoaf and Lium (1976) showed that dimethylsulphoxide (DMSO) is superior to acetone in the extraction of chlorophyll from a wide range of algal species. Hiscox and Israelstam (1979) developed a method by using DMSO to extract chlorophyll from the unmacerated leaves/leaflets by keeping test tubes, containing weighed amount or leaf disc in a known volume of DMSO, at 65°C for 2-3 h. The chlorophyll remains stable in DMSO and spectrophotometeric readings can be taken at convenienece within a day or so.

Materials

Test tubes, pipettes, graduated tube, oven, spectrophotometer, DMSO and leaf tissue.

Procedure

One hundred mg of leaf tissue (fragmented) is placed in a vial containing 7 ml DMSO and chlorophyll is extracted into fluid without grinding at 65°C till the tissue became chlorophyll free (2-4 h). Heavily cutinized and thick leaves will take more time than the thin leaves with light cutin layer. The liquid extract is transferred to a graduated tube and the volume made to 10 ml with DMSO. Absorbance is recorded immediately or sample-containing vials can be stored at 4°C.

Determination of chlorophyll and carotenoids content

A 3.0 ml aliquot of chlorophyll extract is transferred to a cuvette and the abosrbance (A) values are recorded at 480, 645 and 665 nm against DMSO as blank using a spectrophotometer. If the absorbance values are greater than 0.7, dilute the extract with DMSO. Calculate the concentrations using Wellburn's (1994) equations.

Chlorophyll a ($\mu g\ ml^{-1}$) = 12.19 A 665 - 3.45 A 645 = X

Chlorophyll b ($\mu g\ ml^{-1}$) = 21.99A645 - 5.32A665 = Y

Carotenoids ($\mu g\ ml^{-1}$) =(1000 A480-2.86 ChI a-129.9 ChI b)/221= Z

From these values pigment amount can be expressed on µmol basis using equations of Hendry and Price (1993).

ChI a nmol cm^{-2} or g^{-1} or $leaf^{-1}$

$$= \frac{X \times 10 \times 1.119}{Leaf\ area\ \left(cm^{-1}\right) or\ 0.1\ g\ or\ No.\ of\ leaf}$$

ChI b nmol cm^{-2} or g^{-1} or $leaf^{-1}$

$$= \frac{Y \times 10 \times 1.102}{Leaf\ area\ \left(cm^{-1}\right) or\ 0.1\ g\ or\ No.\ of\ leaf}$$

Carotenoids nmol cm^{-2} or g^{-1} or $leaf^{-1}$

$$= \frac{Z \times 10 \times 1.102}{Leaf\ area\ \left(cm^{-1}\right) or\ 0.1\ g\ or\ No.\ of\ leaf}$$

Advantages

(i) Large number of samples can be estimated easily as this method does not involve the grinding and centrifugation.

(ii) Chlorophyll is more stable in DMSO, the samples can be stored for 2-4 days at 0-4°C and absorbance is recorded at convenience.

Precautions

(i) Ensure that all chlorophyll has been extracted from the leaf sample.

(ii) Heating temperature should not exceed beyond 65°C.

(iii) Heating period should not be more than 4 h.

Estimation of carotenoids

Materials

Fresh, healthy and leaf spot infected groundnut leaves

Blender or pestle and mortar

Buchner funnel

Whatman No. 42 filter paper

Beakers, 100 ml

Separatory funnel

Pipettes, 10 ml

Rotary evaporator

Hot-plate

Colorimeter

Reagents

Acetone or methanol

Peroxide-free ether

Ethanol

Aqueous 60% KOH

Method

Chop the freshly collected leaves weighing about 2 g and transfer to a mortar along with 20 ml of either dist acetone or methanol. Homogenize thoroughly in a mortar with pestile. Decant the extract and filter it on a Buchner funnel through Whatman No. 42 filter paper. Repeat the extraction twice or until the tissue is free from the pigments. Pool the filtrates and partition with equal quantity of peroxide-free ether thrice using a separatory funnel. Add water if necessary to produce two layers during the initial ether extraction. The ether phase contains the carotenoids. Evaporate the combined ether extract under reduced pressure at 35°C in a rotary evaporater or on a hot water bath. Dissolve the residue in minimum quantity of ethanol. Add 60% aqueous KOH at the rate of 1 ml for every 10 ml of ethanol extract to saponify it. This would remove chlorophylls and interfering lipids and also cleave the esterified carotenoids. Keep this mixture in the dark in presence of nitrogen or boil it for 5-10 min or leave overnight at room temperature. Add equal amount of water and partition twice with peroxide free ether. Evaporate the ether under reduced pressure and dissolve the residue in minimum volume of ethanol. Measure the absorbance of this solution at 450 nm in a colorimeter. Calculate the carotenoid content using a calibration curve constructed using B-carotene.

Carotenoid in crude plant extracts

An estimate of total carotenoid of extracts containing negligible amounts of chlorophylls may be obtained directly.

Method

Make suitable dilutions of acetone or methanol extract of the plant tissue and measure absorbance at 450 nm and 670 nm in a colorimeter. If A_{450} is 10 times that of A_{670} the carotenoids present in the extract may be estimated from A_{450} values using the following formula:

$$C = \frac{D \times V \times f \times 10}{2500}$$

where, C = total amount of carotenoids in mg,

D = absorbance at 450 nm in a 1.0 cm cell,

V = Volume of the original extract in ml

f = dilution factor, and

2500 = average extinction coefficient of the pigments.

Thin-layer Chromatography of Chloroplast Pigments

Principle

The purpose of this exercise is to extract the chloroplast pigments from leaves of spinach, separate the pigments by thin-layer chromatography and to identify the pigments by their chromatographic behaviour and absorption spectra. The chlorophylls as well as carotenoids are extractable in organic solvents like ethanol (80%), acetone (80%) ether and DMSO. The chloroplasts of all higher plants contain chlorophylls (`a' and `b') and carotenoids. The carotenoids are of two fundamental types : the carotenes (C-40 hydrocarbon chains) or xanthophylls (C-40 oxygenated hydrocarbon chains). A typical representative of carotenes is b-carotene and that of xanthophylls is lutein. Chlorophylls and carotenoids occur in close association with each other and are constituents in grana of chloroplasts with the protein and lipid membrane.

Materials

Pipettes, measuring cylinders, Erlenmeyer flasks, tissue homogenizer, Buchner funnel, rotary film evaporatro, TLC glass plates, TLC gel applicator, TLC-plate developing chamber. Spinach leaves, n-hexane, ethyl ether acetone silica gel G.

Procedure

Slurry of silica gel G is prepared by vigorously shaking 30 g of the silica gel with 60 ml of distilled water in a stoppard 250 or 500 ml Erlenmeyer flask. The slurry must be prepared quickly and used immediately, because silica gel contains $CaSO_4$ which acts as binder. Using the special mounting board and applicator provided, coat a numbe of 5 x 20 cm glass plates with slurry of silica gel G. The applicator should set to apply a layer of absorbent 250 mm thick. The plantes are placed in a drying rack and stored in desiccators until the time to apply the samples to be chromatographed. In the same way 20 x 20 cm glass plates can be prepared.

Chloroplast pigments are extracted from spinach leaves by grinding approximately 25 g of fresh leaves (or approximately 2.5 g of dried leaves) in approximately 100 ml of acetone or 80% acetone (if dried leaves are used) in an homogenizer. When the acetone is turned deep green, add several grams (1-2.5 g) of anhydrous Na_2SO_4 to remove water from the extract and filter the preparation on a Buchner funnel. Evaporate the acetone extract in a rotary film evaporator; redissolve the dried pigments in approximately 10 ml of chloroform.

On the glass plates, 5 x 20 or 20 x 20 cm coated with absorbent, a sheet of tissue paper is run around the margins to remove excess or flanking gel. On the end of the gel, to which sample is to be applied, small scrateches are made on both sides at a distance of 1.5 cm from the end. Then scratch a small mark at a distance of 15 cm on each side of the plate above the original marks. The two sets of markes described above mark the origin and the position to which the solvent will be allowed to migrate, respectively.

In the next step, using a 0.01 or 0.1 ml pipette or a special micro burette apply a small aliquot of the extract between the two marks. Be sure the sample is spotted near the center of the origin end of the plate. A coo-air fan/hair dryer may be used to hasten evaporation of the solvent if repeated application of dilute extract is necessary.

When the sample spot is thoroughly dry, place the plagte in a developing chamber containing solvent (hexane-diethyl ether-acetone, 60 : 30 : 20 v/v). The depth of the solvent is critical, since the sample spot must not dip into the solvent.

Observe the development process and differential migration of the various pigments on the chromatogram. When the solvent front has advanced 15 cm above the origin, remove the chromatogram from the developing tank and allow it to dry.

Observations

Make a scaled diagram of the chromatogram, showing all visible pigments. Calculate Rf value for each pigment.

$$\text{Rf value} = \frac{\text{Distance moved by the pigment}}{\text{Distance moved by the solvent}}$$

TLC on MN30 cllulose in petroleum (b.p 60-80°C) - acetone-n-propanol (90 : 10 : 0.45) for 30 min. run on 20 x 20 cm glass plates. Separation can also be achieved on microscopic slides (5 min run).

As soon as inspection of chromatograms is completed, bands of silica gel containing the individual pigments are scraped using a clean sharp razor blade. The scraped silica gel is extracted with 5 ml of 80% acetone by vortexing and centrifuging. The supernatant from each tube is removed and used for measuring the absorbance with the help of spectrophotometer between 400-700 nm at 10 nm intervals. The absorbance is plotted against wavelength to obtain absorption spectrum.

Table 1. Rf values and colours of chlorophylls a and b on thin layer chromotography.

S.N.	Pigments	Rf value	Colour in day light	Nature of artifacts
1.	Phaeophytin a	93	Grey	Magnesium free
2.	Phaeophytin b	80	Yellow-brown	chlorophylls
3.	Chlorophyl a	60	Blue-green	–
4.	Chlorophyl b	35	Yellow-brown	–
5.	Pheophorbide a	18	Grey	Magnesium-free
6.	Pheophorbide b	07	Yellow-brown	chlorophyllides
7.	Chlorophyllide a	03	Blue-green	Chlorophylls without
8.	Chlorophyllide b	02	Yellow-green	phytyl side chains

Precautions

(i) To minimize degradation, the pigments should be eluted as soon as possible and if the extracts are to be stored, they should be stored at low temperature.

(ii) The dry chromatogram should be covered with a glass plate and the edges taped air tight, if the pigments cannot be eluted within 10-15 min. following drying.

(iii) Development of TLC plates should be carried out under dim light.

CHLOROPHYLL STABILITY INDEX

Chlorophyll stability index (CSI) is a measure to know the extent of chlorophyll pigments decomposition at higher temperature. CSI is measured by determining the difference between the chlorophyll value of fresh material and that of the same material when heated to 65°C for 1 h (Murthy and Majumdar, 1962). Lower the value of CSI, greater is the stability of chlorophyll at higher temperature. Thus, chlorophyll stability is a function of temperature and this property of chlorophyll stability correlates well with drought resistance.

Materials

Large size graduated glass tubes of 30 ml capacity, 80% buffered acetone (80 ml acetone made up to 100 ml with 20 ml 2.5 mM sodium phosphate buffer, pH 7.8), distilled water, Balance, UV visible spectrophotometer, water bath with thermostatic control.

Procedure

(i) Place 100 mg of leaf sample (small pieces) in graduated clean glass tuves containing 25 ml 80% buffered acetone and keep it as control.

(ii) Place another 100 mg of this leaf sample in graduated glass tube containing 10 ml distilled water and subjected it to high temperature of 65°C for 1 h in water bath.

(iii) After 1 h, drain out the distilled water and add 25 ml of 80% buffered acetone.

(iv) Keep both the tubes (control and high temperature treated) in a dark place at 4°C for overnight.

(v) Shake the contents of the tubes occasionally to accelerate the pigment extraction.

(vi) After the incubation period, filter the extract liquid through glass wool to remove leaf pieces and trasnfer to another graduated tube and total volume is made 25 ml with 80% buffered acetone.

(vii) The chlorophyll content is spectrophotometrically analysed in a UV visible spectrophotometer using 3 ml glass cuvettes at wave length 648 nm and 665 nm

(viii) The chlorophyll content can be calculated following the equation proposed by Barnes *et al.* (1992). The equation for quantitative determination of chlorophyll is as under :

$$\text{Total Chlorophyll} = 7.49A^{665} + 20.34A^{648}$$

(ix) CSI is the difference between two readings from heated and unheated samples.

Comments

Dimethyl sulfoxide (DMSO) can also be used as an extraction solvent but the recovery of chlorophyll is 7% more in acetone incubation method (Krishnan *et al.,* 1996). Excised leaf samples should be brought to be laboratory in polythene bags lined with moist filter paper inside. It is important to check the period of incubation at high temperature as this period may vary with different leaf types.

CHLOROPHYLL FLUORESCENCE

Photosynthesis is the process by which plants, algae, cyanobacteria and photosynthetic bacteria convert radiant energy into a chemically stable form. The pathway of this energy transduction is complex, involving several physical and chemical mechanisms and many components. The process is initiated when light is absorbed by the antenna molecules within the photosynthetic membrane. Light energy is absorbed by chlorophyll within plant tissue and used to drive photochemistry of photosynthesis and thus become chemical energy available to the plant for growth. Changes in chlorophyll content occur due to nutrient deficiencies, exposure to environmental stresses, exposure to certain herbicides and differences in light environment during the plant growth. Thereby chlorophyll contents can be used to manage nutrient optimization programmes that both improve crop yield and help to protect the environment.

Energy in excess of that used for photochemistry must be effectively dissipated by non-photochemical processes such as emission of heat and re-emission as longer wavelength red/far-red light energy. This re-emission of light is termed Chlorophyll Fluorescence, Although fluorescence emission from whole leaf system is too weak to be viewed with the naked eye, but it can be observed from illuminated extracts of chlorophyll solution. Peak fluorescence occurs in the red region of spectrum (685 nm) and extends into the infrared region up to 800 nm.

Chlorophyll fluorescence is a rapid and reliable non invasive probe of photochemistry. There are two separate photochemical steps during the photosynthesis associated with different groups of pigments; these are known as photo system I and photo system II . At physiological temperatures, approximately 95% of the chlorophyll fluorescence signal observed is derived from chlorophyll molecules associated with photo system II. Kinetics, analysis and interpretation of Chlorophyll a fluorescence provide information about the photosynthetic capacity and the vitality of the plant material.

The fluorescence signal from illuminated tissue/leaf can be detected with opto-electronic equipment, which separates the red fluorescence emissions from ambient light on the leaf. Instruments measuring this fluorescence are classified into two categories: (a). continuous excitation and (b). modulated fluorescence systems.

Handy PEA (Plant Efficiency Analyser) measures the fluorescence continuously from 1-120 seconds. PEA has high time resolution (10 μs) and a data acquisition capacity thus fast rise of the ms time resultion and 12 bit signal resolution. Saturating levels of illumination (3000 $\mu mol\ m^{-1}\ s^{-1}$) permit accurate monitoring of the time dependent fluorescence induction kinetics observed when a sample is strongly illuminated after a period of darkness. The high resolution of the fluorescence rise allows observing several steps, Fo, K, J, I, P etc. Typical measurements take only 1 second, providing a very high capability for large scale screening of samples. Dark adaptation of samples is achieved by plastic leaf clips, which are suitable for use on a range of samples including broad and small leaves plants. Shutter is closed and after about 2 min the Handy PEA sensor unit is held over the clip and the shutter is opened and single button press activates the high intensity LED array within the sensor, which provides required light intensity, ensuring accurate determination of the fluorescence maximum (Fm). The fluorescence parameters Fo, Fm, Fv, Fv/Fm. Tmax, area over fluorescence, curve between Fo and Fm are automatically calculated and immediately displayed on Handy PEA. All the standard fluorescence parameters can be trasnferred to the PC software for further analysis of 48 different parameters. A fluorescence rise provides the following basic information.

(i) Ratio of the extrema (Fo/Fm)

(ii) Fluorescence intensity at any time (Ft)

(iii) Shape of Fluorescence kinetics (Vt)

(iv) Time to reach the maximal fluorescence (tF_{max})

(v) Area above the Fluorescence transient (Area)

Information related to structure and functions of PSII and the activity of the electron transport after Q_A and to the activity of photosynthetic metabolism. This forms the basis of OJIP (fluorescence rise steps Fo, K, J. I, P) analysis, and following information can be easily selected.

I. Technical Fluorescence Parameters

(i) Fo = Fluorescence Intensity at 0 μs.

(ii) F_{150} = Fluorescence Intensity at 150 μs.

(iii) F_{300} = Fluorescence Intensity at 300 μs.

(iv) F_J = Fluorescence Intensity at the J step at 2 ms.

(v) F_M = Maximal fluorescence Intensity.

(vi) t F_M = Time to reach F_M in ms.

(vii) $V_J = F_J - F_o / F_M - F_O$.

(viii) Area, $(dV/dt)_o$, S_m, and N.

II. Flux Ratios per RC

ABS/RC

TR_O/RC

ET_O/RC

DI_O/RC

III. Flux Ratios or yield or Quantum efficiencies

TR_O/ABS

ET_O/ABS

ET_O/TR_O

IV. Density of RCs

RC/CS_O

RC/CS_M

V. Phenomenological fluxes or Phenomenological activities

ABS/CS_O

TR_O/CS_O

ET_O/CS_O

DI_O/CS_O

RC/CS_O

ABS/CS_M

TR_O/CS_M

ET_O/CS_M

Where,

ABS = Photons absorbed by the antenna pigment Chlorophyll; TR = Trapping Flux, RC = Reaction Centers; EC = Electron transport; CS = Cross section of leaf.

Applications

(i) Fluorescent transient of a given plant material is used to analyze the response to chemical changes in the environment using inhibitors, fertilizers, gases or any other chemical stressor.

(ii) Fluorescent transient of a given plant material is used to analyze the response to physical changes in the environment like quality and intensity of light, temperature (Srivastava and Strasser, 1996; Srivastava *et al.,* 1997).

(iii) Fluorescent transient of a given plant material is used to analyze the response to both chemical and physical changes (Quzounidou *et al.,* 1997).

(iv) Fluorescence transient of a given plant material is used to analyze the vitality, productivity, sensitivity or resistance to stresses, structure and functions (Van Renseburg *et al.*, 1996).

(v) Fluorescent transient is also used to address the effects of herbicides, pesticides, hormones, nodulation, N_2 fixation, drought, heat, cold and salt stress on crop productivity (Strasser *et al.*, 2000; Schmitz *et al.*, 2001; Dudeja and Chaudhary, 2005).

Procedure

(i) After properly placing the leaves in the plastic clips, then shutter is to be closed for dark adaptation of the leaves.

(ii) Care is taken to provide the maximum saturating levels of light for accurate monitoring of the fluorescence

(iii) Perform the Memory test for the Handy PEA at regular intervals.

(iv) Adjust the number of flashes and duration of flash as per your requirement; usually one flash is set per one second.

(v) Gain of the Handy PEA is to be adjusted in such a way that the Fv/Fm is around 0.8, which will indicate proper and optimized measurement of the fluorescence.

(vi) Number each and every observation and save the files in Handy PEA and download the file after recording the observations, to avoid loss of data.

(vii) Use Handy PEA properly to avoid the occurrence of bad sectors in the memory card of the instrument.

MEMBRANE THERMO-STABILITY INDEX

Introduction

A major impact of plant environmental stress is on cellular membrane modification, which results in its perturbed function or total dysfunction. The exact structural and functional modification caused by stress is not fully understood. However, the cellular membrane dysfunction due to stress is well expressed in its increased permeability for ions and electrolytes, which can be readily measured by the efflux of electrolytes. Hence, the estimation of membrane dysfunction under stress by measuring cellular leakage from affected leaf tissue into an aqueous medium is finding a growing use as a measure of cell membrane stability test and as a screening parameter for stress resistance.

Measuring Membrane Thermo-Stability Index

Stress induced damage of plant membrane can be determined using membrane thermo-stability (MTS) assay, which measures electrolyte leakage from leaf tissue after exposure to high temperature. This is the fastest method to screen plants for temperature tolerance and measures the increased electrolyte diffusion resulting from heat induced cell membrane permeability (Sullivan, 1972). Membrane thermo-stability of plant is expressed in terms of electrical conductance. The electrolyte leakage is measured by bathing the heat stressed tissue in deionised water and reading electrical conductance.

Materials

Conductivity bridge with conductivity cell, thermo-statically controlled water bath, autoclave, long test tube to accommodate conductivity cell for measurements, aluminium foil, measuring cylinder and de-ionised water.

Procedure

(i) Collect the upper most fully expanded leaf for assaying MTS.

(ii) Remove the mid rib from the middle portion of the leaf and take approximately 7 cm^2 of leaf and cut it into 1 cm pieces.

(iii) Put leaf pieces into 16 x 150 mm test tube and wash thrice with deionized water to remove unwanted ions from leaf pieces. After final wash, drain water from test tube.

(iv) Cover the tubes with aluminum foil and incubate in a water bath at 49°C for 30 min.

(v) After the heat treatment period, the tubes are kept over night at room temperature.

(vi) Conductance of solution is measured with an electrical conductivity meter after calibration with a 1N KCl solution.

(vii) Autoclave the test tubes at 120°C and 0.10 MPa for 10 min. Cool the tubes at 25°C and measure the conductance again.

(viii) Membrane thermal stability (MTS) is expressed in % as the reciprocal of relative leakage and can be calculated using the following formula :

$$MTS\ (\%) = (1 - T_1/T_2)$$

Where, T_1 is the initial conductivity reading after heat treatment and T_2 is the final conductivity reading after autoclave. Whitlow *et al.*, (1992) suggested improvement in expressing results of membrane stability by using two formulae : 1st is index of injury (Id) = 100 (Rt-RO)/(1-RO)

Where, Rt = Ec initial/Total Ec for stressed plant, RO = same ratio of EC for control plant. Initial Ec is after a given period of incubation at room temperature and total Ec is Ec after heat killing of the tissues.

2nd is Tissue ionic conductance (GTi = Jj/A(CO-Ci); Jj = dsj/dt the flux of solute species j/unite time. A is surface area through which diffusion occurs and CO-Ci is difference between outside (higher) concentration and the inside (lower) concentration of ions. This method gives better representative of permieability damage.

Steps to use Electrical Conductivity Meter

(i) Switch on the instrument and clean and conductivity cell, dry it and connect it in the conductivity input socket.

(ii) Adjust the display equal to cell constant value with CELL CONST knob using in KCl.

(iii) Keep the ``RANGE" switch to max. range, i.e. 1000 mhos position.

(iv) Dip the conductivity cell inot the solution whose conductance is to be measured. The cell must not contain trapped air. Tap the cell to remove any bubbles and dip it 2 or 3 times to assure proper wetting.

(v) Bring the function switch to `NORM' Switch.

(vi) Bring the `RANGE' switch to the position where maximum resolutions obtained.

(vii) The display will indicate the conductance value in mhos cm^1.

REFERENCES

Arnon, D.I. 1949. Copper enzymes in isolated chloroplast. Polyphenoloxidase. *Plant Physiology* 24 : 1-15.

Barnes, J.D., Balaguer, L., Manrique, E., Eqvira, S. and Davison, A.W. 1992. A preappraisal of the use of DMSO for the extraction and determination of chlorophyll `a' and `b' in lichens and higher plants. *Environmental and Experimental Botany* 32: 85-100.

Bruinsma, J. 1963. The quantitative analysis of chlorophylls a and b from plants extract. *Photochemistry and Photobiology* 2 : 241-249.

Dudeja, S.S. and Poonam Chaudhary 2005. Fast chlorophyll fluorescence transient and nitrogen fixing ability of chickpea nodulation variants. *Photosynthetica* 43 : 253-259.

Graan, T. and Ort, D.R. 1984. Quantitation of rapid electron donors to P700, the functional plastoquinone pool and the ratio of the photo-systems in spinach chloroplasts. *Journal of Biological Chemistry* 259 : 14003-14010.

Hendry, G.A.F. and Price, A.H. 1993. Stress indicators : Chlorophylls and carotenoids. In: *Methods in Comparative Plant Ecology* (Eds. G.A.F. Hendry and J.P. Grime). pp. 148-152. Chapman and Hall, New York.

Hiscox, J.D. and Israelstam, G.F. 1979. A method for extraction of chlorophyll from leaf tissue without maceration. *Canadian Journal of Botany* 57 : 1332-1334.

Krishnan, P., Ravi, I. and Nayak, S.K. 1996. Methods for determining the leaf chlorophyll content of rice. *Indian Journal of Experimental Biology* 34 : 1030-1033.

Murthy, K.S. and Majumdar, S.K. 1962. Modification of the technique for determination of chlorophyll stability index in relation to studies of drought resistance in rice. *Current Science* 31 : 470-471.

Quzounidou, G., Moustakas, M. and Strasser, R.J. 1997. Sites of action of copper in the photosynthetic apparatus of maize leaves : Kinetic analysis of chlorophyll fluorescence, oxygen evolution, absorption changes and thermal dissipation as monitored by photoacoustic signals. *Australian Journal of Plant Physiology* 24 : 81-90.

Schmitz, P., Maldonado-Rodriguez, R. and Strasser, R.J. 2001. Evaluation of the nodulated status of *Vigna unguiculata* probed by the JIP test based on the chlorophyll a fluorescence rise. In: *Proceedings, 12th International Congress on Photosynthesis,* Chapter S36-012 : pp 1-4. Brisbane, Australia, August 18-23, 2001.

Shoaf, T.W. and Lium, B.W. 1976. Improved extraction of chlorphyll a and b from algae using dimethylsulphoxide. *Limnology Oceanography* 21: 926-928.

Slack, K.V., Averett, R.C., Greenson, P.E. and Lipscomb, R.G. 1973. Methods for collection and analysis of aquatic biological and microbiological samples. In U.S. *Geological Survey, Technical Water Resource Invest Book* 5.

Srivastava, A. and Strasser, R.J. 1996. Stress and stress management of land plants during a regular day. *Journal of Plant Physiology* 148 : 445-455.

Srivastava, A., Guisse, B., Greppin G. and Strasser, R.K 1997. Regulation of antenna structure and electron transport in PSII of *Pisum sativum* under elevated temperature probed by the fast polyphasic chlorophyll a fluorescence transient. OKJIP. *Biochemica et Biophysica Acta* 1320:95-106.

Strain, H.H. and Snec, A. 1966. Extraction separation and isolation of chlorophylls. In: *The Chlorophylls* (Eds., L.P. Vernon and G.R. Seeley). pp. 22-66. Academic Press, New York.

Strasser, R.J., Srivastava, A. and Tsimilli-Michael, M 2000. The fluorescence transient as a tool to characterize and screen photosynthetic samples. In : *Probing Photosynthesis* : *Mechanism, Regulation and Adaptation,* (Eds. M. Yunus, U. Pathre and P. Mohanty). pp 445-483. Taylor & Francis, New York.

Sullivan, C.Y. 1972. Mechanisms of heat and drought resistance in grain sorghum and methods of measurement. In: *Sorghum in the Seventies* (Ed. N.G.P. Rao and N.R. House) pp. 247-267. Oxford and IBH Publishing Co., New Delhi, India.

Van Renseburg, L., Kruger, G.H.J., Eggenberg, P. and Strasser, R.J. 1996. Can screening criteria for drought resistance in *Nicotiana tabacum* L. be derived from the polyphasic rise of the chlorophyll a fluorescence transient (OJIP). *South African Journal of Botany* 62 : 337-341.

Wellburn, A.R. 1994. The spectral determination of chlorophyll a and b, as well as total carotenoids using various solvents. *Journal of Plant Physiology* 144 : 307-313.

Whitlow, T.H., Bassuk, N.L., Ranney, T.G. and Reichest, D.I. 1992. An impr-oved method for using lectrolyte leakage to assess membrane compe-tence in plant tissues. *Plant Physiology* 98: 198-205.

CHAPTER - 5

ANALYSIS OF ORGANIC MOLECULES IN PLANT TISSUES

INTRODUCTION

Organic and inorganic molecules affect the osmotic potential of plant cells. The inorganic solutes i.e. Na^+, K^+, CO_3^-, SO_4^- and Cl^- occurs predominantly in the vacuoles whereas in the cytoplasm organic molecules i.e. sucrose, amino acids, glycerol and glycine-betaine help in adjustment of osmotic potential. Both organic and inorganic solutes play major role in osmoregulation. Therefore, these organic molecules play an important role in stress conditions. This chapter describes analytical methods for the estimation of important organic metabolites in plants.

CARBOHYDRATES / SUGARS

Carbohydrates are important constituents of all the living cells and these are defined as aldehydes or ketones of polyhydric alcohol and also include those bropolymers which yield these compounds on hydrolysis. Among carbohydrates, sugars such as glucose and fructose are the major source of energy whereas starch and glycogen function as storage polysaccharides in plants and animals.

ESTIMATION OF CARBOHYDRATE (SOLUBLE SUGAR) IN PLANT SAMPLES

Materials

Ethanol (80% V/V), Anthrone (0.2% in concentration sulphuric acid), Glucose (for preparation of standard). Test tube, Pipette, Test tube stand, Beaker, Water bath, Centrifuge tubes, Distilled water, Measuring cylinder, Bottles etc.

Athrone reagent : 0.2 g anthrone in 100 ml concentrated sulphuric acid.

Principle

Monosaccharides are dehydrated to furfural and its derivatives which react with anthrone to give blue-green complex with absorbance at 630 nm (Yemm and Willis, 1954).

Procedure

Homogenize 100-200 mg plant sample in 10 ml of 80% ethanol and centrifuge at 4000 rpm for 20 minutes. Collect the supernatant and re-extract the residue twice with 5 ml hot 80% ethanol and centrifuge at 4000 rpm for 20 minutes. Combine all the supernatants and make volume to 20 ml. Evaporate a known amount of ethanol extract (0.1 - 0.2 ml) to dryness in a test tube on water bath and cool it to room temperature. Add one ml of distilled water to each tube and mix thoroughly. To each add 4.0 ml of anthrone reagent and heat on a water bath at 100°C for 15 minutes. Cool rapidly under ice cold water and measure the absorbance at 630 nm against reagent blank. Prepare a standard curve by using graded concentrations of glucose.

Precautions

1. All the glasswares should be neat and clean.
2. Anthrone reagent should be prepared afresh preferably. Put the flask on ice bath while adding sulphuric acid. Use fume hood.
3. Anthrone reagent is added by the side of the tubes and mixed gently.
4. After colour development absorbance should be read within half an hour.
5. Avoid spilling and contact to skin since sulphuric acid is corrosive.

ESTIMATION OF STARCH CONTENT

Reagents

(i) Anthrone (0.2% in conc. H_2SO_4)

(ii) Perchloric acid 0.46 N. Dilute 10 ml of the 4.6 N $HClO_4$ to 100 ml with distilled water.

(iii) Glucose stock solution. Prepare a 100 ppm glucose stock solution in distilled water.

Prepare each of the standard (given below) by placing the amount of 100 ppm solution indicated in test tube. Then add 0.6 ml 0.46 N $HClO_4$ to each tube and make up to 5-ml with distilled water.

Glucose standards	Glucose stock solution
mg/tube	(ml)
0	0
0.1	1.0
0.2	2.0

Note : Glucose stock solution and standards should be prepared daily and kept cool.

Principle

Perchloric acid hydrolyses glycosidic bonds of starch molecules and produce monosaccharides which are dehydrated to furfural and its derivatives. Furfural reacts with anthrone and produces blue green complex with absorbance maximum at 630 nm (Macare, 1971).

Procedure

(A) Starch Extraction

The residue left after sugar extraction is oven dried at 80ºC for starch extraction. Add 2 ml of distilled water to the centrifuge tube containing residue. Put the tube in a boiling water bath for 15 minutes and stir occasionally. Allow the tubes to cool and add 2 ml 9.2 N perchloric acid while stirring constantly for 15 minutes. The suspension is then made to about 10 ml with distilled water and centrifuged. Collect the supernatant and add 2 ml 4.6 N perchloric acid to the residue. Stir this suspension for 15 minutes and then make up the volume up to 10 ml with distilled water. Centrifuge and then combine the two supernatant. Make up the volume to 50 ml with distilled water. Analyse starch in this starch extract.

Note : Use only 50 mg of dried samples when analyzing starch in panicles at harvest or after the dough stage.

(B) Starch Analysis

Put 5 ml of starch extract into a 50 ml volumetric flask and make up to volume with distilled water. Put 5 ml of this diluted starch extract into test tube, and then put this tube and tubes containing the standards into an ice bath. To each tube slowly add 10 ml of anthrone reagent, allowing the reagent to run down the side of the test tube. Stir slowly with a glass by vortaxing.

Put the tubes into a boiling water bath for exactly 7.5 minutes. Then immeidately cool in ice cold water and measure the absorbance at 630 nm.

Note

(i) This method can be used when starch and sugar concentrations are between 5 and 20 per cent. If the concentration is outside this range, dilute the extract and record computation.

(ii) The size and thickness of the test tubes influence the colour development. It is essential, therefore, to use similar test tubes for all samples and standard.

(iii) Anthrone gives only a very slight colour reaction with pentoses.

Starch Estimation After Enzymatic Hydrolysis

1. Take about 10-50 mg accurately weighed dried tissue and perform the ethanol extraction procedures same as in acid hydrolysis procedures.
2. To the residue, after alcohol extractions, add 4 ml of deionised water and place the tube in a boiling water bath for 20 min, to allow traces of alcohol to escape. Fit an air condensor to tube. Let the tube remain in water bath for further 2 hours.
3. Remove the tube from water bath, cool and add buffer solution (5 ml, 0.2 M acetate buffer pH 4.5) and 0.5 ml of amyloglucosidase suspension (10 units/ml) Place a layer of liquid paraffin (0.5 cm thick) over the liquid, to minimise loss of glucose due to oxidation. Incubate the contents at 55ºC for about 40 hrs.
4. Cool the contents to room temperature and thoroughly mix the contents which are below the paraffin layer. Remove 1 ml aliquot of extract.
5. Deproteinize this aliquot by adding 5 ml water, 2 ml (0.3 N) NaOH and 2 ml of 5% $ZnSO_4$.
6. Filter and remove 1 ml of filterate. This filterate contains glucose formed by the enzymatic hydrolysis of starch. The glucose content can be measured by

   (a) Somogyi-Nelson method for reducing sugars.

   (b) Dinitrosalicylic acid reagent.

   (c) By coupled glucose oxidase-peroxidase system.

Out of these enzymatic method (c) is most sensitive and procedure has been given in this chapter.

7. Measure the glucose concentration in filterate by any of 3 methods described above and calculate the glucose content of tissue as follows :

$$\text{Starch content} = \frac{\textit{Glucose conc. determined } (\mu g/ml) \times V \times 10 \times 0.91}{1000 \times \textit{wt. of starting tissue } (mg)} \times 100\%$$

DETERMINATION OF REDUCING SUGARS BY THE NELSON METHOD

This method is very commonly used for the estimation of reducing sugars content of plant tissues, and is based on the principle that when sugar is heated with an alkaline solution of copper-tartarate, cuprous-oxide is produced which reacts with arseno-molybdate to give molybdenum blue (Nelson, 1944). The intense blue colour is then measured in the colorimeter. Before estimation of reducing sugars, the plant extracts should be deproteinised.

Prepare the reagents as follows :

Nelson's Reagent A. Dissolve 12.5 g of Na_2CO_3 (anhydrous), 12.5 g of potassium tartarate, 10 g of $NaHCO_3$, and 100 g of Na_2SO_4 (anhydrous) in 350 ml of H_2O and dilute to 500 ml.

Nelson's Reagent B. Dissolve 7.5 g of $CuSO_4$. 5 H_2O in 50 ml of water and add one drop of concentrated H_2SO_4.

Arsenomolybdate reagent. Dissolve 25 g of $(NH_4)_6$ MO_7O_{24}-$4H_2O$ in 450 ml H_2O and add 21 ml of concentrated H_2SO_4. Dissolve 3 g of Na_2H As O_4. $7H_2O$ in 25 ml of water and add to the acid molybdate above. Store in brown bottle for 24 hr at 37°C before use.

Glucose standard. Dissolve 100 mg of glucose (dextrose) in 1000 ml of water and store in cold room.

Procedure

1. Prepare Nelson's alkaline copper reagent by mixing 12.5 ml of Nelson's reagent A with 0.5 ml of Nelson's reagent B (make larger volume if necessary).
2. Add 1 ml (if you use less than 1 ml, make up difference with water) of a solution containing about 100 mg of reducing sugar to a test tube and add 1.0 ml of Nelson's alkaline copper reagent.
3. Shake well and place the tubes in a boiling water bath (a 500 ml beaker or larger if bath is not available) and heat for exact 20 min.

4. Remove the tubes simultaneously and place them in a beaker of cold water to cool.
5. When the tubes are cool (25°C), add 1 ml of arseno-molybdate, reagent to each, and shake well several times during a 5 min period to dissolve the Cu_2O and to reduce the arsenomoly-bdate.
6. Add 7.0 ml of water to each tube, and mix thoroughly.
7. Read optical density in a spectrophotometer at 540 nm.
8. Use a blank sample (no sugar) to set absorbance to zero.
9. Before determining the reducing sugar of the unknowns, it would be desirable to run a standard curve using glucose. During each assay a glucose standard should also be included.

GLUCOSE OXIDASE METHOD FOR GLUCOSE DETERMINATION

The estimation of activity of disaccharidases such as maltase, cellobiase and trehalase involve necessary application of this method for glucose estimation. Nelson method or dinitrosalicylic acid methods for reducing sugars cannot be used in assay of these enzymes because substrates (maltose, cellobiose and trehalose) and product (glucose units) are all reducing sugars. Glucose oxidase assay thus provide extremely specific method for product estimation. Moreover this method is used in estimation of glucose, released by enzymatic hydrolysis of starch.

Glucose oxidase is an enzyme obtained by extraction of fungus, *Aspergillus niger*. This enzyme catalyses the oxidation of a-D-glucose to D-glucono-1, 5 lactone with the formation of hydrogen peroxide. The enzyme is specific for a-D-glucose, but most enzyme preparations contain mutarotase, which catalyses the interconversion of the a and b forms. Peroxidase and o-dianisidine are incorporated in reaction mixture, so that oxygen is liberated from the hydrogen peroxide and reacts with the o-dianisidine to give a red chromophore product.

$$Glu\cos e + O_2 \ \ - Glu\cos e\ Oxidase - \ \ H_2O + gluconic\ acid$$

H_2O_2 + O-dianisidine - Peroxidase - chromophoric product.

This assay allows accurate determination of glucose even in the presence of other monosaccharides, reducing substances, and the many other components of plant tissues. Following are the practical instructions for the method.

1. Take 25 mg of o-dianisidine and dissolve it completely in one ml of methanol.
2. Add 49 ml of 0.1 M phosphate buffer, pH 6.5.
3. Dissolve 5 mg of peroxidase and 5 mg of glucose oxidase in above prepared o-dianisidine solution.
4. Take in a test tube 0.5 ml of deproteinized plant extact (deproteinization is not necessary in samples with very low protein contents).
5. To this extract add 1 ml of glucose-oxidase, peroxidase reagent. Incubate for 40 mins at 35°. Terminate the reaction by addition 2.0 ml of 6 N HCl. Record optical density at 540 nm.
6. Prepare a calibration curve by using standard glucose solution (10-100 μg).

ESTIMATION OF PROTEIN CONTENT IN PLANT TISSUES

The estimation of protein content is the fundamental need in any enzymological study. In most of the studies there is a need for relating activity of a particular enzyme with total protein content i.e. to calculate specific activity of an enzyme. The most well known method for protein is Folin-Lowry assay. This method is based on the principle that protein reacts with Folin-Ciocalteau reagent to give a coloured complex (Lowry *et al.*, 1951). The colour so formed is due to the reaction of alkaline copper with the protein to give a violet coloured complex. The depth of the colour obtained is a measure of the number of peptide bonds and also due to the reduction of phosphomolybdate by tyrosine and tryptophan present in the protein. The intensity of the colour depends upon the amount of these aromatic amino acids present and will thus vary for different protiens. This method has been indiscriminately used in the protein estimation in plant tissue homogenate. But this method has large number of limitations. Estimation of protein content directly in the plant tissue homogenates by this method gives erroneous results due to

(i) Components of common laboratory reagents such as K^+, Mg^{++}, Tris, EDTA and thiol compounds interfer in Lowry protein assay

(ii) Since most plant extracts contain phenolic compounds which react with Folin's reagent. Therefore, the Lowry procedure gives exaggerated values.

(iii) Some buffers like HEPES also interfere in Lowry protein determination.

Therefore, this method should only be used when non-proteinaceous interfering substances are removed from the tissue homogenates. This is done by precipitating proteins immediately after extraction. Proteins can be very easily precipitated by addition of trichloroacetic acid (TCA) to plant extracts. The precipitated proteins are dissolved in NaOH solution and then this suspension is used for protein estimation. Following are the practical instructions for protein estimation by Lowry's method.

Lowry Protein Estimation

Take 0.5 ml samples of tissue homogenate and precipitate the protein with 1 ml of cold 10% trichloroacetic acid (TCA). Centrifuge at 5,000 to 10,000 g for 10 min and decant the supernatant. Dissolve the precipitate in 5 ml of 0-1 N NaOH and use 0.1 ml for protein estimation.

Prepare following reagents

(a) 2.0% Na_2CO_3 in 0.1 N NaOH.

(b) 0.3% $CuSO_4$. 5 H_2O in 1% sodium potassium tartrate,

(c) Phenol reagent (Follin-Ciocalteau) 1.0 normal.

Use commercially available reagent or prepare fresh in the laboratory.

(d) Standard protein solution : bovine serum albumin (BSA) at 0.2 mg/ml in 0.1 N NaOH.

(e) Mix reagents (a) and (b) (50 : 1) shortly before assay.

Procedure

To test tubes add 0.1 ml protein suspension (in NaOH) and add 5 ml of reagent (e) and let stand for 10 min. Then add 0.5 ml of reagent (c) and mix immediately. Allow the (sample to stand at room temperature (about 25°) for 30 min and read the absorbance at 750 nm. Prepare a reference curve by using 0.05 ml to 0.5 ml solution of BSA in test tubes, carry through whole procedure and determine the account of color development.

Dye-Binding Method for Protein Estimation

As already discussed Lowry's procedure is not suitable for protein determination in plants extracts unless proteins are precipitated immediately after precipitation. This precipitation step poses serious practiceal problems.

In most of cases the investigator may be possibly dealing with mgs of plant material such as fungal spores, fern spores, pollen grains, epidermal peels, etc.

This implies that very small amount of homogenate can be made and application of TCA precipitation procedure do not remain feasible.

This problem of dealing with very small amounts of plant tissues can be solved by using dye binding methods for proteins estimation (Bredford, 1976). Some dyes like amido black or coomassie blue quantitatively bind to protein molecules and the amount of dye bound is directly proportional to amount of protein present. In practice, this method is done by spotting very small amount of tissue extract on filter-paper dipping it in the dye solution, removing the excess dye and then or ultimately eluting the protein bound dye in a suitable solvent. The absorbance of this released dye is measured and the amount of protein present in tissue extract is estimated. Other laboratory reagents and non-proteinaceous material i.e. phenolic compounds do not interfere in this procedure. However, drawback with this method is that different proteins bind different amounts of dye. However, this method has been recently used in many studies in plant physiology.

Following are the practical instructions for dye-binding protein assay according to the method described by Bramhall *et al.* (1969).

1. Spot plant tissue extract on a 2 X 2 cm. filter paper (for example Whatman 42). Dry the papers, possibly in a stream of warm air. Many samples can be applied to a single piece of filter paper previously marked into labelled squares with a soft pencil. After drying, these squares can be cut apart. Mark the sample no. with lead pencil.
2. Put these small squares into cold 7.5% trichloroacetic acid to fix the proteins into the paper. Many papers can be put into a single Petri dish containing TCA.
3. Then place the petridish containing papers in an oven set at 80ºC for half an hour to remove non-proteinaeous material.
4. Take out these paper squares with forceps and wash with ether/ethanol (1/1 v/v) and then ether to remove lipids and TCA. Then dry the papers.

Next steps involve the use of specific protein binding dyes. Two dyes can be used, one for dilute samples (5-25 μg proteins) and second for conc. plant extracts (which possible have 50-200 μg proteins).

A : *For dilute samples*

1. Put the dried squares in a dye solution (brilliant cyanin G, 10 mg/ml in 7% acetic acid) for 15 min at 50°C with occasional swirling. About 2 ml dye should be used for each paper.
2. Remove excess dye with hot (50-60°C) 7% acetic acid, until the background is almost white. Drain the papers dry.
3. Place each square in a separate test tube. Add 5 ml of destain solution (66 ml methanol, 34 ml water and 1 ml conc ammonia). Shake and then read the abosrbance of the released dye at 610 nm, against a blank prepared by carrying 2 x 2 cm. paper through whole procedure.
4. Prepare a reference curve by spotting known amounts (5-50 mg) of standard protein (BSA) and carrying through the whole procedure. From this reference curve calculate the value of protein content in plant extract.

B : *For conc. extracts*

1. Put the papers in a dye solution (Amido black 10 mg/ml in acetic acid/methanol/water, 1/4/5) at 50°C for 15 min.
2. Then rinse the paper squares with hot (50°C) 7% acetic acid several times to remove excess dye until background becomes almost white. Discard the free liquid.
3. Put papers in individual test tubes. Add 5 ml of 0.05 N NaOH in each tube. Read the absorbance of the released dye at 620 nm.
4. Prepare a reference curve by spotting varying amounts (40-250 μg) of Bovine serum albumin on paper squares and then carrying through the whole procedure. From this reference curve read the values of protein in plant extracts.

COLORIMETRIC ESTIMATION OF DNA AND RNA

Nucleic acids are nitrogen containing high molecular weight compounds that are present in all living organisms. There are two types of nucleic acids that occur in cells, deoxyribonucleic acid (DNA) and ribonucleic acid (RNA). Hydrolysis of DNA and RNA under controlled conditions yields nucleotides which are the basic monomeric units of nucleic acids. Each nucleotide consists of a heterocyclic nitrogenous base, a sugar and phosphoric acid. There are two types of nitrogenous bases in nucleic acids, namely, purines and pyrimidines. The two purine bases are adenine (A) and guanine (G) whereas pyrimidine bases are thymine (T), uracil (U) and cytosine (C). RNA differs from DNA in that, firstly, it contains uracil as one of the pyrimidine bases in place of thymine, and secondly, the pentose sugar in RNA is D-ribose instead of 2-deoxyribose (Adams *et al.*, 1992).

Most DNAs occur in a doublue helical form though some viral DNAs exist as single strands. Double stranded DNA consists of two polynucleotide chains wound together into a right handed helix and are held together by H-bonding between complementary bases (A pairs with T and G with C). The A-T base pair has two H-bonds whereas G-C base pair has three H-bonds.

DNA is the hereditary material in all living organisms except in a few virus which contain only RNA as their genetic material. The function of DNA is to store and express the genetic information in an organism. It serves as a template for synthetic of various types of RNAs which participate in synthesis of protein. RNA present in the cells are of three different types i.e. Messanger RNA (mRNA), ribosomal RNA (rRNA) and transfer RNA (tRNA). All the three types of RNAs exist as single polyribonucleotide strands and play an important role in protein synthesis. Besides these three forms, certain RNA's are associated with specified proteins to form ribonucleoproteins that participate in post-transcriptional processing of other RNAs (Parish, 1972; Wilson and Walker, 1996).

ESTIMATION OF NUCLEIC ACIDS

Extraction of Total Nucleic Acids from Plant Tissues

Principle

Nucleic acids are extracted by grinding the tissues with ice cold trichloroacetic acid (TCA) followed by centrifugation to separate them from the acid soluble compounds (nucleosides and nucleotides etc.). The precipitated nucleic acids are then extracted with ethanol-ether to remove the interfering lipid material. RNA is selectively solubilized from the residue with alkali and DNA subsequently recovered. Extracted DNA and RNA are then quantitatively estimated by their colour reactions with diphenylamine and orcinol, respectively, as outlined later on (Cherry, 1962; Burton, 1968).

Materials and Reagents

1. Cooling centrifuge
2. Boiling water bath at 37°C
3. Plant tissue-Any tissue such as leaves, roots or embryos from generating seeds can be used.

4. Trichloroacetic acid (10%, w/v)
5. Ethanol : Ether (2 :1)
6. 0.5 N NaOH
7. 5% $HClO_4$

Procedure

1. Macerate 2.5 g plant tissue in 10 ml of ice cold 10% trichloroacetic acid (TCA).
2. Centrifuge the homogenate at 3000 rpm for 10 minutes.
3. Discard the supernatant and suspend the pellet in 5 ml of cold 10% TCA.
4. Centrifuge the suspension at 3000 rpm for 10 min, discard the supernatant and suspend the precipitate in 5 ml of ethanol. ether mixture.
5. Again centrifuge as in step 4 and suspend the precipitate in 5 ml of 0.5 N NaOH. Mix well and keep it for 18 h at 37°C.
6. Centrigue this alkali treated preparation at 3000 rpm for 10 min. Decant the supernatant, which contains RNA in the hydrolyzed form as well as proteins, in a test tube.
7. To the precipitate, which contains DNA and some proteins, add 1 ml of 5% $HClO_4$ and after mixing the contents well, heat it in a boiling water bath for 1 hour. Cool the preparation and centrifuge it at 3000 rpm for 10 min. Collect the supernatant, which contains deoxyribo nucleotides released from hydrolysis of DNA, makeup to a known volume with a 5% $HClO_4$ and using diphenylamine reaction estimate the amount of DNA in a suitable aliquot of this preparation.
8. To the supernatant obtained in step 6 (which contains RNA), add an equal volume of 10% TCA. Centrifuge at 3000 rpm for 10 min. Collect the supernatant. Make upto known volume with 5% $HClO_4$ and estimate RNA using orcinol method.

Estimation of DNA by Diphenylamine Reaction

Principle

This is a general reaction given by deoxypentoses. The 2-deoxyribose of DNA, in the presence of acid, is converted to w-hydroxylevulinic which reacts with diphyenylamine to form a blue coloured complex with a absorbance maxima at 600 nm (Burton, 1968). Compounds such as furfuryl alcohol and arabinal, which can be converted into w-hydroxylevulinic aldehyde will also give this reaction. In DNA, since only deoxyribose of purine nuclotides is released, the value obtained represents half of the total deoxyribose in the sample.

Materials and Reagents

1. Spectrophotometer
2. Standard DNA solution : Dissolve 0.5 g DNA in 50 ml of 1 N $HClO_4$.
3. Working standard of DNA : (Conc. 100 ug/ml) Dilute 1 ml of stock standard to 100 ml with 1 N $HClO_4$.
4. 1.6% Acetaldehyde : Dissolve 1 ml of ice cold acetaldehyde in 50 ml of distilled water.
5. Diphenylamine solution : Dissolve 1.5 g diphenylamine in 100 ml glacial acetic acid. Add 1.5 ml conc. H_2SO_4 to it.
6. Diphenylamine reagent : Mix 0.5 ml of 1.6% acetaldehyde to 100 ml diphenylamine solutions. Prepare fresh just before use.

Procedure

1. Take 2.0 ml of extract sample in which DNA has to be estimated in a test tube (For blank, take 2 ml of 0.5 N $HClO_4$).
2. Add 4 ml diphenylamine reagent (Reagent No. 6) and mix the contents properly.
3. Keep at room temperature in dark for 16-18 hours or overnight.
4. Record the absorbance at 600 nm.

Result is expressed as mg DNA/g fresh wt. of tissue. For standard curve make different dilutions of standard DNA solution ranging from 20-100 μg DNA/ml using 0.5 N $HClO_4$. Draw a standard curve of A_{600} VS DNA concentrations. From absorbance of the sample, determine the amount of DNA in it. Expressthe results as mg of DNA/g fresh weight of tissue.

Determination of RNA by Orcinol Method

Principle

This is a general method for estimation of pentoses. Acid hydrolysis of RNA releases ribose which in presence of strong acid undergoes dehydrotion to yield furfural. Orcinal, in the presence of ferric chloride as a catalyst, reacts with furfural producing a green coloured compound with absorbance maxima at 665 nm. The reaction leading to the formation of a green coloured complex (Cherry, 1962).

Materials and Reagents

1. Spectrophotometer
2. Boiling water bath
3. 5% $HClO_4$
4. Standard RNA solution : Dissolve yeast RNA (500 mg/ml) in 5% $HClO_4$. Make different dilutions to obtain solutions containing 100-500 ug RNA/ml with 5% $HClO_4$.
5. Orcinol reagent-Dissolve 100 mg of ferric chloride ($FeCl_3$. $6H_2O$) in 100 ml of conc. HCl and then add 3.5 ml of 6% solution of orcinol prepared in alcohol.

Procedure

1. Take 2ml of test sample in which RNA has to be estimated in a test tube.
2. In another set of test tubes take 2.0 ml solution of each of dilutions of RNA standard solutions. In one of the test tube take 2.0 ml of 5% $HClO_4$ as a reagent blank.
3. Add 3.0 ml orcinol reagent to all the tubes and mix properly.
4. Keep the test tubes in a boiling water bath for 20 min.
5. After cooling, take absorbance at 665 nm against blank.
6. Plot a graph between A_{665} VS amount of RNA and from this standard curve determine the amount of RNA in the provided sample. Calculate the amount of RNA in 1 g fresh weight of the sample.

Determination of DNA by UV Method

Principle

Both DNA and RNA exhibit strong absorption of ultra violet light due to presence of conjugated double bonds of the constituent purine and pyrimidine bases (Parish, 1972; Adams *et al.*, 1992). These have a characteristics absorbance maxima at 260 nm which is linearly related with concentration of DNA or RNA in solution up to an O.D. value of 2.0. This method can be employed for judging purity of DNA and RNA preparations. Proteins are usually the major contiminants in nucleic acid extracts and these have absorption maxima at 280 nm. The ratio of absorbance at 260 and 280 nm, hence, provides a rough idea about the extent of contaminants in the preparations. A ratio between 1.8-2.0 is indicative of fairly pure DNA and RNA preparations but values less than 1.8 signify presence of proteins as impurities.

Materials and Reagents

1. UV spectrophotometer
2. Saline sodium citrate (SSC) solution: Prepare 0.015 M solution of sodium citrate (pH 7.0) and dissolve NaCl so that its final concentration in solution is 0.15 M.

Procedure

1. Switch `ON' the spectrophotometer and allow it to warm up for about 10 min. Adjust the wavelength at 260 nm and put `ON' the UV-lamp.
2. Set the instrument at zero absorbance with SSC solution.
3. Read absorbance of solution of the provided sample. If O.D. is too high, approximately dilute the sample solution with SSC and again take the reading.
4. Calculate concentration of DNA and RNA in the sample from the following formula.

(i) For double standard DNA

Concentration of DNA = 50 x A_{260} x Dilution factor

sample solution (µg/ml).

(ii) For RNA

Concentration of RNA = 40 x A_{260} x Dilution factor in sample solution (µg/ml)

AMINO ACIDS

The amino acids are organic compounds, contains amino and carboxyl group. Proteins are polymers of amino acids and contains about 22 different types of amino acids. Among them proline and hydroxyproline are called imino acids. General formula of an amino acids is given as :

$$R - \underset{\displaystyle H}{\overset{\displaystyle NH_2}{\underset{|}{\overset{|}{C}}}} - COOH$$

Amino acids are the basic bulding blocks of protein. In proteins, amino acids are linked by peptide (CONH) bond, formed by condensation of a-COOH of one amino acid with a-NH_2 group of the next amino acid.

Quantitative Analysis of Free Amino Acid Content in Plant Tissue by Ninhydrin Method

Principle

Ninhydrin (triketohydrinadene hydrate) a powerful oxidising agent react with all amino acids between pH 4 and 8 to give a purple coloured complex with absorbance at 570 nm (Yemm and Cocking, 1955).

Materials and Reagents

(1) Centrifuge

(2) Boiling water bath

(3) Colorimeter/Spectrophotometer

(4) Germinating seed

(5) Surface sterilize by deeped the pea seeds in 0.1% mercuric chloride for 1-2 minutes. Wash the seeds several time with distilled water and keep them at 25°C in petriplates lined with moistened whatman No. 1 filter paper, if necessary add water periodically till they start germinating.

(6) Ethanol 70% (V/v)

(7) KCN-acetone ninhydrin solution : Take 50 ml of 5% ninhydrin solution in acetone and mix it with 250 ml of KCN-acetone solution (prepared by diluting 5 ml of 0.01 M KCN in 60% ethanol to 250 ml with acetone). Alternatively ninhydrin reagent may be prepared by dissolving 2 g of ninhydrin in 25 ml methyl cellosolve and then adding 25 ml of 0.2 M acetate buffer (pH 5.5).

(8) 0.2 M citrate buffer (pH 5.0) : Dissolve 21g citric acid in 200 ml water and mix with 200 ml of 1 N NaOH. If necessary adjust the pH to 5.0 and make volume to 500 ml with D.W.

(9) Standard amino acid solution : prepare a solution containing 100 μg glycine/ml of 0.2 M citrate buffer. (Reagent No. 8).

Procedure

(1) Free amino acids are extracted by grindling 0.5 to 1.0 g of germinating seed in 10 ml of 70% ethanol in a pestle and mortar. Centrifuge the homogenate at 5000 rpm for 10 minutes. Decant the supernatant. Repeat the extraction 4-5 time and combine the supernatant and make the volume to 50 ml with 70% (V/v) ethanol.

(2) Take an appropriate volume (5-10 ml) of this ethanolic extract of germinating seeds and evaporate it to dryness on the boiling water bath and dissolve the residue in 5 ml of 0.2 M citrate buffer (pH 5.0).

(3) Pipette 2ml of above sample preparation in a test tube, for a regent blank take 2.0 ml of 0.2 M citrate buffer (pH 5.0) in place of sample preparation. In another set of tubes take graded concentration of glycine (0-100 μg) and make the total volume to 2 ml with 0.2 M citrate buffer (pH 5.0). Add 1 ml of KCN acetone ninhydrin reagent that are prepared in methyl cellosolve to each tube and mix thoroughly.

(4) Keep the test tubes in boiling water bath for 20 minutes. Cool under running tap water and make volume to 10 ml with D.W.

(5) Note the absorbance at 570 nm after setting the instrument with reagent blank.

(6) Prepare the standard curve by plotting A_{570} against amount of glycine.

(7) Find the amount of free amino acids content in the sample test tubes from the standard curve.

(8) Express the free amino acids content in seeds in term of mg of glycine equivalent per g of seeds.

ESTIMATION OF FREE PROLINE

It is measured by spectrophotometrically according to the method of Bates *et al.* (1973).

Reagents

a) Acid Ninhydrin

It is prepared by dissolving 1.25 g ninhydrin in 30 ml glacial acetic acid and 20 ml 6-M ortho phosphoric acid with continuous tirring, dissolved and stored at low temperature before use.

(b) 3% sulphosalicylic acid (W/v) is prepared in distilled water

Procedure

200 mg of plant sample is homogenized in 5 ml of 3% aqueous sulphosalicylic acid and centrifuged at 4000 rpm for 20 minutes. The residue re-extracted with 5 ml of 3% sulphosalicylic acid and volume made to 10 ml. 2.0 ml of this aliquot is transferred into test tube and 2.0 ml each of ninhydrin and acetic acid are added. The mixture is heated on a boiling water bath for one hour, after which reaction is terminated by placing the tube in ice-box. Thereafter, the reaction mixture is shaken vigorously with 4.0 ml toluene and kept for several hours at room temperature. Chromatophore is thus extracted into toluene phase which is separated and its absorbance is measured at 520 nm using toluene as blank.

Standard curve to be prepared with graded concentration of DL Proline.

EXTRACTION AND ESTIMATION OF ASCORBIC ACID

1. Titrimetric Method Using 2,6 Dichlorophenol Indophenol : (Aberg, 1958)

(A) *Extraction :* Cut plant materials into 2-3 mm pieces. Place them in mortar and add oxalic acid (0.4% oxalic (w/v); use fresh reagent) at the rate of 4 ml/l g tissue. Crush the tissues thoroughly for 5 minutes. Filter the extract through 2 layers of cheese cloth and centrifuge the filterate at 1,000 g for 20 min. Make up the final volume with the oxalic acid reagent to represent 1 g tissue in 5 ml of the solvent. Acid washed sand may be added to improve the grinding. If the ascorbic acid concentration is low, extract the tissues with minimal amount of oxalic acid.

(B) *Estimation :* Ascorbic acid is easily estimated by the visual titration method based on the reduction of 2, 6-dichlorophenol indophenol dye.

1. Prepare standard ascorbic acid solution by dissolving 50 mg of ascorbic acid to 50 ml of 0.4% oxalic acid in a 250 ml volumetric flask and make up the volume to 250 ml with oxalic acid. One ml of this solution contains 0.2 mg of ascorbic acid. Use fresh solution for each standardization.
2. Indophenol reagent can be prepared by dissolving 50 mg of sodium 2, 6-dichlorophenol indophenol in 150 ml of glass

distilled water in a 200 ml volumetric flask. Warm the flask gently in a hot water bath to dissolve the dye. Add 42 mg of $NaHCO_3$ cool the solution and make up the volume to 200 ml with glass distilled water. Store the reagent in dark glass at 2°C and use it within one week.

3. The indophenol reagent must be standardized before use. Pipette 5 ml of the standard ascorbic acid solution to a white proclain dish and titrate against the indophenol dye kept in a burette until the solution becomes pink which should persist for at least 15 seconds.
4. For the estimation of unknown, pipette 5 ml of oxalic acid extract to a white porcelein dish and titrate against the standardized indophenol reagent until the solution becomes pink which should persist for at least 15 seconds.
5. Calculate the ascorbic acid content of the extract by formula.

$$I \times S \times D/A \times 100/W = \text{mg of ascorbic acid/100 g tissue.}$$

Where,

I = ml of indophenol reagent used in the titration.

S = mg of ascorbic acid reacting with 1 ml of the reagent.

D = Volume of the extract in ml,

A = the aliquot titrated in ml,

W = the weight of the sample in g,

2. Alternative Titrimetric Method

The method for quantitative estimation of ascorbic acid using DCPIP has been very widely used in the quantitative estimation of ascorbic acid in plant tissues. However, this method is not exactly specific for L-ascorbic acid. In view of this Barkat *et al.* (1973) have proposed a new titrimetric method for determination of ascorbic acid. By this method a minimum of 25 mg of L-ascorbic acid can be estimated. This method involves the titration of L-ascorbic acid with standard cupric salt solution e.g. cupric sulfate or cupric acetate.

L-ascorbic acid reacts with exactly two equivalents of cupric ions either in the form of cupric sulfate or cupric acetate. Cupric salts are known to react with an excess of potassium iodide to form cupric iodide. The latter is unstable and spontaneously decomposes to give cuprous iodide and free iodine. The iodine liberated then oxidises

$$2CuSO_4 + 4KI \rightarrow 2CuI_2 + 2K_2SO_4$$

$$2CuI_2 \rightarrow Cu_2I_2 + I_2$$

L-ascorbic acid to L-dehydroascorbic acid and hydrogen iodide.

The reaction continues until all the L-ascorbic acid present in the solution is oxidised by iodine released. The slightest trace of iodine in excess will develop a blue colour with starch indicator. The end point is sharp and easily detected. Following are the practical instructions for this method of ascorbic acid determination.

A. Extraction of ascorbic acid

Take 2-5 g of plant material and homogenise it in metaphosphoric acetic acid solution (This solution can be prepared by dissolving 15 g of metaphosphoric acid in a mixture containing 40 ml of acetic acid and 450 ml of distilled water).

Filter the homogenate and transfer an aliquot of clear filterate to an Erlenmeyer flask.

B. Titrimetic estimation

Prepare following reagents :

(a) 0.01 M solution of cupric sulfate or cupric acetate

(b) Potassium iodide 30% (W/V)

(c) 1% (W/V) solution starch.

To the flask in which sample has been added, add 10 ml of 30% potassium iodide (or 5% in case of very dilute samples) and a few drops of starch indicator.

From a 5 ml microburette run dropwise standard cupric sulfate solution. Titrate by shaking vigorously until a blue colour first appears. Note the volume of titer. Run a blank experiment using only metaphosphoric-acetic acid solution instead of homogenate. Substrate the readings of blank from sample titer before calculations.

$$\text{L-ascorbic acid} = V \times C$$

(mg/aliquot)

V = ml 0.01 M $CuSO_4$ $5H_2O$ solution.

C = 0.88 mg (L-ascorbic acid equivalent to 1 ml 0.01 M $CuSO_4.5H_2O$ solution).

3. Colorimetric Micro-Method (Glick *et al.*, 1953)

1. Take a glass-stoppered reaction vessel with a 0.25 ml capacity, add 20 µg of 0.5% oxalic acid. Add the tissue sample to this, using a glass needle.

2. Allow to stand at room temperature for 30-40 minutes without stirring.
3. Then add 50 µl of n-amyl alcohol to form a layer over the solution.
4. To the extract below the amyl alcohol layer, add 32 ml of 2, 6-dichlorophenol indophenol (5 mg in 100 ml). Add the dye rapidly, stir the mixture by bubbling air through it. Stopper the tube andshake vigorously by hand for 15 seonds to extract the residual dye.
5. Centrifuge to separate liquid layers.
6. Transfer an aliquot of clear amyl alcohol layer to a cuvette, and measure the optical density at 546 nm.
7. Prepare a blank by repeating the whole procedure but replacing the dye solution with water.
8. Measure the optical density of the sample at 546 nm, using the blank n-amyl alcohol extract for the control. For the standard curve use ascorbic acid of known conc. prepared in the same manner as the control.

ESTIMATION OF NITRATE CONTENT IN PLANT TISSUES

The method is based on nitration of the 2, 6-xylenol in the presence of a high concentration of the sulphuric acid. In order to make the measurements in visible spectrum, the nitrophenol is converted to the alkali metal 4-nitro-2, 6 xylenol is then separated by solvent extraction. Mercuric sulphate is added to prevent interference due to chloride ions.

Reagents

1. 2, 6 xylenol reagent Dissolve 1 g of 2, 6 xylenol (dimethyl-phenol) in 100 ml of glacial acetic acid.
2. Mercuric sulphate stock solution : Dissolve 1 g of mercuric sulphate in 200 ml of water and 10 ml of conc. H_2SO_4.
3. Sulphuric acid reagent : Add 158 ml of concentrated sulphuric acid to 42 ml of the mercuric sulphate stock solution with stirring. Allow the mixture to cool. Prepare fresh every time before use.
4. Sodium hydroxide solution : Dissolve 80 g of sodium hydroxide in water, making final volume to one litre.
5. Toluene.

6. Potassium nitrate (for preparing calibration curve) 72 mg KNO_3 make final volume 100 ml. This solution contains 100 mg of Nitrate-N in one litre.

Procedure

Take about 50 mg of plant material in a small conical flask, add 5 ml of deionised water and 50 mg activated charcoal.

Heat on boiling water bath for 10 min. After cooling centrifuge the contents at 8000 x g and take an aliquot for nitrate estimation. To one ml of sample add 3 ml of sulphuric acid reagent. Cool the contents by keeping the test tubes at low temperature. Then add 0.5 ml of xylenol reagent and heat the contents on water bath at 35°C for 30 min. Pour the contents in a separating funnel and add 2 ml of deionised water and 4 ml of toluene. Mix the contents thoroughly by vigorous shaking and then keep the funnel for about 10 minutes to allow the two layers to separate out. Discard lower layer. Wash toluene layer with 3 ml of water. then in the toluene layer, add 4 ml of sodium hydroxide and shake. After 20-25 minutes, measure the optical density in toluene layer at 432 nm.

ESTIMATION OF FATTY ACIDS AND OILS

Estimation of Free Fatty Acids

A small quantity of free fatty acids is usually present in oil along with the triglycerides. The free fatty acid content is known as acid number/acid value. It increases during storage. The keeping quality of oil therefore relies upon the free fatty acid content.

Principle

The free fatty acid in oil is estimated by titrating it against KOH in the presence of phenolphthalein indicator. The acid number is defined as the mg KOH required to neutralize the free fatty acids present in 1g of sample. However, the free fatty acid content is expressed as oleic acid equivalents (Cox and Pearson, 1962).

Reagents

1% phenolphthalein in 95% ethanol.

0.1 N potassium hydroxide

Neutral Solvent : Mix 25 ml ether, 25 ml 95% alcohol and 1 ml of 1% phenolphthalein solution and neutralize with N/10 alkali.

Procedure

1. Dissolve 1-10 g of oil or melted fat in 50 ml of the standard solvent in a 250 ml conical flask.
2. Add a few drops of phenolphthalein.
3. Titrate the contents against 0.1 N potassium hydroxide.
4. Shake constantly until a pink color which persists for fifteen seconds is obtained.

Calculation

$$\text{Acid value (mg KOH/g)} = \frac{\textit{Titrate value} \times \textit{Normality of KOH} \times 56.1}{\textit{Weight of sample (g)}}$$

The free fatty acid is calculated as oleic acid using the equation 1 ml N/10 KOH = 0.028 g oleic acid

Note

To find out the exact strength of KOH, prepare 0.1 N oxalic acid solution (630 mg in 100 ml water) and titrate against KOH with phenolphthalein as indicator. Calculate the strength of KOH by the formula $V_1N_1 = V_2N_2$.

Estimation of Oil in Oilseeds

Fats and fatty acid esters of glycero. Fat as liquid is called oil. Seeds like gingely, groundnut, castor, sunflower etc. contain oil as reserve food material for the embryo.

Principle

Oil from a known quanity of the seed is extracted with petroleum ether. It is then distilled off completely, dried, the oil weighed and the % oil is calculated.

Material

1. Petroleum Ether (40-160°C).
2. Whatman No. 2 Filter paper.
3. Absorbant Cotton.
4. Soxhlet Apparatus.

Procedure

1. Fold a piece of paper in such a way to hold the seed meal. Wrap around a second filter paper which is left open at the

top like a thimble. A piece of cotton wool is placed at the top to evenly distribute the solvent as it drops on the sample during extraction.

2. Place the sample packet in the butt tubes of the Soxhlet extraction apparatus.
3. Extract with petroleum ether (140 drops/min) for 6 h without interruption (for castor beans use hexane) by gentle heating.
4. Allow to cool and dismantle the extraction flask. Evaporate the ether on a steam or water bath until no order of ether remains. Cool at room temperature.
5. Carefully remove the dirt or moisture outside the flask and weight the flask. Repeat heating until constant weight is recorded.

Calculation

$$\text{Oil in ground sample \%} = \frac{\textit{Weight of Oil}\ (g)}{\textit{Weight of sample}\ (g)} \times 100$$

$$\text{Oil to dry weight basis} = \frac{\textit{\% oil in ground sample}}{\textit{100\% moisture in whole seed}}$$

Sample preparation varies depending upon the materials.

Peanut

Place about 50 g of Kernels in a drying dish and dry at 130°C for not more than 20 min in a forced draft oven. Cool to room temperature and then pass through the nut slicer. Care is to be taken to prevent expressing of any oil while slicing. Mix the sliced sample well. Weight accurately 2 g into the filter paper fold.

Sunflower

Weigh about 50 g of sunflower seeds into a large beaker. Add an equal weight of Hyflow supercel or equivalent (dried at 130°C overnight before use). Mix well. Grind the mixture immediately.

For sunflower seeds oil in ground sample, % =

$$\frac{\textit{Weight of Oil}\ (g) \times 200}{\textit{Weight of Sample}\ (g)}$$

since equal quantity of Hyflow supercel is added.

Estimation of Volatile Acids

Volatile fatty acids are water-soluble and they can be distilled at atmospheric pressure. These fatty acids are intermediates in the process of methane production and many other allied reactions. These acids can be removed by steam distillation, in spite of high boiling points, because of their high vapor tension. This group includes water-soluble fatty acid having up to six carbon atoms.

Principle

Straight distillation or steam distillation after acidifying the sample with 1+1 H_2SO_4 results in the evaporation of volatile acids which on condensation yield liquefied volatile acids. These acids could be titrated against 0.1 N alkali with phenolphthalein indicator to find out the amount of volatile fatty acids. The steam distillation method is more tedious, requires at least 4 h to complete, but results in the recovery of 92% to 98% volatile acids from sewage sludge. The steam distillation is conducted on a $MgSO_4$ saturated, acidified, sludge-free liquor produced by chemical treatment with ferric chloride and filter-aid followed by vaccum filtration. The effects of varying concentrations of dissolved solids and mineral acids during distillation are avoided (Rand *et al.,* 1976).

Reagents

1. Dilute Sulphuric Acid (1 + 1) + Mix one volume of conc. H_2SO_4 with one volume of water.
2. Ferric Chloride Solution : Dissolve 82.5 g $FeCl_3$ $6H_2O$ in one liter distilled water.
3. Diatomaceous-Silica Filter Aid.
4. Magnesium Sulphate ($MgSO_4$, $7H_2O$)
5. Standard Sodium Hydroxide, titrant 0.1 N : Dissolve 4 g sodium hydroxide in 1 *l* of distilled water and standardize against any standard acid.
6. Phenolphthalein Indicator : 0.1% in 95% alcohol.

Procedure

1. Adjust the sample containing volatile fatty acids (sludge and water) 200 to 1,000 ml to pH 3.5 with (1 + 1) H_2SO_4.
2. Add 6 ml of $FeCl_3$ solution/l sample (equivalent to 500 mg/l).
3. Add 50 g filter aid/l, and mix well.

4. Filter by suction, using a Buchner funnel containing a filter paper freshly coated with a thin layer of filter-aid.
5. Wash the residue thoroughly three or four times with water, and adjust the filtrate to pH 11 with NaOH solution.
6. Concentrate by evaporation to 150 ml and cool in a refrigerator.
7. Adjust the cooled filtrate to pH4 with dilute H_2SO_4 and add it quantitatively and quickly to the distilling flask.
8. Add magnesium sulphate to a slight excess of saturation.
9. Apply heat with a small flame to the flask until rapid evolution of volatile acids commences. This will prevent excessive increase in the volume of the mixture.
10. Steam distill slowly so that about 200 ml of distillate is collected in 25 min.
11. Increase the rate of distillation and continue until a total of 600 ml is collected.
12. Titrate the distillate against 0.1 N NaOH using phenolphthalein indicator.

Calculation

$$\text{mg/L volatile acids as acetic acid} = \frac{ml\ 0.1\ N\ NaOH \times 6{,}000}{ml\ sample}$$

(molecular weight of acetic acid is 60)

Identification and Quantification of Fatty Acids

To understand the quality and composition of fat, identification and quantification of fatty acids present are important. Though fatty acid esters could be identified by TLC, quantification is best done with gas-liquid chromatography (GLC).

Principle

Fatty acids are made volatile by converting them into methyl esters. The conversion of fatty acids in methyl esters is carried out directly by trans esterification i.e., from glycol ester to methyl ester. Esterification can also be done after saponification. The esters are identified and quantified by injecting into GLC and comparing with a set of standard esters (Metcalfe *et al.*, 1966; Munshi, 1987).

Materials

1. 10% BCl_3 or 14% BF_3 in methanol
2. Saturated NaCl Solution
3. Na_2SO_4 (anhydrous)
4. Hexane
5. Nirogen Gas
6. GLC
7. Column Pretested 10% silar 10°C on gas chrom Q 100 to 200 mesh (6' x 1/4" - Internal diameter (4mm)
8. Detector - FID
   Injector temp - 280°C
   Detector temp - 250°C
9. Carrier gas - Nitrogen at 50 ml/min
10. Column/oven temp - 165°C

Procedure

1. To 150 to 300 mg of oil, in a culture tube add 3 ml of 10% BCl_3 or 14% BF_3.
2. Add boiling chips, cover the tube with aluminum foil and heat at 83°C for 6 min.
3. Transfer the contents to a 30 ml separating funnel and also collect the washing of the tube into the separating funnel. Washing is done four times with 1 ml portions of hexane.
4. Shake and allow to separate.
5. Add 4 ml saturated NaCl solution.
6. Shake and collect hexane layer over anhydrous Na_2SO_4.
7. Rinse funnel with 2 ml hexane, collect the hexane layer and combine the hexane extracts.
8. Filter the hexane extract through filter paper (Whatman No. 4).
9. Reduce the volume of the filtrate to 2-3 ml by drying with a low stream of nitrogen.
10. Inject an aliquot to pre-conditioned GLC.
11. Inject standard methyl esters separately and calculate the retention time for individual esters.

Calculations

Identify peaks by relative position on the chart. Esters appear in order of increasing number of carbon atoms and of increasing unsaturation for the same number of carbon atoms. Calculate the peak area of each acid in the sample and compare with the standards. From the quantity injected and the attenuation, the amount of fatty acid content can be calculated. The latest computerized GLC models give concentration directly.

Notes

1. If free fatty acids alone are to be determined, incorporate methyl urea (520-540 mg) in the tube to prevent transecterification of fatty acids of glycerides. This will ensure reduction of trans-esterification to 0.02-0.05%.
2. An alternative method for trans-esterification of fat is given below :

Take 5-50 mg of lipid sample in 15 x 45 mm culture tube with a cap (Evaporate the solvent if the sample is in the solvent). Add 0.25/0.5 ml of 0.4 N sodium methoxide in anhydrous methanol and cover with a cap. Immerse the capped tube to a depth of 12 mm for 30 sec in a water bath at 65°C. Esterification is considered complete when the mixture of lipid and reagent becomes homogeneous. Continue heating for another 90 sec. Remove the tube and cool to room temperature. Quickly add to this, 1 g mixture of silica gel G and anhydrous $CaCl_2$ (1 : 1 w/w). Mix the contents thoroughly with a glass rod. Add 5 ml of hexane/carbon disulphide and shake for 1-2 min. Centrifuge for a short time. Withdraw the clear solvent layer. Wash with the solvent to have a good recovery. Evaporate the solvent at low heat to 0.5 ml final volume. This material now contains the methyl esters of fatty acids that can be injected into a GLC for identification and quanti-fication.

3. Instead of trans-esterification, saponification followed by esterification can also be done as follows. Saponify 1 g fat with 25 ml of 4% methanolic KOH for 30 min (Reflux with an air condenser over a boiling water bath). After saponification, remove methanol in a flash evaporator below 50°C. Dissolve the residue in sufficient water. Transfer to a separating funnel. Add petroleum ether (60 - 80°C), shake, and remove the petroleum ether layer to get rid of unsaponifiable matter. Hydrolyze the saponified material with HCl and extract the free fatty acid with petroleum ether four times. Pool the ether extracts and dry over nhydrous Na_2SO_4. Evaporate to dryness

under vacuum at 50°C. Add 20-fold methanol-HCl and reflux for 2h over a water bath. Remove excess methanol by rotary evaporation. Add petroleum ether and trasnfer to a separating funnel. Shake with 0.5 N sodium carbonate and remove the aqueous layer for washing. The resultant petroleum ether extract contains the methyl esters of fatty acids which are then injected into a GLC for quantification.

4. Any unesterified liquid component will spoin the column in most cases and hence the esterification must be checked before injecting into GLC. The check is done by carrying out the separation by TLC (5 x 20 cm glass plates) using silica gel G in a solvent system of petroleum ether-diethyl ether-acetic acid 80:20:1 v/v), the prominent spot for esters and the absence of spots for triglycerides confirm the completion of esterification.

Saponification Value

Saponification is the process by which the fatty acids in the glycerides of the oil are hydrolyzed by an alkali. Saponification value is the amount (mg) of alkali required to saponify a definite quantity (1g) of an oil or fat. This value is useful for a comparative study of the fatty acid chain length in oils.

Principle

A known quantity of oil is refluxed with an excess amount of alcoholic KOH. After saponification, the remaining KOH is estimated by titrating it against a standard acid (William, 1975).

Materials

1. Hydrochloric acid 0.5 N, accurately standardized.
2. Alcoholic KOH - Dissolve 40 g KOH in one liter of distilled alcohol keeping the temperature below 15.5°C while the alkaline is being dissolved. This solution should remain clear.
3. Phenolphthalein Indicator - 1% in 95% alcohol.
4. Air Condenser.

Procedure

1. Melt the sample if it is not already liquid and filter through paper to remove any impurities and the last traces of moisture. The sample must be completely moisture-free.

2. Weigh 4-5 g sample into the flask. Add 50 ml of alcoholic KOH from burette by allowing it to drain for a definite period of time.
3. Prepare a blank also by taking only 50 ml of alcoholic KOH allowing it to drain at the same duration of time.
4. Connect air condenser to the flasks and boll them gently for about 1h.
5. After the flask and condenser get cooled, rinse down the inside of the condenser with a little distilled H_2O and then remove the condenser.
6. Add about 1 ml of indicator and titrate against 0.5 N HCl until the pink color just disappears.

Calculations

$$\text{Saponification value} = \frac{28.05 \times (\textit{titer value of blank} - \textit{titer value of sample})}{\textit{Weight of sample } (g)}$$

Notes

1. Alcohol is inflammable. Use electrical heating.
2. Alcohol should not get dried up during saponification. Effective cooling of alcohol vapor is essential.
3. Clarity and homogeneity of the test solution are indicators of complete saponification.

Determination of Iodine Value of Oil

The iodine value is a measure of the degree of unsaturation in an oil. It is constant for a particular oil or fat. Iodine value is a useful parameter in studying oxidative rancidity of oils since higher the unsaturation the greater the possibility of the oils to go rancid.

Principle

The oils contain both saturated and unsaturated fatty acids. Iodine gets incoporated into the fatty acid chain wherever the double bond exist. Hence, the measure of the idonie absorbed by an oil, gives the degree of unsaturation. Iodine value/number is defined as the `g' of iodine absorbed by 100 g of the oil (William, 1975).

Materials

1. Hanus Iodine Solution: Weigh 13.6 g of Iodine and dissolve in 825 ml glacial acetic acid by heating, and cool. Titrate 25 ml

of this solution against 0.1 N sodium thiosulphate. Measure another portion of 200 ml of glacial acetic acid and add 3 ml of bromine to it. To 5 ml of this solution add 10 ml of 15% potassium iodine solution and titrate against 0.1 N sodium thiosulphate. Calculate the value of bromine solution, to double halogen content of the remaining 800 ml of the above iodine solution as follows :

2. X = B/C, where X = ml of bromine solution required to double the halogen content, B = 800 x thiosulphate equivalent of 1 ml of iodine solution and C = thiosulphate equivalent of one ml of bromine solution.
3. 15% potassium Iodine Solution
4. 0.1% Sodium Thiosulphate
5. 1% Starch

Procedure

1. Weigh 0.5 or 0.25 g of oil into an iodine flask and dissolve in 10 ml of chloroform.
2. Add 25 ml of Hanus iodine solution using a pipette, draining it in a definite time. Mix well and allow to stand in dark for exactly 30 min with occasional shaking.
3. Add 10 ml of 15% KI, shake thoroughly and add 100 ml of freshly boiled and cooled water, washing down any free iodine on the stopper.
4. Titrate against 0.1 N sodium thiosulphate until yellow solution turns almost colorless.
5. Add a few drops of starch as indicator and titrate until the blue color completely disappears.
6. Towards the end of titration, stopper the flask and shake vigorously so that any iodine remaining in solution in $CHCl_3$ is taken up by potassium iodine solution.
7. Run a blank without the sample.

Calculation

The quantity of thiosulphate required for blank minus the quantity required for sample gives thiosulphate equivalent of iodide adsorbed by the fat or oil taken for analysis.

$$\text{Iodine number} = \frac{(B-S)\times N\times 12.69}{Weight\ of\ Sample(g)}$$

where,

B = ml thiosulphate for blank

S = ml thiosulphate for sample

N = normality of thiosulphate solution

Amount of fat/oil taken should be adjusted such that the excess iodine in the added 25 ml of Hanus iodine solution has about 60% of excess iodine of the amount added, i.e., if (B–S) is greater than B/2, repeat the smaller amount of sample.

Determination of Peroxide Value

Rancidity is brought about by the action of air (oxidative rancidity or by microorganisms (Ketonic rancidity) in oil. In oxidative rancidity oxygen is taken up by the fat with the formation of peroxides. The degree of peroxide formation and the time taken for the development of rancidity differ among oils.

Principle

Peroxide value is a measure of the peroxide contained in the oil. The peroxides present are determined by titration against thiosulphate in the presence of KI. Starch is used as indicator (Cox and Pearson, 1962).

Materials

1. Solvent Mixture : Mix two volumes of glacial acetic acid with one volume of chloroform.
2. 5% potassium Iodine Solution.
3. 1% Starch Solution.
4. N/500 Sodium Thiosulphate Solution. Prepare N/10 solution and dilute to N/500 on the day of use.

Procedure

1. Weigh 1 g of oil or fat into a clean dry boiling tube and add 1 g of powdered potassium iodide and 20 ml of solvent mixture.
2. Place the tube in boiling water so that the liquid boilds within 30 seconds and allow to boil vigorously for not more than 30 seconds.
3. Transfer the contents quickly to a conical flask containing 20 ml of 5% potassium iodide solution.
4. Wash the tube twice with 25 ml water each time and collect into the conical flask.

5. Titrate against N/500 sodium thiosulphate solution until yellow color is almost disappeared.
6. Add 0.5 ml of starch, shake vigorously and titrate carefully till the blue color just disappears.
7. A blank should also be set at the same time.

Calculation

Peroxide value (milliequivalent peroxide/kg sample) =

$$\frac{S \times N \times 1000}{Weight\ of\ sample\ (g)}$$

where,

S = ml $Na_2S_2O_3$ (Test-blank) and

N = normality of $Na_2S_2O_3$

PHOSPHATASES

(Orthophosphoric-monoesterphosphoydrolase, alkaline medium 3.1.3.1)

(Orthophosphoric-monoesterphosphohydrolase, acid medium 3.1.3.2)

Phosphatases liberate inorganic phosphate from organic phosphate esters. Acid phosphatase (3.1.3.2) hydrolyzes a number of phosphomonoesters and phosphoproteins. Alkaline phosphatase catalyses the hydrolysis of numerous phosphate esters, such as esters of primary and secondary alcohols, sugar alcohols, phenols and amines. Phosphodiesters are not hydrolyzed by ether of them.

Principle

The enzyme phosphatase hydrolyzes p-nitrophenol phosphate. The released p-nitrophenol is yellow in color in alkaline medium and is measured at 405 nm. The optimum pH for acid and alkaline phosphatases are 5.3 and 10.5 respectively (Fiske and Subba Row, 1925; Lowry *et al.*, 1954).

(i) Acid Phosphatase

Materials

1. *Sodium Hydroxide 0.085 N*

Dissolve 0.85 g sodium hydroxide in 250 ml water.

2. Substrate Solution

Dissolve 1.49 g EDTA, 0.84 g citric acid and 0.03 g p-nitrophenyl phosphate in 100 ml water and adjust to pH 5.3.

3. Standard

Weigh 69.75 mg p-nitrophenol and dissolve in 5.0 ml distilled water (100 mM).

4. Enzyme Extract

Homogenize 1g fresh tissue in 10 ml of ice-cold 50 mm citrate buffer (pH 5.3) in a pre-chilled pestle and mortar. Filter through four layers of cheese cloth. Centrifuge the filtrate at 10,000 g for 10 min. Use the supernatant as enzyme source.

Procedure

1. Incubate 3 ml of substrate solution at 37°C for 5 min.
2. Add 0.5 ml enzyme extract and mix well.
3. Remove immediately 0.05 ml and mix it with 9.5 ml of sodium hydroxide 0.085 N. This corresponds to zero time assay (blank).
4. Incubate the remaining solution (substrate + enzyme) for 15 min at 37°C.
5. Draw 0.5 ml sample and mix it with 9.5 ml sodium hydroxide solution.
6. Measure the absorbance of blank and incubated tubes at 405 nm.
7. Take 0.2 to 1.0 ml (4 to 20 mM) of the standard, dilute to 10.0 ml with NaOH solution. Read the color and draw the standard curve.

Calculations

Specific activity is expressed as m moles p-nitrophenol released per min per mg protein.

Notes

For alkaline phosphatase extract the enzyme in 50 mM glycine NaOH buffer pH 10.4.

Alkaline phosphatse functions optimally at about pH 10.5. The assay procedure is similar to that for acid phosphatase, except for the substrate solution. Prepare the substrate solution as follows.

Dissolve 375 g glycine, 10 mg magnesium chloride, 165 mg p-nitrophenyl phosphate in 42 ml of 0.1 N sodium hydroxide and dilute to 100 ml. Adjust to pH 10.5.

(ii) Assay using Glycerophosphate

Acid and alkaline phosphatases can also be assayed by measuring the amount of inorganic phosphorus released from sodium glycerophosphate. Phosphorus is estimated by Fiske-Subbarow method as described below.

Materials

1. Substrate : 0.1 M solution of /3-glycerophosphate 3.153 g in 100 ml water.
2. 10% TCA
3. Ammonium Molybdate (2.5%)

Dissolve 25 g ammonium molybdate in 400 ml of water. Add 500 ml of 10 N H_2SO_4 and make up the volume to one liter with water.

4. *1-Amino 2-Naphthol 4-Sulphonic Acid (ANSA) Reagent*

Dissolve 30 g sodium metabisulphite, 6 g sodium sulphite and 5.00 mg ANSA separately in small quantities of water. Combine all the solutions and make up to 250 ml with water. Allow to stand overnight and filter. Store refrigerated in an amber-colored bottle. Prepare fresh reagent every fortnight.

5. *Standard Phosphate Solution:* Dissolve 439 mg potassium dihydrogen phosphate in water, add 10 ml of 10 N H_2SO_4 and make up to one liter with water (1 ml = 0.1 mg P).

Add 0.5 ml chloroform as preservative.

Dilute 10 ml of the above stock solution to 50 ml with water and use as working standard (1 ml = 20 μg P).

6. *0.2 M solution of magnesium acetate* - 4.289 g in 100 ml water.

Procedure

1. Pipette out 2 ml of respective (acid or alkaline) buffer in a test tube.
2. Add 0.5 ml magnesium acetate solution.
3. Add 2 ml of b-glycerophosphate solution.

4. Make up the volume to 7 ml with the buffer.
5. Add 1 ml enzyme extract and incubate for 1 h at 37°C.
6. Stop the reaction by the additon of 2 ml of 10% TCA.
7. For the control add the enzyme after incubation.
8. Centrifuge at 10,000 g for 10 min.
9. Pipette out different known aliquots of supernatant of test, and control into test tubes. Make up the volume in each tube to 5 ml by adding water.
10. 10 Add 1 ml of ammonium molybdate solution to each tube and mix thoroughly.
11. Add 0.4 ml of ANSA reagent to each tube and mix well. After standing for 10 min dilute to 10 ml. Read the intensity of blue color at 660 nm against a reagent blank.
12. Pipette out different aliquots of phosphate solution (5-50 µg P) into test tubes and treat as described above. A reagent blank is also prepared with water in place of phosphate solution. Plot absorbance versus phosphorus concentration. Read the phosphorus content from the calibration curve and express enzyme activity as the amount of phosphorus released in unit time per mg protein.

Notes

1. Any traces of phosphate impurity in the distilled water interferes with the determination. This is easily noticed from the reagent blank which should actually be colorless.
2. After developing, the blue color intensifies with the progress of time. It is preferable to read the tubes 10 min after adding ANSA reagent.

LIPASE

(Triacylglycerol acylhydrolase EC 3.1.1.3)

Lipase hydrolyses triglycerides to release free fatty acids and glycerol.

Triglycerides + H_2O – Glycerol + Fatty Acids

During germination of oilseeds, lipases play an important role in hydrolysing the stored oils so that the required energy for growth and carbon skeleton for synthesis of new compounds are produced. The germinating seeds of castor bean, sunflower, groundnut are good sources of lipases.

Principle

The quantity of fatty acid released in unit time is measured by the quantity of NaOH required to maintain pH constant. he milliequivalent of alkali consumed is taken as a measure of the activity of the enzyme (Jayaraman, 1981).

Materials

1. Substrate : Take 2 ml of any clear vegetable oil, neutralize to pH 7.0, if necessary, and stir well with 25 ml of water in the presence of 100 mg bile salts (sodium taurocholate) till an emulsion is formed. Addition of 2g gum arabic hasten emulsification.
2. 0.1 N NaOH
3. Enzyme Source : Grind a known quantity of sample with a mortar and pestle. Homogenize the tissue with twice the volume of ice-cold acetone. Filter and wash the powder successively with acetone, acetone : ether (1 : 1) and ether. Air dry powder. This acetone powder, can be stored in a refrigerator. Extract 1 g of the powder in 20 ml ice-cold water of suitable buffer. Centrifuge at 15,000 rpm for 10 min and use the supernatant as enzyme source.
4. 50 mM phosphate buffer (pH 7.0).

Procedure

1. Take 20 ml of substrate in a 500 ml beaker. Add 5 ml of phosphate buffer (pH 7.0).
2. Set the beaker on top of a magnetic stirrer cum hot plate and stir the contents slowly. Maintain the temperature at 35°C. Dip the electrodes of a pH meter in the reaction mixture. Note the pH and adjust it 1 7.0.
3. Add enzyme extract (0.5 ml), immediately record the pH and set the timer on. Let it be pH at zero time.
4. At frequent intervals (say 10 min) or as the pH drops by about 0.2 unit add 0.1 NaOH to bring pH to the initial value. Continue the titration for 30-60 min period.
5. Note the volume of alkali consumed.

Calculation

The enzyme activity is defined as the amount of enzyme which releases one milliequivalent of free fatty acid per minute per g sample. Specific activity is expressed as milliequivalent / min/ mg protein.

$$\text{Activity meq/min/g sample} = \frac{\textit{Volume of alkali consumed} \times \textit{Strength of alkali}}{\textit{Wt. of sample in g} \times \textit{Time in} \text{ min}}$$

REFERENCES

Aberg, B. 1958. Ascorbic acid. In: *Encyclopedia of Plant Physiology* 6: 479-494. (Ed. H. Burton) Springer-Verlag, Berlin.

Adams, R.L.P., Knowler, J.T. and Leader, D.P. 1992. *The Biochemistry of Nucleic Acids.* 11th edition, Chapman and Hall Ltd., London.

Barkat, M.Z., Shenab, S.K., Darwish, N. and El-Zohery, A. 1976. A new titrimethic method for the determination of vitamin C. *Analytical Biochemistry* 53: 245-250.

Bates, L.S., Waldren, R.P. and Teare, I.D. 1973. Rapid determination of free proline for water stress studies. *Plant and Soil* 39: 295-297.

Bramhall, S., Noack, N., Wu, M. and Loewenberg, J.R. 1969. A simple colorimetric method for the determination of protein. *Analytical Biochemistry* 31: 146-148.

Bredford, M. 1976. A rapid and senstive method for the quantification of microgram quantities of protein utilizing the principle of protein-dye binding. *Analytical Biochemistry* 72: 248-258.

Burton, K. 1968. Determination of DNA concentration with dippenylamine. In: *Methods in Enzymology* (Grossman L. and Moldave K., eds.), Vol. 12, Part B, pp. 163-167, Academic Press, New York.

Cherry, J.H. 1962. Nucleic acid determination in storage tissue of higher plants. *Plant Physiology* 37: 670-678.

Cox, H.E. and Pearson, D. 1962. *The Chemical Analysis of Foods.* Chemical Publishing Co. Inc. New York, 421 p.

Fiske, C.H. and SubbaRow, Y. 1925. *Journal of Biological Chemistry* 66 : 575.

Glick, D., Alpert, M. and Stecklein, H.R. 1953. Studies in histochemistry: XXVII. The determination of L-Ascorbic acid and dehydro-L-Ascorbic acid plus Diluto-L-Gulonic acid in microgram quantities of tissue. *Journal of Histochemistry and Cytochemistry* 1: 326-335.

Jayaraman, J. 1981. In : *Laboratory Manual in Biochemistry*. Wiley Eastern Limited New Delhi p. 133.

Lowry, O.H., Rosenbrogh, N.J., Farr, A.L. and Randall, R.J. 1951. Protein measurement with Folin-phenol reagent. *Journal of Biological Chemistry* 193: 265-275.

Lowry, O.H., Roberts, N.R., Mei-Ring, W.U. Hixon, W.S. and Crawford, E.J. 1954. *Journal of Biological Chemistry* 207 : 19.

Macare, J.C. 1971. Quantitative measurement of starch in very small amounts of leaf tissue. *Planta* 96: 101-106.

Metcalfe, L.D. Schmitz, A.A. and Pelke, J.R. 1966. *Analytical Chemistry* 38 : 514.

Munshi, S.K. 1987. In: *Manual in Modern Analytical Techniques in Agricultural Biochemistry* (Ed. Sudarshan Singh). Punjab Agricultural University, Ludhiana. 226 p.

Nelson, N. 1944. A photometric adaptation of the Somogy method for the deterimination of glucose. *Journal of Biological Chemistry* 153: 375-380.

Parish, J.H. 1972. *Principles and Practise of Experiments with Nucleic Acids*, Longman Group Ltd., London.

Rand, M.C., Greenberg, A.E. and Taras, M.J. (eds.) 1976. In: *Standard Methods for the Examination of Water and Waste Water* (14th Ed) American Public Health Association Washington p. 529.

William Horowitz (ed.) 1975. *Official Methods of Analysis of AOAC Association of Official Analytical Chemists*. Washington 12th Ed. 190 p.

Wilson, K. and Walker, J. 1996. *Practical Biochemistry-Principles and Techniques,* 4th edn., Combridge University Press, Combridge, London.

Yemm, E.W. and Willis, A.J. 1954. The estimation of carbohydrates in plant extracts by anthrone. *Biochemical Journal* 57: 508-514.

Yemm, E.W. and Cocking, E.C. 1955. The determination of amino acids with ninhydrin. *Analyst* 80: 209-230.

CHAPTER-6

pH AND BUFFER SOLUTIONS

ACIDS AND BASES

The modern concept of acids and bases developed by Bronsted, Lowry and others defines acids as proton donors and bases as proton acceptors. Thus an acid is in equilibrium with its conjugate base as under :

$$\text{Acid} = \text{Base} + H^+$$

The older view of an acid is as a compound yielding hydrogen ions and a base as one giving hydroxyl ions. Some ionic species in fact act both as acid and base. Such compounds are known as ampholytes or are said to be amphoteric (Chopra and Kanwar, 1976).

pH : In 1909 Sorenson introduced the term pH as a convenient way of expressing hydrogen ion concentration. He defined pH as the negative logarithm of the hydrogen ion concentration. pH is more strictly defined as the negative logarithm of the hydrogen ion activity, although in practice the activity is nearly equal to the concentration except in strongly acid solutions (Metlock, 1961).

BUFFER, pH And Life

Animal Cell

Cells are mostly insulated from the external environment and are bathed in fluid. The composition of pH of this fluid is maintained within quite a narrow range by the organism, and ay radical departure from the normal range results in the death of the cell. In human being the plasma pH lies within the very narrow limits of 7.35 to 7.45 at 37°C. At pH 7.0 or below, death results from acidotic coma and at pH 7.8 or above death occurs due to tetany. From this, it is clear that there must be an efficient homeostatic mechanism for the maintenance of the plasma pH within limits for life and even narrowr limits for health.

This is effected by the buffering systems of the plasma, bicarbonate, protein and phosphate, of which bicarbonate is the most important.

$$H_2CO_3 = HCO_3^- + H^+$$

From the Henderson-Hasselbalch equation :

$$pH = pK + \log_{10} \frac{[HCO_3^-]}{[H_2CO_3^-]}$$

The plasma pH, therefore, depends on the ratio of the bicarbonate to carbonic acid and not the absolute concentrations. Any tendency for the pH to change is buffered and can be corrected by adjusting this ratio.

Quite a lot is known of the buffering systems and their function in extracellular fluid,but relatively little about the pH within the cell. The values of most cells examined lies near pH 6.8. This figure has been calculated from a knowledge of the concentration of bicarbonate and other ions within the cell and also measured by the use of indicators and microelectrodes.

Plant and Bacterial Cells

The cell sap of the majority of land plants lies within the range of pH 5.2-6.5, although indicators suggest that the cytoplasmic pH is about 6.8. Some plant juices are very acidic (Orange, grapefruit, pH=3). The acid seems to be present in vacuoles and, therefore, segregated from the rest of the cytoplasm.

The internal pH of bacteria examined appear to be around neutrality although many forms grow well at pH 6 or 9, a thousand fold difference. Thiobacilli will grow at pH 0, which is that of N sulphuric acid, while some fungi grow at pH 11.

Clearly, the simpler forms of life appear to be able to survive a wide pH range of the external medium although the internal pH does not vary to that extent. Each living system has a powerful buffering system to maintain the internal cell pH.

SOLUTIONS

The use of solutions is common in experimental procedures of plant physiology. It is very important for a student to be familiar with their preparation. The substance dissolved in a solution is called solute and the medium in which it is dissolved is known as solvent.

Solutions of two liquid (e.g. alcohol in water) and solid in liquid (e.g. NaCl in water) are most common (Plummer, 1971). There are different ways to express the composition of a solution in terms of quantity of solute present in the solvent. These are as follows :

1. Mole Concentration

The basic SI unit of quantity is the mole which gives the amount of a substance present irrespective of the volume. Mole is defined as molecular weight in grams.

One mole = molecular weight in grams = 6.023 x 1023 molecules (Avogardo's number). Thus 1 mole of glucose (molecular weight 180) is 180 g.

Molarity (M) of a solution is the number of moles of a solute dissolved per liter of a solution, thus to prepare 1 M solution of glucose we dissolve 180 g of glucose in water and make the volume 1 liter. Molality of a solution is the number of moles of a solute dissolved in 1 kg of a solvent thus to prepare 1 molal solution of glucose dissolve 180 g of glucose in 1 kg of water (1000 ml water).

2. Mass Concentration

Sometimes the molecular weight of a compound is unknown or the substance to be dissolved is an impure mixture (e.g. casein hydrolysate, extract of protein, cell sap etc.) In such cases concentration has to be expressed in terms of weight per unit volume (w/v) or volume per unit volume (v/v) on per cent basis. Even in cases where the molecular weight of a solute is known per cent solutions are sometimes more straight forward to deal with.

One g of NaCl dissolved in water to make the volume 100 ml (1 g NaCl + 99 g water) shall make 1% NaCl (w/w) solution. Similarly, 1 ml of alcohol dissolved in water to make the volume 100 ml shall make 1% alcohol (v/v) solution.

One mg of a substance dissolved in 1 kg of a solvent (1000 ml of water) written as 1 mg kg-1 (1 mg 1 -1 of water) is another convenient measure for very dilute solutions. Since 1 mg forms 1 millionth part of a kg by weight this forms 1 part per million (1 ppm).

3. Solutions of Acids and Alkalies

Commonly used acids like hydrochloride acid and sulphuric acid are not pure and for making their solutions their actual strength has to be taken into account. Relevant parameters of some common acids are given in Table 1.

Table 1. Parameters of some commonly used commercially available acids and alkalies.

Name of acid or alkali	Purity (% by wt.)	Specific gravity at 30°C
Hydrochloric acid	36	1.19
Sulphuric acid	96	1.84
Nitric acid	69.5	1.42
Glacial acetic acid	99.5	1.05
Phosphoric acid	85	1.71
Perchloric acid	60	1.54
Ammonium hydroxide	28	0.9
Sodium hydroxide (saturated)	50	1.5

To prepare a solution of desired normality the following formula is applied :

$$V1 = \frac{Equivalent\ weight\ of\ acid \times V2 \times Normality \times 1000}{1000 \times specific\ \ gravity \times purity\ \%}$$

Where,

V1 = required volume of the commercial acid

V2 = volume of the acid solution of desired normality to be prepared

Thus, to prepare 800 ml of 0.1 N sulphuric aid from commercial concentrated acid (purity, 96%; specific gravity, 1.84; molecular weight, 98; equivalent weight 49 (sulphuric acid is diprotic 49) we compute

$$V1\ = \frac{49 \times 800 \times 0.1 \times 100}{1.84 \times 96} = 2.2\ ml$$

Take 2.2 ml of commercial concentrated sulphuric acid and dilute it to 800 ml with distilled water to get 0.1 N solution of sulphuric acid.

4. Buffer Solutions

Buffer resists a change in pH on the addition of acid or alkali. Certain purification procedures or reaction processes that require a particular pH are to be conducted in appropriate buffer solutions. In natural systems all biochemical reactions of the cells are catalysed by enzymes whose stability as well as activity is highly dependent on the constant pH of the system. Buffer solutions have conjugate acid-base pairs. Most commonly a buffer solution consists of a mixture of a weak acid and its conjugate base, e.g., a mixture of acetic acid and sodium acetate is a buffer solution. Composition of some commonly used buffers are given in Methods in Enzymology, 1955, 1984 and 1990.

COMMON BUFFERS

(1) Phosphate Buffer

Stock Solutions

A : 0.2 m solution of monobasic sodium phosphate (NaH_2PO_4) 27.6 g in 1000 ml.

B : 0.2 solution of dibasic sodum phosphate (53.65 g) of Na_2HPO_4 7 H_2O or 71.7 x ml of A + y ml. of B, diluted to a total of 200 ml gives 0.2 M Buffer.

x	y	pH	x	y	pH
93.5	6.5	5.7	45.0	55.0	6.9
92.0	8.0	5.8	39.0	61.0	7.0
90.0	10.0	5.9	33.0	67.0	7.1
87.7	12.3	6.0	28.0	72.0	7.2
85.5	15.0	6.1	23.0	77.0	7.3
81.5	18.5	6.2	19.0	81.0	7.4
77.5	22.5	6.3	16.0	84.0	7.5
73.5	26.5	6.4	13.0	87.0	7.6
68.5	31.5	6.5	10.5	90.5	7.7
62.5	37.5	6.6	8.5	91.5	7.8
56.5	43.5	6.7	7.0	93.0	7.9
51.0	49.0	6.8	5.3	94.7	8.0

(2) Malate Buffer

Stock Solutions

A : 0.2 M solution of acid sodium malate (8 g of NaOH + 23.2 g of maleic acid or 19.6 g of maleic anhydride in 1000 ml)

B : 0.2 M NaOH

50 ml of A + x ml of B, diluted to a total of 200 ml.

x	pH	x	pH
7.2	5.2	33.0	6.2
10.5	5.4	38.0	6.4
15.3	5.6	41.6	6.6
20.8	5.8	44.6	6.8
26.9	6.0		

(3) Tris-Malate Buffer [(Tris (hydroxymethyl) aminomethane maleate Buffer]

Stock Solutions

A : 0.2 M solution of Tris acid malate (24.2 g of tris (hydroxymethyl) amino methane+ 23.2 g of maleic acid or 19.6 g of maleic anhydride in 1000 ml)

B : 0.2 M NaOH

50 ml of A + x ml of B, diluted to a total of 200 ml.

x	pH	x	pH
7.0	5.2	48.0	7.0
10.8	5.4	51.0	7.2
15.5	5.6	54.0	7.4
20.5	5.8	58.0	7.6
26.0	6.0	63.5	7.8
31.5	6.2	69.0	8.0
37.0	6.4	75.0	8.2
42.5	6.6	81.0	8.4
45.0	6.8	86.5	8.6

A buffer grade tris can be obtained from the sigma chemical CO, st. louis 3, Mo. or from Matheson coleman and Bell, East Ruthee foed, N.J.

4. Tris (hydroxymethyl) amino methane (Tris) Buffer

Stock Solutions

A : 0.2 M solution of Tris (hydroxymethyl) amino methane (+24.2 g. in 1000 ml)

B : 0.2 M HCl

50 ml of A + x ml of B, diluted to a total of 200 ml.

x	pH	x	pH
5.0	9.0	26.8	8.0
8.1	8.8	32.5	7.8
12.2	8.6	38.4	7.6
16.5	8.4	41.4	7.4
21.9	8.2	44.2	7.2

5. Boric Acid-Borax Buffer

Stock Solution

A : 0.2 M Solution of Boric acid (12.4 g in 1000 ml).

B : 0.05 M solution of Borax (19.05 g in 1000 ml, 0.2 M in terms of sodium borate)

50 ml of A + x ml of B, diluted to a total of 200 ml.

x	pH	x	pH
2.0	7.6	22.5	8.7
3.1	7.8	30.0	8.8
4.9	8.0	42.5	8.9
7.3	8.2	59.0	9.0
11.5	8.4	83.0	9.1
17.5	8.6	115.0	9.2

6. Glycine-NaOH Buffer

Stock Solution

A : 0.2 M solution of Glycine (15.01 g in 1000 ml).

B : 0.2 M NaOH

50 ml of A + x ml of B, diluted to a total of 200 ml.

x	pH	x	pH
4.0	8.6	22.4	9.6
6.0	8.8	27.2	9.8
8.8	9.0	32.0	10.0
12.0	9.2	38.6	10.4
16.8	9.4	45.5	10.6

7. Borax-NaOH Buffer

A : 0.05 M solution of borax (19.05 g in 1000 ml; 0.02 M in terms of sodium borate)

B : 0.2 M NaOH

50 ml. of A + x ml of B, diluted to a total of 200 ml

x	pH
0.0	9.28
7.0	9.35
11.0	9.4
17.6	9.5

23.0	9.6
29.0	9.7
34.0	9.8
38.6	9.9
43.0	10.0
46.0	10.1

8. Glycine-HCl Buffer

Stock Solutions

A : 0.2 M solution of glycine (15.01 g in 1000 ml)

B : 0.2 M HCl

50 ml. of A + x ml of B, diluted to a total of 200 ml

x	pH	x	pH
5.0	3.6	16.8	2.8
6.4	3.4	24.2	2.6
8.2	3.2	32.4	2.4
11.4	3.0	44.0	2.2

9. Carbonate-Biocarbonate Buffer

Stock solution

A : 0.2 M solutions of anhydrous sodium carbonate (21.2 g in 1000 ml)

B : 0.2 M solutions of sodium bicarbonate (16.8 g in 1000 ml)

x ml of A + y ml of B, diluted to a total of 200 ml.

x	y	pH
4.0	46.0	9.2
7.5	42.5	9.3
9.5	40.5	9.4
13.0	37.0	9.5
16.0	34.0	9.6
19.5	30.5	9.7
22.0	28.0	9.8
25.0	25.0	9.9
27.5	22.5	10.0
30.0	20.0	10.1
33.0	17.0	10.2

35.5	14.5	10.3
38.5	11.5	10.4
40.5	9.5	10.5
42.5	7.5	10.6
45.0	5.0	10.7

10. Citrate Buffer

Stock Solution

A : 0.1 M Solutions of citric acid (21.01 g in 1000 ml)

B : 0.1 M solutions of sodium citrate (29.41 g $C_6 H_5 O_7 Na_3 2H_2O$ in 1000 ml, the use of the salt with 5½ HO is not recommended).

x ml of A + y ml of B, diluted to a total of 100 ml.

x	y	pH
46.5	3.5	3.0
43.7	6.3	3.2
40.0	10.0	3.4
37.0	13.0	3.6
35.0	15.0	3.8
33.0	17.0	4.0
31.5	18.5	4.2
28.0	22.0	4.4
25.5	24.5	4.6
23.0	27.0	4.8
20.5	29.5	5.0
18.0	32.0	5.2
16.0	34.0	5.4
14.9	35.1	5.5
13.7	36.3	5.6
11.8	38.2	5.8
9.5	41.5	6.0
7.2	42.8	6.2

11. Citrate-Phosphate Buffer

Stock Solutions

A : 0.1 m solution of Citric acid (19.21 g in 1000 ml)

B : 0.2 M solution of dibasic sodium phosphate (53.65 g of $Na_2 HPO_4 7H_2O$ or 71.7 g of $Na_2 HPO_4$, 12 H_2O in 1000 ml)

x ml of A + y ml of B, diluted to a total of 100 ml.

x	y	pH
44.6	5.4	2.6
42.2	7.8	2.8
39.8	10.2	3.0
37.7	12.3	3.2
35.9	14.1	3.4
33.9	16.1	3.6
32.3	17.7	3.8
30.7	19.3	4.0
29.4	20.6	4.2
27.0	22.2	4.4
26.7	23.3	4.6
25.2	24.8	4.8
24.3	25.7	5.0
23.3	26.7	5.2
22.2	27.8	5.4
21.0	29.0	5.6
19.7	30.3	5.8
17.9	32.1	6.0
16.9	33.1	6.2
15.4	34.6	6.4
13.6	36.4	6.6
9.1	40.9	6.8
6.5	43.6	7.0

12. Acetate Buffer

Stock Solutions

A : 0.2 M solutions of Acetic acid (11.55 ml in 1000 ml).

B : 0.2 M solution of sodium acetate (16.4 g of $C_2H_3\ O_2\ Na$ or 27.2 g of $C_2H_3O_2\ Na$, 3 H_2O in 1000 ml).

x ml of A + y ml of B, diluted to a total of 100 ml.

x	y	pH
46.3	3.7	3.6
44.0	6.0	3.8
41.0	9.0	4.0
36.8	13.2	4.2
30.5	19.5	4.4
25.5	24.5	4.6

20.0	30.0	4.8
14.8	35.2	5.0
10.5	39.5	5.2
8.8	41.2	5.4
4.8	45.2	5.6

13. Succinate Buffer

Stock solutions

A : 0.2 M solution of succinic acid (23.6 g in 1000 ml.)

B : 0.2 M NaOH

25 ml. of A + x ml of B, diluted to a total of 100 ml.

x	pH	x	pH
7.5	3.8	26.7	5.0
10.0	4.0	30.3	5.2
13.3	4.2	34.2	5.4
16.7	4.4	37.5	5.6
20.0	4.6	40.7	5.8
23.5	4.8	43.5	6.0

REFERENCES

Chopra, S.L. and Kanwar, J.S. 1976. *Analytical Agricultural Chemistry.* Kalyani Publishers, New Delhi. 514 p.

Matlock, G. 1961. *pH Measurement and Titralion.* Haywood and Co. Ltd., London.

Plummer, D.T. 1971. *An Introduction to Practical Biocemistry.* Tata McGraw Hill Publishing Company Ltd., Bombay-New Delhi. 369 p.

Methods in Enzymology 1955. I: 138-146.

Methods in Enzymology 1984. 104. 404-417.

Methods in Enzymology 1990. 182: 24-38.

CHAPTER - 7

ENZYMES : EXTRACTION AND ASSAY

INTRODUCTION

Enzymes are biological catalysts which in a very minute concentrations bring about a large transformations in chemical reactions. They are by and large protein in nature. Enzymes change the rate of chemical reactions but do not affect the final equilibrium. Living organisms are able to transform and use energy very rapidly due to the presence of enzymes.

Enzymes have very high specificity i.e. they catalyse only specific reactions in certain well defined pH, and temperature range. Each enzyme has certain pH and temperature optimal at which the enzyme activity is at its maximum. Enzyme acts on some specific substance called as substrate and forms an unstable enzyme substrate complex which is finally converted into product. At the end of the reaction the enzyme is released or set free to start a new cycle of reaction.

$$substrate\ (S) + Enzyme\ (E)\ \frac{K_{+1}}{K_{-1}}\ ES \rightarrow K_2 \quad Product\ (P) + E$$

The reaction rate of an enzyme thus, may be measured either by the disappearance of the substrate or the appearance of the reaction products.

All the known enzymes are classified and arranged in a systematic manner in order to have a scientific nomenclature. This system is known as Digital System of Enzyme classification. Enzymes are classified on the basis of the type of reaction catalysed. Each enzyme has a name and number to indicate its group and sub-group. Mainly six groups are laid out in order to accomodate all the enzymes. They are,

1. Oxido-reductases
2. Transferases
3. Hydrolases
4. Lyases

5. Isomerases

6. Ligases (Synthetases)

The nomenclature for each enzyme recommended by the International Union of Biochemistry is given in the publication ``Enzyme Nomenclature 1965".

In this chapter, an attempt has been made to describe the assay procedure for some of the common enzymes of nitrogen metabolism, oxygen scavenging system, and alcoholic fermentation. Quantitative measurement of enzyme activity could be done in two conditions :

(i) *in vivo* (ii) *in vitro*

In the former case the activity is measured *in situ* conditions without disturbing the state of enzymes in the living organism. However, in the latter case the enzymes are extracted out of the living organism and the activity is measured under ideal conditions in the external medium. Care is taken to simulate the *in vivo* condition in order to maintain the original level of enzyme activity.

Extraction and Purification

The enzyme extraction is accomplished in a well buffered system at low temperature containing all the essential cofactors and conditions present under natural conditions. The necessary procedure should liberate the enzyme protein in soluble form from the intact cell or tissue structure without causing loss of activity. Usually mechanical blending or homogenization of animal tissues is used to break cell membranes and release the cell contents, which then be assayed for their content of desired enzyme protein. For bacteria, plant cells and yeasts much more vigorous procedures are necessary to break the tough cell walls, e.g. sonic radiation, grinding with sand or fragmentation in a high pressure press.

Next step is to ascertain whether the enzyme in question is localized in one of the major sub cellular organelles, such as nucleus, mitochondria, chloroplast or the soluble portion of the cytoplasm. If the enzyme protein is present in specific organelle then the organelle needs to be fractionated using differential centrifugation and finally enzymes are isolated from the purified organelle. In all cases the enzyme proteins are to be obtained in soluble form before actual assay is performed.

The soluble fraction is commonly known as crude enzyme extract. A number of enzymes can be assayed in crude extract. However, to be more precise, the crude extract needs to be purified using standard techniques.

Among the separation procedures used are isoelectric focussing (precipitation), fractionation by salting out, or solvent precipitation. At each fractionation step the enrichment of enzyme activity factor and the recovery per cent is to be determined. These methods are further followed by more refined chromatographic procedures in order to obtain better purification. If the recovery per cent of enzyme is quite high, it can also be crystallized and finally redissolved to obtain pure enzyme preparation for assay. During purification, the yield which is expressed as percentage of the original extract, will decrease whereas the specific activity expressed as units of activity per mg protein should incease. The results of a purification are best recorded in a tabular form as under :

- Volume (ml)
- Enzyme activity (units/ml)
- Total enzyme activity (units/ml x extraction volume)
- Protein (mg/ml extract)
- Specific activity (units/mg protein)
- Yield (recovery) percentage of original total activity.

Assay Conditions

The amount of an enzyme in a given solution or tissue extract can be assayed quantitatively in terms of the catalytic effect it produces. For this purpose, it is necessary to know :

(i) The overall stoichiometry of the reaction catalyzed.

(ii) Whether the enzyme requires cofactors

(iii) Km of both substrate and cofactors.

(iv) Its optimum pH.

(v) Its optimum temperature.

(vi) A simple analytical procedure for determining the disappearance of the substrate or the appearance of the reaction products.

Normally, enzymes are assayed in test systems at optimum pH and temperature and at high substrate concentration (at saturating concentration) so that the initial reaction rate is zero order for substrate. Under these conditions, the initial reaction rate is proportional to enzyme concentration alone. When enzymes require cofactors, such as metal ions or coenzymes, they must also be added in concentration near to saturation.

Usually the rate of the formation of end product measurement is more accurate than measurement of the disappearance of the substrate.

Fixed Time Assay

For this type of assay the enzyme is incubated with the substrate at fixed temperature and pH, samples are withdrawn at suitable time intervals and analysed. In addition to the test mixture ``controls" are always prepared in which either enzyme is omitted or boiled enzyme is used. Controls are used to compare changes observed due to non-enzymatic transformation in the assay mixture.

Rate Kinetics

In this type of measurement, the rate of disappearance of substrate or formation of end product is continuously monitored over a time range in which the reaction rate is constant and linear.

Expression of Enzyme Activity

Measured enzyme activity can be expressed in any one of the following ways :

(i) Fresh weight basis

(ii) Dry weight basis

(iii) Protein basis

Of these, expression on protein basis is most appropriate and meaningful.

QUANTITATIVE DETERMINATION OF ENZYMES

Nitrogen Metabolism

1. Nitrate Reductase (EC 1.6.6.1)

Nitrate reductase (NR) activity can be assayed according to the method of Jaworski (1971). Nitrate reductase catalyses the conversion of nitrate to nitrite which in turn is converted to hyponitrous acid by nitrite reductase and finally to ammonium ion.

$$NO_3^- \xrightarrow[NR]{2e^-} NO_2^- \xrightarrow[NiR]{2e^-} HNO_2^- \xrightarrow{4e^-} NH_4^+$$

Reagents

1. Phosphate buffer 0.1 M, pH 7.5
2. Potassium nitrate, 0.02 M
3. Propanol, 5%
4. Sulphanilamide, 1% in 3 N HCl
5. N-1 naphthyl ethylene diamine hydrochloride 0.02%.

Procedure

250 mg of sample is suspended in screw cap vials having 4.5 ml medium containing 0.1 M phosphate buffer (pH 7.5), 0.02 M KNO_3, 5% propanol. A control is kept omitting plant sample. The vials are capped and kept in dark at 30°C for desired incubation period. Nitrite released into the medium is determined by treating 0.4 ml aliquot with 0.3 ml each of sulphanilamide and N-1 naphthyl ethylene diamine hydrochloride. After 20 minutes the solution is diluted with distilled water to make the volume upto 5 ml and the absorbance is measured at 540 nm using reagent blank.

Standard curve to be prepared using KNO_2 or $NaNO_2$ solution graded concentrations of up to 100 ppm and the amount of NO_2 produced by the activity of nitrate reductase in the assay medium is calculated.

Calculation for Nitrate Reductase

0.25 g plant sample suspended in 4.5 ml reaction mixture for 1 h in dark

0.4 ml aliquot +0.3 ml Sulfanilamide + 0.3 ml of NEDH + 4 ml dH_2O (total volume is 5 ml)

Since, 0.4 ml aliquot produces X µg nitrite/unit time (derived from standard curve)

so, 4.5 ml aliquot produces $\frac{X \times 4.5}{0.4}$

This activity is in 0.25 g/fr. wt.

Now 0.25 g fresh weight gives $\frac{4.5\ X}{0.4}$ µg activity

Hence 1 g gives = 45 X

Dilution factor in the present assay condition is 45

where X is the µg activity derived from standard curve.

2. Nitrite Reductase (EC 1.6.6.4)

The nitrite reductase (NiR) is assayed according to the method of Ferrari and Varner (1971).

Reagents

1. Phosphate buffer 0.1 M, pH 7.5
2. Sodium nitrite, 2mM
3. Dimethyl sulphoxide
4. Sulphanilamide, 1% in 3 N HCl
5. N-1 napthyl ethylene diamine hydrochloride, 0.02%
6. Chloramphenicol, 0.5 mg/ml.

Procedure

The nitrite reductase activity is assayed by measuring nitrite disappearance after placing 250 mg sample, separately in 4.5 ml medium containing phosphate buffer, chloramphenicol and sodium nitrite (2 mM). Nitrite is added to start the assay.

0.1 ml aliquot is immediately removed to provide a measure of initial nitrite concentration of the reaction mixture (A_1) in the presence of tissue pieces. After incubation for the desired period, 0.1 ml aliquot of the medium is again removed for NO_2 determination. The difference between the two NO_2 concentration, indicates the amount of NO_2 taken up (A_2) by the tissues. Dimethyl sulphoxide is then added to the medium (final conc., 50% v/v) and the flasks are kept on hot plate until the medium started boiling (20-30 sec). Under these conditions, the NO_2 of the tissue rapidly leakes out into the medium. After cooling, 0.1 ml aliquot is removed for NO_2 determination (A_3). The difference between the initial concentration and this nitrite concentration (A_1-A_3) gives the amount of nitrite reduced. Residual NO_2 is determined colorimetrically.

Calculation for Nitrite Reductase

0.25 g plant sample suspended in 4.5 ml reaction mixture

0.1 ml aliquot + 0.3 ml sulphanilamide + 0.3 ml of NEDH + 4.3 ml d H_2O (total volume is 5 ml)

0.1 ml aliquot reduces X μg nitrite (derived from standard curve)

So, 4.5 ml will reduce $= \dfrac{X \times 4.5}{0.1} = 45\,X$

This activity comes from 0.25 g tissue fresh weight

Now, 0.25 g fr. wt. gives 45 X activity / unit time

hence, 1 g gives = 180 X

Where, X is the µg activity derived from standard curve

3. Glutamine Synthetase (EC 6.3.1.2)

Glutamine synthetase (GS) activity is assayed by the modified r-glutamyl transferase method of Elliott (1955)

Principle

Glutamine synthetase brings about a reaction in which amino acid glutamate functions as an ammonia acceptor to produce the amide glutamine. The enzyme has a very high affinity for ammonia and is, thus, capable of incorporating ammonia even if present in very low concentration.

Reagents

1. Tris-HCl buffer (0.1 M, pH 7.5)
2. ATP 10 mM, pH 7.2
3. Glutamate 100 mM, pH 7.2
4. Mg SO_4 1 M
5. NH_4OH 1 M, pH 7.0
6. Cysteine 1 M pH 7.2
7. $FeCl_3$ reagent; equal volumes of 10% $FeCl_3$ $6H_2O$ in 0.2 N HCl, 24% TCA and 50% (v/v) are mixed together.

Extraction

A known weight of sample (250 mg) is homogenised separately in 10 ml 0.1 M Tris-HCl buffer (pH 8.0) containing 0.1 mM each of cysteine and EDTA and centrifuged at 9000 x g in a cooling centrifuge for 25 minutes. The supernatant is collected and retained for enzyme assay as crude extract. Protein in this extract is estimated by the method described by Lowry *et al.* (1951). All operations unless otherwise stated are done at 4°C.

Assay

Pre-incubation is carried out at 35° for 5 minutes in 12 ml centrifuge tubes containing the following mixtures. Buffer, 0.5 ml;

ATP, 0.5 ml; Glutamate, 1.5 ml; Mg SO_4 0.1 ml; cysteine, 0.1 ml; enzyme extract 0.1 ml + H_2O to final volume of 4 ml. The reaction is started by the addition of 0.1 ml NH_2OH and after incubation for 20 minutes, 0.1 ml of ferric chloride reagent is added to each tube, and the protein precipitated is removed by centrifugation. The hydroxamine in supernatant is estimated by measuring the optical density at 540 nm in Spectrophotometer against reagent blank. Standard curve to be prepared using graded concentration of r-glutamyl hydroxamate (Sigma chemicals). Enzyme activity is expressed as mM r-glutamyl hydroxamate produced per mg protein per 20 min (Specific actvitiy). Total activity is calculated on per g tissue fresh weight.

Calculations

0.250 g sample-extract 10 ml

0.1 ml extract used for assay

since, 0.1 ml extract produces X μ g r-glutamyl Hydroxamate

so, 10 ml extract will produce $\frac{X \times 10}{0.1} \mu g$

This value is equivalent to 0.250 g plant sample

Now, 0.250 sample gives - 100 X μg

so, 1 g sample will give = $\frac{100\ X \times \mu g}{0.250}$

= 400 X μg r-glutamyl Hydroxamate

4. Aspartate Amino Transferase (GOT) (EC 2.6.1.1)

The GOT activity of sample is assayed by the method of Wootton (1964) in the sampe extract which is used for GS assay.

Reagents

(i) Substrate

DL-aspartate (200 mM) and L-Ketoglutarate (2mM) in phosphate buffer (0.1 M) pH 7.5 (2.66 g of DL-aspartate and 0.0292 g of L-Keto-glutarate to be dissolved in 100 ml of 0.1 M phosphate buffer, pH 7.5.

(ii) Developer

1 mM 2,4 dinitrophenyl hydrazne solution containing hydrochloric acid is prepared by dissolving 0.0198 of dinitrophenyl hydrazine in 10 ml of conc. HCl and volume made to 100 ml with distilled water.

(iii) Standard

Standard curve to be prepared using graded concentrations of pyruvate dissolved in 0.1 M phosphate buffer, pH 7.5.

(iv) NaOH

0.4 N NaOH to be prepared in distilled water.

Procedure

The incubation mixture is prepared by mixing the reagents in the proportion given below.

Ingredients	Analysis	Reagent blank
Substrate	0.5 ml	0.5 ml
	Pre-incubation for 10 minutes at 37°C	
Enzyme Extract	0.2 ml	--
	Incubation for 30 minutes at 37°C	
Developer	0.5 ml	0.5 ml

After mixing thoroughly, it is allowed to stand for 20 minutes at room temp.

Distilled water	0.3 ml	0.5 ml
0.4 N-NaOH	5.0 ml	5.0 ml

After thorough, mixing, the solution to be kept for a period of 10 minutes at room temperature.

Optical density is measured at 550 nm against the reagent blank. The amount of pyruvate formed is then determined from the standard curve and the values expressed as mg pyruvate released per μg protein/30 minutes. Total activity is expressed as μg pyruvate released per gm tissue fresh or dry weight in 30 minutes at 37°C. A standard curve to be prepared using 1-7 μ mole pyruvate in 5.0 ml of buffer.

5. Alanine Amino Transferase (GPT) (EC 2.6.1.2)

This enzyme activity is assayed by the method of Wootton (1964).

Extraction

The enzyme extract which is used for GOT assay can also be used for GPT

Reagents

L-alanine (200 mM) and L-Ketoglutarate (2mM) in 0.2 mM phosphate buffer, pH 7.5 (1.780 g of L-alanine and 0.0292 g of L-Keto-glutarate to be dissolved in 100 ml of 0.2 M phosphate buffer, pH 7.5.

The rest of the reagents are same as for GOT enzyme assay.

Procedure

The procedure for enzyme assay is similar to that of GOT except that alanine is used as substrate instead of aspartate. The activity is calculated and expressed as in case of GOT for a standard curve prepared using 1.7 μ mole pyruvate in 5.0 ml buffer.

Calculation for GOT and GPT

Crude Extract : 0.25 g plant sample extracted in 10 ml Tris-buffer.

Assay : 0.2 ml enzyme extract used for assay.

Since, 0.2 ml enzyme extract gives X μg pyruvate

So, 10 ml enzyme extract will give = 50 X

Since, 0.250 g sample gives 50 X activity

So, 1 g sample will give = $\dfrac{50 \times X_1}{0.250} = 200\,X$

OXYGEN SCAVENGING SYSTEM

6. Catalase (EC 1.11.1.6)

Catalase activity can be assayed colorimetrically according to method given in Analystical Biochemistry (Sinha, 1972). Catalase facilitates the dismutation of H_2O_2 to water and O_2 according to the reaction.

$$H_2O_2 \xrightarrow{\text{Catalase}} H_2O + 1/2\ O_2$$

The enzyme plays an important role in association with SOD as well as in photorespiration and glycolate pathway.

Reagents

1. Phosphate buffer, 0.1 M, pH 7.0
2. Potassium dichromate acetic acid (5% $K_2Cr_2O_7$ + glacial acetic acid in 1:3 ratio.
3. H_2O_2 (0.2 M)

Procedure

200 mg plant sample is homogenized with 10 ml of phosphate buffer 0.1 M (pH 7.0) and centrifuged at 10,000 rpm at 40°C for 30 minutes. Collect the supernatant and store at low temperature (4°C). Use supernatant for enzyme assay and estimate the enzyme activity as given below :

Reaction Mixture

Test	Blank	Reagents
1.25 ml	--	H_2O_2
0.50 ml	0.50 ml	Enzyme extract
3.25 ml	4.50 ml	Phosphate buffer
5.00 ml	5.00 ml	

Take reaction mixture in conical flask and mix rapdily at 37°C. After 3 minutes withdraw 2.0 ml reaction mixture and add 2.0 ml potassium dichromate acetic acid reagent. Keep on boiling water bath for 10 minutes and record O.D. at 570 nm against blank after cooling.

Express result as enzyme unit per gram fresh or dry weight or per mg protein basis.

Calculation

0.2 g sample extracted with 10 ml, buffer

0.5 ml enzyme extract + 1.25 ml H_2O_2 + 3.2 ml buffer (5 ml assay mixture)

(Total material is 5.0 ml)

After 3 minutes of reaction, 2 ml mixture was withdrawn for assay.

2.0 ml rection mixture + 2.0 ml $K_2Cr_2O_7$ acetic acid reagent.

Now, 2 ml reaction mixture gives X. O.D.

So, 5 ml reaction mixture gives $\frac{5 \times X\ O.D.}{2}$

0.5 ml enzyme extract was used for assay, therefore

0.5 ml enzyme extract = $\frac{5 \times X\ O.D.}{2}$

So for 10 ml enzyme extract = 50 X O.D.

Which is equivalent to 0.2 g plant sample

Since, 0.2 g plant sample gives = 50 X O.D.

So, 1 g fresh wt. will gives = 250 X O.D. /3 minutes

Therefore, the Dilution factor = 250/3 minutes

Since 0.1 O.D. = 1 enzyme unit

Therefore, X OD = 10X

Now enzyme unit/g fresh weight/3 minute can be calculated by multiplying by dilution factor of 250 i.e 250 x 10 X

Where, X is the observed OD.

Finally, enzyme unit is calculated on the basis activity/g fr. wt/min. which is as under :

$$\frac{250 \times 10X}{3}$$

7. Peroxidase (EC 1.11.1.7.)

Peroxidase activity can be assayed colorimetrically according to method given in Plant Physiology 34:416-418 (Curne and Galston, 1959). The enzyme catalyses the oxidation of a substrate by removal of hydrogen which combines with H_2O_2.

$$AH_2 + H_2O_2 \ - \ 2H_2O + A$$

Reagents

1. Phosphate buffer, 0.1 M (pH 6.0)
2. Pyrogallol (0.1 N)
3. H_2O_2 (0.02%)

Procedure

Take 200 mg plant sample and homogenize with 10 ml of Phosphate buffer 0.1 M (pH 6.0). Centrifuge at 10,000 rpm at 4°C for 30 minutes. Collect the supernatant and store at low temperature. Use the supernatant for enzyme assay and estimate the enzyme activity as given below :

Reaction Mixture

Test	Blank	Reagents
2.0 ml	2.0 ml	Enzyme extract
2.0 ml	3.2 ml	Phosphate buffer
1.0 ml	--	Pyrogallol
0.2 ml	--	H_2O_2
5.2 ml	5.2 ml	

Shake the mixture well and keep it at 37°C on water bath for 10 minutes for the formation of purpurogallin. Measure the activity at 430 nm and express result as enzyme unit per gram fresh or dry weight or per mg protein basis.

Calculation

0.2 g sample extracted with 10 ml buffer

2 ml enzyme extract + 2 ml Buffer + 0.2 ml H_2O_2 + 1 ml Pyrogallol - OD measured at 430 nm

2 ml enzyme extract gives ---- XOD

So, 10 ml enzyme extract gives - - - - 5X

0.2 g plant sample is equivalent to - - - - 5 X 0D

Therefore, 1 g plant sample is equivalent to -- - 25 X OD

Hence, the dilution factor = 25

0.1 OD = 1 enzyme unit

So, X OD = 10X

Now total activity/g/fresh wt. / minute = 10 X x 25

where, X is the observed OD.

Superoxide Dismutase (EC 1.15.1.1)

Superoxide dismutase (SOD) occur in all aerobic organisms and play essential role in the survival of these organisms. They protect the tissues from the deleterious effect of oxygen free raicals formed in various enzyme reactions in which a single electron is transmitted to O_2 (Fridovich, 1975; Giannopolitis and Ries, 1977).

$$O_2 + e \ - \ O_2 \text{ (superoxide)}$$

$$O_2^- + O_2^- + 2H^+ \xrightarrow{\text{SOD}} + H_2O_2 + O_2$$

$$2\,H_2O_2 \xrightarrow{\text{Catalase}} 2\,H_2O + O_2$$

Mechanism of assay

Superoxide dismutase (SOD) is assayed on the basis of its ability of inhibit the photochemical reduction of nitro blue tetrazolium (NBT) into blue formazan which has a wavelength maximum at 560 nm (Asada *et al.*, 1974; Beauchamp and Fridovich, 1971). SOD essentially inhibits the formation of the blue formation and can be quantitated on this basis. The chemical reactions involved are

1. Photoreduction-oxidation reaction between riboflavin and methionine with the flavin being reduced.
2. Generation of superoxide (O_2) from reoxidation of reduced flavins in air.
3. a. Reduction of nitro blue tetrazolium (NBT) by superoxide into blue formazan with wavelength maxima at 560 nm.

   b. Inhibition of step 3a by SOD dismutation of the superoxide into hydrogen peroxide and molecular oxygen.

Extraction

1. Weigh about 0.2 g FW leaf sample (cut into small pieces).
2. Add a pinch of acid-washed sand and 2 ml cold phosphate buffer pH 7.0 (0.05 M) Grind in clean, dry mortar and pestle in ice bath.
3. Transfer to centrifuge tubes and wash 2 x with buffer (1 ml each of the buffer pH 7.0 and pH 7.8 which were used in initial trial extraction a total of 2 ml per washing) to make a total of 8 ml.
4. Centrifuge at 30,000 x g for 30 minutes at 0.4°C (about 15,000 RPM in our refrigerated centrifuge, be sure to put tubes of more or less the same weight in exactly opposite positions to balance). Collect the clear soup and keep at 0.4° for assay.

Reagents

1. 0.05 M phosphate buffers (pHs 7.0 and 7.8).
   a. Dissolve 13.4 g Na_2 $HPO_4.7H_2O$ (MW = 268) and 6.9 g NaH_2 $PO_4.H_2O$

      (MW = 137.99) in sterile dH_2O to total volume of 1 liter.
   b. Take an aliquot of solution la and adjust the pH to 7.8 Record the initial pH (about 3.0 ml per 500 m buffer).
   c. Adjust the pH (about of another aliquot of solution la as in step 1b adjust the pH to 7.0. Dissolve 1 g polyvinyl pyrrolidone per 100 ml buffer (1% w/w PVP final concentration).
   d. Keep the buffers in reagent bottles labelled with phosphate pH 7.0 and 7.8, respectively.
2. 0.15 M EDTA. 20 mM methionine, and 0.12 nitro blue tetrazolium (NBT) in buffer pH 7.8.

a. Weigh 27.92 mg Na_2 EDTA.$2H_2O$ (MW = 372.2), 1.4921 g methionine (MW = 149.21) and 49.06 mg NBT (MW = 8.176) and dissolve together in 100 ml buffer pH 7.8. This stock solution gives the following concentrations : 0.75 mM EDTA, 100 mM methionine, and 0.6 mM NBT.

b. Aliquot 10.0 ml of solution instep 2a and dilute to a total of 50 ml with buffer H 7.8. This gives the following concentration : 0.15 mM EDTA, 20 mM methionine, and 0.12 mM NBT.

c. Use 2.0 ml of solution 2b per 3-ml assay total volume with the following concentrations : 0.1 mM EDTA, 13 mM methionine, and 0.08 NBT.

3. 12 μM riboflavin in buffer pH 7.8 (to be prepared fresh prior use)

a. Weigh 4.5 mg riboflavin (MW - 37637) and dissolve in 400 ml buffer pH 7.8. This gives riboflavin concentration of 29.89 μM.

b. Aliquot 20.0 ml of solution in step 3a and dilute to a total of 50 ml with buffer pH 7.8. This gives riboflavin concentration of 11.96 μM.

4. Use 0.5 ml of solution 3b (11.96 μM riboflavin) per 3-ml assay total volume with a final concentration of 1.993.

Procedure

Presuming that 50% inhibition of NBT reduction occurs with 0.2 ml enzyme extract, a 3 ml reaction mixture contains the following :

Description	Blank	Reference	Test
Buffer pH 7.0	0.3 ml	0.5 ml	0.3 ml
Solution 2 b (Methionine, NBT, EDTA)	2.0 ml	2.0 ml	2.0 ml
Riboflavin (Solution 3b)	0.5 ml	0.5 ml	0.5 ml
Enzyme extract	0.2 ml	-	0.2 ml
Total volume	3.0 ml	3 ml	3 ml
Light Treatment	Dark	Light	Light

Use blank to set instruments ZERO and measure Absorbance of Reference (V) and TEST (v) at 560 nm.

1. A-3 ml reaction mixture should contain the following

(a) 0.05 M phosphate buffer (pH 7.8) ml

(b) 13 mM methionine ml

(c) 75 μM nitro blue tetrazolium ml

(d) 100 nM EDTA ml

(e) 0-200 ul enzyme crude extract (component 1 a above should be added to dilute the enzyme extract.

2. Shake all the tubes and place the sample tubes 30 cm below a light bank consisting of two 30-watt-fluorescent bulbs (reference tube must be placed in the dark)
3. Start the reaction by switching on the light and allow to run for 20 min.
4. Terminate the reaction by switching off the light with the tubes covered with a black cloth. Tubes are then singly wrapped with used carbon paper or aluminium foil.
5. Read absorbance against the blank (reference tube) at 560 nm. Volume of enzyme extract would range from 0 mixture of which gives ABS_1 (V) to 200 ml (mixtures of which give ABS_2 (V) then onwards).
6. Plot the ratio of ABS_1/ABS_2 in the y-axis against volume of the enzyme extract in the reaction mixture. The volume of enzyme extract producing 50% inhibition of the reaction (that is with ABS_1/ABS_2 ratio 2 of is to be derived from the graph. The volume of extract which gave 50% inhibition is defined to have one unit of SOD activity.

Calculations

Use blank to set instruments ZERO and measure absorbance of reference (V) and test (v) at 560 mm. Calculate enzyme activity as under :

SOD units/ml = [(V/v) - 1] x (dilution factor)

- Dilution factor 1 = 5 (for/ml activity).
- Dilution factor 2 = 40 (for/0.2 g fr. wt. equivalent to 8 ml extract)

(since 0.2 g fr. wt. was extracted with 8 ml buffer solution)

Final dilution factor for per g fr. wt. therefore will be = 40 x 5 = 200 (units/g fr. wt.)

9. Ascorbate Peroxidase (EC 1.11.1.11)

Photoproduction of superoxide by the univalent reduction of molecular oxygen in PSI is an inevitable reaction in chloroplasts. The superoxide radicals are disproportionated ty hydrogen peroxide and molecular oxygen by superoxide dismutase contained in chloroplasts. Oxygen evolution on addition of hydrogen peroxide to illuminated chloroplasts, and the stoichiometry of the oxygen evolution and the decomposition of H_2O_2 suggest that the H_2O_2 in chloroplasts is reduced to water by a peroxidase using a photoreductant as the electron donor. Conclusive evidence for this scavenging system of hydrogen peroxide has been provided by the observations that illuminated intact chloroplast evolve $^{16}O_2$ on addition of $H_2\,^{18}O_2$ with the same stoichiometry as that of $^{16}O_2$ evolution without an accumulation of $H_2\,^{18}O_2$.

Ascorbic peroxidase activity can be assayed colorimetrically according to method given by Nakano and Asada (1987).

$$2O^{-2} + aSe + 2H \xrightarrow{\text{Ascorbate}} 2H_2O_2 + \text{DHase}$$

$$2\text{DHasc} + 4\,H_2O - \xrightarrow{\text{Ascorbate peroxidase}} -2H_2O_2 + 2\,\text{aSc}$$

Reagents

(i) 1 M and 50 mM Potassium phosphate buffer, (pH 7.8)

(ii) 10 mM H_2O_2

(iii) 1 mM and 10 mM ascorbate in sterile dH_2O

Extraction

Take 700 mg plant sample and homogenize with 5 ml of 5 mM potassium phosphate buffer pH 7.8 with 1 mM ascarbate. Centrifuge at 10,000 rpm at 4°C for 30 minutes. Collect the supernatant and store at low temperature. Use the supernatant for enzyme assay and estimate activity as given.

Enzyme assay

Ascorbate peroxidase is measured by decrease in absorbance A) at 290 mM due to ascorbate oxidation. Run a time course at wavelength 290 using 1 ml curette with sterile dH_2O in the reference cell. The activity is measured in reaction mixture (1 ml) containing 100 μL, 1 M potassium phosphate buffer (pH 7.8), 700 μL sterile dH_2O, 50 μL 10 mM aqueous ascorbate, 100 μL of crude extract. Then mix thoroughly and cover the sample compartment of the spectrophotometer. Start scan to observe the straight line while continuously scanning add immediately 50 μL of 10 mM H_2O then mix with plastic laddle and cover the sample compartment of the spectrophotometer. Once a straight line is formed after the addition of 10 mM H_2O_2, a curette containing the next sample is introduced to replace the older sample curette.

Note

The total assay volume is each cuvette must be equal to 1000 μL or 1 ml. In case where the aliquot of crude enzyme extract has to be changed a corresponding change in aliquot of sterile dH_2O must be done in order to maintain the total assay volume of 1000 ml or 1 ml.

Keep all graphing papers in a folder for data collection. Record the slope (Change in OD per unit time). The molar absorptivity constant of oxidized ascarbate product measured by UV method is 2.8 optical density (mM^{-1}) ascrobate $(cm)^{-1}$. Express activity of ascorbate peroxidase in terms of moles ascorbate oxidized per mg protein per minute.

10 Ascorbate Content Determination

Ascorbate content can be assayed colorimetrically according to method given by Roe and Kuether (1943) and Shigeoka *et al.* (1979).

Reagents

(i) 5% metaphosphoric acid : 5 grams of metaphosphoric acid in dH_2O to give a total final volume of 100 ml.

(ii) Ascorbic acid : Purpose 1 µg per µL standard stock solution. Dissolve 10 mg ascorbate acid in 5% metaphosphoric acid to a final volume of exactly 10 ml.

(iii) 3 mM DCIP : 43.52 mg of 2, 6-DCIP sodium salts dissolve in 50 ml of dH_2O.

(iv) 1.0% thiourea

(v) 20% H_2SO_4

(vi) 10 mM DNPH (2, 4-dinitro phenyl hydrazine, MW = 198.1). Dissolve 99.05 mg of 2, 4-DNPH in 50 ml of 20% H_2SO_4.

(vii) 85% H_2SO_4

Procedure

1. Homogenize leaf sample (1 g per 20 ml extractant, solution 1 above) with 5% metaphosphoric acid in a mortar and pestle.
2. Centrifuge at 15,000 rpm in refrigerated centrifuge for 30 minutes. Then discard the residue and collect supernatant for ascorbate determination.

3. There are two assays to be done at the same time for each sample the total ascorbate assay and the oxidized ascorbate assay. Both assays are described in steps number 4a and 4b which follow below.
4. (a) For total ascorbate, aliquot 1 ml of the supernatant and add 0.5 ml of 3 mM DCIP, to oxidized ascorbic acid into dehydroascorbic acid quantitatively. Follow step 4 a in the assay of the standard aliquots prepared in solutions 2 above.

4b. For the oxidized ascorbate aliquot 1 ml of the supernatant and 0.5 ml sterite dH_2O.

5. Prepare a reference tube containing 1 ml of the extract to be used and 0.5 ml of 3 mM DCIP to be used in the assay of standards and both the total and oxidized ascorbate samples. Note too that the total final volume of the reference tube at this time is 1.5 ml also.
6. Let all the tubes in steps 4 and 5 to stand for 20 minutes after which add 1 ml of 1% thourea in 5% metaphorphoric acid, solution 4 above to every tube (note that each tube contains a total of 2.5 ml this time.
7. Add 1 ml of 20% H_2SO_4 for the reference tube only. But for the rest of the tubes (Standard, total ascorbate and oxidized ascorbate), add 1 ml 10 mM DNPH dissolve in 20% H_2SO_4 to form the 2, 4-DNPH derivative of dehydroascorbic acid (each tube must have a total of 3.5 ml this time).
8. All tubes will be incubated at 50°C for 1 hour.
9. After the incubation, they are allowed to cool in ice bath and after the first 15 minutes still on ice bath add very gradually 2.5 ml of 85% H_2SO_4, solution 7 (addition has to finish in about 1 minute per tube so as not to allow the temperature of the mixture to be raised appreciably as this will bring about charring of sugars or other organic matter which interferes in the assay). This gives a total volume of 6.0 ml each tube.
10. When the 2, 4-DNPH derivative of dehydro ascorbic acid is treated with 85% H_2SO_4, a reddish product is formed which absorbs maximally at 500-550 and 350-380 nm. Each tube must be in ice-bath for total of 30 minutes than proceed to next step.
11. Finally, add 1 ml 20% H_2SO_4 to tubes for standards, total ascorbate, and oxidized ascorbate. But the reference tube add 1 ml of 10 mM DNPH dissolved in 20% H_2SO_4 to give a total of 7 ml for each tube.

12. Measure absorbance at 520 nm against the reference. The colored product is stable for 1 hour.

11. Glutathione Reductase (EC 1.6.4.2)

The simple tripetide glutathione (GSH) is found in almost all living cells and takes part in numerous biochemical reactions as well as helping to stabilize certain enzyme by preventing oxidation of their thiol groups. The oxidized form of glutathione (GSSG) may be converted back to the reduced glutathione reductase as under :

$$GSSG + NADPH^{+} + H^{+} \rightarrow 2GSH + NADP$$

GSG and glutathione reductase have been reported to be present in the chloroplast, mitochondria and cytoplasm of plant cells.

Glutathione reductase activity can be assayed colormetrically according to method given by Halliwell and Foyer (1978).

Reagents

(i) M & 50 mM potassium phosphate buffer (pH 7.8)

(ii) 10 mM NADPH

(iii) 10 mM oxidized glutathione

(iv) 1 mM ascarbate

Extraction

Take 700 mg plant sample and homogenized in 5 ml of 50 mM potassium phosphate buffer, (pH 7.8) with 1 mM ascarbate as the extraction buffer. Centrifuge at 10,000 rpm at 4°C for 30 minutes. Collect the supernatant and stored at low temperature. Use the supernatant for enzyme assay and estimated activity has given below.

Enzyme assay

Glutathione reductase was measured by decrease in absorbance at wavelength 340 nm using 1 ml curettle with sterile dH_2O in the reference cell. Reaction mixture containing 100 μL of 1 M potassium phosphate buffer, (pH 7.8), 20 μL 10 mM NADPH, 810 μL sterile dH_2O, 50 μL of crude extract then mix with plastic laddle and cover the sample of the spectrophotometer.

Start scan to observe a straight line and while continuously scanning add immediately 20 μL of 10 mM oxidized glutathione then mix with plastic laddle and coner the sample compartment of the spectrophotometer. Once a straight line is formed after the addition of 10 mM oxidized glutathione, a curette containing the next sample is introduced to replace the older sample curette.

Note : The total assay volume in each curette must be equal to 1000 μL or 1 ml. In case where the aliquot of crude enzyme extract has to be changed a corresponding change in aliquot of sterile dH_2O must be done in order to maintain the total assay volume of 1000 μL or 1 ml. Some time and paper by adjusting the total scanning time to combne 3-4 sample scans, or more if the number of cuvettes permits, in one piece of graphing paper. Keep all graphig papers in a folder for data collection. Record the slope (Change in OD per unit time).

The molar absorptivity constent of NADPH oxidized measured by uv method at 340 nm is 6.22 x 10^{-3} optical density $(mM)^{-1}$ NADPH $(cm)^{-1}$. Express the activity of glutathione reductase in terms of moles NADPH oxidize per mg protein per minute.

12. Reduced Glutathione Content Determination

Reduced glutathione content can be assayed colorimetrically according to method given by Tietze (1969), Griffith (1980) and Law *et al.* (1983).

Reagents

(i) 55 metaphorphoric acid

(ii) 1 M tricine in 6 M NaOH

(iii) 2-Vinyl pyridine (pure)

(iv) 1 M sodium phosphate buffer in 40 mM EDTA (pH 7.5). Prepare 1 M NaH_2PO_4 and 1 M Na_2 HPO_4 separately then pour off NaH_2PO_4 into the Na_2HPO_4 solution until the pH becomes 7.5.

(v) 6 mM 5, 5 dithiobis (2 nitro benzoic acid) in 0.1 M sodium phosphate buffer (pH 7.5).

(vi) 10 mM NADPH in 0.1 M sodium phosphate buffer (pH 7.5)

(vii) 1 U glutathione reductase GR.

Extraction

Take 1 g of leaf sample and plant tissue were grinded in liquid nitrogen by pestle and mortar. Crude homogenates are prepared by homogenizing the leaf sample in 10 ml of 5% metaphosphoric acid solution land centrifuge mixture at 15,000 rpm at 4°C for 30 minutes. Collect the supernatant and discard the residue. Desired the pH (60.0-7.0) of supernatant was created by adding 1 M tricine (reagent 2) very gradually into a tube 60 μL of reagent 2 for 1 ml of supernatant.

Assay

Prepare two tubes of this mixture. One tube for total and other for oxidized glutathione. For the total glutathione at 0.5 ml of dH_2O while for oxidized glutathione 0.5 ml of 2-venyl pyridine (Reagent 3) to remove the reduced form of glutathione. Further, mixture of solution were centrifuged at maximum speed for 1 minute for the tube meant for oxidized glutathione remove the upper layer sharply and very carefully not to remove bit of the lower layer and also not to retain any of the upper layer. Glutathione content is assayed in 1000 μL of assay mixture containing 100 μL of 1 M sodium phosphate in 4 M EDTA (pH 7.5), 10° μL of 6 mM 5, 5-thio bis (2-nitro-benzoic acid), 25 μL of 1° mM NADPH, 615 μL dH_2O, 150 μL crude extract. Initiation of reaction started by adding of 10 μL, 1 unit of glutathione reductase. Record the increase in absorbance at 412 nm for 1 minute. Reduced glutathione would be computed by difference between total glutathione and oxidized glutathione.

Standard curve

3 repeat measurements of 0 to 10 μg GSH [or 0 to 20 μg GSSG] standard in the assay buffer per ml of standard assay volume and perform the same procedure as in the total glutathione preparation and assay.

REFERENCES

Asada, K., Takahasi, S.M. and Nagate, M. 1974. Assay and inhibitors of spinach superoxide dismutase. *Agricultural Biological Chemistry* 38(2): 471-473.

Beauchamp, C. and Fridovich, I. 1971. Superoxide dismutase : Improved assays and an assay applicable to acrylamide gels. *Analytical Biochemistry* 44 : 276-287.

Bergameyer, H.N. 1974. *Methods of Enzymatic Analysis* Vols. 1-4. Springer.

Elliott, W.H. 1955. Isolation of glutamine synthetase and glutamate transferase from green peas. *Journal of Biological Chemistry* 201: 661-669.

Ferrari, T.E. and Varner, J.E. 1971. Intact tissue assay for nitrite reductase in barley alcurone layers. *Plant Physiology* 47: 790-794.

Fridovich, I. 1975. Superoxide dismutases. *Annual Review of Biochemistry* 44 : 147-159.

Giannopolitis, C.N. and Ries, S.K. 1977. Superoxide dismutases I: Occurrence in higher plant. *Plant Physiology* 59 : 309-314.

Griffith, O.W. 1980. Determination of glutathione and glutathione disulphide using glutathione reductase and 2-vinylpyritine. *Analytical Biochemistry* 106 : 207-212.

Halliwell, B. and Foyer, C.H. 1978. Properties and physiological function of glutathione reductase purified from spinach leaves by affinity chromatography. *Planta* 139 : 9-17.

Jaworski, E. 1971. Nitrate reductase assay in intact plant tissues. *Biochemical Biophysical Research Communication* 43: 1274-1279.

Law, M.Y., Charles, S.A., Halliwell, B. 1983. Glutathione and ascorbhic acid in spinach (*Spinacia oleracea*) chloroplasts - the effect of hydrogen peroxide and paraquat, *Biochemical Journal* 210 : 899-903.

Lowry, O.H., Rosebrough, N.J., Farr, A.L. and Randall, R.J. 1951. Protein measurements using the 'Folin Phenol reagent. *Journal of Biological Chemistry* 193: 265-275.

Nakano, Y and Asada, K. 1987. Purification of ascorbate peroxidase in spinach chloroplast in activation in ascorbate depleted medium and reactination by monodehydeo ascorbate radicals. *Plant and Cell Physiology* 28 : 131-140.

Roe, J.H. and Kuether, C.A. 1943. The determination of ascorbic acid in whole blooc and urine through the 2,4-dinitro phenyl hydrazine derivative of dehydro ascorbic acid. *Journal of Biological Chemistry* 147 : 399-407.

Shigeoka, S., Yokato, A., Nakano, Y. and Kitaoka, S. 1979. The effect of illumination of the 1-ascorbic acid content in Euglena gracilis. *Agricultual Biological Chemistry* 43 : 2053-2058.

Sinha, S.K. 1972. Colorimetric assay of catalase. *Analytical Biochemistry* 47: 389-394.

Tietze, F. 1959. Enzymic method for quantitative determination of nanogram amounts of total and oxidized glutathione: Applications to mammalian blood and other tissues. *Analytical Biochemistry* 27: 502-522.

Wooten, I.D.P. 1964. *Microanalysis in Medical Biochemistry*. J.A. Churchil Ltd., London.

CHAPTER-8

ASSAY OF SOME KEY ENZYMES OF PLANTS

INTRODUCTION

All the known enzymes till date are classified and arranged in a systematic manner in order to have a scientific nomenclature. This system is known as digital system of enzyme classification. Enzymes are classified on the basis of the type of reaction catalysed. Each enzyme has a name and number to indicate its group and sub-group. Mainly six groups are laid out in order to accomodate all the enzymes. They are,

1. Oxido-reductases
2. Transferases
3. Hydrolases
4. Lyases
5. Isomerases
6. Ligases (Synthetases)

The nomenclature for each enzyme recommended by the ``International Union of Biochemistry" is given in the publication ``Enzyme Nomenclature 1965".

In this chapter, an attempt has been made to describe the assay procedure for some of the key enzymes under each of the above groups.

OXIDO-REDUCTASES

1. Ascorbic Acid Oxidase

Ascorbic acid oxidase, a copper containing enzyme is of wide-spread occurrence in plant tissues. The activity of this enzyme has been reported in both cytoplasmic as well as cell wall fractions. The role of this enzyme has been proposed to be important in regulating the levels of reduced and oxidized glutathione and NADPH because of the existence of a system linking NADPH to oxygen.

Ascorbic acid oxidase can be measured by following the rate of disappearance of ascorbate at 265 nm in a UV spectrophotometer.

Enzyme Reaction

$$\text{Ascorbic acid} + \frac{1}{2}O_2 \rightarrow \textit{Dehydroascorbic acid} + H_2O$$

Procedure

Add to a clean quartz cuvette of a UV spectrophotometer.

Phosphate buffer, 0.2 M, pH 6.2	1.0 ml
Ascorbic acid (1mM)	0.2 ml
Enzyme extract	0.1 or 0.2 ml

Add distilled water to bring volume to 3.0 ml. Insert cuvette immediately in chamber and record initial absorbance (A_o), and then read optical densities after 30 sec. intervals upto 3 minutes.

Calculations

From the linear phase of reaction, compute the change in optical density per minute.

2. Indole-3-Acetic Acid Oxidase

The plant growth hormone indole acetic acid can be oxidised by peroxidase and IAA-oxidase (Hare, 1964). The activity of IAA-Oxidase can be altered by many naturally occurring compounds such as phenols, manganese salt and plant acids. Investigations with phenols generally support the view that monophenols act as cofactors of IAA-oxidase while ortho-and paradi-hydroxy phenols and polyphenols are inhibitors of this enzyme (Endo, 1968). IAA-Oxidase activity can be assayed in plant extracts by three methods :

1. By estimating colorimetrically the amount of residual IAA-left after enzyme assay.
2. Colorimetric estimation of IAA-oxidation products.
3. By following oxygen uptake manometerically.

However, only, colorimetric methods are given here.

Estimation of Residual IAA

In this method, enzyme extract is incubated with IAA solution and after that residual IAA is estimated by using Salkowski's reagent. This reagent is prepared by mixing 0.5 M ferric chloride and 35% perchloric acid in the ratio of 1 : 50.

Procedure

For assay of IAA-oxidase activity prepare reaction mixture by adding following constituents in a test tube; 1 ml IAA, solution (10 mg in 100 ml containing 0.5 mM $MnCl_2$); 0.25 ml of 0.1 mM, 2, 4-dichlorophenol and 3.5 ml of 0.05 M phosphate buffer, pH 6.5 and 1 ml of enzyme extract. Incubate at 30°C for one hour in the dark.

After incubation remove two ml of above reaction mixture in a test tube and add 2 ml of Salkowski's reagent. Incubate for 30 minutes and record absorbance at 530 nm. For controls withdraw samples at zero time and add Salkowski's reagent.

Calculations :

Since in the controls the enzyme is not given time to oxidise IAA, therefore absorbance values in these tubes will be higher.

For calculations substract absorbance values obtained in the reaction tubes from that of control tubes. This will give the measure of indole-acetic acid oxidised by the enzyme action. Express enzyme activity in terms of μg IAA oxidised per mg protein per hour.

Estimation of IAA-oxidation Products

The oxidiation products produced by IAA-oxidase activity react with p-dimethylaminocinnamaldehyde to give a red violet coloured compound, and this reaction forms the basis of 1AA oxidase assay by this method (Meudt and Gaines, 1967).

Procedure :

For assaying 1AA-oxidase activity by this method, prepare reaction mixture by adding following constituents to test tubes. 0.2 ml of 10 mM 1AA; 0.1 ml, 1 mM 2, 6-dichlorophenol; 0.1 ml, 1 mM $MnCl_2$ and 0.4 ml, 0.1 M, phosphate buffer, pH 6.0. Start reaction by the addition of 0.2 ml of tissue homogenate. Incubate for two hours at 30°C and then add 1 ml p-dimethylaminocinnamaldehyde (0.5% solution in 1 N HCl). Keep for one hour and record 0.D at 562 nm. For controls add paradimethylaminocinnamaldehyde solution at zero time.

Calculations :

For calculating IAA-oxidase activity substract absorbance values obtained in control tubes from that of enzyme assay tubes. Express enzyme activity in terms of μg 1AA oxidised per mg protein per hour.

3. Peroxidase

Peroxidases are widely distributed in plant tissues and are of immense physiological interest because of their association with numerous catalytic functions. Numerous functions have been proposed and most important among these is the ability to oxidise indole-3-acetic acid (Siegel and Galston, 1960). Peroxidase can be assayed quite easily due to its high activity and stability and the availability of large number of hydrogen donors which can be used for assaying its activity (Pooviah and Rasmussen, 1973).

Enzyme-Reaction :

$$H_2O_2 - \xrightarrow{Peroxidase} - H_2O + O$$

$$O - dianisidine + O \rightarrow Oxidised\ O \rightarrow dianisidine\ (coloured)$$

The reaction is followed as rate of increase of absorbance at 430 nm.

Procedure

To a clean dry colorimeter cuvette with a 1 cm light path add -

Phosphate buffer, pH 6.5, 0.1 M	3.50 ml
Plant tissue homogenate	0.20 ml
O-dianisidine solution (1 mg/ml in methanol)	0.10 ml

Bring the mixture to 28-30°C in a constant temperature bath and then transfer the cuvette to spectrophotometer set at 430 nm.

Then add to cuvette 0.2 ml hydrogen peroxide (0.2 M). Simultaneously, start a stop watch. Read the initial absorbance A_o and then after 30 sec. intervals record subsequent absorbance values upto 3 minutes (A_1 to A_6). If the rate of increase of absorbance is too fast then repeat the procedure by using less homogenate.

Calculations

Plot the graph of different increasing absorption values against time as described earlier. From the linear phase of initial reaction read the change in optical density per minute. Express the enzyme activity in terms of rate of increase of absorbance per unit time per mg protein or tissue weight as the case may be.

4. Polyphenol Oxidase

The polyphenol oxidase category refers to two enzymes in this category, laccase or p-diphenol oxygen oxidoreductase and catechol oxidase or o-diphenol oxygen oxidoreductase. Catechol oxidase which is also referred to as phenolase, polyphenol oxidase, tyrosinase or catecholase. Both catechol oxidase and laccase oxidize phenolic substrates. The presence of phenolic compounds in plants, their oxidation following injury, either mechanical or due to infection and the relatively high toxicity of the oxidation products have long drawn attention (Mayer and Harel, 1979). The possible relationship of these properties to plant resistance to disease has prompted many research workers to ascribe a role to catechol oxidase in disease resistance. Catechol oxidase has been found to increase following infection by virus, bacteria fungi or mechanical injury.

The most convenient method for assaying the activity of catechol oxidase is to follow the initial rate of formation of quinone (product) spectrophotometerically (Mayer *et al.,* 1965).

Enzyme Reaction:

$$\text{Catechol} \xrightarrow{\text{Catechol oxidase}} \text{Quinone}$$

Procedure :

To a clean cuvette of a spectrophotometer add 3.0 ml buffered catechol solution (0.01 M catechol, freshly prepared in 0.1 M phosphate buffer, pH 6.0). Set the spectrophotometer to read absorbance at 495 nm. Insertte the curvet in cuvette chamber of spectrophotometer and adjust absorbance to zero. Add 1 ml enzyme extract to cuvette, mix immediately, place the cuvette in the spectrophotometer and record changes in absorbance for every 30 seconds upto 5 minutes.

Calculations :

Plot the increasing absorbance values and from the linear phase of the curve, read the increase in optical density per minute.

DEHYDROGENASES

5. Glucose-6-Phosphate Dehydrogenase

It has been known that glucose-6-phosphate dehydrogenase is a key enzyme of the oxidative pentose phsophate (PP) pathway in microorganisms and animals. The regulatory role of this enzyme in PP pathway has also been suggested in case of plant tissues (Ashihara and Komamine, 1976). The enzyme next to glucose-6-phosphate dehydrogenase in pentose phosphate pathway is 6-phosphogluconate dehydrogenase and both of these enzymes have been demonstrated in many plants (Gahan *et al.*, 1979).

The assay of this enzme can be done by following the reduction of NADP with the changes in the absorbance studied at 340 nm (Gusling and Ross, 1979).

Enzyme Reaction

$$\text{Glucose - 6 - phosphate} \xrightarrow[\text{G - 6 - PDH}]{\text{NADP} \quad \text{NADPH}} \text{6 - phosphogluconate}$$

The rate of formation of NADPH is followed by absorbance increase at 340 nm.

Procedure

To assay glucose-6-phosphate dehydrogenase, prepare the reaction mixture by adding the following constituents to a glass cuvette.

1.6 ml, 0.05 M Tris- HCl buffer, pH 7.5; 0.5 ml, 20 mM $MgCl_2$; 0.1 ml, 4 mM NADP and 0.2 ml of tissue homogenate. These constituents should previously have been incubated in a water bath at 30°C to ensure proper temperature for the assay. Initiate the reaction by adding 0.3 ml, 10 mM glucose-6-phosphate and follow the change in optical density at 340 nm.

Calculations

From the linear phase of initiasl enzyme velocity calculate the change in absorbance per minute. Express enzyme activity as change in O.D per mg protein per min or alternatively convert the data n moles substrate converted per mg protein per min using molar extinction coefficient of reduced NADP to be 6220.

6. Glutamate Dehydrogenase

Glutamate dehydrogenase is considered as most important enzyme for the incorporation of ammonia into complex organic molecules. This enzyme has been found to be very active in the germinating seeds. It is present in the mitochondria which gives it a close access to its substrates, NADH (or NADPH) and a-keto-glutarate, which arises from the TCA cycle, means that forward reaction is favoured (Lea and Miflin, 1974). Glutamate has been found to be a primary product of ammonia assimilation in photosynthetic tissues, although other amino acids such as alanine and aspartic acid have been demonstrated to be important products when nitrate is utilised as nitrogen source.

Glutamate dehydrogenase activity in plant tissue extracts can be conveniently assayed by following the oxidation of NADH to NAD spectrophotometrically at 340 nm. This oxidation of NADH accompanies the reductive amination of α-ketoglutarate to glutamic acid in the presence of ammonium ions.

Enzyme Reaction :

α-ketoglutarate + NADH + NH_4^+ → Glutamate + NAD

Procedure :

To assay glutamate dehydrogenase, prepare the reaction mixture by adding the following constituents to glass cuvettes : 0.2 ml, 0.2 M a-ketoglutarate; 2.3 ml, 0.1 M Tris-HCl buffer, pH 8, 0; 0.1 ml 5 mM NADH and 0.2 ml enzyme extract. Equilibrate for about 5 minutes at 30°C. Transfer the cuvette to the chamber of spectophotometer set at 340 nm. Initiate the reaction by addition of 0.2 ml, 1.5 M ammonium sulphate. Record absorbance at intervals of 30 seconds upto 3-5 min.

Calculations :

From the linear phase of change in absorbance per unit time graph, calculate change in optical density per min. Calculate the enzyme activity on tissue fresh weight basis and also determine the specific activity.

7. Malate Dehydrogenase

Malate dehydrogenase activity from higher plants has been the subject of large number of studies. It has been proposed that different isozymes of malate dehydrogenase (NAD specific) are localized in different cellular compartments. Such as mitochondria, peroxisomes cytoplasm and even chloroplasts (Rocha and Ting, 1971; Yang and Scandalios, 1974). In both C_3 and C_4 plants, NADP-specific MDH has been reported. This enzyme is concerned with the respiratory process and also carboxylation phenomena in plants, particularly with the reaction catalysed by PEP-carboxylase. The oxaloacetate produced by the carboxylase enzyme can be reduced to malate by malate dehydrog-enase. The highest activity of this enzyme has been reported in pollen (Malik and Singh, 1976, 1980).

Malate dehydrogenase activity can be determined with respect to reduction of oxalo-acetic acid with coupled oxidation of NADH to NAD.

Enzyme Reaction

Oxaloacetate + NADH → Malate + NAD

NADH absorbs light at 340 nm while NAD does not. It implies that formation of malate from oxaloacetate will be accompanied by a decrease in O.D at this wavelength.

Procedure

To assay malate dehydrogenase activity prepare reaction mixture by adding following constituents to a glass cuvette :

2.8 ml, 0.1 M Tris-HCl buffer, pH 7.8;

0.10 ml NADH (7 mg/ml) and

0.10 ml tissue homogenate.

Equilibnate for about 5 minutes at 30°C and than initiate reaction by addition of 0.10 ml oxaloacetate (1 mg/ml). Follow the decrease in absorbance at 340 nm for three minutes.

Calculations

Calculate decrease in absorbance per min. from the linear phase of initial reaction velocity. Express enzyme activity in terms of change in optical density per mg protein per min.

8. Succinate Dehydrogenase

This enzyme is an essential part of Kreb's-tricarboxylic acid cycle and so is formed in all aerobic cells. It is firmly bound to the structure of the mitochondrion but has been isolated and purified. The isolated enzyme contains flavine-adenine-dinucleotide and its pH optimum is about 7.6. This enzyme catalyses the reversible oxidation of succinate to fumarate when phenazine methosulfate is used as an electron acceptor. Since this enzyme is present in mitochondria, so it is often assayed by taking mitochondrial fraction by centrifugation of homogenate. For the estimation of its activity the reduction of flavine-adnine-dinucleotide (FAD) is coupled to a redox-dye, dichlorophenol-indophenol.

Enzyme Reaction

$$\text{Succinate} \xrightarrow[\text{FAD FADH}_2]{\text{SDH}} \text{Fumarate}; \quad \text{DCPIP} \longrightarrow \text{DCPIPH}_2$$

The enzyme reaction is followed by decrease in absorbance at 600 nm.

Procedure

To assay succinate dehydrogenase, prepare the reaction mixture by adding the following constituents to glass cuvette.

2.6 ml, 0.1 M phosphate buffer, pH 7.6 :

0.1 ml, 1.2 M sodium succinate; 1.0 ml of 10^{-3} M 2, 6-dichlorophenol indophenol (2.6 DCSIP); Equilibriate at 30°C in a water bath then add 0.1 ml of mitochondrial preparation (or curde tissue homogenate). Initiate the reaction by adding 0.1 ml of phenazine methosulphate (10 mg/ml) to the cuvette. Measure the rate of reduction of 2, 6 DCPIP at 600 nm during the interval between 10 and 60 sec. after initiating.

Calculations

Express enzyme activity in terms of change in absorbance at 600 nm per mg protein per min.

9. Nitrate Reductase

In green algae, fungi and higher plants the reduction of nitrate to nitrite is catalysed by nitrate reductase, an enzyme complex of high molecular weight which contains FAD, cytochrome and molybdenum as prosthetic group. The next step in the pathway of nitrate assimilation involves the activity of nitrite reductase which converts nitrite to ammonia. Many factors have been reported to effect nitrate reductase activity in higher plants and the visible light is most important because it controls both its synthesis and its activity.

Nitrate reductase activity can be assayed by following the reduction of nitrate to nitrite in the presence of plant homogenate *in vitro* or by using intact tissues in an *in vivo* assay (Jaworski, 1971). Both of these two types of assays are described below :

In vitro Assay

For the assay of nitrate reductase activity in plant extract prepare the reaction mixture by adding following constituents in a test tube.

1 ml phosphate buffer, 0.1 M, pH 7.2; 0.5 ml NADH (1.5 mM); 0.5 ml distilled water and 1 ml enzyme extract. Equilibrate the tubes to 25°C and then initiate the enzyme reaction by addition of 0.5 ml potassium nitrate, 0.1 M. After incubating for 30 min terminate the reaction by addition of 0.8 ml zinc acetate 1 M. Centrifuge and retain 3 ml of supernatant. To this add 2 ml colour developing reagent (1% sulphanilamide in 1.5 NHCl and 0.2% NEDH, mix in equal volume). After 10 min record O.D at 540 nm. For keeping the control use the enzyme extract inactivated by boiling.

Calculations

Prepare a reference curve by taking known amounts of nitrite and react it with the colour developing reagent. From this reference curve read the amount of nitrite produced by the enzyme action. Express enzyme activity in terms of μ moles nitrite formed per unit time.

In Vivo Assay :

For *in vivo* assay of nitrate reductase take about 200 mg of fresh plant material, and cut it into the slices. Suspend this chopped plant tissues in a 5 ml reaction mixture which comprises 55 propanol and 0.02 M potassium nitrate in 0.1 M phosphate buffer, pH 7.5. Cap the test tubes and incubate in dark at 30°C for two hours. After incubation remove 0.4 ml aliquot of reaction mixture and mix it with 0.2 ml of 1% sulphanilamide in 3 N HCl and 0.2 ml of 0.2% N-napthylene diamine hydrochloride (NEDH). After 20 min add 4 ml of distilled water to each tube. Read adsorption at 570 nm.

Calculations :

Calculate the quantity of nitrite released from the standard curve which is prepared by using potassium nitrite solution. Express the enzyme activity in terms of μg nitrite released per unit time.

TRANSFERASES

10. Alanine Amino Transferase

Alanine amino transferase (L-alanine : 5-oxoglutarate amino-trasnferase) has been reported in many plant tissues (Matheron and Moore, 1973). It has been proposed that this enzyme plays a major role in the synthesis of the alanine which occurs in cotyledons of many seeds during germination. This alanine is then transported from cotyledons to the axis of the germinating seeds. Alanine could be synthesized from pyruvate and glutamate by alanine amino transferase. Alanine amino transferase activity is also known to increase during the greening of leaves of plants having the C_4 photosynthesis pathway.

The activity of this enzyme can be assayed based on the transamination of L-alanine and a-ketoglutaric acid to give pyruvic acid and glutamic acid. The pyruvic acid forms a hydrozone with 2, 4-dinitrophenlyhydrazine, and this hydrazone changes to a deep brownish colour on the addition of the alkali.

Enzyme Reaction

Alanine + a-ketoglutarate → Pyruvic acid + Glutamic acid

Pyruvic acid + 2, 4 DNPH → Pyruvic acid hydrazone (brown at alkaline pH)

Procedure

Prepare two reagents A and B as follows :

Reagent A

Mix 6 ml, 0.2 M Tris-HCl buffer, pH 8.7; 1.25 ml, 0.5 M, a-Alaline; 0.75 ml, 0.1 M a-ketoglutarate and 0.035 ml 10% bovine serum albumin.

Reagent B

This mixture is used for the blanks, and is prepared as A except that alanine is replaced with water.

Prepare the assay tubes by adding 0.5 ml of reagent A or B and 0.1 ml of the tissue homogenate. Incubate samples as well asd blanks at 35°C for 30 min. Stop the reaction by adding 0.1 ml of 80% TCA to erach tube. Add 0.5 ml, 2,4-dinitro-phenyl hydrazine (0.1% in 2 NHCl). After 5 min of incubation at 35°C extract the reaction mixture with 1 ml of water saturated toluene. Add 3 ml of alcoholic KOH to 0.6 ml of the toluene layer and record the absorbance at 530 nm.

Calculations

Read the amount of pyruvic acid formed from the standard curve. Express enzyme activity in terms of μ moles pyruvate formed per mg protein per hour.

Spectrophotometric Method

This enzyme can be assayed spectrophotometrically by using lactate dehydrogenase as an auxiliary enzyme. The pyruvate formed due to aminotrasnferase activity is converted to lactate by lactate dehydrogenase and an equivalent amount of NADH is oxidised. This implies that amino transferase activity can be followed by recording rate of decrease of optical density at 340 nm.

Enzyme Reaction

$$\text{Alanine} + \alpha\text{-ketoglutarate} \xrightarrow[\text{NADH NAD}]{} \text{Pyruvate} + \text{Glutamate}$$

$$\text{Pyurvate} \xrightarrow[\text{LDH}]{} \text{Lactate}$$

Procedure

To assay alanine amino trasnferase, prepare the reaction mixture by adding the following constituents to a glass cuvette. 2.2 ml phosphate buffer, 0.1 M, pH 7.5; 0.1 ml, 1.5 mM NADH; 0.1 ml, 19 mM α-ketoglutarate; 0.1 ml lactate dehydrogenase (4 units) and enzyme extract in a final volume of 3 ml. Record the decrease in absorbance at 340 nm.

Calculations

Plot the graph of absorbance vs. time and from this find out the initial reaction velocity. Express enzyme activity in terms of rate of decrease in O.D per min.

11. Starch Phosphorylase

Plants contains starch phosphorylase that catalyses the reaction :

$$\alpha\text{-(1-4)-glucan (n)} + \text{Pi} - \alpha\text{ (1-4)} \rightarrow \text{glucan (n-1)} + \text{glucose - 1- phosphate}$$

Starch phosphorylase can catalyse the successive phosphorytic removal of glucose units from the non-reducing end of amylose. The enzyme attacks amylopectin readily to give 33 to 45% as glucose-1-phosphate and rest as a limit dextrin. It has been proposed that in case of pea seeds phosphorylase plays a major role in the break-down of starch during seed germination. This enzyme can be assayed from plant tissue homogenates by incubating with glucose-1-phosphate. In the presence of prior starch molecule, *in vitro* synthesis of starch will occur due to addition of glucose units from glucose-1-phosphate. There is simultaneous release of inorganic-phosphate in the reaction medium and to determine phoosphorylase, this released phosphate is measured.

Enzyme Reaction

$$\text{Glucose-1-phosphate} + \underset{\text{(PRIMER)}}{(1, 4\text{ glucan})_n} \rightarrow (1, 4\text{ glucan})_{n+1} + \text{Pi}$$

This phosphate released is estimated by the method of Fiske and Subbarow (1925).

Procedure

Incubate 0.2 ml enzyme extract with 0.6 ml Tris-maleate buffer (pH 6.5) containing ImM sodium flouride. Then add 0.2 ml of 0.05 M glucose-1-phosphate. Incubate for about one hour at 30°C and then terminate the reaction by adding 0.5 ml cold 5% TCA. Centrifuge to settle protein precipitate and use supernatent for the estimation of inorganic phosphate. In addition to the reaction tubes, maintain control tubes in which the enzyme extract is denaturated by boiling for 2 to 3 minutes. For measuring the inorganic phosphate take 0.5 ml of supernatent add to it 3.3 ml distilled water and 1 ml ammonium molybdate reagent (1.5 g ammonium molybdate + 30.3 ml conc. HCl, diluted to 100 ml with distilled water).

Shake the tubes well and after about 5 minutes add 0.2 ml Fiske and Subbarow reagent (Prepare this reagent by dissolving 1.45 g sodium metabisulphite, 50 mg sodium sulfite and 25 mg 1-amino-2 - napthol-2-sulphonic acid in 5 ml water and then making final volume to 10 ml). Keep for 15 min at about 30°C and then record absorbance at 660 nm.

Calculations

Prepare reference curve for estimation of inorganic phosphate by Fiske and Subbarow method and from this determine the amount of phosphate present in reaction as well as in control tubes. Substract the control values from reaction values and this excess inorganic phosphate will give a measure of phosphorylase activity.

HYDROLASES

12. Acid Phosphatase

Orthophosphoric monoester phosphohydrolase, commonly known as `phosphatase' catalyzes a variety of biochemical reactions and plays a vital role in regulating the plant cell metabolism through inorganic phosphorus level (Tuubol *et al.,* 1957). It has been demonstrated that carbohydrate metabolism of plants may be indirectly controlled by availability of inorganic phos-phate which in turn is regulated by acid phosphatase activity level. Murray and Collier (1977) have shown that a specific isoenzyme of acid phosphatase present in seed coats of *Pisum* seeds regulates the movement of phosphorus from seed coat to developing embryos. The activity of this enzyme is usually assayed by using p-nitrophenyl phosphate as an artificial substrate (Gazan and McClann, 1967).

Enzyme Reaction

$$\text{p-nitrophenyl phosphate} \xrightarrow{\text{Acid Phosphatase}} \text{p-nitrophenol} + \text{Pi}$$

p-nitrophenol is yellow at alkaline pH and its amount is measured by reading absorbance at 410 nm.

Procedure

For the assay of acid phosphatase activity, prepare reaction tubes by adding following constituents : 0.5 ml of the substrate solution (50 mg p-nitrophenyl phosphate in 10 ml water + 2.5 ml acetate buffer, 0.1 M, pH 4.8), and 0.1 ml enzyme extract. Incubate for 30 min at 35°C. Terminate the reaction by adding 3.4 ml of 0.1 N NaOH. Read absorbance at 410 nm.

Calculations

Prepare a reference curve by taking known amounts of p-nitrophenol and adding NaOH and then reading absorbance at 420 nm. express enzyme activity in terms of μ moles p-nitrophenol released per unit time.

13. Amylases

α- and β-amylases can hydrolyse α-1, 4 bonds of starch. α-amylase from plat tissues catalyses the hydrolyis of the glycosidic bonds in α-1, 4-D-glucan (Bernfeed, 1955). The linkage in maltose is not attacked. a-amylase from plants contains 1g atom of calcium per mole of enzyme. In barley and wheat multiple forms of a-amylase have been reported (Kruger, 1972). The enzyme is synthesized in and secreted by the aleurone layer and the overall process is dependent upon gibberellin (Chrispeels and Varner, 1967). Infact, induction of the release of α-amylase by gibberellin is one of the most commonly employed bioassay for this hormone.

For estimation of α-amylase activity the homoegenization medium for enzyme extraction should contain high concentration of calcium ions. Infact, most of the workers grind the plant tissues in buffered calcium chloride solution. From the homogenate the α-amylase activity can be estimated by measuring decrease in colour of iodine-dextrin complex. This assay is done by variable time method.

13a. α-amylase

Most commonly used substrate is soluble starch, however, use of β-limit dextrin gives more accurate results. In the following, a method for α-amylase is described which involves use of soluble starch. For β-limit dextrin method one can refer to the paper by Davis (1977).

Enzyme Reaction

$$\text{Starch} \xrightarrow{\alpha\text{-amylase}} \text{- short dextrins + reducing sugars}$$

The decrease in starch is measured by taking colour of iodine dextrin complex at zero time and after variable time intervals.

Procedure

For assay of a-amylase activity prepare a reaction mixture by adding following constituents. 1 ml of starch solution (boil 150 mg potato starch, 600 mg KH_2PO_4, 20 ml of anhydrous $CaCl_2$ and 100 ml of distilled water for one minute and cool, filter and use the clear supernatent) and 0.5 ml enzyme extract. At zero time, remove 0.2 ml aliquot of reaction mixture and add it to 3 ml IKI solution (254 mg I_2 and 4.0 g; KI in one litre distilled water).

Record optical density at 620 nm. Then after every 30 min, remove the aliquot and repeat the colour developing process.

Calculations

Express enzyme activity in terms of decrease in optical density at 620 nm per unit time.

13b. β-amylase

β-amylase has only been reported from cells of higher plants and there are reports of multiple-forms of this enzyme. Most of the studies of β-amylase has been done on sweet potato enzyme. The enzyme is readily inactivated by sulphydryl reagents. In cereals, most of the β-amylase in the seeds is present in a latent or inacticve form that appears to be converted to an active form during germination. This enzyme catalyses the successive removal of maltose units from the non-reducing end of α-1, 4-D glucan. Thus, β-amylase can catalyse the complete conversion of amylose to maltose. The action of this enzyme on amylopectin results in the formation of maltose, and of a β-limit dextrin that constituents above 43% of the origina lamylopectin. In starch breakdown, during germination of cereal seeds β-amylase is also likely to be involved and its main role may be hydrolysis of shorter chains produced by the action of α-amylase.

The activity of β-amylase is estimated by measuring the increase in reducing groups in the presence of EDTA. This chelating compound is added to inhibit the a-amylase activity. The reducing groups are estimated by heating with dinitrosalicylic acid reagent.

Enzyme Reaction

$$\text{Starch} \xrightarrow{\beta\text{-amylase}} \text{- Maltose units}$$

Since maltose is a reducing sugar, it can be estimated by colorimetric method.

Procedure

For assaying the activity of β-amylase prepare reaction mixture by adding the following constituents :

0.7 ml starch solution (0.25 in 0.067 M phosphatre buffer, pH 6.0), 0.1 ml EDTA solution (0.1 M) and 0.1 ml enzyme extract. Incubate for 30 min at 30°C. Stop the reaction by the addition of 1 ml dinitrosalicylic acid reagent. Boil the tubes in water bath for 20 min. Cool and add 3 ml of distilled water to each tube, and then record the optical density at 560 nm. For the controls, keep the reaction mixture similar to as in reaction tubes but terminate the reaction at zero time.

Calculations

Prepare reference curve by heating known amounts of maltose sugar with dinitrosalicylic acid reagent. Express enzyme activity in terms of maltose released per unit time.

14. Cellulase

Cellulose is one of the main constituent of plant cell walls and chemically it is a linear polymer of glucose units with 1, 4 linkages. This macromolecule can be degraded by plant enzyme known as cellulase. Cellulase activity has been correlated with leaf and fruit abscission (Greenberg *et al.,* 1975) and fruit ripening (Hinton and Pressey, 1974). Multiple molecular forms of cellulase have also been reported. This enzyme has been shown to be present in both soluble as well as membrane fraction of plant cell walls (Koehler *et al.,* 1976). Cellulase activity from plant cell extracts can be assayed by viscometery or colorimeteric methods (Lewis and Koehler, 1979). In the former method enzyme extract is incubated alongwith carboxymethylcellulose and decrease in the viscosity due to cellulose breakdown is reduced. In the colorimetric method, the reducing sugars produced by cellulase action are measured by dinitrosalicylic acid reagent or Nelson's reagent. However, only colorimeteric method is described here.

Enzyme Reaction

$$\text{Cellulose} \xrightarrow{\text{Cellulase}} \text{Cellobiose + glucose}$$

Both the products are the reducing sugars and are measured colorimetrically.

Procedure

For the assay of cellulase activity prepare reaction tubes containing following constituents.

0.5 ml acetate buffer, 0.1 M, pH 5.5; and 1 ml enzyme extract. Equilibriate the mixture at 30°C and then add 1.5 ml, 1% carboxylethyl cellulose. Incubate for two hours at 30° and then add 3 ml dinitrosalicylic acid reagent. Boil the tubes for 10 min. Cool and record absorbance at 560 nm. For keeping the controls boil the enzyme for 3 to 5 min prior to the addition of carboxymethyl cellulose.

The dinitrosalicylic acid can be prepared as follows :

To 50 ml of distilled water add 2.5 g dinitrosalicylic acid. Dissolve completely and add 4 g. of sodium hydroxide. After dissolving add 75 g of sodium potassium tartarate and make total volume 250 ml with distilled water.

Calulations

Prepare standard curve by heating known amounts of glucose with dinitrosalicylic acid reagent and from this standard curve, calculate the amount of reducing sugars released, by cellulase activity. Express enzyme activity in terms of nmoles glucose equivalent released in one hour.

15. β-Glucosidase

Glycosidases are enzymes which catalyze the hydrolysis of glycosidic linkages. Plant glycosidases are optimally active at slightly acidic pH values. The multiple forms of glycosidases in single tissue have been reported. It has been proposed that glycosidases play a direct role in cell wall modification and cell expansion (Evans, 1974). The enzyme activity has been reported in cell walls also. An increase in β-glucosidase activity was shown to be roughly correlated in time with the decrease in cell wall glucan and an increase in groweth rate of hypocotyls of seedlings of *Phaseolus vulgaris.* Murray and Bandurski (1975) found highest specific activity of β-glucosidase in cell walls from the most rapidly elongatng region of pea epicotyls. Most of the workers have used synthetic substrates i.e. nitrophenyl glycosides in studying glycosidase activities from plant cells (Cohen *et al.*, 1952).

Enzyme Reaction

p-nitrophenyl-β-D-glucoside- $\xrightarrow{\beta\text{-Glucosidase}}$ -p-nitrophenol+glucose

p-nitrophenol gives intense yellow colour at alkaline pH and amount of colour gives an accurate measure of glycsidase activity.

Proceudre

To assay β-glucosidase prepare the reaction mixture by adding the following constituents to the test tubes :

0.1 ml substrate (15 mg p-nitrophenyl-β-glucoside dissolved in 2 ml water).

0.8 ml 0.1 M citrate phosphate buffer, pH 4.8.

Incubate these constituents in a water bath 30ºC, to ensure proper temperature for the assay. Then initiate the reaction by addition of 0.1 ml enzyme extract. After 30 min terminate the reaction by the additon of 2 ml 1 M sodium carbonate and read absorbance at 410 nm. At the pH of the enzymatic reaction there is no substrate breakdown in the absence of enzyme, therefore, for correcting any initial absorbance by the reaction mixtures, use non-enzymatic blanks to which substrate and sodium carbonate is added prior to the addition of enzyme. The presence of light causes the breakdown of the substrate under alkaline conditions therefore, after termination keep the reaction mixtures in the dark until O.D. is taken.

Calculations

Prepare reference curve by using known concentrations of p-nitrophenol (10 to 100 μg/ml), adding 2 ml 1 M sodium carbonate and recording O.D. at 410 nm. From this reference curve read the amount of p-nitrophenyl released due to action of β-glucosidase on p-nitrophenyl glucoside.

16. Invertase

Invertase is one of the most important enzymes in plant tissues because of its ability to catalyse the breakdown of sucrose. The invertases may be divided into acid-invertases, alkaline or neutral invertases. However, acid-invertase activity is of most widely spread occurrence in plants, and this enzyme is a b-fructo furanosidase. Inverstase has been reported both from cytoplasmic as well as cell-wall fractions. It has been proposed that membrane and wall-bound invertases help in facilitating the uptake of sucrose by plant cells.

Studies with mature storage tissue of sugar cane show that movement of exogenous sucrose into the cell is atleast partially dependent upon sucrose hydrolysis and there is evidence that some of this hydrolysis occurs in the free space of plant-cells. This implies that acid-invertase in such tissues play an important role in sucrose uptake. Inverstase can be very easily assayed from plant tissue extracts i.e. by measuring the formation of reducing sugars by the action of this enzyme on sucrose.

Enzyme Reaction

$$\text{Sucrose} \xrightarrow{\text{Invertase}} \text{Glucose + Fructose}$$

The formation of products can be measured by either (a) estimation of total reducing sugars (Glucose + Fructose) by Dinitro salicylic acid reagent (Sumner, 1935) or by Nelson method (1944) or (b) estimation of glucose by coupled glucose-oxidase peroxidase reaction.

Procedure

In clean test tubes add 0.4 ml of 0.2 M acetate buffer (pH 4.8), 0.25 ml of 0.4 M sucrose. To this reaction mixture add appropriate amount of enzme extract to give a total volume of one ml. In the control tubes add sucrose solution only when enzyme has been inactivated by boiling for about 5 min. After incubation at 30ºC for 30 min add 1 ml of 3, 5 di-nitrosalicylic acid reagent. Boil the tubes for 10 min then dilute it to 5 or 10 ml and record optical density at 560 nm.

Calculations

Calculate the amount of reducing sugars formed from the reference curve prepared by heating varying amount of anhydrous glucose (10 to 100 μg) with dinitro-salicylic acid. Express invertase activity in terms of μg reducing sugars formed per mg protein.

17. Leucine-Amino-Peptidase

Leucine aminopeptidase is the term for an aminopeptidase which has a preference for leucine. The activity of this enzyme is not, however, limited to n-terminal leucine and also does not show a preferential hydrolysis of l-leucyl peptides, e.g. in comparison to its activity with l-alanylpeptide; in addition, amino acid amides and arylamides are hydrolyzed. Amino peptidases and amino acid arylamidases have for several years been included in aminopeptidases group of enzymes. Aminoacyl-pnaphthylamides have been widely used as substrates for the histochemical localization and quantitative estimation of leucine-amino peptidase.

Enzyme reaction

$$\text{Leucyl-}\beta\text{-naphthylamide} \xrightarrow{\text{LAP}} \text{leucine} + \beta\text{ - naphthylamine}$$

$$\beta\text{-naphthylamine} + \text{ Diazonium} \xrightarrow{\text{Salt}} \text{Azo dye (coloured)}$$

Procedure

The colorimetric assay based on the reaction of β-naphthylamine with fast blue BB is recommended. The reaction mixture should contain 0.8 ml of 25 mM Tris-maleate buffer, pH 6.8; 0.1 ml substrate solution (3 mM leucyl-β-naphthylamide-hydrochloride) and 0.1 ml enzyme extract. Incubate reaction mixture for 30 min at 28°C. After which 2 ml of 5% Tween-20 (or Triton X-100) and 2 ml 0.1% Fast blue BB salt (in 0.1 M acetate buffer, pH 4.2) is added. Read at 525 nm against a control tube in which distilled water is added instead of enzyme extract.

Calculations

The enzyme activity expression should be based on the amount of β-naphthylamine released per unit time. The amount of β-naphthylamine is read from reference curve prepared by coupling of known amounts (10 to 100 μg) of β-naphthylamine with Fast blue BB.

18. Phytase

Phytic acid (Myoinositol hexaphosphate) is most important phosphate storing form in developing seeds. During the germination of seeds, dephosphorylation of phytate is mediated through enzyme phytase (Mandal and Biswas, 1970). Goel and Sharma (1979) have demonstrated that polyphenols which are present in the seeds are known to inhibit seed germination can inhibit the activity of phytase. This enzyme can be assayed by using phytic acid as the substrate and by measuring the amount of inorganic phosphate released.

$$\text{Myoinositol hexaphosphate} \xrightarrow{\text{Phytase}} \text{Myoinositol pentaphosphate} + \text{Pi}$$

The inorganic phosphate released is mesured by the method of Fiske and Subbarow (1925).

Procedure

For assay of phytase, prepare reaction mixture containing the followings in a volume of 2 ml : 0.64 mM of phytic acid; 0.17 mM of acetate buffer (pH 4.8) and enzyme extract. Incubate at 45°C for 60 min and terminate and reaction by adding one ml cold 10% TCA. From the supernatant estimate inorganic phosphate by the method of Fiske and Subbarow (1925).

Calculations

Prepare reference curve by taking known amounts of inorganic phosphate (KH_2PO_4) and read the values of inorganic phosphate released by phytase action. Express enzyme activity in terms of μ moles of inorganic phosphate released per unit time under the assay conditions.

19. Polygalactouronase

This enzyme hydrolyses the glucosidic bonds present in the pectic substances causing the release of galacturonic acids. Riov (1974) reported that polygalacturonase plays important role in the abscission of citrus leaf explants. Fungal pathogens also secrete this enzyme; such a release is responsible for causing soft-rots of plant tissues. This enzyme is assayed by using sodium polypectate or pectic acid as substrates. After incubating the substrate with the enzyme, the reducing groups released are estimated by using dinitrosalicylic acid reagent.

Enzyme Reaction

$$\text{Sodium polypectate} \xrightarrow{\text{PG}} \text{Galactouronic acid}$$

Procedure

For estimating the activity of polygalactouronase activity prepare assay mixture by adding the following constituents :

0.5 ml enzyme extract and 0.5 ml substrate solution. The substrate solution should contain 0.25% sodium polypectate, 1 mM sodium hydrosulphite in 0.1 M acetate buffer, pH 5.0. Incubate for 6 hours at 37°C with constant shaking. Terminate the reaction by the addition of 0.5 ml dinitrosalicylic acid reagent. Prepare this reagent as for cellulase. For keeping controls prepare a reaction mixture to which active enzyme solution is added at the end of the incubation period.

Calculations

Prepare reference curve by using known amounts of galacturonic acid and heating with dinitrosalicylic acid reagent. Express enzyme activity in terms of nmoles of galacturonic acid per unit time.

20. Protease

Proteases have been classified as endopeptidases (with acidic, sulfahydryl serine and metal dependent sub-classes) and amino-and carboxy-peptidases (Sopenen and Lauriers, 1976). The endopeptidases have been found to be sulfahydryl dependent and optimum activity is seen under acidic conditions (Basha and Beevers, 1975; Salmia *et al.*, 1978). The endopeptidase activity plays a prominent role in the mobilization of reserve proteins during germination of seeds. The increased activity of proteases have been demonstrated in senescing leaves. The endopeptidases, by themselves can not hydrolyse proteins to amino acids; these convert proteins into small peptides which are further degraded by amino and carboxy peptidases (Salima and Mikola, 1976). The role of these enzymes is also important in host-parasite relationship because parasite needs to secrete proteolytic enzymes to degrade the proteins in the host cell walls. The degradation products i.e. amino acids serve as nitrogen and carbon sources for the utilisation by the parasite.

The assay of protease activity is generally done by incubating enzyme extract with proteins like casein or haemoglobin. The TCA soluble products released by protein hydrolysis are estimated colorimetrically by using ninhydrin or Folin-phenol reagent.

Enzyme Reaction

Proteins $\xrightarrow{\text{Protease}}$ peptides (TCA - soluble) amino acids
(TCA- insoluble)

The amino acids released are measured colorimetrically.

Procedure

To measure proteolytic activity prepare the following reaction mixture :

0.5 ml casein solution (1% in, 0.1 M, pH 6.0) and 0.1 ml enzyme extract. Incubate at 45°C for one hour and then terminate the reaction by the addition of 0.1 ml TCA (40%). Centrifuge and retain the TCA soluble component. Measure the amino acid content of TCA soluble component after reaction with folin-phenol reagent. For keeping the controls terminate the reaction at zero time by the addition of TCA.

Calculations

Prepare reference curve by reacting known amounts of tyrosine with folin-phenol reagent. Express enzyme activity in terms of mg amino acids released per unit time.

21. Ribonuclease

Plants possess many different kinds of ribnonucleases, but little is known about their specific functions. Out of nuclease category a number of RNA-ase have been characterized in plant tissues. This enzyme can be conviently assayed because of very high activity present in plant cells.

Enzyme Reaction

RNA $\xrightarrow{\text{RNA-ase}}$ Nucleoside monophosphate (Acid insoluble) + Oligonucleotide + CoreRNA (Acid soluble)

The amount of RNA-ase activity is determined by measuring an increase in acid soluble material. This is done by reading the absorbance of PCA-soluble material released by the action of the ribonuclease.

To assay for ribonclease activity prepare assay tubes each containing the followings :

0.4 ml of RNA (0.2%) solution prepared in 0.1 M citrate-phosphate buffer (pH 5.0). tart the reaction by the addition of 0.2 ml of tissue homogenate. After incubation for one hour at 30°C stop the reaction by adding 0.1 ml uranyl acetate solution (0.75% prepared in 25% perchloric acid). Place the tubes at 4°C for about one hour after which centrifuge to remove the precipitate. Remove 0.1 ml of the supernatent and add it to 2.9 ml of distilled water and measure extinction at 260 nm. For keeping the controls use an equivalent amount of enzyme extract but add uranyl acetate solution at zero time and then proceed further as done in enzyme assay tubes.

Calculations : Express ribonuclease activity in terms of increase in the optical density of reaction tubes over control at 260 nm per mg, protein.

LYASES

22. Isocitrate-Lyase

Many plant seeds and pollen can convert fat into carbohydrates as they germinate. Kornberg and Beevers (1957) demonstrated that extracts from germinating castor seeds contain the enzyme necessary for the conversion of fat into carbohydrates via acetyl COA as in following reaction sequences :

Fat → Acetyl COA → malate → carbohydrate

The above pathway is described as *glyoxalate cycle*. The key enzymes of this cycle are isocitrate lyase and malate synthetase

(Singh and Malik, 1976). In one turn of the glyoxalate cycle, 2 molecules of acetate are used and one molecule of succinate is produced. The succinate can be used for synthesis of carbohydrates.

Enzyme Reaction:

$$\text{Isocitrate} \xrightarrow{\text{Isocitita lyase}} \text{Glyoxalate + Succinate}$$

The glyoxalate thus produced can be measured colorimetrically by forming its complex with phenythydrazine and then measuring absorbance at 517 nm.

Procedure

Prepare the following reaction mixture :

Phosphate buffer, 0.2 M, pH 7.5	0.2 ml
$MgCl_2$ solution (10 mM)	0.1 ml
Cysteine hydrochloride	0.1 ml
Enzyme extract	0.1 ml

Equilibrate for 5 min at 30°C and start the reaction by the addition of 0.1 ml isocitrate solution. Incubate for 10 min and then terminate the reaction by the addition of 0.2 ml trichloroacetic acid (15%). Clear the mixture by centrifugation and to the supernatent add 1 ml of 0.5 M phosphate buffer (pH 7.0). Maintain the mixture at 25°C for 10 min and then add 0.1 ml (2% w/v) Phenylhydrazine hydrochloride. After two min. of keeping at this temperature add 1.2 ml conc. HCl (ice-cold), then after further one min add 0.1 ml (10% w/v) potassium ferricyanide. After each addition mix the contents of the tubes thoroughly. Keep the reaction mixture for 2-3 min at 25°C after the last addition and measure absorbance at 517 nm.

Calculations

A blank containing all the components except the homogenate should be carried through the same procedure. Substract the absorbance of this sample from that of test samples and read the amount of glyoxalate produced from the reference curve. The reference curve can be prepared by taking known amounts of pure glyoxalic acid in place of enzyme extract.

SYNTHETASE

23. Sucrose-Synthetase

Sucrose-synthetase catalyses the reaction :

Sucrose + nucleoside - diphosphate → nucleoside - diphospho glucose + fructose

This enzyme is of wide occurrence in plants. Most of the studies on purification and characterization have been done in *Phaseolus aureus.* This enzyme is reversible and can catalyse the synthesis of nucleoside-diphosphoglucose from sucrose. The nucleoside in the above reaction may be uridine, adenine, thymidine, cytosine or guanidine. However, ADP-glucose or UDP-glucose have been found to be important sugar nucleosides for sucrose synthesis *in vivo.* The properties of this enzyme indicate that it plays an important part in converting sucrose to sugar-nucleotides. There is evidence that sucrose-synthetase activity is high in tissues in which there are heavy demands for sugar nucleosides for the synthesis of structural or storage polysaccarides. Most of sucrose-sythetase activity may be associated with membrane systems in plant cells.

Reaction

The enzyme can be assayed in the direction of sucrose synthesis.

UDPG + Fructose = UDP + Sucrose

For the estimation of enzyme activity homogenate from plant tissues is incubated in presence of UDPG and sucrose synthesized is measured by using resorcinol reagent. In addition to above, control for each homogenate is also maintained by incubating an equivalent amount with fructose only in the incubation medium. After incubation the tubes are heated at 100°C for about 10 minutes after addition o sodium hydroxide solution. This treatment destroys free fructose Then to all controls and reaction tubes resorcinol reagent is added and subsequently heated. The sucrose formed in reaction tubes by th action of sucrose-synthetase gives red colour with this reagent. Nov the absorbance values obtained in control tubes are substrated fron those in reaction tubes to get correct values for sucrose synthesize enzymatically.

Procedure

Prepare two reaction mixtures :

1. To 9 ml of 0.05 M HEPES buffer, add 11 mg fructose and 1 mg UDP-glucose.
2. To the same volume of buffer add 11 mg of fructose only. Tak 0.15 ml of each incubation mixtures in separate test tube and add 0.1 ml of homogenate to it. Incubate for 15 minute at 35°C, after which add 0.3 ml (0.3 N) NaOH solution. He

in boiling water bath for 10 min. Cool to room temperature and add 0.5 ml of resorcinol reagent A (50 mg resorcinol dissolved upto 50 ml with 95% ethanol) and 0.5 ml of resorcinol reagent B (50 ml of conc. HCI is added carefully to 10 ml of water). Heat the tubes at 80°C for 10 min. Cool and add 1.5 ml ethanol. Record O.D. at 490 nm.

Calculations

Calculate the amount of sucrose formed from a reference curve prepared by treating varying amounts of sucrose (0.1 μM to 0.5 μM) with resorcinol reagent.

ISOMERASE

24. Fructose Diphosphate Aldolase

Aldolase is a significant enzyme of glycolytic pathway. There is enough evidence to support the views that the glycolysis is universal in higher plants. Fructose-diphosphate aldolase causes conversion of fructose-1, 6-diphosphate to glyceraldehyde 3-phosphate and dihydroxyacetone phosphate. This enzyme can be assayed by measuring the products formed by using colorimeter or through coupled enzymatic reactions.

Colorimetric estimation : Aldolase present in the plant extracts catalyzes the breakdown of fructose-1, 6-diphosphate to dihydroxyacetone 3-phosphate and glyceraldehyde 3-phosphate. Hydrazine is added to form the hydrozones, and the phosphates are hydrolyzed with sodium hydroxide to give methyl glyoxal. Reaction of methyl glyoxal with 2, 4 dinitrophenyl hydrazine gives a coloured hydrazone which is read at 540 nm.

Coupled enzymatic reaction : In this method the products are estimated by using commercially available triosephosphate isomerase and glyceraldehyde 3 phosphate dehydrogenase as auxilliary enzymes. Triose phosphate isomerase converts dihydroxyacetone phosphate to glyceraldehyde 3 phosphate dehydrogenase reduced NAD. Thus aldolase may be assayed by measuring the kinetics of apperance of NADH at 340 nm in the presence of excess of fructose-1, 6-diphosphate, glyceraldehyde-3-phosphate dehydrogenase, triosephos-phate isomerase.

Reaction

Fructose-1, 6-diphosphate = Dihydroxyacetone phosphate + glyceraldehyde-3-phosphate

$$\text{Dihydroxyacetone phosphate} \xrightarrow[\text{NAD} \;\; \text{NADH}]{\text{TPI}} \text{glyceraldehyde-3-P}$$

Glyceraldehyde-3-phosphate → 1, 3 diphosphoglyceric acid

Thus from the above reaction sequence it is clear that during the breakdown of one molecule of fructose 1, 6-phosphate two molecules of NADH will be produced.

Procedure

For colorimetric assay of aldolase activity, incubate 0.5 ml enzyme extract with 2 ml of substrate (10 mM Fructose-1, 6-diphosphate and 0.1 M hydrazine in Tris-HCl buffer, 0.05 M, pH 8.5). After 30 min of incubation at 25°C terminate the reaction by adding 2 ml of 10% TCA. Centrifuge and retain one mol of supernatant for further estimations. To this supernatant add 1 ml of 0.75 N NaOH solution. Keep for 10 min at room temperature and then add 1 ml 2,4 dinitrophenyl hydrazine (0.125% in N HCl). Keep at 35°C for 10 min. Then add 7 ml of 0.75 N NaOH and read the optical density at 540 nm. For controls add 2 ml of 10% TCA prior to the addition of enzyme extract.

Calculations

Express enzyme activity in terms of increase in optical density at 540 nm per unit time.

REFERENCES

Ashihara, H. and Komamine, A. 1976. Characterization and regulatory properties of glucose-6-phosphate dehydrogenase from black gram (*Phaseolus mungo*). *Physiologia Plantarum* 36: 52-56.

Basha, S.M.M. and Beevers, I. 1975. The development of proteolytic activity and protein degradation during germination of *Pisum sativum*. *Planta* 124: 77-87.

Bernfeed, P. 1955. α and β Amylases. *Methods in Enzymology* 1: 149-158.

Chrispeels, M.J. and Varner, J.E. 1967. Gibberellic acid enhanced synthesis and release of α-amylase and ribonuclease by isolated barley aleurone layers. *Plant Physiology* 42: 378-383.

Cohen, R.B., Tsou, K.C., Rutenbergs, H. and Seligman, A.M. 1952. The colorimetric estimation and histochemical demonstration of B-gluco-sidase. *Journal of Biological Chemistry* 195: 239-249.

Davis, B.D. 1977. Occurrence of amylase in the axis of germinating peas. *Plant Physiology* 60: 513-518.

Endo, T. 1968. Indole acetate oxidase activity of horse raddish and other plant peroxidase isozymes. *Plant and Cell Physiology* 9: 333-338.

Evans, M.L. 1974. Evidence against the involvement of galactosidase or glucosidase in auxin- or acid-stimulated growth. *Plant Physiology* 54: 213-215.

Farkas, G.L. and Kiraly, T. 1958. Enzymological aspects of plant diseases. I-oxidative enzymes. *Phytopathology* 31: 251-272.

Fiske, G.H. and Subbarow, Y. 1925. The colorimetric determination of phosphorus. *Journal of Biological Chemistry* 66: 375-400.

Gahan, P.B., Auderset, G. and Greppin, H. 1979. Pestose phosphate pathway activity during floral induction in *Spinacia oleracea. Annals of Botany* 44: 121-124.

Gazan, P.B. and McClann, J. 1967. Acid phosphatse in roots of *Vicia faba. Biochemical Journal* 102: 47-54.

Goel, M. and Sharma, C.B. 1979. Multiple forms of phytase in germinating cotyledons of *Cucurbitia maxima. Phytochemistry* 18: 1939-1942.

Gusling, P.G. and Ross, J.D. 1979. Characterization of glucose-6-Phosphate dehydrogenase and 6-phosphogluconic acid dehydrogenase for hazel cotyledons. *Phytochemistry* 18: 1441-1449.

Greenberg, J., Garen, R. and Riov, J. 1975. The role of cellalose and polygalacturonase in absicission of young and mature shanouli organic fruits. *Physiologia Plantarum* 34: 1-7.

Hare, R.C. 1964. Indole acetic acid oxidase. *Botanical Review* 30: 129-164.

Hinton, D.M. and Pressey, R. 1974. Cellulase activity in peaches during ripening. *Journal of Food Science* 39: 783-785.

Jaworski, E.G. 1971. Nitrate reductase in intact plant tissues. *Biochemical Biophysical Research Communications* 43: 1274-1279.

Koehler, D.E., Leonard, R.T., Linkins, A.E. and Lewis, L.N. 1976. Association of latent cellulase activity with plasma membrane from Kidney bean abscission zone. *Plant Physiology* 58: 324-330.

Kornberg, H.L. and Beevers, H. 1957. A mechanism of conversion of fat to carbohydrate in caster beans. *Nature* 180: 35-36.

Kruger, J.E. 1972. Changes in the amylase of hard red sping wheat during growth and maturation. *Cereal Chemistry* 49: 379-390.

Lea, P.J. and Miflin, B.J. 1974. Alternate route for nitrogen assimilation in higher plants. *Nature* 251: 614-616.

Lewis, L.N. and Koehler, D.E. 1979. Cellulase in kidney bean seedlings. *Planta* 146: 1-5.

Malik, C.P. and Singh, M.B. 1976. Fluctuations in the activity of some dehydrogenases during pollen germination in *Calotropis procera. Biochemistry Physiology Pflanzen* (BPP) 169: 583-588.

Malik, C.P. and Singh, M.B. 1980. *Plant Enzymology and Histo-Enzymology - A Text Manual.* Kalyani Publishers, New Delhi.

Mandal, N.C. and Biswas, B.B. 1970. Metabolism of inositol phosphatases. I. phytase synthesis during germination in cotyledons of mungbeans. *Phaseolus aureus. Plant Physiology* 45: 4-7.

Matheron, M.E. and Moore, T.C. 1973. Properties of an aminotransfic rases of pea (*Pisum setivum*). *Plant Physiology* 52: 63-67.

Mayer, A.M. and Harel, E. 1979. Polyphenol oxidates in plants. *Phytochemistry* 18: 193-215.

Mayer, A.M., Hant E. and Ben-Shaul, R. 1965. Assay of catechol. oxitase: A critical comparasion of methods. *Phytochemistry* 5: 783-789.

Meudt, W.J. and Gaines T.P. 1967. Studies on the oxidation of indole-3-acetic acid by peroxidase enzymes. I. colorimetic determination of indole-3-acid oxidation products. *Plant Physiology* 42: 1395-1399.

Murata, T., Akazawa, T. and Fukuchi, S. 1968. Enzymic mechanism of starch breakdown in germinating rice seeds. I. Analytical Study. *Plant Physiology* 43: 1899-1903.

Murray, A.K. and Bandurski, R.J. 1975. Correlative studies in cell wall enzymes and growth. *Plant Physiology* 56: 143-147.

Murray, D.R. and Collier, M.D. 1977. Acid phosphatese activities in developing seeds of *Pisum sativum*. *Australian Journal of Plant Physiology* 4: 843-848.

Pooviah, B.W. and Rasmussen, H.P. 1973. Peroxidase activity in the absicission zone of bean leaves during abscission. *Plant Phsyiology* 52: 263-267.

Riov, J. 1974. A polygalacturonase from citrus leaf explants: role in abscission. *Plant Physiology* 53: 312-316.

Rocha, V. and Ting, I.P. 1971. Malate dehydrogease of leaf tissue from *Spinacea oleracea* : properties of three isoenzymes. *Archives of Biochemisty and Biophysics* 14: 114-119.

Salmia, M.A. and Mikola, J.J. 1976. Localization and activity of napthylamidases in germinating seeds of scots pipe. *Physiologia Plantarum* 38: 73-77.

Salmia, M.A., Nyman, S.A. and Mikola, J.J. 1978. Characterization of proteins present in germinating seeds of scots pine, *Pinus sylvestrus*. *Physiologia Plantarum* 42: 157-162.

Siegel, B.Z. and Galston, A.W. 1960. The isoperoxidase of *Pisum sativum*. *Plant Physiology* 42: 221-226.

Singh, M.B. and Malik, C.P. 1976. Isocitrate lyase - a new enzyme for pollen. *Indian Journal of Experimental Biology* 15: 357-359.

Sopanen, T. and Lauriers, C. 1976. Activity of various peptidases in the leaf of wheat. *Physiologia Plantarum* 36: 251-254.

Summer, J.B. 1935. A more specific reagent for the determination of sugar in urine. *Journal of Biological Chemistry* 69: 393-407.

Tuubol, K.K., Wiener, G. and Hudsen, P.B. 1957. Acid phosphatase VII. Yeast phosphomonoestrase, isolation, procedure and flability characteristics. *Journal of Biological Chemistry* 224: 621-622.

Yang, N.S. and Scandalions, J.G. 1974. Purification and biochemical properties of genetically defined malate dehydrogenase in maize. *Archives of Biochemistry and Biophysics* 161: 335-340.

CHAPTER - 9

PHOTOSYNTHETIC ENZYMES AND IMPORTANT METABOLITES

INTRODUCTION

Photosynthetic fixation of CO_2 produces assimilates and storage carbohydrates needed for production of energy (thorugh mitochondorial respiration). Since enzymes catalyse the production/utilization of these intermediates, changes in their levels would reflect the changes in the overall process of photosynthesis. It is therefore, essential to measure the activities of CO_2 fixing enzymes and levels of important metabolites for exploiting the photosynthetic potential of the plants. Any change in the production and utilization of these assimilates would adversely affect the growth and development vis-a-vis yield of the test plant. This is addressed in the present chapter.

CO_2 FIXING ENZYMES

The reducing power (reduced pyridine nucleotide NADPH) and ATP generated during the light phase of photosynthesis convert CO_2 into carbohydrates by three different mechanisms: (a) Calvin cycle (also called C_3 pathway or photosynthetic carbon reduction (PCR) cycle), (b) C_4 pathway and (c) the crassulaean acid metabolism (CAM) pathway. These mechanisms involve a series of enzyme catalyzed reactions for photosynthetic CO_2 fixation in higher plants. However, net fixation of CO_2 into carbohydrates is essential for the growth of the plants and it is achieved by Calvin cycle (Walker, 1974). Since C_4 and CAM pathways also use the Calvin cycle to convert CO_2 into carbohydrates, therefore, these three mechaniss are considered as one basic mechanism (C_3 pathway) with two variants (C_4 and CAM pathways). However, the two variants also involve additional enzyme catalyzed steps whose nature, the time during the day when they occur and their cellular location distinguish C_4 and CAM pathways from each other.

Evidence that these three pathways exist in higher plants, rests on the detection and on the results of $^{14}CO_2$ (radiolabelled carbon dioxide) feeding experiments. In other words, measurement of the enzyme activities has become one of the major approaches for the elucidation of various pathways of CO_2 assimilation. Thirteen enzyme catalyzed reactions constitute the Calvin cycle of photosynthesis, however, here we emphasis only key enzymes viz., RuBP carboxylase, NAD(P) glyceraldehydes-3-phosphate dehydrogenase, ribulose-5-phosphate kinase, phosphoglycerate kinase and fructose 1, 6-bisphosphatase. Similarly, PEP carboxylase, NAD(P) malate dehydrogenase, NAD(P) malic enzyme, glutamate oxaloacetate transaminase (GOT) and glutamate pyruvate transaminase (GPT), pyruvate diP dikinase and PEP carboxykinase have been selected from C_4 and CAM pathways. Different assay methods have been used by different workers for measurement of these enzyme activities.

Principle

The enzyme activity is determined spectrophotometerically at 340 nm by coupling with commercially available standard enzymes and following either the oxidation of NAD(P)H or the reduction of NAD(P). The amount of product formed is then calculated from a standard curve prepared using NAD(P)H.

Materials

Plant material under study, chemicals of high purity analytical grade, glassware, pestle, mortar, high speed refrigerated centrifuge, spectrophotometer preferably Cal Biochem LC-340B photometer (which measures absorbance only at 340 nm) or any spectrophotometer having time scanning.

Procedure

Preparation of the enzyme extract : Plant material (0.5 g) is hand homogenized at 0-4°C in a pre-chilled mortar and pestle using acid washed quartz sand with 50 mM imidazole-HCl bufer (pH 8.0) containing 10 mM DTT, 20 mM $MgCl_2$, 2 mM EDTA and 1% (w/v) PVP. The resulting homogenate is centrifuged at 12000 x g for 30 min at 0-4°C. The supernatant is further passed through Sephadex G-25 column and final volume made to 5.0 ml. This enzyme extract is used for determination of enzyme activities.

Enzyme assays

An enzyme assay is the reaction mixture in which the substrate is converted to an end product under optimum physiological conditions. Depletion of the substrate or the formation of the end product is measured to determine the activity of a particular enzyme.

(1) ENZYME OF C_3 PATHWAY

C_3 pathway is the basic mechanism for photosynthetic fixation of CO_2 to carbohydrates and those plants which use it exclusively are called C_3 plants. Such plants generally grow in temperate regions of the world and their productivity is lower than C_4 plants. Thirteen enzymes catalyze sequential reactions in C_3 pathway. However, RuBP carboxylase, NAD(P)-glyceraldehyde-3-phosphate dehydrogenases, 3-PGA kinase and fructose 1-6-bisphosphatase are key enzymes of Calvin cycle. The assay methods for these are discussed here.

1. RuBP carboxylase (EC 4.1.1.39) : RuBP carboxylase catalyzes the key reaction in Calvin cycle in which CO_2 reacts with RuBP to form two molecules of PGA. RuBP carboxylase constitutes about 50% of the soluble proteins in the leaves and as such must be the most abundant protein in nature. It was originally distinguished from the other soluble leaf proteins by analytical ultra centrifuguation and was called Fraction I protein.

Reactions involved

$$\text{R-5-P} \xrightarrow{R-5-P\ isomerase} Ru-5-P$$

$$Ru-5-P+ATP \xrightarrow{Ru-5-P\ kinase} RuBP + ADP$$

$$RuBP + {}^{*}CO_2 \xrightarrow{RuBP\ carboxylase} 3-PGA$$

(*in the form of $NaHCO_3$)

$$\text{3-PGA+ATP} \xrightarrow{\text{3-PGA kinase}} \text{1, 3-diPGA + ADP}$$

$$\text{1, 3-diPGA + NADH} \xrightarrow{\text{GAP dehydrogenase}} \text{GAP + NAD + Pi}$$

The assay mixture (Lilley and Walker, 1974) in a total volume of 1.5 ml contain : 50 mM HEPES buffer (pH 7.8), 10 mM KCl 1 mM EDTA, 15 mM $MgCl_2$, 5 mM ATP, 5 mM DTT, 0.2 mM NADH, 20 mM $NaHCO_3$, 2 units each of GAP dehydrogenase, R-5-P isomerase and 3-PGA kinase and an approximate amount of the enzyme extract. The reaction is started by adding R-5-P. However, the reaction mixture for radiochemical assay (Bjorkman, 1968) in a final volume of 0.42 ml contains (umoles) : $NaH^{14}CO_3$ = 2.5, GSH = 1.25, EDTA = 0.1, tris-HCl (pH 5.2) = 1.0, $MgCl_2$ = 2.5 and RuBP = 0.15. The reaction is started by adding a suitable aliquot of the enzyme extract and terminated after 10 min of incubation by adding 0.5 ml 2 N HCl. A suitable aliquot of the reaction mixture is then dried on a steel planchet under infra red lamp and radioactivity is determined.

2. NAD-glyceraldehyde-3-P dehydrogenase (EC 1.2.1.12) : 3-PGA resulting from RuBP carboxylase reaction is then phosphorylated to give 1,3-DiPGA. ATP is the phosphorylating agent and 3-PGA kinase is the enzyme to catalyze the reaction. 1, 3-DiPGA is then reduced to glyceraldehydes -3-P by NAD-linked glyceraldehyde-3-dehydrogenase.

$$\text{3-PGA+ATP} \xrightarrow{\text{3-PGA kinase}} \text{1, 3-diPGA+ADP}$$

$$\text{1, 3-diPGA + NADH} \xrightarrow{\text{GAP dehydrogenase}} \text{GAP + NAD + Pi}$$

The assay mixture (Latzko and Gibbs, 1968) in a total volume of 1.5 ml contain : 10 mM tris-HCl buffer (pH 7.8), 10 mM $MgCl_2$, 5 mM GSH, 5 m ATP, 3-PGA kinase (2 units), 0.2 mM NADH, 20 mM 3-PGA and the enzyme extract. The reaction is initiated by adding 3-PGA.

3. NADP-glyceraldehydes-3-P dehydrogenase (EC 1.2.1.13) : It is similar to above enzyme in action, except that it uses NADP to reduce 1, 3-DiPGA.

$$\text{3-PGA+ATP} \xrightarrow{\text{3-PGA kinase}} \text{1, 3-diPGA + ADP}$$

$$\text{1, 3-diPGA + NADPH} \xrightarrow{\text{GAP dehydrogenase}} \text{GAP + NADP + Pi}$$

The composition of the reaction mixture of this enzyme is the same as that of NAD-glyceraldehyde-P-dehydrogenase except that 0.2 mM NADH is replaced by 0.2 mM NADPH.

4. Ribulose-5-P kinase (EC 2.7.1.19) : This enzyme catalyzes the last reaction of Calvin cycle, i.e., regeneration of RuBP by phosphorylation of Ribulose-5-P, which is produced during either ribulose phosphate 3-epimerase or ribose-5-phosphate isomerase catalysed reaction.

$$\text{R-5-P} \xrightarrow{\text{R-5-P isomerase}} \text{Ru-5-P}$$

$$\text{Ru-5-P+ATP} \xrightarrow{\text{Ru-5-P kinase}} \text{RuBP + ADP}$$

$$\text{ADP + PEP} \xrightarrow{\text{Pyruvate kinase}} \text{Pyruvate + ATP}$$

$$\text{Pyruvate + NADH} \xrightarrow{\text{LDH}} \text{Lactate + NAD}$$

This assay mixture (Racker, 1957) in a total volume of 1.5 ml contain : 100 mM Tris-HCl buffer (pH 7.8), 10 mM $MgCl_2$, 20 mM KCl, 1 mM ATP, 1.5 mM PEP 0.2 mM NADH, 10 mM DTT, Ribose-5-P-isomerase (3 units), 3 mM R-5-P and the enzyme extract. The reaction is started by adding R-5-P.

5. Phosphoglycerate kinase (EC 2.7.2.3) : 3-PGA, the primary product of photosynthetic CO_2 fixation, is phosphorylated by 3-PGA kinase using ATP generated during light phase of photosynthesis. 1, 3-DiPGA so produced is then measured by coupling with another enzyme catalyzes reaction through 3-PGA kinase catalyses the freely reversible reaction but continuous production of ATP during light phase drives the reaction in the direction of 1, 3-DiPGA.

$$\text{3-PGA+ATP} \xrightarrow{\text{3-PGA kinase}} \text{1, 3-diPGA + ADP}$$

$$\text{1, 3-diPGA + NADH} \xrightarrow{\text{GAP dehydrogenase}} \text{GAP + NAD + Pi}$$

The assay mixture (Latzko and Gibbs, 1968) in a total volume of 1.5 ml contain 150 mM Tris-HCl (pH 7.4), 15 mM $MgCl_2$, 7.5 mM GSH, 7.5 mM ATP, 0.3 mM NADH, GAP dehydrogenase (2 units), 3 mM 3-PGA and the enzyme extract. The reaction is started by adding 3-PGA.

6. Fructose-1, 6 bisphosphatase (EC 3.1.3.11) : Fructose-1, 6-bisphosphate is produced by conversion of glyceraldehyde-3-P and Dihydroxyacetone phosphate catalyzed by aldolase enzyme. It is then dephosphorylated by Fructose 1-6-bisphosphatase to yield fructose-6-Phosphate. This fructose-6-P is then estimated by coupling with two other enzyme catalyzed reactions as shown here :

$$\text{F-1, 6-BP} \xrightarrow{\text{F-1, 6 bisphosphatase}} \text{F-6-P + Pi}$$

$$\text{F-6-P} \xrightarrow{\text{Phosphoglucoisomerase}} \text{G-6-P}$$

$$\text{G-6-P + NADP} \xrightarrow{\text{G-6-P dehydrogenase}} \text{6-P-gluconate + NADPH}$$

The assay mixture (Latzko *et al.*, 1974) in a total volume of 1.5 ml contain : 100 mm tris-HCl (pH 8.8 for chloroplast FBPase), 10 mM $MgCl_2$, 2 mM EDTA, 0.2 mM NADP, 2 units each phosphoglucoisomerase and glucose -6-P dehydrogenase and the reaction is started by adding F-1, 6-BP.

(2) ENZYMES OF C_4 AND CAM PATHWAYS

Although, the C_4 mode of photosynthesis was discovered in two tropical grasses (sugarcane and maize), but it occurs in many other species of gramineae, some species of Cyperaceae and in some species of dicotyledons families, Aizoaceae, Amaranthaceae, Chenopodiaceae, Compositeae, Euphorbiaceae etc. These C_4 plants grow optimally at high light intensities and at day temperatures of 30-35°C. The first proof that the mechanism of photosynthetic CO_2 fixation in such plants is different from that of C_3 plants came in 1965 when it was shown that the C_4 dicarboxylic acids (L-malic acid and L-asparatic acid were labeled first when sugarcane leaves were allowed to photosynthesize in $^{14}CO_2$ and that there was a distinct lag phase before ^{14}C appeared in 3-PGA and the remaining Calvin cycle intermediates. On the other hand, plants which exhibit crassulacean acid metabolism (CAM) are succulents. Many belong to dicotyledonous family Crassulaceae but other belong to monocotyledonous families.

7. PEP carboxylase (EC 4.1.1.31) : CO_2 present in the air in leaf intercellular spaces gets dissolved and ionizes in the cytoplasm of mesophyll cells under the catalytic influence of carbonic anhydrase. The HCO_3 so produced is then used by PEP carboxylase to carboxylate PEP with the formation of oxaloacetic acid, which is then estimated by coupling with MDH catalyzes reaction.

$$\text{PEP} + \text{HCO}_3 \xrightarrow{\text{PEP carboxylase}} \text{OAA} + \text{Pi}$$

$$\text{OAA} + \text{NADH} \xrightarrow{\text{Malate dehydrogenase}} \text{Malate} + \text{NAD}$$

The assay mixture (Hatch, 1972) in a total volume of 1.5 ml contain : 50 mM tris-HCl (pH 8.1), 5 mM $MgCl_2$, 2 mM DTT, 0.2 mM NADH, 20 mM $NaHCO_3$, MDH (2 units), 5 mM PEP and the enzyme extract. The reaction is initiated by adding PEP.

However, the radiochemical assay mixture in a total volume of 1.6 ml contain (mmoles) : $NaH^{14}CO_3$ = 0.5, 2-mercaptoethanol = 1.0, tris-HCl (pH 5.2) = 1.0, $MgCl_2$ = 0.5 and PEP = 0.2. The reaction is started by adding an aliquot of the enzyme extract and terminated after 10 min of incubation by adding 0.1 ml 10% TCA. A suitable aliquot of the reaction mixture is then dried on a steel planchet under infra-red lamp and radioactivity determined.

8. NADP-malate dehydrogenase (EC 1.1.1.82) : Oxaloacetic acid produced in PEP carboxylase reaction is passed into mesophyll chloroplasts and is rapidly converted to malate by NADP malate dehydrogenase using NADP produced in light reaction. NADP malate dehydrogenase is activated in presence of light.

$$\text{OAA} + \text{NADPH} \xrightarrow{\text{NADP} - \text{MDH}} \text{Malate} + \text{NADP}$$

The assay mixture (Hatch and Slack, 1966) in a total volume of 1.5 ml contain : 50 mM tris-HCl (pH 8.0), 1 mM EDTA, 0.2 mM NADPH, 0.5 mM OAA and the enzyme extract. The reaction is started by adding oxaloacetate.

9. NADP-Malic enzyme (EC 1.1.1.40) : In some C_4 plants (NADP-Malic enzyme species), the malic acid passes from chloroplasts of mesophyll cells to chloroplasts of bundle sheath cells probably by way of cytoplasm of plasmodesmata which connect the two types of cells. It is then decarboxylated by NADP specific malic enzyme to yield pyruvate and CO_2. The CO_2 then becomes the substrate for RuBP carboxylase.

$$\text{Malate} + \text{NADP} \xrightarrow{\text{NADP-ME}} \text{Pyruvate} + CO_2 + \text{NADPH}$$

The assay mixture (Johnson and Hatch, 1970) in a total volume of 1.5 ml contain : 50 mM tricine buffer (pH 8.0), 1.0 mM EDTA, 0.25 mM NADP, 2.5 mM malate, 5.0 mM $MgCl_2$ and the enzyme extract. The reaction is started by adding malate.

10. NAD Malate dehydrogenase (EC 1.1.1.37) : In some other C_4 plants (NAD-malic enzyme species) oxaloacetic acid in the mitochondria is reduced by the NAD-specific malate dehydrogenase to malic acid.

$$\text{OAA} + \text{NADH} \xrightarrow{\text{NAD-MDH}} \text{Malate} + \text{NAD}$$

NAD-malate dehydrogenase is estimated in the same way as NADP-MDH except that 0.2 mM NADPH is replaced by 0.2 mM NADH.

11. NAD-Malic enzyme (EC 1.1.1.39) : In some C_4 plants (NAD-malic enzyme species), the malic acid in the mitochondria is decarboxylated by mitochondrial NAD species malic enzyme to yield pyruvate and CO_2. The CO_2 diffuses from mitochondria to chloroplasts where it is incorporated into carbohydrates by Calvin cycle.

$$\text{Malate} + \text{NAD} \xrightarrow[\text{Mn}]{\text{NAD-ME}} \text{Pyruvate} + CO_2 + \text{NADH}$$

The assay mixture (Hatch and Kagawa, 1974) in a total volume of 1.5 ml contain : 50 mM imidazole-HCl buffer (pH 7.8), 1.0 mM EDTA, 4.0 mM DTT, 8.0 mM $(NH_4)_2SO_4$, 2 mM NAD, 5.0 mM malate, 5.0 mM $MnCl_2$ and the enzyme extract. The reaction is started by adding malate.

12. Glutamate oxaloacetate transaminase (EC 2.6.1.1) : In NAD-malic enzyme type C_4 plants oxaloacetic acid produced in the cytoplasm of mesophyll cells by PEP carboxylase reaction is transaminated by cytoplasmic glutamate oxaloacetate transaminase to aspartate which is passed into mitochondria of bundle sheath cells via plasmodesmata. There oxaloacetic acid is reduced by the reversal of GOT reaction and is reduced by NAD-specific malate dehydrogenase as mentioned above.

$$\text{Asparatate} + \alpha\text{-ketoglutarate} \xrightarrow{\text{GOT}} \text{OAA} + \text{Glutamate}$$

$$\text{OAA} + \text{NADH} \xrightarrow{\text{MDH}} \text{Malate} + \text{NAD}$$

The assay mixture (Bergmeyer, 1970) in a total volume of 1.5 ml contain : 100 mM tris-HCl (ph 7.8), 35 mM neutralized aspartate, 0.2 mM NADH, malate dehydrogenase (2 units), 7 mM α-ketoglutarate and the enzyme extract. The reaction is started by adding α-ketoglutarate.

13. Glutamate pyruvate transaminase (EC 2.6.1.2) : The pyruvate produced as a result of decarboxylation of malate in the mitochondria of NAD-specific malic enzyme diffuses into the cytoplasm, where it is transminated to alanine by glutamate pyruvate transaminase. The alanine then passes from cytoplasm of bundle sheath cell to that of mesophyll cell via plasmodesmata, where it is converted back to pyruvate by reversal of glutamate pyruvate transaminase reaction. This pyruvate then passes into mesophyll chloroplast for regeneration of PEP.

$$\text{Alanine} + \alpha\text{-ketoglutarate} \xrightarrow{\text{GPT}} \text{Pyruvate} + \text{Glutamate}$$

$$\text{Pyruvate} + \text{NADH} \xrightarrow{\text{LDH}} + \text{Lactate} + \text{NAD}$$

The assay mixture (Hatch and Mau, 1973) in a total volume of 1.5 ml contain : 100 mM HEPES (pH 7.8), 2 mM EDTA, 0.2 mM NADH, 10 mM alanine, lactate dehydrogenase (4 units), 5 mM α-ketoglutarate and an approximate amount of enzyme extract. The reaction is started by adding α-ketoglutarate.

14. Pyruvate Pi dikinase (EC 2.7.9.1) : In NAD-malic enzyme type C_4 plants, the pyruvate is produced by NADP-malic enzyme passes back from bundle sheath chloroplasts to mesophyll chloroplast. It is then converted into PEP by enzyme pyruvate Pi dikinace. PEP so produced then diffuses from chloroplasts into the cytoplasm of mesophyll cells and becomes the substrate for PEP carboxylase.

$$\text{Pyruvate} + \text{Pi} + \text{ATP} \xrightarrow{\text{PPDK}} \text{PEP} + \text{AMP} + \text{PPi}$$

$$\text{PEP} + \text{HCO}_3 \xrightarrow{\text{PEP carboxylase}} \text{OAA}$$

$$\text{OAA} + \text{NADH} \xrightarrow{\text{MDH}} \text{Malate} + \text{NAD}$$

The assay mixture (Andrews and Hatch, 1969) in a total volume of 1.5 ml contains 100 mM tris-HCl (pH 8.0), 10 mM $MgCl_2$, 0.5 mM EDTA, 5 mM DTT, 20 mM $NaHCO_3$, 0.2 mM NADH, 1.25 mM pyruvate, 2.5 mM K_2HPO_4, 1.25 mM ATP, 2 units each of MDH and PEP carboxylase and the enzyme extract. The reaction is started by adding ATP.

15. PEP carboxykinase (EC 4.1.1.32) : In some C_4 plants (PEP carboxykinase type C_4 plants), oxaloacetic acid in the cytoplasm of bundle sheath cells is decarboxylated by PEP carboxykinase to produce PEP and CO_2. The CO_2 diffuses into the bundle sheath chloroplast and is used as substrate for Calvin cycle. PEP is thought to reach mesophyll chloroplasts via pyruvate through alanine, where pyruvate is converted into PEP by pyruvate Pi dikinase reaction as stated above.

$$PEP + HCO_3 + ADP \xrightarrow{\text{PEP carboxykinase}} OAA + ATP$$

$$OAA + NADH \xrightarrow{\text{MDH}} Malate + NAD$$

The assay mixture (Dittrich *et al.,* 1973) in a total volume of 1.5 ml contain : 100 mM HEPES (pH 7.6), 10 mM $MgCl_2$, 10 mM $MnCl_2$, 20 mM $NaHCO_3$, 0.2 mM NADH, 2.0 mM PEP, malate dehydrogenase (2 units), 10 mM ADP and an approximate amount of enzyme extract. The reaction is started by adding ADP.

Calculations

Let 10 ml be the final volume of enzyme extract prepared from 0.5 g of plant material and 0.1 ml of it gives an absorbance change of 0.04 min^{-1} at 340 nm. From the standard curve of NAD(P)H let an absorbance of 0.08 corresponded to 25 nmoles.

Now 0.08 absorbance = 25 nmoles

0.04 absorbance = 12.5 nmoles

If one mole of NAD(P)/NAD(P)H reacts with one mole of substrate to give one mole of product then total enzyme activity in 10 ml of enzyme extract would be

12.5 x 10 x 10 nmoles min^{-1} 0.5 g^{-1} fresh wt.

or 1250 nmoles min^{-1} 0.5 g^{-1} fresh wt.

or 2500 nmoles min^{-1} g^{-1} fresh wt.

or 2.5 μmoles min^{-1} g^{-1} fresh wt.

If we want to express the enzyme activity on protein basis (specific activity) then protein content in the enzyme extract is measured. If 12.5 mg is protein content in the 10 ml of enzyme extract. Then the specific activity of the particular enzyme would be

0.1 μmoles min^{-1} mg^{-1} protein

or 100 nmoles min^{-1} mg^{-1} protein

Similarly, the enzyme activity can also be expressed on chlorophyll content basis.

Enzyme units

If we want to express the enzyme activity in units then select the enzyme activity unit arbitrarily. Let 1 unit of enzyme activity is the amount of enzyme which catalyses the formation of 1 nmole of product then the specific activity as above can be expressed as under :

100 units min^{-1} mg^{-1} protein.

COMMENTS

Three types of assays viz., radiochemical, colorimetric and spectrophotometric are in common use. Radiochemical assays are used particularly for carboxylases. Assays utilizing $H^{14}CO_3$ as substrate and containing radioactivity in the product after acidification of the reaction mixture are often convenient but may be imprecise because of dilution of $H^{14}CO_3$ with endogenous HCO_3^- It is possible to count radioactive OAA directly but this compound decarboxylates at a substantial rate and it is, therefore, preferable to convert OAA to a stable phenyl hydrazone following completion of the reaction (Goatly and Smith, 1974). Use of 2, 4-dinitrophenyl hydrozone is less desirable because of quenching problems. Alternatively OAA can be converted to malate during carboxylation by addition of NADH and coupling the reaction by standard malate dehydrogenase and the oxidation of NADH is recorded spectrophoto metrically at 340 nm. The coupling assay methods are most convenient, less time consuming and generally preferred now a days.

(3) METABOLITES

In plant cells, groups of enzymes work together in sequential chains or systems to carry out a given metabolic process such as fixation of CO_2 to produce sucrose and starch. In such multienzyme systems, the reaction production of the first enzyme becomes the substrate of the next and so on. Three different pathways (viz., Calvin cycle, C_4 pathway and CAM pathway, each of which involves a series of enzyme catalyzed reactions), have been recognized for photosynthetic CO_2 fixation. These metabolic pathways, in the plant cell, have to be strongly controlled as otherwise considerable amount of ATP, triose phosphates etc. present in chloroplast in dark would also permit CO_2 fixation in dark (Heldt *et al.,* 1980).

In each metabolic pathway, there is at least one enzyme `the pace maker' that sets the rate of overall sequence because it catalyzes the slowest or rate limiting step. In most multienzymes systems, the first enzyme of the system is the pacemaker enzyme. Such pacemaker enzymes which are also regulatory in nature, not only have a catalytic function but their activity is often allosterically regulated i.e. specifically affected by low concentration of certain key metabolites which do not have structural resemblance to the substrates for the enzyme. For example, a number of metabolites such as Fru-6-P, FBP, NAD(P)H etc. stimulate and/or inhibit RuBP carboxylase - a key enzyme catalyzing the primary carboxylation reaction of Calvin cycle (Chu and Bassham, 1975; Ladzina, 1978; Tabita and Colletti, 1979; Vacter and Salnikow, 1978; Whiteman *et al.,* 1979). Similarly, glu-6-P, α-glycero phosphate, FBP, 3-PGA and Pi have been reported to be the activators of PEP carboxylase - an enzyme of central importance to C_4 and CAM pathways (Coombs *et al.,* 1973; Ting and Osmond, 1973; Wang and Davies, 1973; Uedan and Sugiyama, 1976; Mukherji, 1977), whereas, oxaloacetate and malate inhibit the enzyme activity (Ting, 1968; Lowe and Slack, 1971; Huber and Edwards, 1975; Bhagwat and Sane, 1976; Singal and Singh, 1986). Thus metabolites play particularly important role in controlling the activity of key enzymes of metabolic pathways.

Principle

Metabolites are more conveniently estimated spectrophotometrically at 340 nm by making use of commercially available standard enzymes ad the assay is based on either the oxidation of NAD(P)H or reduction of NAD(P). However, for the estimation of those metabolites which are present in extremely low concentration in plant preparations, a quantitative enzymic-optical assay is not feasible. Such metabolites are estimated by radioisotopic methods.

Materials

To estimate different metabolites of CO_2 fixation pathways, we need (i) Plant material from which the metabolites are to be estimated, (ii) Biochemicals including standard auxiliary enzymes and other chemicals of high purity analytical grade, (iii) glassware including pestle mortar, (iv) High speed refrigerated centrifuge, (v) spectrophotometer preferably Cal Biochem LC-340B photometer which measures absorbance only at 340 nm and (vi) $NaH^{14}CO_3$ and liquid scintillation counter for measurement of radioactivity.

Procedure

Preparation of the extract

Different workers have used different extraction procedures and the one described here is that used by Latzko and Gibbs (1972).

One g of plant material is hand homogenized in at 0-4°C in a prechilled mortar and pestle using acid washed quartz sand with 20% (w/v) $HClO_4$. After removal of the debris by centrifugation for 5 min at 7000 x g at 0-4°C and the supernatant titrated with 5 M K_2CO_3 to pH 5.5. The precipitated $KClO_4$ is removed by centrifugation for 3 min at 3000 x g at 0-4°C. The clear supernatant is used immediately for estimation of various metabolites.

Estimation of Metabolites

Estimation of metabolites is essential to support the enzyme activity profile. Different metabolites can be estimated using standard enzymes, which act on them.

1. Ribulose-1, 5-bisphosphate (RuBP) : Ribulose-1, 5-bisphosphate is the substrate for the very first enzme RuBP carboxylase of Calvin cycle. It reacts with CO_2 to form two molecules of 3-phosphoglyceric acid.

$$\text{RuBP} + {}^{14}CO_2 + H_2O \xrightarrow{\text{RuBP carboxylase}} 2\ 3\text{ - PGA} + {}^{14}C$$

Since the concentration of RuBP is extremely low in most of plant preparations, it is estimated using radioisotopic assay by incorporating ${}^{14}CO_2$ (in the form of $NaH^{14}CO_3$) in the presence of RuBP carboxylase according to the method of Witz *et al.* (1980). The assay mixture in a final volume of 250 µl contain : 50 mM Tris-HCl (pH 8.1), 20 mM $MgCl_2$, 2 mM $NaHCO_3$, 1 mM DCPIP (included to inhibit phosphoribulokinase that might be present in RuBP carboxylase preparation), 4 µCi $NaH^{14}CO_3$ and 50 mg RuBP carboxylase. Then 625 ml 10 N HCOOH is added and the reaction mixture evaporated to dryness at 50-60°C, redissolved in 100 µl 1 M HCl, reevaporated and to it is added 10 ml scintillation fluid (prepared by mixing 6 g butyl PBD, 600 ml toluene and 400 ml methylglycol). Then radioactivity is counted. In the blank, extract is replaced by equal volume of $KClO_4$ solution saturated at 0°C.

2. Glucose-6-P, glucose-1-P, fructose-6-P and fructose-1, 6-P_2 : Glucose-6-P can be converted into glucose-1-P or fructose-6-P using mutase or isomerase respectively, whereas, fructose-6-P_2 is produced by phosphorylation of fructose-6-P under the catalytic influence of fructose-6-P Kinase

$$\text{Glucose-6-P} + \text{NADP} \xrightarrow{\text{G-6-P dehydrogenase}} \text{6-phospho-gluconate} + \text{NADPH}$$

$$\text{Glucose-1-P} \xrightarrow{\text{Phosphoglucomutase}} \text{Glucose-6-P}$$

$$\text{Fructose-6-P} \xrightarrow{\text{Hexose-P isomerase}} \text{Glucose-6-P}$$

$$\text{Fructose-1, 6-P}_2 \text{ - } \xrightarrow{\text{Fru-1,6-BPase}} \text{Fructose-6-P + Pi}$$

The combined assay of above metabolites is based on the reduction of NADP by glucose-6-P dehydrogenase reaction (Latzko and Gibbs, 1972). The assay mixture in a final volume of 1.0 ml contains 0.1 ml 1 M tris-HCl (pH 8.5), extract, 0.05 ml 5 mM NADP, 0.02 ml 0.5 M $MgCl_2$ and 1 unit glucose-6-P dehydrogenase. Then change in absorbance at 340 nm is noted which would be due to the presence of G-P-P. From a standard curve the amount of metabolite is estimated. After that following components are added in sequence and change in absorbance at 340 nm is noted.

1 unit of hexose-P-isomerase ($(\Delta A_{340} = fructose-6-P)$, 1 unit of phosphogluco mutase (Δ *A_{340} = glucose-1-P) and 0.02 ml of 0.1 M EDTA and 1 unit of FBPase (Δ A_{340} = fructose-1, 6-BP).

In the blank cuvette the extract is replaced by an equal volume of $KClO_4$ solution saturated at 0°C.

3. 3-phosphogluceric acid : 3-Phosphoglyceric acid is the primary product of photosynthetic carbon fixation via Calvin cycle. After phosphorylation by 3-PGA kinase, it is reduced to glyceraldehydes-3-P using NADH generated during light phase of photosynthesis.

$$\text{3-PGA + ATP} \xrightarrow{\text{3-PGA kinase}} \text{Glycerate 1, 3-P}_2 \text{ + ADP}$$

$$\text{Glycerate 1, 3-P}_2 \text{ + NADH} \xrightarrow{\text{Glyceraldehyde-3-P dehydrogenase}} \text{Glyceraldehyde-3-P + NAD}$$

3-PGA is estimated according to Latzko and Gibbs (1972) by coupling two glycolytic back reactions to reduce 3-PGA to glyceraldehydes-3-P consuming stoichiometric amount of ATP and NADH and the oxidation of NADH is followed at 340 nm. The standard assay mixture in a final volume of 1.0 ml contains : 0.1 ml tri-EDTA buffer (0.4 M in 0.2 M EDTA, pH 7.6), test extract, 0.02 ml 0.005 M NADH, 0.1 ml 0.05 M ATP, 0.02 ml 0.5 M $MgCl_2$, 1U Gly-3-P DH and 1U of 3-PGA kinase. In blank the extract is replaced by $KClO_4$ solution saturated at 0°C.

4. Phosphoenol pyruvate (PEP) and pyruvate: Phosphoenol pyruvate and pyruvate play a pivotal role in carbohydrate metabolism. PEP accepts CO_2 in the mesophyll of all C_4 plants. It can be converted to pyruvate via pyruvate kinase reaction, which may be converted to lactate using stoichiometric amounts of NADH.

$$\text{PEP + ADP} \xrightarrow{\text{Pyruvate kinase}} \text{ATP + Pyruvate}$$

$$\text{Pyruvate + NADH} \xrightarrow{\text{Lactate DH}} \text{Lactate + NAD}$$

PEP and pyruvate are measured by converting then enzymatically to lactate with the oxidation of NADH according to Czok and Lamprecht (1974). The oxidation of NADH is proportional to the substrate converted to product and is measured spectrophotometrically at 340 nm. The assay mixture in a final volume of 1.1 ml contain : 0.3 mM Tris-HCl buffer (pH 7.6), 0.1 mM NADH, 3 mM Tris-HCl buffer (pH 7.6), 0.1 mM NADH, 3 mM EDTA, sample extract, 3 U of lactate (DH (ΔA_{340} = Pyruvate) and 2 U of pyruvate kinase (ΔA_{340} = PEP).

In blank, the extract is replaced by $KClO_4$ saturated at 0°C.

5. Oxaloacetate: In C_4 plants carboxylation of PEP via PEP carboxylase produces oxaloacetate, which is very rapidly converted to malate under the catalytic influence of malate dehydrogenase thereby consuming stoichiometric amount of NADH.

$$\text{Oxaloacetate + NADH} \xrightarrow{\text{Malate DH}} \text{Malate + NAD}$$

Oxaloacetate is measured by method of Wahlefeld (1974). The assay mixture in a final volume of 1.5 ml contains : 0.13 M tris-HCl buffer (pH 7.6), 13 mm EDTA, sample extract, 0.06 mm NADH and 2 units malate DH. The reaction is started by adding NADH.

6. Malate : Malate produced by reduction of oxaloacetate is decarboxylated to release CO_2 for refixation by RUBP carboxylase in bundle sheath chloroplsts. The decarboxylation may take place either in chloroplasts (NADP-ME enzyme C_4 plants) or in mitochondria (NAD-ME type C_4 plants).

$$\text{L-malate + NAD} \xrightarrow{\text{Malate DH}} \text{Oxaloacetate + NADH}$$

Malate is estimated by following the method of Gutmann and Wahlefeld (1974). The assay mixture in a final volume of 1.5 ml contains : 0.1 M hydrazine/glycine buffer (pH 9.0), 2.75 mM NAD, sample extract and 4 U of malate DH. The reaction is started by adding NAD.

7. Aspartate : In NAD-ME type C_4 plants asparatate produced by transamination of oxaloacetate is again converted to oxaloacetate in mitochondria bundle sheath cells. Then it is reduced to malate, which is then decarboxylated. In PEP carboxylkinase type C_4 plants asparatate is converted to OAA in cytoplasm of bundle sheath cells through transamination reaction. Oxaloacetate so produced is directly decarboxylated to PEP, which reaches mesophyll cytoplasm via pyruvate and alanine.

$$\text{Aspartate} + \alpha\text{-ketoglutarate} \xrightarrow{\text{GOT}} \text{OAA} + \text{Glutamate}$$

$$\text{OAA} + \text{NADH} \xrightarrow{\text{MDH}} \text{Malate} + \text{NAD}$$

Aspartate is determined according to Bergmeyer *et al.* (1974). The assay mixture in a final volume of 1.5 ml contains : 0.19 mM NADH, 3.1 mM a-ketoglutarate, 0.5 U MDH, sample extract and 0.5 U glutamate oxaloacetate transaminase (GOT). The reaction is started by adding α-ketoglutarate (α-KG).

COMMENTS

Stitt *et al.* (1983) have used a different extraction procedure for the estimation of various metabolites. The plant material is frozen in liquid nitrogen, powdered and extracted for 15 min in 0.6 ml 0.05 M HEPES (pH 8.5) containing 2 mM EDTA, 4 ml methanol and 1.5 ml chloroform and then 4 ml H_2O is added, shaken for 20 sec and centrifuged at 1500 x g for 5 min. Supernatant is dried down at 35 to 40°C in a rotary evaporator, redissolved in 500 ml H_2O and treated with 6 mg activated charcoal. This extract is then used for estimation of metabolites.

REFERENCES

Andrews, T.J. and Hatch, M.D. 1969. Properties and mechanism of action of pyruvate phosphate dikinase from leaves. *Biochemical Journal* 114 : 117-125.

Bergmeyer, H.U. 1970. Glutamate oxaloacetate transaminase from pigheart. In : *Methods of Enzymatic Analysis,* (Ed. H.U. Bergmeyer). pp. 462-463. Academic Press, New York.

Bergmeyer, H.U., Bernt, E., Mollering, H. and Pfleiderer, G. 1974. L-aspartate and L-asparagine. In : *Methods of Enzymatic Analysis* (Ed., H.U. Bergmeyer). pp. 1696-1700. Academic Press, New York.

Bhagwat, A.S. and Sane, P.V. 1976. Studies on enzymes of C_4 pathway; partial purification and kinetic properties of Maize phosphoenol pyruvate carboxylase. *Indian Journal of Experimental Biology* 14 : 155-158.

Bjorkman, O. 1968. Carboxy dismutase activity in shade adapted and of sun-adapted species of higher plants. *Physiologia Plantarum* 21 : 1-10.

Chu, D.K. and Bassham, J.A. 1975. Regulation o ribulose 1, 5-bisphosphate carboxylase by substrate and other metabolites. *Plant Physiology* 55 : 720-726.

Coombs, J., Baldry, C.W. and Bucke, C. 1973. The C_4 pathway in *Pennisetum purpureum*. *Planta* 110 : 95-107.

Czok, R. and Lamprecht, W. 1974. Pyruvate phosphoenolpyruvate and D-glycerate-2-phosphate. In : *Methods of Enzymatic Analysis,* (Ed., H.U. Bergmeyer). pp. 1446-1451. Academic Press, New York.

Dittrich, P., Wilbur, H. and Black, C.C. 1973. Phosphoenolpyruvate carboxylase in plants exhibiting crassulacean acid metabolism. *Plant Physiology* 52 : 357-361.

Goatly, M.D. and Smith, H. 1974. Differential properties of phosphoenolpyruvate carboxylase from etiolated and green sugarcane. *Planta* 117:67-73.

Gutmann, I. and Wahlefeld, A.W. 1974. L-malate determination with malate DH and NAD. In: *Methods of Enzymatic Analysis* (Ed., H.U. Bergmeyer). pp. 1585-1589. Academic Press, New York.

Hatch, M.D. 1972. Synthesis o L-malate-4-C^{14} and determination of label in C_4 carboxyl of malate. *Analytical Biochemistry* 47 : 174-183.

Hatch, M.D. and Kagawa, T. 1974. NAD-malic enzyme in leaves with C_4 pathway of photosynthesis. *Archives of Biochemistry and Biophysics* 160 : 340-349.

Hatch, M.D. and Mau, S.L. 1973. Activity, location and role of aspartate amino transferase and alanine amino transferase in leaves with C_4 pathway of photosynthesis. *Archives of Biochemistry and Biophysics*. 156:195-206.

Hatch, M.D. and Slack, C.R. 1966. Photosynthesis by sugarcane leaves : A new carboxylation reaction and pathway of sugar formation. *Biochemical Journal* 101 : 103-110.

Heldt, H.W., Laing, W., Lorier, G.H., Stitt, M. and Wirtz, W. 1980. On the regulation of CO_2 fixation by light. Proc. 5th *International Congress on Photosynthesis,* Greece.

Huber, S.C. and Edwards, G.E. 1975. Inhibition of phosphoenolpyruvate carboxylase from C_4 plants by malate and aspartate. *Canadian Journal of Botany* 53 : 1925-1933.

Johnson, H.S. and Hatch, M.D. 1970. Properties and regulation of leaf NADP malate dehydrogenase and malic enzyme in plants with C_4 dicarboxylic acid pathway of photosynthesis. *Biochemical Journal* 119 : 273-280.

Latzko, E. and Gibbs, M. 1968. Distribution and activity of enzymes of reductive pentose phosphate cycle in spinach leaves and in chloroplasts isolated by different methods. *Zhurnal Pflanzen Physiologie* 59 : 184-194.

Latzko, E. and Gibbs, M. 1972. Measurement of the intermediates of the photosynthetic carbon reduction cycle using enzymatic methods. In : *Methods in Enzymology* (Ed. A.S. Pietro). 24 : 261-268. Academic Press, New York.

Latzko, E., Zimmerman, G. and Feller, W. 1974. Evidence for a hexose diphosphatase from cytoplasm of spinach leaves. *Hoppe Seylerts Physiological Chemistry* 355 : 321-326.

Ladzina, K.J. 1978. Activation of ribulose 1, 5-bisphosphate carboxylase by chloroplast metabolites in a reconstituted spinach chloroplast system. *Planta* 143 : 291-296.

Lilley, M.C.R. and Walker, D.A. 1974. An improved spectrophotometeric assay for ribulose diphosphate carboxylase. *Biochemica and Biophysica Acta* 358 : 226-229.

Lowe, J. and Slack, C.R. 1971. Inhibition of maize leaf PEP carboxylase by oxaloacetate. *Biochemica Biophysica Acta* 235 : 207-209.

Mukherji, S.K. 1977. Corn leaf phosphenolpyruvate carboxylase : Inhibition of $^{14}CO_2$ fixation by SO_3 and activation by glucose-6-phosphate. *Archives of Biochemistry and Biophysics* 182 : 360-365.

Racker, E. 1957. The reductive pentose phosphate cycle. I. Phosphoribulokinase and ribulose diphosphate carboxylase. *Archives of Biochemistry and Biophysics* 69 : 303-310.

Singal, H.R. and Singh, R. 1986. Purification and properties of phosphoenolpyruvate carboxylase from immature pods of chickpea (*Cicer arietinum* L.). *Plant Physiology* 80 : 369-373.

Stitt, M., Gerhardt, R., Kurzel, B. and HJeldt, H.W. 1983. A role for fructose-1, 6-bisphosphate in the regulation of sucrose synthesis in spinach leaves. *Plant Physiology* 72 : 1139-1142.

Tabita, F.R. and Colleti, C. 1979. Carbondioxide assimilation in Cyanobacteria. Regulation of ribulosae-1, 5-bisphosphate carboxylase. *Journal of Bacteriology* 140 : 452-458.

Ting, I.P. 1968. CO_2 metabolism in corn roots : Inhibition of phosphoenolpyruvate carboxylase by L-malate. *Plant Physiology* 43 : 1919-1924.

Ting, I.P. and Osmond, C.B. 1973. Activation of plant phosphoenolpyruvate carboxylase by glucose-6-P. A particular role in crassulacean acid metabolism. *Plant Science Letters* 1 : 123-128.

Uedan, K. and Sugiyama, T. 1976. Purification and characterization of phosphoenol-pyruvate carboxylase from maize leaves. *Plant Physiology* 57 : 906-910.

Vacter, J. and Salnikow, J. 1978. Identification of two binding sites of the D-ribulose-1, 5-bisphosphate and effectors of carboxylation reaction. *Archives of Biochemistry and Biophysics* 194 : 190-197.

Wahlefeld, A.W. 1974. Oxaloacetate : UV spectrophotometric determination. In : *Methods of Enzymatic Analysis* (Ed., H.U. Bergmeyer). pp. 1604-1608. Academic Press, New York.

Walker, D.A. 1974. Some characteristics of primary carboxylation mechanism. In : *Plant Carbohydrate Biochemistry* (Ed. J.B. Pridhan) pp. 7-25. Academic Press, New York.

Wang, K.F. and Davies, D.D. 1973. Regulation of phosphoenolpyruvate carboxylase of *Zea mays* by metabolites. *Biochemical Journal* 131 : 451-45.

Whiteman, W.B., Martin, M.N. and Tabita, F.R. 1979. Activation and regulation of ribulose bisphosphate carboxylase-oxygenase in absence of small subunits. *Journal of Biological Chemistry* 254 : 10184-10189.

Witz, W., Stitt, M. and Heldt, H.W. 1980. Enzymic determination of metabolites in sub-cellular compartments of spinch protoplasts. *Plant Physiology* 66 : 187-193.

Chapter - 10

ENZYMES OF INFECTED PLANTS AND PARASITES

INTRODUCTION

Different extraction procedures are followed to extract enzymes from tissues : gridning in (i) cold distelled water (ii) chilled buffer and (iii) chilled acetone. The `enzyme extract' prepared by these methods in fact contains a mixture of enzymes. The first method should be avoided if possible, as some amount of inactivation may occur. The second method involves numerous trials and errors to select the appropriate buffer and optimum pH for maximum enzyme activity. This method is better than the first, however, the extract should be used within 1-2 days of preparation. Acetone powder offers many advantages. The powder can be stored at –20°C for a long time without any loss of enzyme activity. The enzyme can be purified from the powder by dissolving it in buffers of different pH. Finally several enzymes may be assayed in presence of appropriate substrates.

Extraction of Enzymes from Plant Tissues

Many methods have been evolved to extract enzymes from plants. Some of them are complex while a few are simple, requiring only buffers. With the availability of protein purifying gels and columns, isolation and purification of enzymes have become simple.

Extraction of Enzymes in Buffer

Materials

Leaves of experimental plants

Blender or pestle and mortar

Pipettes 10 ml, 5 ml

Cheese cloth

Refrigerated centrifuge

Reagents

Phosphate buffer, pH 6.6 (use suitable buffers with selected pH depending upon the enzyme).

Method

Weigh the tissues; cut into 1-2 mm pieces, transfer to a blender and add chilled (2-4°C) phosphate buffer; use 5 ml for 1 g of tissue. Grind at low speed for 2-3 min; stir the contents and grind at high speed for 2-3 min. Squeeze the extracts through 3 layers of cheese cloth to remove the pulp and centrifuge for 30 min at 2000 g at 4°C. Decant the supernatant and use the clear extract as enzyme source.

Estimate the protein content of the extract.

Preparation of Acetone Powder

Materials

As above

Buchner funnel

Reagents

Acetone

Diethyl ether

Phosphate buffer, pH 6.6 (use suitable buffers with selected pH depending upon the enzyme)

Method

Weigh the tissues; cut into pieces of 1-2 cm, transfer to a blender, add chilled acetone (–20°C) enough to cover the tissues, blend for 3-5 min at high speed, filter through Buchner funnel using Whatman No. 1 filter paper and wash the powder with chilled acetone in the Buchner funnel at least 3 times. Wash again with cold diethyl ether. Dry the powder on the funnel with suction, spread the powder on Whatman No. 1 filter paper and air dry for about 1 h Store the powder in containers with tight caps in a freezer. This powder can be used for a period of 6 months without loss of activity.

Just before assay, weigh 0.1 g of the powder, grind in 5 ml of phosphate buffer (0.1 M) at 4°C for 10-15 min in a mortar with pestle. Centrifuge the extract at 2000 rpm for 30 min at 4°C and use the supernatant as the enzyme source.

Comment

If no blender is available, use mortar and pestle to grind the tissues. Do all the work in the cold. If no cold room is available, chill the buffer, blender, beakers and flasks in a refrigerator. Do the grinding in an insulated box with ice.

Estimation of Protein in Plant Tissues and Enzyme Extracts

Proteins present in plant tissues and enzyme extracts need to be separated by precipitation and clarified from other interfering substances prior to their estimation.

Preparation of samples

Materials

Fresh plant leaves or enzyme extract

Pipettes 1 ml, 5 ml

Pestle and mortar

Centrifuge

Ice water bath

Reagents

Absolute ethanol, 80% ethanol

Ethanol-ether mixture (3 : 1, v/v)

Trichloroacetic acid (TCA), 5%, 10% solutions

Method

Extract the tissues in hot 80% ethanol and macerate in a mortar with pestle. Transfer the homogenate to centrifuge tubes, centrifuge at 2000 g for 20 min and discard the supernatant. Suspend the pellet in a suitable volume of 5% TCA in an ice bath for 15 min. While processing enzyme extracts for the estimation of buffer soluble proteins, pipette 1 ml aliquots into centrifuge tubes and add 1 ml of 10% TCA to precipitate the proteins. Allow to stand for 15 min in the ice water bath. Centrifuge and discard the supernatant. Repeat this process twice. Re-extract the pellet once with hot absolute ethanol and twice with ethanol-ether mixture, every time discarding the supernatants after centrifugation. This pellet contains proteins and nucleic acids.

Lowry's Method for Protein Estimation

The method is based on the principle that different proteins contain different amounts of aromatic residues which react with Folin-phenol reagent giving a blue colour and is read in a colorimeter (Lowny *et al.*, 1951).

Materials

Protein sample

Pipettes, 1 ml, 5 ml

Test tubes

Water bath with heater

Colorimeter / spectrophotometer

Reagents

Reagent A

Alkaline sodium carbonate solution (2% Na_2CO_3 in 0.1 N NaOH)

Reagent B

Copper sulphate-sodium potassium tartrate solution (0.5% $CuSO_4.5H_2O$ in 1% sodium potassium tartrate; prepare fresh)

Alakline copper reagent : Mix 50 ml of reagent A and 1 ml of reagent B.

Prepare fresh.

Folin-Phenol reagent, either commercially available or prepared in the laboratory.

Dilute the reagent with equal volume of water just before use.

1 N NaOH

Bovine serum albumin or Casein

Method

Place the protein samples in 1 ml of NaOH at 100°C for 4 to 5 min. Add 5 ml of alkaline copper reagent and allow it to stand at room temperature for 10 min (minimum time) or longer. Add 0.5 ml of follin-phenol reagent rapidly and mix immediately. After 30 min, measure the absorbance at 750 nm in a colorimeter. Calculate the amount of protein in the samples with a standard curve prepared using bovine serum albumin or casein solution of different concentrations.

Comment

The time to hydrolyse the sample with the alkali should be kept constant.

Bradford Method for Protein Estimation

In the recent years, the method developed by Bradford has replaced Lowry's method. The method is based on the ability of proteins to bind to the dye Coomassie brilliant blue G-250 and forms a complex whose extinction coefficient is greater than the free dye.

Materials

As in Lowry's method.

Micropipettes

Reagents

Dye solution

Mix 10 mg of Coomassie brilliant blue G-250 with 10 ml of 88% phosphoric acid and 4.7 ml of absolute alcohol, dilute the mixture to 100 ml with dist water. Check absorption at 550 nm. It should be 1.18. If this value is not reached, adjust as required with dye or water. Since the dye solution is almost saturated, a precipitate may form with storage for long period. If this happens, filter the solution through whatman No. 1 filter paper and recalibrate the dye concentration.

Method

Prepare tubes containing 50 uL of increasing dilution of the sample. Add 950 uL of the dye solution to all tubes. Incubate the tubes at room temperature for 5 min. Read the absorbance at 595 nm against a blank containing 50 uL of protein buffer and 950 uL of dye solution.

Use a standard curve prepared with 0.1 to 5 ug of bovine serum albumin to get a final volume of 50 uL.

Hydrolytic Enzymes

The macromolecules of tissues are cleaved by various hydrolytic enzymes and these are important in pathogenesis. Studies on the occurrence of these enzymes during disease development and stress are helpful in understanding symptom development and plant responses.

Pectic Enzymes of Plant Parasites and of Plant Tissues

Most necrotrophic parasites produce pectic (pectinolytic) enzymes in culture media and in the inoculated plants, compared with biotrophs. Parasites secrete the enzymes either inductively in presence of substrates or noninductively. Efforts have been made to correlate pectic enzyme production and pathogenesis. Attempts have also been made to correlate that pectic enzymes have nothing to do with pathogenesis (Mahadevan, 1979).

Pectic substances are found in primary cell walls and in the middle lamella. The simplest monomer of pectin is galacturonic acid and it is linked in an (1-4) fashion. Pectic substances are classified according to the degree of polymerization of galacturonic acids. The simplest classification of pectic substances is : pectic acid, pectin and protopectin. Pectic acid is composed of about 100 galacturonic acid units, their large number of carboxyl groups react with polyvalent cations such as Ca. Pectin (pectinic acid) is made up of about 200 polygalacturonic acid units and the carboxyl groups are esterified with ethoxyl groups, the degree of methoxylation varies from 50-80%. Protopectin is the major component of pectin. It is composed of ca 1,000-2,000 units of esterified carboxyl groups and has bound calcium and phosphate.

Pectin as such does not diffuse into cells but galacturonic acids readily penetrate and are utilized as a carbon by pathogens. Hence, cleavage of pectic substances into smaller units of galacturonic acids is the first act of pathogen and this cleavage is mediated by both hydrolytic and nonhydrolytic enzymes (Mahadevan and Kathirvelu, 1967). In the hydrolytic attack cleavage takes place at the -1, 4-glycosidic linkages resulting in galacturonic acids of different size. Depending upon substrate specificity, the enzyme that cleavage pectin acids is called polymethyl galacturonase (PMG) while polygalacturonase (PG) cleavage pectic acids and derivatives. The enzymes that remove the methoxyl groups are called pectin methyl esterases or pectinesterases (PME).

The nonhydrolytic break down of pectic substances involves the participation of two enzymes, pectin trans eliminase (pectate lyase, PTE) if it acts on pectin of polygalacturonate trans eliminase or pectic acid trans eliminase (pectate lyase PATE) acting on pectic substances. The enzymes act by way of removal of H atom a C-5 position of galacturonic acid chain, formation of double bond between C-4 and C-5 and elimiation of glycosidic residue at C-4, resulting in the formation of a single unsaturated monouronic acid, 4 deoxy-1-threo-5-hexosulose uronic acid and a homologous series of oligouronides.

These enzymes can be individually assayed by employing their appropriate substrates. So far PG, PME, PTE and PATE have been detected in the culture filtrates of plant parasites and in the infected plants (Mahadeven and Kathirvelu, 1967; Mahadevan, 1979).

Preparation of Pectic Enzymes of Parasites

In the presence of pectice substances, *Botrytis, Cladosporium, Erwinia, Fusarium, Pythium, Rhizoctani* and *Verticillium* produce pectic enzymes in culture media.

Materials

Erlenmeyer flasks, 500 ml

Centrifuge

Buchner funnel

Pipettes, 5 ml

Whatman No. 1 filter paper

Cellophane tubing

An isolate of *Fusarium* or *Pythium*

Method

Inoculte 100 ml of Czapek's medium containing 2-4% pectin adjusted to pH 6.5 in 500 ml Erlenmeyer flasks with 2 ml of spore suspension (ca, 4000 spores/ml) of *F. oxysporum F. vasinfectum* or *Pythium* sp. Incubate the flask at 24-26°C for 10-15 days; collect the culture filtrates and centrifuge at 2000 g for 30 min to remove spores. Use the supernatant. Dialyze ther filtrate in cellophane tubing against dist water in a refrigerator for 24 h changing the water every 3 h. Add a few drops of toluene to the dialyzed filtrate and use as enzyme source. Keep the filtrate at 2-4°C when not in use.

Pectic Enzymes from Disaesed Plants

Materials

Wilt-infected cotton or beans inoculated with *Rhizoctonia solani* showing typical symptoms.

Blender or mortar and pestle

Pipettes, 5 ml

Centrifuge

Cheese-cloth

Reagents

NaCl, 0.15 M

Toluene

Method

Weigh 20 g tissue; cut into pieces of 1-2 cm, transfer to a blender or mortar and add NaCl at the rate of 3-5 ml for every g tissue. Blend for about 3-5 min, filter through two layers of cheese-cloth, centrifuge at 2000 g for 30 min and dialyze the supernatant in cellophane tubing against several volumes of dist water for about 24 h at 0-2°C. Change the water at least two times during dialysis. Use the dialysed solution as crude pectic enzyme. Add a new drops of toluene and store at 4°C. Do not use the enzyme extract for more than one week.

Measurement of Polygalacturonase (PG)

PG cleaves hydrolytically the glycoside bonds of pectic substances, release galacturonic acids of different size. The most frequently used substrates are polypectate and pectic acid (Dube and Bordia, 1983).

The enzyme is assayed by a physical method, based upon the loss in viscosity the substrate, or chemically measuring the amount of galacturonic acids. The physical method is rapid but not specific for PG. Other enzymes, particularly PATE, contribute to viscosity reduction.

Viscosity assay

Materials

Viscometer size 300 Ostwald - Fanske or locally fabricated viscometer with minimum efflux time of 10 sec for water is equally good.

Pipettes, 5 ml

Beakers, 250 ml

Water bath, 30 ±1°C

Stop watch

Blender

Cheese-cloth

pH Meter

Reagents

Sodium polypectate substrate : Dissolve 750 mg of the chemical in 100 ml acetate buffer, pH 5.2, heat to 50-60°C and place in a blender. Blend for 2-3 min at low speed and to high speed for another 3 min. Pass through two layers of cheese-cloth. Check the pH of the solution in a pH meter and adjust using HCl or 1 N NaOH. Add a few drops of toluene or 10 ml of 1% merthiolate and keep at 4°C, if not in use. Prepare fresh substrate every week.

Sodium acetate-acetic acid buffer, pH 5.2

Sodium fluoride, 4% aqueous

Method

Pipette 4 ml of substrate into the viscometer. Add 1 ml of acetate buffer, 1 ml of NaF (against bacterial contaminants if the mixture is to be incubated for several hours and 2 ml of the enzyme source. Gently mix the contents by drawing air rapidly through the large arm of the viscometer by suction. Apply suction to viscometer through the small arm and measure the efflux time of the mixture, between A & B. This is the zero time. Measure the efflux time of the mixture at various intervals.

Calculate the per cent reduction in viscosity of the substrate from the formula :

$$V = \frac{T_o - T}{T_o - T_w} \times 100$$

Where, V = Per cent loss in viscosity

T_o = flow time in seconds at zero time

T = flow time of the reaction mixture at time T

T_w = flow time of dist water

Comment

If the per cent loss in viscosity is 95-100 within 2 h, omit NaF, if the per cent loss is rapid within 15 min of the assay, dilute the enzyme preparation.

Estimation of Reducing Groups

The release of galacturonic acids can be estimated during enzymatic cleavage of pectic acid.

Materials

Enzyme source

Erlenmeyer flasks, 100 ml

Glass stoppered Erlenmeyer flask, 100 ml

Pipettes, 5 ml

Burette, 25 ml

Water bath, 30±1°C

Reagents

Pectic acid, 0.714% dissolved in sodium acetate-acetic acid buffer pH 5.2

1 M Na_2CO_3 (anhydrous)

0.1 N Iodine

2 M H_2SO_4

0.1 N Sodium thiosulphate

Method

Add 15 ml of the enzyme adjusted to pH 5.2 to 35 ml of pectic acid dissolved in buffer pH 5.2, placed in Erlenmeyer flasks to give a final conc of the substrate to 0.5%. Incubate the mixture at 30°C in a water bath. Remove aliquots of 5 ml at definite time interval and to 0.9 ml of 1 M Na_2CO_4 in a glass stoppered Erlenmeyer flask followed by 5 ml of 0.1 N iodine solution. After exactly 20 min, acidity the reaction mixture with 2 ml of 2 M H_2SO_4. Titrate the residual iodine with 0.1 N sodium thiosulphate.

Calculate the milliequivalent of reducing groups liberated from pectic acid from standard curve prepared from glucose which has same value as that of galacturonic acid monohydrate. Express the enzyme activity in milliequivalents of reducing group liberated from pectic acid/min/ml of the enzyme exctract or mg protein.

Comment

If pectic acid is not available, use sodium polypectate as the substrate.

Maceration of potato medullary cells

Pectic enzymes macerate parenchymatous tissues with relative ease and for such studies, potato medullary tissue is ideal. For a long time, it has been thought the maceration of tissues is brought by a group of enzymes called 'protopectinase' 'maceration enzymes'. Recent evidences reveal that tissue macefration is caused by diverse groups of other enzymes such as cellulolytic, proteolytic and lignolytic enzymes. The recommended approach is to test the end products of macerated cells and determine the enzymes causing maceration. Both hydrolytic and trans eliminative pectic enzymes contribute to cellular maceration. A simple method is described to assess the maceration of potato cells due to hydrolytic enzymes (Mahadevan, 1979).

Materials

Enzyme source

Potato medullary discs of 100 μ thick, 700-800 μ diameter.

(These can easily be cut by means of a hand microtome. Use fresh discs)

Sterile 4-cm Petri dishes

Test tubes

Glass rods

Pipettes, 5 ml, 10 ml

Whatman No. 42 filter paper

Centrifuge

Water bath with heater

Colorimeter

Reagents

Sodium fluoride, 5% aqueous

0.01 M Thiobarbituric acid (TBA)

0.5 N HCl

Method

Pipette 8 ml of the enzyme extract adjusted to pH 5.2 into a Petri dish and 1 ml of NaF; place 5 potato discs and incubate at room temperature or 30°C. Examine from time to time up to 24 h for separation of cells by gently touching the discs with a glass rod. Rank the solutions 1 through 6 in order of the amount of maceration.

Filter the extract through Whatman No. 42 filter paper. Centrifuge at 2,000 g for 30 min to remove the debris. To 3 ml of the clear supernatant add 10 ml of 0.01 M TBA and 5 ml of 0.5 N HCl. Place the tubes in a boiling water bath for 1 h. Remove, cool, make up the volume and measure the abosrbance between 480-580 nm in a colorimeter.

The PG type enzyme hydrolyses pectin derivatives to galacturonic acid residues, which upon reaction with TBA exhibit maximum absrobance at 510 nm. Use samples drawn at zero time and boiled enzyme extract on potato medullary discs as controls.

Comment

If the maceration is due to PG type of enzymes, the release of TBA reaction substances may be quantitatively followed by measuring the absorbance at 510 nm. The assay must include NaF to prevent bacterial contamination.

Polymethyl Galacturonase (Pectin depolymerase)

Polymethyl galacturonse (PMG) hydrolyses esterified pectic substances more rapidly than pectic acid.

Pectin → Polygalacturonide methyl esters

Pectic acid → Polygalacturonic acids

The enzyme action leads to the accumulation of polygalacturonates polygalacturonides.

Like PG, PGM activity can be followed by measuring the viscosity loss of substrate or the reducing groups released from pectin or pectic acid. As pectic acid is also hydrolysed by PG, it is advisable to use pectin as the substrate to assay PG.

Viscosity assay

The basic method to assay PGM activity is the same as that of PG, except 1% pectin as the substrate which is prepared in the following manner : Dissolve 1 g pectin in 100 ml of acetate buffer, pH 5.2 at 50-60°C, kept in a blender. Blend 2-3 min at low speed and at high speed for another 3 min. Pass through two layers of cheese-cloth. Check pH of the solution and adjust if necessary with 1 N HCl or NaOH. Use a viscometer of size 150 or any viscometer with minimum efflux time 20 sec for water.

Estimation of Reducing Groups

Follow the procedure given for PG, but use pectin as the substrate.

Pectin Methyl Esterase (Pectinesterase)

Pectin methyl esterase (PME) cleaves pectin resulting in pectic acid and methanol.

Pectin → Pectic acid + Methanol

Either methanol or increase in free carboxyl groups can be measured. Estimation of methanol is cumbersome, while the release of free carboxyl groups is easily monitored by a pH meter.

Measurement of Free Carboxyl Groups

Materials

Enzyme source

Beakers, 50 ml

Pipettes, 10 ml

Burette, 25 ml

Water bath, 30° + 1°C

pH meter

Reagents

1% Pectin dissolved in 0.15 M NaCl

0.02 N NaOH prepared from N NaOH on the day of use

Method

Pipette 20 ml of pectin solution in a 50-ml beaker, adjust the pH to 7 in a pH meter. Add 10 ml of the enzyme solution and adjust the pH meter immediately to 7 adding NaOH. This is the zero time. Place the beaker in the water bath at 30°C. Maintain control containing boiled enzyme. Every 15 min, check the pH and if it falls below the reference point, add the alkali from a burette while strring the contents. Use 0.02 N NaOH to adjust the pH and note the volume of the alkali at each interval. This is known as the ``continuous titration method''.

Alternatively, the enzyme-substrate can be incubated for 24 h and the pH is adjusted with 0.02 N NaOH. In this method only the enzyme activity at the end of the assay period is measured.

An indicator can also be used. Methyl red is the frequently used indicator. Published reports indicate weak PME activity with methyl red which may be actually high as the indicator is not sensitive enough to detect traces of the enzyme. If at all indicators have to be used, avoid apple pectin which imparts a colour to the indicator.

Determination of Pectin Content of Plant Tissues

The pectin content of plant tissues may be determined by estimating individual methanol and uronic acid content. From these data it is possible to estimate the methylation of pectic substances present in plant tissues.

Materials

Cotton plants infected by *F. oxysporum* f. sp. *vasinfectum* Use stem.

40-Mesh sieve

Distillation flasks, 10 ml, 50 ml
Pipettes, 5 ml
Water bath with ice
Water bath
Colorimeter

Reagent

1 N HCl
Conc H_2SO_4
Dilute H_3PO_4
1 N NaOH
5% $KMnO_4$
Saturated solution of $NaHSO_3$
2% Aqueous chromotropic acid

Method

Dry the plant tissue and grind to pass through 40-mesh sieve. Place 100 mg sample in 50 ml distillation flasks. Add 4 ml of 1 NaOH and allow to stand for 30 min to saponify the pectin. Then add 4.25 ml of 1 N HCl to each flask and distill. Collect the distill in a 10 ml volumetric flask and dilute to 80 ml with water. Transfer 1 ml aliquots of sample to 10 ml volumetric flasks. Add 3 drops of dilute H_3PO_4 and 5 drops of 5 $KMnO_4$. Allow the mixture to stand at room temperature with occasional shaking to facilitate oxidation of CH_3OH to CH_3O. Add drop wise saturated $NaHSO_3$ to reduce the excess $KMnO_4$. Place the flasks in an ice bath and add 4 ml of conc H_3SO_4 following 5 drops of chromotropic acid. Warm the samples at 60°C in a water bath for 15 min. Cool the flasks in an ice bath and make up the volume to 10 ml with water. Thaw the samples and determine the absorbance in a spectrophotometer at 580 nm. Calculate the methanol content from a standard curve and express in meq/g dry weight tissue.

Estimation of Uronic Acid in Diseased Plant Tissues

Materials

Cotton plants infected by *F. oxysporum* f. sp. *vasinfectum* : use wilted plants

40-Mesh sieve

Beakers, 200 ml, 500 ml

Erlenmeyer flasks, 100 ml

Pipettes, 5 ml, 10 ml

Whatman No. 1 filter paper

Ice bath

Water bath with heater

Hot air oven

pH Meter

Colorimeter

Reagents

Ethanol, 70%, 95%

0.5% Versene (EDTA, disodium salt)

Acetic acid

Conc H_2SO_4

1 N NaOH

Carbazole reagent

Pectinase (from Sigma)

Method

Dry the tissue and grind it to pass through 40-mesh sieve. Allow the dry powder to stand in 70% ethanol (100 ml/g) for about 18 h at room temperature. Filter through Whatman No. 1 filter paper and wash the residue with one fourth the original volume of 70% ethanol, Dry the tissue at 80°C. Place 0.5 to 1 g samples of the tissue in 200 ml beakers and moisten with 95% ethanol. Add 200 ml of 0.5% versene to each sample to sequestor multivalent cations. Adjust the pH of the sample to 11.5 with 1 N NaOH. Allow to stand at room temperature to permit saponification of the pectin. Adjust the pH to 5.0 - 5.5 with acetic acid and add 0.1 g of pectinase to each sample to solubilize the pectic substances. Stir the samples for 1 h and dilute to 250 ml with water. Filter the sample through Whatman No. 1 filter paper. Discard the first few ml of the filtrate and collect the rest. Dilute 2 ml of this filtrate with water to 10 ml and use 2 ml portions of this aliquot for the measurement of uronic acid.

Place 12 ml conc H_2SO_4 in Erlenmeyer flask and cool it in an ice bath. Add 2 ml samples containing ca 5 to 80 μg of deesterified uronide to each tube and mix the contents thoroughly. Cool the mixture to below 5°C in an ice bath and place it on a boiling water bath for 10 min. Add 1 ml of 0.15% carbazole reagent after cooling the mixture to about 20°C and mix the contents thoroughly. After 30 min, measure the absorbance of the samples at 520 nm. Calculate the amount of anhydrouronic acid using a standard curve. Express the results as meqanhydrous galacturonic acid present /g dry tissue.

Pectin Trans Eliminase (pectin lyase; PTE)

PTE splits pectin releasing unsaturated galacturonic acid residues. The rate of release of unsaturated galacturonic acids is estimated by the loss in viscosity of pectin and formation of a chromogenic complex with thiobarbituric acid and HCl. Further confirmatory test is the detection of compounds absorbing at 232 nm. PTE is particularly functional at alkaline pH when compared with PG and PMG which are functional at a pH range of 4.4-5.6 (Dube and Bordia, 1983).

Viscosity assay

Materials

Enzyme source

Viscometer, size 150 or any viscometer with a minium efflux time of 15-20 sec for water.

Pipettes, 5 ml

Stopwatch

Blender

Cheese-cloth

Water bath at 30±1°C

Reagents

Pectin 1% dissolved in boric acid-borax buffer at pH 8.7. This is prepared by dissolving 1 g pectin in buffer at 50-60°C kept in a blender. Blend at low speed for 3 min. Filter through two layers of cheese-cloth and adjust the pH, if necessary. Add 1 ml of 1% merthiolate and ketp at 4°C. It may be used for 1 week. Boric-acid-borax buffer at pH 8.7.

Method

Pipette 4 ml of the substrate, 1 ml of the buffer and 2 ml of the enzyme into the viscometer. Gently apply suction through the large arm of the viscometer to mix the contents. Now apply suction to the small arm and determine the viscosity of the mixture (zero time).

Measure the efflux time of the mixture at various intervals and calculate percent loss in viscosity of sodium polypectate by employing the formula given in the earlier section.

Estimation of TBA Reacting Groups

Materials

The reaction mixture of the previous experiment incubated for 2h.

Pipettes, 5 ml, 10 ml

Test tubes

Centrifuge

Water bath with heater

Colorimeter

Reagents

0.01 M Thiobarbituric acid (TBA)

0.5 N HCl

Method

Pipette 3 ml of the reaction mixture into a 25 ml test tube, add 10 ml of 0.01 M TBA and 5 ml of N HCl and place in a boiling water bath for 60 min. Cool under a running tap and adjust the volume of the solution to 18 ml with dist water. If the resulting solution is cloudly, centrifuge at 2000 g for 30 min. Measure the absorbance of the supernatant between 480-580 nm in a colorimeter. The absorbance of the solution should be at 547 nm. Use enzyme-substrate mixture at zero time and boiled enzyme as blanks.

Comment

A progress curve of enzyme action can be plotted by measuring the absorbance of TBA-HCl system at 547 nm. PTE synthesis in diseased plants can be followed employing this method.

Mecaration of Potato Medullary Cells

Both hydrolytic and trans eliminative pectic enzymes separately cause meceration of potato medullary cells and it is recommended to analyze the end product of enzyme activity. The trans eliminase releases derivatives from pectic substance which react with thiobarbituric acid-HCl giving a chromogenic derivative which exhibits an absorbance maximum at 547 nm. This test is reliable to differentiate eliminase from PG types of enzyme.

Materials

Enzyme source

Potato medullary discs of 100 u thick, 700-800 u diameter; use fresh discs.

Sterile 4-cm Petri dishes

Test tubes

Glass rods

Pipettes, 10 ml

Whatman No. 2 filter paper

Centrifuge

Water bath with heater

Colorimeter

Reagents

Sodium fluoride, 4%

0.01 M Thiobarbituric acid (TBA)

Method

Pipette 10 ml of the enzyme solution adjusted to pH 8.7 to a petri dish, add 1 ml NaF and place 5 potato discs. Incubate the plates at room temperature or at 30ºC. Examine from time to time up to 24 h for separation of the cells by gently touching discs with a glass rod. Rank the solution 1 through 6 in order of amount of maceration.

Filter the extract through Whatman No. 42 filter paper and centrifuge at 2000 g or 30 min to remove the debris. To 3 ml of the clear supernatant add 10 ml of 01 M TBA and 5 ml of 0.5 N HCl. Place the tubes in a boiling water bath for 60 min, under a running tap, make up the volume, and measure the absorbance between 480 and 580 nm in a colorimeter. If the absorbance maximum is at 547 nm, the Maceration of potato discs is due to trans eliminase types of enzyme. Use samples drawn at the start and boiled enzyme preparation on potato discs as controls.

Comment

The assay system must include NaF or any other preservative. The assay should be performed in a pH range of 5.8-5.6 to determine the participation of hydrolytic enzymes in macerating the cells.

Release of ultra-violet absorbing substances

Materials

Enzyme source

Test tubes

Pipettes, 5 ml

Water bath, 30° ± 1°C

Centrifuge

Spectrophotometer

Reagents

Polygalacturonic acid 0.2% dissolved in boric acid-borax buffer, pH 8.7

NaF 4%

Trichloroacetic acid (TCA) 5%

Method

To 4 ml of polygalacturonic acid at pH 8.7 in a test tube, add 2 ml of the enzyme adjusted to pH 8.7 and 1 ml of NaF. Incubate the mixture in the water bath for 24 h. Add 2 ml of TCA to precipitate the proteins and keep the mixture at room temperature for about 3 h. Centrifuge the extract at 2000 g for 30 min to sediment the precipitate. Decant the clear supernatant and measure the absorbance in a wavelength of 200-300 nm. Use a clarified aliquot withdrawn at the beginning of the reaction to adjust the absorbance to zero.

Plot absorbance against wavelength and a peak at 232 nm is an indication of PTE activity. Maintain zero time and boiled enzyme controls.

Polygalacturonate Trans Eliminase

Polygalacturonate trans eliminase (PATE) also referred to as pectic acid trans eliminase) trans eliminatively splits pectic acid, releasing unsaturated galacturonic acids. The rate of release of galacturonic acids can be followed by the various tests for PTE using sodium polypectate or pectic acid as the substrate.

For viscosity assay, use 1-2% sodium polypectate dissolved in hot (50-50°C) boric acid borax buffer with blending. Viscometer size 300 or any other make, with a minimum efflux time of 10 sec for water may be used.

Procedures for TBA test and release of UV absorbing product are the same as used for PTE.

Cellulase of Plant Pathogens and Plant Tissues

Cellulose is the major constituent of plant cell walls, in particular the secondary cell walls. Cellulose is a linear polymer of D-glucose units with B 1.4 linkages.

Hydrolysis of cellulose yields glucose which is an important energy source for parasites. Cellulases secreted by parasites are important in pathogenesis as by way of cellulose hydrolysis, cell walls lose their strength and then collapse (Bostawade *et al.,* 1977; Maheshwari *et al.,* 1994).

The mechanisms involved in cellulose degradation has undergone radical changes in the recent years. The hydrolytic degradation of cellulose takes place through the concerted action of (i) five endo-1, 4 B-glucanase attacking at random the 1, 4 B-glucosidic linkages along the cellulose chain, (ii) one exo-1, 4 B-gluconase, splitting off cellobiose or glucose units from the non-reducing end of the cellobiose and (iii) one or several 1.4 B-glucosidases hydrolyzing cellobiose to glucose and gluconolactone. In addition to the hydrolytic enzymes, an oxidative enzyme, has been demonstrated to be of importance for *in vitro* degradation of cotton cellulose. Cellobiose oxidase has been located in *Sporotrichum pulverulentum.* Ramasamy (1980) presented the following sequence :

Cellulose $\xrightarrow{\text{1,4B-endoglucanase}}$ Smaller polymers, Soluble oligomers

$\xrightarrow{\text{1,4 B-endoglucanase 1,4 B-exoglucanase}}$ Cellobiose -

$\xrightarrow{\text{Cellobiase, aryl B-glucosidase}}$ Glucose

Preparation of Cellulase of Plant Parasites

Many necrotrophic fungi and a few bacteria produce cellulase in presence of cellulose.

Materials

F. oxysporum f. vasinfectum

Pipettes, 5 ml

Buchner funnel

Whatman No. 1 filter paper

Centrifuge

Dialysis tubing

Czapek's medium

Reagents

Carboxymethyl cellulose, 1-2% solution

Method

Inoculate 100 ml of Czapek's medium containing 1-2% carboxymethyl cellulose (CMC) or any other cellulosic substrate (cellulose powder, filter paper pulp etc.) adjust to pH 6.5 in 500 ml) of *F. oxysporum, F. vasinfectum*. Incubate the flasks at room temperature for 7-10 days; filter through Buchner funnel using Whatman No. 1 flter paper and centrifuge at 2000 g for 40 min to sediment the spores. Collect the supernatant and dialyze in cellophane tubing against dist water at 2-4°C for 24 h. changing the water at 8 h interval. Add a few drops of tolerene to the dialysate and use it as the enzyme source. When not in use, store it at 2-4°C.

Enzyme source

Pipettes, 5 ml

Whatman No. 1 filter paper

Water bath, 30° + 1°C

Stopwatch

Blender

Viscometer size 150 or any other viscometer with a minimum efflux time of 15-20 sec for water.

Reagents

Carboxy methyl cellulose (CMC) solution. Dissolve 0.5 g of CMC in 100 ml of sodium acetate acetic acid buffer at pH 5.2 at 50-60°C kept in a blender. Blend for 3-5 min at low speed and again blend at high speed for 3-5 min. Filter through Whatman No. 1 filter paper and add 1 ml of 4% NaF as preservative. Store the substrate at 4°C when not in use.

Method

Pipette 4 ml of CMC, 1 ml of the buffer and 2 ml of enzyme into the viscometer kept in a water bath. Mix the contents by drawing air gently through the large arm of the viscometer. Apply suction to the small arm and measure the efflux times of the mixture.

Comment

The enzyme may be extracted in acetic acid-acetate buffer at pH 5.2 or suitable buffer. Chill the buffer before use. Work at low temperature (2-4°C) if possible Dialysis is known to reduce the cellulase activity, since part of the enzyme absorbed onto the dialysis tubing, composed of cellulose nitrate.

Measurement of Cellulase (Endoglucanase, Cx)

Cellulase (Cx, endoglucanase) activity is measured by loss in viscosity of cellulosic substrate (Bostawade *et al.,* 1977).

Viscosity measurement

Materials

Enzyme source

Pipettes, 5 ml

Whatman No. 1 filter paper

Water bath, 30° + 1°C

Stopwatch

Blender

Viscometer size 150 or any other viscometer with a minimum efflux time 15-20 sec for water.

Reagents

Carboxyl methyl cellulose (CMC) solution. Dissolve 0.5 g of CMC in 100 ml sodium acetate-acetic acid buffer at pH 5.2 at 50-60°C kept in a blender Blend for 3-5 min at low speed and again blend at high speed for 3-5 min. Filter through Whatman No. 1 filter paper and add 1 ml of 4% NaF as preservative. Store the substrate at 4°C when not in use.

Method

Pipette 4 ml of CMC, 1 ml of the buffer and 2 ml of enzyme into the viscometer kept in a water bath. Mix the contents by drawing air gently through the large arm of viscometer. Apply suction to the small arm and measure the efflux tims of the mix at pre-fixed interval. Calculate the per cent loss in viscosity of CMC by employing the formula given in the section on PG.

Comment

If CMC is not available, other cellulosic substrates, preferably soluble derivatives may be used. The viscosity loss is solely due to endoglucanase (Cx).

Measurement of Endo-B-1, 4 Glucanase or Carboxyl Methyl Cellulase

Endo-B-1, 4 glucanase attacks the 1, 4-B glucosidic linkages of cellulose molecule randomly, releasing reducing sugars which can be measured with DNS reagent.

Materials

Enzyme source

Test tubes

Pipettes, 5 ml

Water bath at 30°C

(Other requirement as in the estimation of reducing sugars)

Reagents

CMC, 0.5% dissolved in acetate buffer, pH 5.2

Sodium acetate-acetic acid buffer, pH 5.2

(Other requirements as in the estimation of reducing sugars)

Method

Pipette 4 ml of CMC into a test tube followed by 1 ml of buffer and 2 ml of the enzyme. Incubate the tube in a water bath, withdraw aliquots of 1 ml from each tube at pre-fixed interval and determine the amount of reducing sugars released. Express the enzyme activity as the amount of glucose released/ml of the enzyme extract/unit time.

Exo-1, 4 B-Glucanase

Exo-1, 4 B-glucanase splits off cellobiose or glucose units from the non-reducing end of the cellulose molecule. The soluble sugars released by the enzyme react with orcinol reagent and is measured.

Materials

Enzyme source

Pipettes, 1 ml

Glass rods

Test tubes

Water bath with heater

Centrifuge

Ice

Colorimeter

Reagents

Sodium acetate-acetic acid buffer, pH 5

Avicel, 1% dissolved in the buffer

Orcinol reagent : Dissolve 1 g of orcinol in 50 ml of dist water. Gradually add 20 ml of 67% H_2SO_4 with cooling. Raise the volume to 100 ml with dist water.

Method

To 1 ml of the buffer in a test tube, add 0.5 ml Avicel and 1 ml of enzyme source. Adjust the volume to 5 ml and incubate in a water bath at 30°C. Centrifuge at 2000 g for 15 min to deposit the residual Avicel cellulose.

Soluble sugars in the supernatant are measured with the orcinol reagent.

In a chilled test tube, pipette 2 ml of the reaction mixture followed by orcinol reagent. After cooling, add 10 ml of H_2SO_4 solution and stir with the sample cautiously to keep the temperature as low as possible (Sample kept in ice water for at least 12 h without any loss of activity). Heat the sample water bath at 80°C for exactly 20 min. Immediately cool the sample and measure absorbance at 485 nm.

A blank without orcinol should be similarly treated and measured. The solution can be replaced by 2% H_2SO_4.

Cellobiase

The amount of reducing sugars released from cellobiose by cellobiase is used to assay the enzyme.

Materials

Test tubes

Pipettes, 1, 2, 5 ml

Water bath with heater

Colorimeter

Reagents

Sodium acetate-acetic acid buffer, pH 5.8

Cellobiose, 5 mM dissolved in the buffer at pH 5.8

Method

Pipette 1.5 ml of the buffer, 2.5 ml of 5 mM cellobiose and 1 ml of the enzyle solution into a test tube. Incubate at 30°C for 2 h. Terminate the reaction by placing a tube in a boiling water bath for 10 min.

Measure the amount of glucose liberated by the enzyme as per Nelson's method.

Aryl B-Glucosidase

The assay is based on the measurement of chromogenic nitrophenol-B-D- acopyranoside.

Materials

Test tubes

Pipettes 1,2, 10 ml

Water bath set at 30°C

Colorimeter

Reagents

Sodium acetate-acetic acid buffer, 0.05 M, pH 5

p-Nitrophenyl-B-glucoside, 1 mM dissolved in buffer, pH 5

Na_2CO_3, 1 M

Method

To 1 ml of 1 mM p-nitrophenyl-B-D-glucoside in sodium acetate buffer, pour 2 ml of the enzyme and 1 ml of the buffer. Incubate the reaction mixture at 37° for 1 h. Add 20 ml of 1 M Na_2CO_3 and raise the volume to 20 ml with dist. water and measure the amount of p-nitrophenol liberated by the enzyme by measuring absorbance of the solution at 425 nm in a colorimeter.

Express enzyme activity in units, one unit liberating 5 u mole of p-nitrophenol under the assay condition.

Xylanase Activity

Xylanase degrades the polymer xylan, that links the cellulose and rhamnogalacturonan in the three dimensional structure of plant cell wall. Xylanase are of two types : endo-xylanase that degrades the polymer randomly producing oligomers and B-xylosidase which removes the xylose units from the terminal ends. Endoxylanase releases the reducing groups from larch wood xylan and can be measured by DNS method (Prabakaran and Dube, 1985; Prasad *et al.,* 1988).

Preparation of Xylanase of Parasites

A few plant pathogens have been screened for xylanase production.

Materials

Erlenmeyer flasks, 500 ml

Buchner funnel

Whatman No. 42 filter paper

Centrifuge

Flash evaporation unit, with water bath

F. oxysporum f. *vasinfectum/Rhizoctonia solani* grown in Czapek's medium

Method

Inoculate 100 ml of Czapek's medium, containing 2 per cent xylan, pH adjusted to 5.5, with a 15 day old culture of the fungus. Incubate the flasks at room temperature for 15 days, filter through Buchner funnel using Whatman No. 42 filter paper and flash evaporate under vacuum at 20°C to get 1/10 of the original volume. Use this as enzyme source.

Measurement of Endoxylanase Activity

Endoxylanase (EC 3.2.1.8) releases the reducing groups from larch wood xylan and can be measured by using DNS reagent.

Materials

Enzyme source

Test tubes

Pipette 5 ml

Water bath at 30°C

Reagents

Starch wood xylan (Sigma) 0.5% in phosphate buffer at pH 6.8.

Phosphate buffer, pH 6.8

Other requirements as in estimation of reducing sugars.

Pipette 2 ml of wood xylan solution into a test tube followed by 2 ml buffer and enzyme. Incubate the tube in a water bath; withdraw aliquots of 1 ml from tube at hourly interval and estimate the amount of reducing sugars released with reagent.

Express the enzyme activity as the amount of xylan released/ml of the enzyme.

Measurement of B-xylosidase

Xylosidase (E.C. 3.2.1.7) releases para-nitrophenol from para-nitrophenyl-β-D-xylophyranoside and can be measured in a colorimeter.

Materials

Colorimeter

Endo-xylanase

Reagents

Para-nitrophenyl-B-D-xylopyranoside, 20 mM in phosphate buffer, pH 6.8

Other reagents, as in the measurement of total phenols.

Method

Pipette 2 ml of para-nitrophenyl-B-D-xylopyranoside, into a test tube followed by 2 ml buffer and 2 ml of enzyme. Incubate the tube in a water bath; withdraw aliquots of 1 ml from each tube at hourly interval and estimate the phenol present in the extract.

Amylase and Invertase of Pathogens and of Plants

Starch is the principal storage carbohydrates in plant cells. It contains long chains of glucose linked together by oxygen bridges similar to those present in the disaccharide maltose, in which the terminal carbon atom of each residue is linked through an oxygen atom to the fourth carbon atom of the following glucose residue. Starch is made up of amylose and amylopectin. Amylose possesses unbranched chains and is readily soluble in water while amylopectin consists of unbranched chains and is less soluble. A - and -B-Amylse hydrolyse starch. B-Amylase cleaves amylose molecule to the disacch-aride maltose while a-amylose yields dextrins, short chains of glucose molecules with 6 to 12 glucose residues. Sucrose is an early product of photosynthetic reaction. Invertase cleaves sucrose to glucose and fructose.

Plant pathogens produce these enzymes and their role is of interest in studies dealing with carbohydrate metabolism of diseased tissues.

Preparation of Amylase and Invertase of Pathogens

Materials

Culture of *H. oryzae*

Erlenmeyer flasks, 500 ml having 100 ml Czapek's medium

Pipettes, 5 ml

Buchner funnel

Whatman No. 1 filter paper

Centrifuge

Dialysis tubing

Reagent

Toluene

Method

Inoculate 100 ml of Czapek's medium containing either 10 g of potato starch or 20 g of sucrose in 500 ml Erlenmeyer flasks with 2 ml spore suspension (ca. 4000 spores/ml) of *H. oryzae*. Incubate the flasks at 28-30°C for 10-15 days. Filter the mycelium through Buchner funnel using Whatman No. 1 filter paper and centrifuge the filtrate at 2000 g for 30 min to remove the spores. Dialyze the supernatant in dialysis tubing against dist water for 16 to 20 h at 0-4°C, changing the water every 8 h. Add a few drops of toluene to the filtrate and store at 2-4°C when not in use. Use this as enzyme source.

Preparation of Amylase and Invertase of Diseased Plants

Materials

Diseased plant tissues

Volumetric flask, 100 ml

Pipette, 5 ml

Pestle and mortar

Ice water bath

Cheese-cloth

Reagents

Aqueous glycerol, 20%

Toluene

Method

Homogenize the plant tissues (5 g) in a chilled mortar placed in an ice water bath, with pre-cooled 20% aqueous glycerol solution. Filter through 4-5 layers of cheese cloth and make up the volume with 20% glycerol to 100 ml in a volumetric flask. Add 1-2 ml of toluene above the mark to preserve the extract. Store at 0-4°C when not in use.

Measurement of Amylase and Invertase

Amylase and invertase activities are determined by measuring the amount of reducing sugars released by these enzymes from their respective substrates.

Amylase assay

Materials

Enzyme extract

Erlenmeyer flasks, 100 ml

Test tubes

Incubator

(Other requirements as for the estimation of reducing sugars)

Reagents

1 M Acetic acid-sodium acetate buffer, pH 6

1% Soluble starch solution

(Other requirements as for the estimation of reducing sugars)

Method

Place 5 ml of the crude enzyme extract in 100 ml flasks. Add 10 ml of 1 M acetate buffer, pH 6 and 5 ml of 1% soluble starch. Incubate the mixture at 37°C for 24 h. Pipette 1 ml portions of the reaction mixture and stop the reaction by the addition of 1 ml of either Nelson's or DNS reagent. Estimate the reducing sugars released by either Nelson's or DNS method and express the result as the amount of reducing sugars released per g dry weight of the tissue.

Invertase

Materials

As for amylase assay

Reagents

Sucrose solution, 2.5% aqueous

1 M Acetic-acid sodium acetata buffer, pH 5

Other requirements as in amylase assay

Method

Place 5 ml of the crude enzyme extract in 100 ml flasks. Add 10 ml of 1 M acetic acid-sodium acetate buffer, pH 5 and 5 ml of 2.5% sucrose. Incubate at 37°C for 24 h. Pipette aliquots of 1 ml of the reaction mixture and stop the reaction by adding 1 ml of either Nelson's or DNS reagent. Estimate the reducing sugars released by either Nelson's or DNS method and express the results as mg of reducing sugars released per g dry weight of the tissue.

B-Glucosidase Activity of Diseased Plants

B-Glucosidase releases reducing sugars from glucosides or oligosaccharides. This enzyme is particularly important in disease resistance mechanisms where phenols are released from phenolic glucosides. B-Glucosidase-induced accumulation of flavonoid aglucones may be related to disease resistance as phytoalexins in leguminous crops. B-Glucosidases are both of plant and parasitic origin. The specificity of the enzyme is absolute for the B-glucose portion of the substrate but towards the aglucone, it is relative (Mahadevan, 1979).

Measurement of B-glucosidase

The enzyme activity may be assayed by either measuring the amount of glucose released from the substrate or by estimating the glucone, whichever is convenient. Measurement of glucose may be made using either Nelson's or DNS reagent. A simplified method of measuring the aglucone is presented here. Folin-phenol reagent specifically reacts with the phenol moiety and in presence of sodium carbonate, forms a chromogenic complex which is measured in a colorimeter.

Materials

Enzyme extract from plant tissues

Erlenmeyer flasks, 50 ml

Pipettes, 5 ml, 10 ml

Test tubes

Water bath

Colorimeter

Reagents

Salicin, 0.01 M dissolved in sodium acetate-acetic acid buffer, pH 5.2

Sodium acetate acetic acid buffer, pH 5.2

NaF, 4%

Folin-phenol reagent

Sodium Carbonate, 20% aqueous

Method

To 10 ml of salicin substrate placed in 50 ml Erlenmeyer flasks, add 3 ml of enzyme source and 1 ml of NaF. Incubate the mixture in a constant temperature water bath. Withdraw aliquots of 1 ml at fixed intervals and transfer to test tubes. Add 1 ml of Folin-phenol reagent and 2 ml of Na_2CO_3. Shake vigorously and place in boiling water bath for exactly 1 min. Cool the reaction mixture, raise the volume to 10 ml (if the colour intensity is high, dilute to 25 ml) and measure the absorbance at 650 nm in a colorimeter. Maintain similar reaction mixture with boiled enzyme extracts and samples withdrawn at the beginning of the assay as controls.

Plot a standard curve with known concentrations of catechol and calculate the amount of aglucone released in terms of catechol equivalent.

Estimation of Chitinase from Infected Plants

Although chitin per se is not synthesized by vascular plants, chitinases are found in a number of plants. N-acetyl glucosamine forms a constituent of plant polymers such as glycoproteins. In most host-pathogen interaction increase in chitinase is well recognized. Besides the source from host, pathogen secreted chitinase is an attractive possibility (Boller *et al.,* 1983; Balasubramanian and Manocha, 1992).

Colorimeteric Assay

Chitinase (E.C. 3.2.1.14, poly-1, 4-(2-acetamido-2deoxy) - B-D-glucoside glycanohydrolase) hydrolyses polymers of N-acetyl glucosamine (NAG) including tetramers and less effectively trimers. N-acetyl-B-glucominidase (EC 3.2.1.30) is required to complete the hydrolysis of the resulting dimers (chitobiose) and trimers to NAG.

Materials

Enzyme extract from ground nut leaves infected by *R. solani*.

Test tubes

Blender

Centrifuge tubes with glass balls

Centrifuge

Water bath, 37°C

Water bath, with heater

Reagents

Crab shell chitin. It is prepared by grinding crabs with 1 N HCl and treating with 0.8 N NaOH, 0.5% $KMnO_4$ or 0.2 M sodium chlorite $NaClO_3$). After repeatedly washing in water, a clear white paste is obtained, chitin.

Sodium acetate buffer, 0.1 M, pH 5.2.

Potassium tetraboraborate ($K_2B_4O_7$) 0.8 M p-dimethylaminobenzaldehyde (DMAB) 1 g in 100 ml glacical acetic acid containing 1.25% (v/v) 10 N HCl (prepare fresh).

Method

Chitin suspension of 1.8 mg/ml is prepared by dissolving 25 ml chitin in pH 5.2, 0.1 M sodium acetate buffer in a blender : slow speed for 30 sec. The reaction mixture containing 2 ml chitinase in sodium acetate buffer, 4 ml chitin suspension and 0.2/mg/ml BSA is incubated 37°C for 4 h in a water bath.

One ml of sample and 1 ml of dist water are boiled in a glass ball covered centrifuge tube for 10 min and centrifuged at 2,000 g for 30 min. N-acetylglucosamine is estimated in 0.5 ml aliquots or the supernatant by adding 0.1 ml potassium tetraborate solution and boiling for exactly 3 min in a water bath. After cooling, add 3.0 ml of DMAB solution, mix and keep for 20 min at 37°C. Cool the samples and read within 10 min at 585 nm in a colorimeter and convert to NAG by referring to a calibration curve of standard NAG. Prepare a standard of NAG in the range of 0.1-0.4 M. conc.

One unit of chitinase activity is defined as the amount of enzyme that produces 100 ug of NAG/ml chitin reaction mixture in 1 h at 37°C at pH 5.2. Specific activity is defined as units of enzyme activity per mg protein.

Protease of Pathogens and of Plants

For direct cell penetration, the parasite should elaborate besides cellulolytic and pectinolytic enzymes, proteolytic enzymes, to depolymeterize the proteins in the cell wall. The degradation products, amino acids and amides serve as nitrogen and carbon sources to the pathogen. Although many pathogens were screened for pectinolytic and cellulolytic enzyme production. Only a few have been studied for proteolytic enzyme producing ability (Balasubramanian, 1972; Bisen and Agarwal, 1980).

Preparation of Protease of Pathogens

Microorganisms produce protease inductively in presence of a protein substrates is usually substituted for nitrogen. After a sufficient period of incubation of mycelium is removed and the culture filtrate is assayed for enzyme activity (Mahadevan, 1970; Srivastava and Prasad, 1989).

Materials

Erlenmeyer flask, 500 ml

Beakers, 250 ml

Pipettes, 5 ml

Buchner funnel

Whatman No.1 filter paper

Centrifuge

Dialysis tubing

Method

Inoculate 100 ml of Czapek's medium containing casein as nitrogen source have the casein in small volume of 0.1 N NaOH and add to the final medium. Adjust the medium kept in 500 ml Erlenmeyer flask and adjust pH to 7 using 3 N HCl. Inoculate with of spore and mycelial suspension. Incubate at room temperature for 10 days. The Mycelium by filtering through Buchner funnel using Whatman No. 1 filter and centrifuge at 2,000 g for 30 min. Collect the supernatant and dialyze with several volumes of dist water at 0-4°C for 24 h changing the water at least twice. Use the dialyzed filtrate as enzyme source.

Measurement of Protease

Most of the methods to assay protease involve (i) the measurement of the intact protein and (ii) the amino or carboxyl groups liberated from a protein. The formol titration method of Sorensen poses difficulty of reproducibility. Measurement of amino acids with ninhydrin reagent or estimation of aromatic with Folin-phenol reagent is relatively easy, reproducible and a large number can be screened (Mahadevan and Chandramohan, 1967).

Ninhydrin method

Materials

Enzyme source

Test tubes

Pipettes 1 ml, 5 ml

Water bath

(Other requirements as for the estimation of amino acids)

Reagents

Casein 1% dissolved in phosphate buffer at pH 7 (Since casein is sparing soluble in water. Dissolve it in a minimal quantity of 0.1 NaOH and raise the volume to 100 ml with the buffer. Check the pH and if necessary adjust)

0.1 N NaOH

0. 1 M Phosphate buffer, pH 7

(Other requirements as for the estimation of amino acids)

Method

Pipette 10 ml of casein solution into a test tube. Add 5 ml of 0.1 M phosphate buffer, pH 7 and 5 ml of enzyme extract. Mix by pipetting the contents and incubate the mixture at 30°C in a water bath. Use boiled enzyme and zero time controls. Withdraw aliquots of 1 ml at various intervals and stop the reaction by the addition of ninhydrin reagent. Estimate the amino acid content of the solution. Calculate the amount of amino acids released/unit time/g of protein from a standard curve prepared from aspartic acid or glutamic acid standards.

Measurement of Aromatic Amino Acids with Folin-Phenol Reagent

Materials

Centrifuge

Colorimeter

Reagents

Trichloroacetic acid, 5%

Folin-phenol reagent

Sodium carbonate, 20%

Method

Pipette 1 ml of the enzyme substrate mixture (see earlier section) into centrifuge tube and 1 ml of TCA. Allow the mixture to stand at room temperature for 1 h and centrifuge at 2,000 rpm for 20 min. Pipette 1 ml of the supernatant into a test tube and add 1 ml of Folin-phenol reagent and 2 ml of 20% sodium carbonate. Place the tube in a boiling water bath for exactly 1 min, cool under a running tap and raise the volume to 10 ml with dist water at 650 nm. Calculate the amount of amino acids released from a standard curve plotted from phenylala-nine or tyrosine. Use reaction mixture with boiled enzyme and aliquots drawn at zero time as controls.

Comment

Casein in insoluble in acid pH. The use of haemoglobin as the substrate gives consistent results.

Nucleases in Diseased Plant Tissues

Nucleic acid metabolism in diseased plants has attracted the attention of a few investigators especially with reference to obligate parasitism. Plant tissues infected by obligate parasites exhibit enhanced levels of nucleases. In virus-infected tissues, it is believed that the breakdown products of host ribonucleic acids (RNA) may serve as precursors for the newly synthesized virus RNA. However, it is of significance to note that increase in ribonuclease (RNase) is accompanied by increase in total RNA in the rust-infected tissues (Chakravarty and Scott, 1979).

Preparation of Nucleases from Diseased Plants

Materials

Wilt infected cotton leaves

Pestle and mortar

Pipettes, 10 ml

Refrigerated centrifuge

Reagent

Phosphate buffer, 1.0 M pH 6.5

Method

Homogenize 5 g of leaf sections with ice cold phosphate buffer, in a pre-cooled mortar with pestle. Make up the volume of the homogenate to 8 ml with the same buffer and centrifuge at 20,000 g for 30 min at 0-2°C. Use the supernatant for the assay of enzyme activity. Estimate the protein content of the extract.

Measurement of Nucleases

Ribonuclease (RNase) and deoxyribonuclease (DNase) act upon ribonucleic acid (RNA) and deoxyribonucleic acid (DNA), respectively. These reactions release the respective pentoses D ribose from RNA and deoxyribose from DNA nucleotides. The increase in acid-soluble nucleotides due to enzyme activity can be measured in a spectrophotometer (Chakravarty and Scott, 1979).

Deoxyribonuclease

Materials

Enzyme extract

Pipettes, 1 ml, 5 ml

Water bath

U.V. Spectrophotometer

Centrifuge

Reagents

Calf thymus DNA, 0.05% in 0.05 M phosphate buffer, pH 6.5

$MgCl_2$, 0.2%

Perchloric acid ($HClO_4$), 25%

Method

Pipette aliquots of the enzyme extract into test tubes. Add 0.5 ml of calf thymus DNA in 0.05 M phosphate buffer, pH 6.5 ml of 0.2%. $MgCl_2$, incubate the reaction mixture at 37°C for 4 h in a water bath. Add 0.5 ml of chilled 25%. $HClO_4$, to terminate the reaction and precipitate the unhydrolysed nucleic acid. Maintain zero time blank. Cool the contents of the tubes in tap water, and centrifuge at 2,000 g for 15 min. Measure the absorbance of the supernatant (dilute, if necessary) at 260 nm in a spectrophotometer. Consider an increase of 1.0 A_{260} per hour as one unit of DNAse and express the results as DNAse unit per hour.

Ribonuclease

Materials

Enzyme extract
Pipettes 1 ml, 5 ml
Water bath
Spectrophotometer
Centrifuge

Reagents

Phosphate buffer, 0.05 M pH 6.5
Yeast RNA, 0.15% dissolved in 0.05 M phosphate buffer, pH 6.5
Perchloric acid, 25% containing 0.75% uranyl acetate

Method

Pipette aliquots of 0.2 ml of the enzyme extract into test tubes followed by 1 ml of yeast RNA solution in 0.05 M phosphate buffer, pH 6.5 and 0.8 ml of 0.05 M phosphate buffer, pH 6.5. Incubate the tubes containing the reaction mixture in a water bath at 37°C for 2 h. Add 0.5 ml of chilled $HClO_4$-uranyl acetate reagent to terminate the reaction and precipitate the unhydrolysed ribonucleic acid. Similarly. maintain zero time blank. Cool the tubes under a running tap and centrifuge at 2,000 rpm for 15 min. Measure the absorbance of the supernatant (dilute, if necessary) at 260 nm in a spectrophotometer against the zero time blank. Consider an increase of 1.0 A_{260} per hour as 1 unit of RNase and express the results as RNase unit per hour.

Comment

One A_{260} of unit is that amount of soluble nucleotide that has an A_{260} of 1.0 in a volume of 1.0 ml. If the activity is low, express in milli units. Convert the spectrophotometer readings to standard units using the following formula :

$$Unit/ml = \frac{A_{260} \times (ml\ assay\ solution + ml\ precipitating\ agent)\ dilution\ factor}{(ml\ enzyme\ solution\ assayed) \times \text{min}}$$

Use gloves while working, otherwise nucleases from hands influence results.

Detoxifying Enzymes of Pathogens

A successful parasite overcomes the toxic prohibitins and phytoalexins of the host. It is achieved through cetrain enzyme systems capable of detoxifying the toxic barriers of the host either by degrading them or converting them into nontoxic forms. The concept of detoxifying enzymes was proposed by Mahadevan (1974). The glycosidic prohibitins are acted upon by B-glycosidase releasing the nontoxic-aglycones, Phenols and related aromatic substances may be detoxified (i) by formation of polymers (melanin) e.g. phenolase or tyrosinase type which acts on tyrosine, to give 3, 4-dihydroxy-phenylalanine and on further oxidation, a quinone, forming a heteroxycyclic red compound resulting in the formation of an inert polymer known as melanin, (ii) by masking off the reactive OH group, e.g. tyrosinase, laccase and peroxidase which oxidize a variety of mono-, di- and polyphenols in presence of H_2O_2 and (iii) by ring fisson, e.g. oxygenase which catalyses the incorporation of two atoms of atmospheric oxygen into or substrate especially with aromatic substances (Boominalhan *et al.*, 1986).

B-Glycosidase

Fusarium spp. and *Septaria lycopersici* produce extracellular specific B-glycosidase capable of detoxifying the prohibitin, tomatin (present in tomato). B-Glycosidase can be estimated by measuring the release of p-nitrohenol or phenol from p-nitrophenol B-glucoside or salicin.

Materials

Enzyme source (dialyzed culture filtrate)

Pipettes, 1 ml, 10 ml

Test tubes

Water bath

Colorimeter

Reagents

Acetate buffer, pH 5.5

p-Nitrophenyl-B-glucoside, 0.01 M or 0.01 M Salicin

Na_2CO_3, 1 M

Method

Pipette 1 ml of the enzyme extract containing ca 1 mg of protein into a test tube. Add 8 ml buffer at pH 5.5 and 1 ml of substrate. Incubate the reaction mixture in a water bath at 37°C. At prefixed interval, pipette 1 ml sample and mix with 1 ml of 1 M Na_2CO_2 in a test tube. Dilute with 4 ml of dist water and read the absorbance at 420 nm. Express the specific activity as u mole of p-nitrophenol or phenol liberated per mg of protein.

Comment

The intracellular enzyme activity may be assayed using the enzyme preparation obtained by homogenizing the mycelium in the buffer.

Phenol Oxidase

Although the Commission on Enzymes has grouped laccase or p-nitrophenol oxygen oxidoreductase and catechol oxidase or o-diphenol oxygen oxidorductase together, they are considered as two distinct groups.

Laccase

Laccase oxidizes a wide range of substrates including monophenols, o-and p-diphenols, triphenols and ascorbic acid. White rot fungi possess laccase while soft rot fungi do not. Although most fungi possess extracellular laccase, *Neurospora* and *Glomerella* possess both extracellular and intracellular enzyme. Laccase may be estimated in a colorimeter employing guaiacol as the substrate.

Preparation of Enzyme Extract

Materials

Liquid culture of *Aspergillus* grown in Richard's medium

Buchner funnel

Whatman No. filter paper
Pestle and mortar
Refrigerated centrifuge

Reagents

Phosphate buffer, 0.1 M, pH 6.

Method

Filter a 10 day old culture through filter paper using a Buchner funnel and use the culture filtrate as source of extracellular enzyme.

To detect intracellular laccase, grind the mycelium with sand in 5-7 ml of cold phosphate buffer in a mortar with pestle. Centrifuge the homogenate at 17,000 rpm for 20 min in cold to remove sand and cell debris. Use supernatant as enzyme source.

Assay of Laccase

Materials

Enzyme source
Pipettes, 0.1 ml, 5 ml
Incubator at 30°C
Colorimeter

Reagent

Phosphate buffer 0.1 M, pH 6 containing 10 mM guaiacol.

Method

Pipette 5 ml of sodium phosphate buffer containing guaiacol into test tubes and equilibrate at 30°C. Add 0.1 ml of the enzyme source. After 5 min, determine the absorbance at 470 nm. Express the results as enzyme units (one unit) - absorbance change of 0.01 per min). Keep suitable control with heat killed enzyme.

Catechol Oxidase (Phenol Oxidase)

Catechol oxidase is also referred to phenolase, polyphenol oxidase, tyrosinase, catecholase or cresolase. The ability of o-diphenol oxygen oxidoreductase (catechol oxidase) to oxidize p-diphenol is a typical property of the enzyme. In host-parasite interaction, the pathogen produced catechol oxidase may produce highly reactive quinones which rapidly polymerize and lose toxicity. Thus, it may help to detoxify-prohibitins (Subba Rao *et al.*, 1979; Aruna Kumari and

Mahadevan, 1981). The endogeneous activity of the enzyme in fungal mycelium may be estimated by measuring the formation of dopachrome (2-carboxy-2, 3-dihydroindole-5, 6, quinone) from DOPA (3, 4-dihydroxy-L-phenylalanine).

Materials

Shake culture of *F. oxysporum* f. sp. *vasinfectum* grown on potato dextrose broth for 10-15 days (Prepare the endoenzyme as for laccase).

Pipettes, 1 ml, 5 ml

Test tubes

Water bath

Colorimeter

Reagents

Phosphate buffer, 0.1 M, pH 6.0

L-DOPA, 0.017 M

Method

Pipette 3.5 ml of phosphate buffer into test tubes and 0.5 ml of L-DOPA. Equilibrate the mixture in a water bath at 30°C. Pipette 0.5 ml of enzyme preparation into the tubes containing the substrate solution and incubate the reaction mixture in a water bath for 5 min at 30°C. Measure the absorbance of the solution at 475 nm. Maintain appropriate controls with heat killed enzyme. An enzyme unit is defined as change in absorbance of 0.001% per min which is equivalent to 2.7 x 10^{-7} mol of dopachrome formed using the molar extinction coefficient.

Comment

The tyrosinase activity of cultures can be increased by exposing the mycelium to 4°C for 4 h. The extracellular enzyme may be assayed by incubating the mycelium for 30 min in presence of L-dopa as substrate and measuring the amount of dopachrome formed in the medium after removing the mycelium.

Peroxidase

Several wood destroying fungi especially white rot fungi secrete extracellular peroxidase and the enzyme is involved in pathogenesis (Mahadevan, 1979). Peroxidase activity may be measured by using o-dianisidine as H-donor and H_2O_2 as oxidant in the assay system which is sensitive (Vora and George, 1978). Extracellular production of peroxidase by the fungi may be assayed using o-dianisidine with or without H_2O_2, to measure peroxidase plus phenol oxidase activities.

The peroxidase activity may be derived from the difference between the activities of two enzymes (Dohroo, 1989).

Materials

Buchner funnel
Pipettes, 1 ml, 2 ml
Test tubes
Incubator at 25°C
Colorimeter

Reagents

Phosphate buffer, 0.5 M, pH 6
H_2O_2, 0.003 M
o-Dianisidine, 0.003 M

Method

Three week old culture of *F. oxysporum* f. *vasinfectum* grown in Czapek's medium consisting of 2% malt extract, 2% glucose (in place of sucrose) and 1% peptone in dist water, at pH 6.5. Filter the mycelium and use the culture filtrate as enzyme source.

Pipette 1 ml of phosphate buffer into test tubes. Add 1 ml of H_2O_5 and 0.8 ml of o-dianisidine and equilibrate the tube at 25°C. Add 0.2 ml of enzyme source, incubate the reaction mixture at 25°C for 3 min and measure the absorbance at 460 nm. Similarly, perform another assay replacing H_2O_2 with 1 ml of dist water. Maintain controls with heat killed enzyme. Calculate peroxidase activity by noting the difference in absorbance between the two systems. Calculate the reaction rates from the initial slope. A unit of enzyme activity is defined as the amount of enzyme causing a change of 1 absorbance unit per min.

Catechol Oxygenase

Ring cleavage is an important step in the detoxification of secondary plant substances, especially phenols. Both bacteria and fungi produce catechol oxygenase which oxidizes catechol, releasing cis, cis-muconic acid. The cleavage goes through the reactions :

Catechol → cis, cis Muconic acid → Muconolactone → B-Ketoadipate enol-lactone → B-Ketoadipate → TCA cycle.

By secreting oxygenase, phenols and other aromatic substance are degraded (Nagarajan and Mahadevan, 1979; Aruna Kumari and Mahadevan, 1985).

Oxygeanse Indution

Materials

Isolate of *Pseudomonas solanacearum* / F. o.f. sp *vasinfectum*

Erlenmeyer flasks, 100 ml

Pipettes, 1 ml 5 ml

Rotary shaker

Reagents

Modified Dye's medium

Modified Dye's medium suppelemented with succinic acid (10 mM as sole carbon source)

Catechol, 0.1 M, aqueous

Method

Preinduce *P. solanacearum* cells by growing them in modified Dye's medium for 24 h. Inoculate 100 ml flasks containing 25 ml of Dye's medium having succinic acid with 24 h old culture of *P. solanacearum*. Incubate the flasks on a rotary shaker at room temperature for 30 min under aseptic conditions. Add 2.5 ml of catechol (inducer) to get a final conc of 2.5 mM. Incubate the flasks for another 2.5 h.

Enzyme Preparation

Materials

Liquid culture of *P. solanacearum*

Refrigerated centrifuge

Deep freezer

Sonicator/Pestle and mortar with acid washed sand

Reagent

Phosphate buffer, 0.025 M, pH 7.4

Method

After 5 days of growth, collect the cells by centrifuging the liquid broth in cold at 7,000 g for 20 min. Discard the supernatant, suspend the cells in 20 ml of chilled phosphate buffer; centrifuge and repeat the step. Suspend the cells in 4 ml of the buffer and freeze them at 20°C. Thaw the cell suspension and sonicate the cells for 5-6 min at current strength of 4 A at 5°C. Alternatively the cells can be ground in acid washed sand to a fine paste. Centrifuge the suspension at 22,000 g for 30 min at 5°C to remove the cell debris. Use supernatant as the crude enzyme extract.

Enzyme assay

Materials

Enzyme source

Pipettes, 1 ml

Test tubes

Water bath with heater

Reagents

Phosphate buffer, 0.025 M, pH 7.4

Catechol, 0.002 M.

Method

Place aliquots of 1 ml of the enzyme extract in test tubes. Add 1 ml of catechol and 1 ml of phosphate buffer. pH 7.4. Incubate the tubes containing the reaction mixture at 30°C in a water bath for 20 min. Draw an aliquot of 1 ml of the reaction mixture and estimate the residual catechol.

Estimate the protein content of the enzyme extract and express the results as u mole of catechol oxidized per mg of protein in 20 min.

Rothera Rection for Detecting the Mode of Ring Cleavage

The oxidative ring fission of phenol usually occurs at either the meta or ortho postion. A few microorganisms cleave by both routes. The manner of fission can be detected using Rothera reaction which differentiates the type of cleavage and has been employed to distinguish the ring cleavage pattern by pathogenic fungi and bacteria.

Materials

Isolate of *Pseudomonas solanacearum*

Test tubes

Pipettes, 1 ml 5 ml

Incubator

Reagnets

Mineral medium for *Pseudomonas* supplemented with (i) p-hydroxy benzoate, (ii) m-hydroxy benzoate at 2 g/l, (iii) glucose (1 g/l) and yeast extract (0.2 g/l)

Tris (hdyroxymethyl) aminomethane buffer, 0.02 M, pH 8.

Toluene

Protocatechuate, 4 mM

$(NH_4)_5 SO_4$

Sodium nitroprusside 1%

Conc ammonia

Method

Grow the bacterium in the mineral medium, having 2 g per L of p-hydroxy benzoate, m-benzoate or quinate at room temperature. Maintain control cultures in the same basal medium supplemented with glucose (1 g/l) and yeast extract (2.5 g/l). After 2-3 days of growth, transfer a thick loopful of the test organism and suspend it in 2 ml of Tris buffer, pH 8.0. Then add 0.5 ml of toleune and 2.5 ml of 4 mM of protocatechuate (final conc. 2 mM). Shake the tubes vigorously. Formation of a green-yellow colour within 3 min indicates a positive meta cleavage of the substance. If no reaction occurs, incubate the tubes in a shaker for 18 h in a water bath at room temperature and add 1 g of $(NH_4)_4 SO_4$, 5 drops of freshly prepared aqueous 1% sodium nitroprusside solution and 0.5 ml of conc ammonia. Appearance of a deep purple colour within 3 min indicates a positive test for ortho-cleavage of the substrate.

Dehydrogenase of Plant Tissues

Dehydrogenase catalyses the transfer of hydrogen atoms from a substrate to a hydrogen acceptor. Changes in dehydrogenase activity are reported in a few host parasites interactions. In compatible host-parasite relationship, the dehydrogenase may depress the accumulation of oxidized phenols.

Measurement of Dehydrogenase

In the assay method, a dye is included in the reaction mixture and the rate of reaction is measured according to the degree of reduction of the dye, usually methylene blue or triphenyl tetrazoleum chloride (TTC).

Materials

Enzyme extract
Test tubes
Pipettes, 1 ml, 2 ml
Centrifuge
Water bath
Colorimeter

Reagents

Tripheynl tetrazoleum chloride (TTC) 0.1%
Phosphate buffer, pH 7.0, 0.1 M
Sodium succinate, 0.2 M
Acetone, reagent grade

Method

Pipette 2 ml of sodium succinate into a test tube and add 1 ml of phosphate buffer, pH 7, 1 ml of TTC and 2 ml of enzyme extract. Incubate the mixture in a water bath at 30ºC. At various intervals, remove 1 tube; add 7 ml of acetone to stop the reaction and shake vigorously for a few seconds. Centrifuge at 2000 g for 30 min to remove the precipitates. Decant the supernatant the measure the absorbance of the supernatant at 460 nm. Use boiled extract in the control reactions.

Plot a standard curve with TTC reduced by suspending a few crystals of sodium hydrosulphite into the dye solution and express the enzyme activity in terms of ug of dye reduced per unit time per g fresh weight of protein content.

REFERENCES

Aruna Kumari, A. and Mahadevan, A. 1985. Regulatory studies on inducible catechol 1, 2 dioxygenase in *Pseudomonas solanacearum*. *Indian Journal of Biochemistry and Biophysics* 22 : 161-165.

Balasubramanian, K.A. 1972. The possible role of proteolytic enzymes in the pathogenesis of *Rhizopus stolonifer*. *Indian Phytopathology* 25 : 475-477.

Balasubramanian, R. and Manocha, M.S. 1992. Cytosolic and membrane-bound chitinases of two mucroraceous fungi. A comparative study. *Canadian Journal of Microbiology* 38 : 331-338.

Bisen, P.S. and Agarwal, G.P. 1980. *In vitro* production of pectolytic enzymes by *Aspergillus niger* van Tiegh causing soft rot in apples. *Phytopathology* Z. 97 : 317-326.

Boller, T., Gehri, A., Mauch, F. and Vogeli, U. 1983. Chitinase in bean leaves : induction by ethylene, purification, properties and possible func-tion. *Planta* 157 : 22-31.

Boominathan, K., Gurujeyalakshmi, G. and Mahadevan, A. 1986. Detoxification mechanisms of plant pathogens. In : A Varma and J.P. Verma (eds.) *Vistas in Plant Pathology*. pp. 45-60. Malhotra, New Delhi.

Bostawade, K.B., Deshpande, V.V., Jogiekar, A.V., Lakshmi Kantham, B.C., Mishra, C., Phansalkar, S.B., Rao, M., Seeta, R.,. Srinivasan, M.C. and Jagannathan, V. 1977. Cellulolytic enzymes of *Penicillium strain*. In: T.K. Ghosh (ed.) *Bioconversion of Cellulosic Substances into Energy Chemicals and Microbial Protein*. pp. 143-151. Indian Institute of Technology, New Delhi.

Chakravarty, A.K. and Scott, K.J. 1979. Changes in two barley leaf ribonuclease fractions during infection by the powdery mildew fungus. *Physiological Plant Pathology* 14 : 85-97.

Dohroo, N.P. 1989. Peroxidase and polyphenol oxidase activities in rhizome rot of ginger. *Indian Phytopathology* 42 : 167.

Dube, H.C. and Bordia, S. 1983. Differential control of exopolymethyl galacturonase and pectic acid lyase production by *Helminthosporium sacchari*. *Natural Academy Letters* 6 : 317-320.

Komarajah, M. and Reddy, S.M. 1988. Production of cellulases by *Corynespora cassiicola* Wei, a seed borne fungus of methi. *Acta Botanica Indica* 14: 133-138.

Mahadevan, A. 1970. The role of proteolytic enzymes in disease development. In : *Plant Disease Problems*. pp. 179-184. Indian Phytopathology Society, New Delhi.

Mahadevan, A. 1974. Detoxification mechanisms of plant pathogens. *Journal of Science of Indian Research* 33 : 131-138.

Mahadevan, A. 1979. Biochemical aspects of disease resistance. *Biochemistry Review* 69 : 51-56.

Mahadevan, A. 1979. Recent studies in the disease resistance of plants. In : GP Agarwal and K.S. Bilgrami (eds.) *Physiology of Parasitism, Current Trends in Life Sciences* 7 : 45-78. Today Tomorrow's New Delhi.

Mahadevan, A. and Chandramohan, D. 1967. Protease, trans-eliminase and fusaric acid in *Fusarium* wilted cotton plants. *Phytopathologia Mediternnea* 6 : 86-94.

Mahadevan, A. and Kathirvelu, R. 1967. Pectin and polygalacturonate trans eliminase in *Fusarium moniliforme* and *Cephaloporium sacchari*. *Current Science* 36 : 396.

Maheshwari, D.K., Jahan, H. and Gohade, S. 1994. Bioconversion of cellulose -facts, in P.S. Bisen (ed.) *Frontiers in Microbial Technology* pp. 279-289. CBS Publishers, Delhi.

Nagarajan, M. and Mahadevan, A. 1979. Carbohydrate repression and effect of cyclic AMP on the synthesis of catechol oxygenase in *Pseudomonas tabaci*. *Indian Journal of Experimental Biology* 17 : 757- 759.

Prabakaran, K. and Dube, H.C. 1985. Xylanase production by *Penicillium islandicum*. *Microbiology Letters* 28 : 123-126.

Prasad, G.S., Podile, A.R. and Dube, H.C. 1988. Regulation of xylanase synthesis and activity in *Fusarium oxysporum* f sp *udum*. *Indian Phytopathology* 41 : 443-449.

Prasad, N.M., Singh, H.N.P. and Kumari, S. 1995. Changes in level of protein and phenol contents in banana fruits under pathogenesis, in A.K. Roy and K.K. Sinha (eds). *Recent Advances in Phytopathological Research*. pp. 93-98. M.D. Publication, New Delhi.

Ramasamy, K. 1980. On the mechanism of cellulose degradation by a *Pseudomonas* isolated from activated sludge. *Agricultura* 28 : 473-619.

Srivastava, A.K. and Prasad, M. 1989. The production of protease in culture of *Xanthomonas campestris* pv. *phaseoli* and its activities in bean leaf tissues of varied responses. *Z Mikrobiology* 144 : 23-27.

Subba Rao, P.V., Saraswati, C. and Kalghatgi, K.K. 1979. Degradation of phenolic compounds by the pathogenic fungus *Rhizoctonia solani* in : A. Mahadevan (ed.) *Physiology of Host Pathogen Interaction*. pp. 479-495. Today Tomorrow's, New Delhi.

Vora, A.B. and George, V.C. 1978. Catalase and peroxidase activities of Capnodium infected leaves of some common plants. *Science and Culture* 44 : 139-140.

CHAPTER 11

DISEASE RESISTANCE RELATED ESTIMATIONS

INTRODUCTION

Resistance of plants to parasites is largely determined by their ability to inhibit the invader either during penetration or during its development in the tissues. Parasites differ in their degree of tolerance to the inhibitors and this influences the specificity of pathogenic microorganisms (Mahadevan, 1969). The preformed and post-infectionally synthesized inhibitors play a major role in restricting invasion by the parasite and is intrinsic of host resistance. Mahadevan (1969) proposed that the preformed inhibitory substances that confer some degree of protection to the host plants against microorganisms should be called ``prohibitions". These belong to different groups of secondary plant metabolites. The importance of prohibitins in disease resistance has been well amplified in a series of reviews (Mahadevan, 1973, 1979a, 1979b, 1982, Sridhar and Mahadevan, 1979, Verma and Prasad, 1987).

Although a very large number of plants have been screened for prohibitins, there is very little information on the quantity of prohibitins during their development. The effect of nutrients on the prohibitins of plants has not been studied. Even varieties have not been screened for prohibitins. The mode of action of prohibitins on microorganisms, their enzymes and on their toxins has not been studied extensively.

Volatile Prohibitins

Some of the substances emitted by plants are volatile and a few of them are toxic to microorganisms. Cultivated plants differ in the amount and nature of volatile prohibitins that they emit. The volatile prohibitins might interfere with the development of plant pathogens, particularly during their pre-penetrative stages (Mahadevan, 1982). A simple method to detect the volatile prohibitins emitted by lemongrass leaves which is toxic to *Curvularia lunata* is described.

Detection of Volatile Prohibitins

Materials

Freshly collected lemon-grass (*Cymbopogan citratus*) leaves *Curvularia lunata* culture, grown in PDA for 8-10 days.

Petri dishes (10 cm)

Filter paper discs

Microscope slides

Adhesive tape

Microscope

Method

Prepare a spore suspension of *C. lunata* in dist water. Place a drop of spore suspension on a microscope slide kept in 10-cm Petri dishes lined with moist filter paper. Spread five to six, 5-cm apical segments of lemon-grass leaves on the moist filter paper by the side of the slide. Cover the petri dish with its lid and secure tight the lid and bottom portion of the dish with adhesive tape to make the dish air-tight. Incubate the Petri dishes on a laboratory bench at room temperature for 6 to 8 h. Examine the slide for the number of germinated spores and measure the germ tube length of spore in 10 microscope fields selected at random. Express the results as per cent inhibition over control.

Isolation and Assay of Prohibitins from Groundnut

Groundnut (*Arachis hypogea*) roots, leaves and pods contain a spectrum of prohibitins (Kalaichelvan and Mahadevan, 1987, 1988).

Materials

Groundnut shells

Spore suspension (ca 4 x 10 spores/ml) obtained from 5-7 day old cultures of *C. lunata* grown on potato dextrose agar.

Pipettes, 0.1 ml

Microscope

Cavity slides

Water bath at 80°C

Centrifuge

Humidity chamber (Petri dish lined with filter paper moistened with water.

Method

Extract 50 g shells of groundnut in 100 ml of hot water for 3 h on a hot water bath. Filter and centrifuge at 4,000 g for 20 min. Use the supernatant for the assay.

Use spores washed in sterile dist water. Spore suspension is centrifuged for 20 min; the supernatant is decanted and the sedimented spore suspension is suspended in sterile dist water.

Place 0.1 ml of the extract in a cavity slide and add 0.1 ml of spore suspension. Incubate the slide in a moist chamber for germination. Examine the slide for spore germination at 30 min interval for the first 6 h and thereafter once in 3 h. Examine at least 50 spores in 5 fields. Maintain sterile dist water control. Calculate per cent inhibition of spore germination.

Comments

Delay in the germination of spores is an indication of prohibitin activity. Development of malformed germ-tubes is another indication of the effect of prohibitin on germ-tubes (Kalaichelvan and Mahadevan, 1979, 1987).

Organic solvents may be used to extract prohibitins. Other fungi and bacteria may be employed for bioassay. The prohibitin may be sprayed on plants and their effect may be investigated.

Inhibitors of Plant Viruses

Extracts of plants inhibits plant viruses. Plant extracts with anti-viral activity induce systemic resistance in specific host-virus combination. Raychaudhuri and Prasad (1965) demonstrated that radish mosaic virus is inhibited by chilli, Datura and spinach leaf extracts. The method used by them to assess the antiviral property of plant extracts is described.

Extraction of Prohibitins from Chilli Leaves

Materials

Chilli leaves

Pestle and mortar

Cheese cloth

Test tubes

Pipettes, 1 ml, 5 ml

Centrifuge

Reagent

Phosphate buffer, 1 M, pH 7

Method

Macerate fresh leaves (10 g) of chilli in a pestle with mortar using 50 ml of phosphate buffer. Filter the homogenate through layers of cheese cloth and centrifuge at 3000 g for 20 min. The supernatant contains virus inhibitors. Prepare suitable dilutions of the clear supernatant.

Preparation of Virus Inoculum

Materials

Radish plants infected by radish mosaic virus

Pestle and mortar

Cheese cloth

Test tubes

Pipettes, 1 ml, 5 ml

Centrifuge

Reagent

Phosphate buffer, 1 M, pH 7.0

Method

Collect young, diseased radish leaves showing mosaic symptoms and crush them (10 g) in a mortar with adequate quantity (50 ml) of buffer. Filter the homogenate through layers of cheese cloth. Centrifuge the filtrate at 3,000 g for 20 min and decant the supernatant. Dilute the supernatant to 50 ml with the buffer. Use the diluted extract as the inoculum. Mix equal volumes of the extracts containing the virus inoculum and the one containing virus inhibitors for testing the effect of inhibitors on the virus.

Infectivity Titer

The extent of inhibitory activity of the virus prohibitin is determined based on the infectivity titer (counts of the local lesions produced) of the virus inoculum. The infectivity titer can be assessed using a local lesion host. *Brassica juncea* is a suitable host.

Materials

Virus inoculum
Inhibitor extract
Test tubes
Pipettes, 1 ml, 5 ml
Spatula
Brassica juncea plants

Method

Mix the virus inoculum with the extract having prohibitin kept in test tubes in suitable dilutions. Inoculate *B. juncea* leaves either with the virus inoculum alone or with the mixture containing virus inoculum and the extract. Dip a spatula in the inoculum and rub it gently, but firmly on the upper surface of *B. juncea* leaves. Maintain suitable control plants inoculated with the virus inoculum. Examine the plants at 24 h interval. Count the number of lesions formed on the leaves inoculated with the virus alone (control) and on those inoculated with the virus and prohibitin. Express the activity of prohibitin as per cent reduction in lesions.

Solanine from Potato Peel

Alkaloids constitute the largest class of secondary plant substances. They influence the development of parasites. Solanine, the steroid alkaloid of solanaceous plants is a modified terpenoid and is a prohibitin, effective against many fungi and bacteria and a few viruses. The resistance of solanaceous plants has been attributed to alkaloids (Raychaudhari, 1980, 1984).

Extraction of Solanine

Materials

Potato peel
Blender
Cheese cloth
Water bath at 70°C
Centrifuge
pH Meter

Reagents

Acetic acid, 5%

Conc. NH_4OH

Method

Crush the potato peel in acetic acid (20 ml/g tissue) in a blender and filter the extract through cheese cloth. Warm the filtrate to 70°C in a water bath and add conc. NH_4OH drop wise till the pH is 10. Centrigute at 4,000 g for 30 min and discard the supernatant. Wash the residue with 1% NH_4OH and recentrifuge at 4,000 g for 30 min. Collect the precipitate, dry and weigh the crude solanine.

Estimation of Solanine

Materials

Alkaloid extract

Test tubes

Pipettes, 1 ml ,10 ml

Colorimeter

Reagents

Ethanol, 96%

H_2SO_4, 20%, 60%

Formaldehyde 0.5% in 60% H_2SO_4

H_2SO_4, 1 M

Solanine

Method

Dissolve about 50 mg of the crude solanine in a 1 : 1 mixture of ethanol and 20% H_2SO_4. To 5 ml of 60% H_2SO_4 in a test tube, add 1 ml of the alkaloid solution. After 5 min, add 2-3 ml of formaldehyde reagent. Allow to stand for 3 h at room temperature and measure the absorbance at 565-570 nm in a colorimeter. Calculate the amount of solanine in the crude preparation using reference solutions prepared with pure solanine treated similarly with formaldehyde in 1 M H_2SO_4.

Comment

Solanine can be separated by paper chromatogrpahy using ethyl acetate-pyridine water (3 : 1 : 3) or by TLC on silica gel G using acetic acid-ethanol (1 : 3) as solvents (Rf 0.46). Similarly, tomatine from unripe tomatoes can be extracted and estimated. Prohibitin activity of extracts may be assayed by using spore germination test.

Lectins

Lectins are sugar binding proteins or glycoproteins of non immune origin. These are devoid of enzymatic activity towards sugar to which they bind and do not require free glycosidic hydroxyl groups for the binding. Lectins are widely distributed in plants and in microorganisms. Distribution of lectins in plants varies with plant parts and age. Lectins are toxic to a variety of microorganisms (Mahadevan *et al.*, 1982).

Agarwal and Mahadevan (1984) have isolated and quantified lectins from several legume seeds and assessed the effect of seed microlora of lectin content and the effect of lectins on *Colletotrichum capsici* and *Fusarium solani*.

Extraction of Lectins

Materials

Groundnut seeds

Pestle and mortar

Erlenmeyer flask, 250 ml

Soxhlet apparatus

Centrifuge

Reagents

Petroleum ether (b.p 80-100°C)

Ammonium sulphate

Phosphate buffer 0.05 M (pH 7.2)

Ascorbic acid, 50 mM

Method

De-coat the seeds and wash them thoroughly with water. Crush the seeds (50 g) in mortar with pestle and dry them at room temperature. Place the air-dried seed crushings in a Soxhlet apparatus and extract repeatedly with petroleum ether to defatten the material. Extract 10 g of defatted meal with phosphate buffer and ascorbic acid in 250 ml Erlenmeyer flask and precipitate lectin by the addition of ammonium sulphate. Usually ammonium sulphate precipitate can be obtained at 50 and 100% saturation levels. Centrifuge at 3,000 g for 20 min and discard the supernatant. The residue contains the lectin.

Estimation of Lectins

Lectin is measured by the hemagglutination test.

Materials

Rabbit erythrocytes

Human erythrocytes A, O, B

Human leucocytes A, O, B

Lectin extract

Reagent

Saline (NaCl, 0.9%)

Method

The photometric method is based on the observation that rabbit erythrocytes sediment at a rate which is proportional to the conc of hemagglutinin. Hemagglutinating activity is calculated from absorbance measurements of the suspensions of unsedimented cells after 30 min and 2 h.

Collect blood samples from local hospital. Dilute the blood with Saline to get a suspension, containing 1.5 to 1.50% cells. Serially dilute 0.2 ml to 1 ml of the blood suspension to 2 ml with NaCl and read the suspension at 620 nm. Prepare a standard graph.

Prepare different conc of lectin ranging from 10-1000 μg/ml and add 1 ml of blood sample (1.25 to 1.50% cells) and make the final volume to 2 ml with saline. Mix the contents immediately following the addition of blood sample. Incubate at room temperature for 30 min and 2 h. Measure the absorbance at 620 nm.

Comment

Absorabance readings are readily converted to per cent of the cells remaining in suspension from a standard graph. Hemagglutinating activity is expressed as hemagglutinating unit (H.U.) and is defined as that level of test solution which causes 50% of the standard cell suspension to sediment after 2 h and 30 min incubation.

Effect of Lectin on Microorganisms

Materials

Conidial suspension of *Colletotrichum capsici* or *Fusarium* sp. (ca 10^6 conidia/ml)

Test tubes
Cheese cloth
Centrifuge
Microscope
Glass slides

Reagents

Lectin solution (1 mg lyophilized powder/ml) in 0.005 M phosphate buffer (pH 7.2).

Richard's medium diluted to 1%.

Method

Mix 1 ml conidial suspension (ca 10^6 conidia/ml) with 1 ml of lectin solution in test tubes and incubate at 30°C with occassional stirring for 18 h. Observe agglutination conidia.

Wash thoroughly the agglutinated conidia in sterile dist water by centrifugation. Suspend the conidia in drops of 1% Richard's medium placed on a slide for 6 to 8 h. Examine the slide for germination and germ tube growth of the fungus.

Comments

While extracting lectins from legume seeds, add polyvinyl-polypyrolidone (PVP) to the extract to precipitate phenols, before precipitating the lectins with ammonium sulphate. Lyophilize and store the lectins at -10°C. The ammonium sulphate precipitated lectins can be further purified in Sepharose 4B column (15 x 3.5 cm). The fractions can be further purified by dialysis against 0.005 M phosphate buffer.

Lectins from germinated and ungerminated seeds can be isolated and their activities can be evaluated.

Estimation of Hydrocyanic Acid (HCN) released from Plant Tissues

Cyanogenesis is exhibited by about 750 plant species belonging to 60 families. These plants contain cyanogenic glycosides. These glycosides protect the young seedlings from being invaded by pests and diseases. Sorghum seedlings contain high levels of the cyanogenic glycoside dhurrin (p-hydroxy-mandelonitrile-B-D-glucoside).

An active glycosidase is present in sorghum and during pathogenesis, the enzyme activity is enhanced. Hydrolysis of the glycoside releases an unstable aglycone, p-hydroxy mendelonitrile which decomposes to HCN and the aromatic aldehyde p-hydroxybenzaldehyde. Sorghum also possesses the enzyme, oxynitrilase which catalyses the decomposition of the nitrile.

HCN is toxic to fungi and bacteria and hence, its concentration has some significance in disease resistance.

Materials

Sorghum leaves
Polythene bags
Erlenmeyer flask, 500 ml with rubber stopper
Whatman No. 1 filter paper strips (15 x 2 cm)
Colorimeter

Reagents

Sodium carbonate, 2.5%
Picric acid solution, 0.5%

Method

Place leaves in 500 ml flasks and stopper them lightly with rubber stopper.

Soak Whatman No. 1 filter paper strips in sodium carbonate solution and then in picric acid. Suspend a strip into each flask from the rubber stopper. Allow to stand at room temperature for 24 h, transfer the filter paper strips to 100 ml of water and allow them to be bleached completely. Maintain appropriate controls. Measure the absorbance of this solution at 450 nm in a colorimeter. Calculate the conc of HCN from a standard curve.

Comment

Experiments may be conducted with cut tissues, infected leaves and leaves at different ages of maturity. Cassava tuber peels have cyanogenic glycoside and needs investigation. Prohibitin activity of cassava peels may be evaluated as per the procedure for volatile prohibitins.

Folin-phenol Method to Estimate Total Phenols

Estimation of phenols with Folin-phenol reagent is based on the reaction between phenols and an oxidizing agent phosphomolybdate which results in the formation of a blue colour. The intensity of the colour is measured in a colorimeter (Anwar *et al.,* 1995).

Materials

Alcohol extract of plant tissues

Pipettes, 1 ml, 5 ml

Test tubes (graduated at 25 ml)

Water bath with heater

Colorimeter

Ragents

Folin-Phenol reagent : Dissolve 100 g sodium tungstate ($Na_2 MO_4$, H_2O) and 25 g sodium molybdate in 700 ml water in 1*l* flask. Add 50 ml 85% orthophosphoric acid and 100 ml HCl and boil under reflux gently for 10 h. Cool and add 150 g lithium sulphate dissolved in 50 ml water and 4-5 drops of liquid bromine. Boil the mixture without condenser for 15 min to remove the excess bromine. Cool, dilute to volume with water and filter. The reagent should be golden yellow in colour; store in brown bottles. It is stable for many months. Just before use, dilute 1 volume of the stock solution with 2 volumes of water.

Sodium carbonate 20%.

Method

Pipette 1 ml of the extract in a graduated test tube and 1 ml of Folin-phenol reagent followed by 2 ml of Na_2CO_3 solution. Shake the tube and heat in a boiling water bath for exactly 1 min. Cool under a running tap. Dilute the blue solution to 25 ml with water and measure its absorbance at 650 nm in a colorimeter. If any precipitate forms, filter or centrifuge the solution at 2,000 g for 15 min before measuring its absorbance. Read the unknowns from a standard curve made from different concentrations of catechol or guaiacol. A blank containing all the reagents minus plant extract should be used to adjust the absorbance to zero.

Estimation of Ortho-dihydric (O.D.) Phenols

Ortho-dihydric phenols are important in disease resistance reactions. They are oxidized by phenol oxidase and the resulting quinones are highly reactive and toxic to pathogens and to their enzymes (Mahadevan, 1991).

Arrow's Method

Arrow's reagent specifically reacts with O.D. phenols by producing a pink colour; The intensity of which is measured in a colorimeter.

Materials

Alcohol extract of plant tissues

Pipettes, 1 ml, 2 ml

Test tubes (graduated at 25 ml)

Colorimeter

Reagents

Arrow's reagent: dissolve 10 g of sodium nitrite and 10 g of sodium molybdate in 100 ml of dist water. Store the reagent in a brown bottle. It is stable for one year.

HC1, 0.5 N

NaOH, 1 N

Method

Pipette 1 ml of the alcohol extract into a test tube, add 1 ml of 0.05 N HCl, 1 ml of Arrow's reagent, 10 ml of dist water and 2 ml of 1 N NaOH. Soon after the addition of the alkali, the pink colour develops. Maintain a reagent blank without the extract. If the colour intensity is high, dilute to 25 ml and read the absorbance of the solution at 515 nm. Calculate O.D. phenols present in the samples from a standard curve prepared with catechol.

Separation and Detection of Phenols by Paper and Thin-layer Chromatography

Solvents

Paper Chromatography

n-Butanol-acetic acid-water (4 : 1 : 1, v/v)

Iso-Propanol-conc ammonium hydroxide water (20 : 1 : 2, v/v).

Thin-layer Chromatography

Ethyl acetate – iso propanol – water (65 : 24 : 11, v/v).

Benzene – ethyl acetate-formic acid (50 : 40 : 10, v/v).

Spray reagents

Diazotized sulphanilic acid : Dissolve 50 g of sulphanilic acid in 250 ml of 10% KOH, cool and add 200 ml of 10% NaNO, Add the mixture dropwise to 120 ml of HCl (a mixture of 80 ml of conc HCl and

40 ml of water). Filter the precipitated diazonium salt with ice water. Wash the precipitate with ethanol followed by ether and dry in air. Store the dried salt at 0-4°C. It can be used for 10-12 months. Handle the dried salt with extreme care, as it may explode.

For spray, dissolve 50 mg of the salt in 20 ml of 20% Na_2CO_2. After spraying, dry the paper at room temperature. Phenols appear as deep yellow, light red and other colours.

Diazotized p-nitroaniline

Stock solution A : Dissolve 300 mg of p-nitroaniline in 100 ml of 8% HCl.

Stock solution B : Dissolve 5 g of sodium nitrite in 100 ml water.

Just before spray, mix 25 parts of A and 1.5 parts of B.

Comment

It would be a good practice to view the chromatograms before spraying under UV light (short wave 254 nm and long wave 365 nm) and record fluorescence if any.

Extraction and Estimation of Bound Phenols

Occurrence of esterified phenols in bound form with cell-wall components is known. Their possible involvement in disease resistance reactions in rice had been pointed out (Chattopadhyay and Samaddar, 1980). The bound phenols may be liberated by treatment of plant tissues with NaOH at room temperature (Chattopadhyay and Bera, 1980; Mall and Suresh, 1989).

Materials

Rice leaves, either blast or brown spot infected

Pestle and mortar

Pipette, 5 ml

Centrifuge

Colorimeter

Reagents

Sodium lauryl sulphate solution, 3%

Ethanol

Diethyl ether

NaOH, 0.5 M

Method

Cut the leaves into small bits of 0.5 to 1 cm. Grind 100 mg sample in a mortar with pestle and 5 ml of sodium lauryl sulphate. Transfer the suspension to centrifuge tubes and centrifuge at 2,000 g for 5 min. Discard the supernatant and wash the residue successively, once with 5 ml of sodium lauryl sulphate, twice with 5 ml of diethyl ether. After each washing, centrifuge and discard the supernatant. Allow the residue to dry and suspend in 3 ml of 0.5 M NaOH in centrifuge tubes. Allow it to stand at room temperature overnight, centrifuge and collect the supernatant. The alkali extract contains the released phenols, which have absorption peaks at 330 and 290 nm. Dilute 1 ml to 8 ml with 0.5 M NaOH. Measure the absorbance at 290 nm. The concentration of bound phenols can be expressed as absorbance values. Alternatively the concentration may be estimated colorimetrically, using Folin-phenol reagent.

Comment

Save the residue after the cold alkaki extract if lignin and brown pigment contents of the tissues have to be determined.

Extraction and Estimation of Chlorogenic Acid

Chlorogenic acid, a O.D. phenol has been implicated in the resistance of plants against fungi and bacteria. Potato peel and coffee beans are rich in chlorogenic acid (Reddy and Prasad, 1979).

Reagents

Sodium lauryl sulphate solution, 3%

Ethanol

Diethyl ether

NaOH, 0.5 M

Method

Cut the leaves into small bits of 0.5 to 1 cm. Grind 100 mg sample in a mortar with pestle and 5 ml of sodium lauryl sulphate. Transfer the suspension to centrifuge tubes and centrifuge at 2,000 g for 5 min. Discard the supernatant and wash the residue successively, once with 5 ml of sodium lauryl sulphate, twice with 5 ml of diethyl ether. After each washing, centrifuge and discard the supernatant. Allow the residue to dry and suspend in 3 ml of 0.5 M NaOH in centrifuge tubes. Allow it to stand at room temperature overnight, centrifuge and collect the supernatant. The alkali extract contains the released phenols, which have absorption peaks at 330 and 290 nm. Dilute 1 ml to 8 ml with 0.5 M NaOH. Measure the absorbance at 290 nm. The concentration of bound phenols can be expressed as absorbance values. Alternatively, the concentration may be estimated colorimetrically, using Folin-phenol reagent.

Comment

Save the residue after the cold alkali extract if lignin and brown pigment contents of the tissues have to be determined.

Extraction and Estimation of Chlorogenic Acid

Use Folin-phenol reagent to locate the spots on the chromatogram. Alternatively, examine the unsprayed chromatograms under UV light and locate the fluorescence chlorogenic acid. Elute the fluorescent band containing chlorogenic acid in ethanol and measure its absorbance at 324 nm. Calculate the amount of chlorogenic acid present from a standard curve constructed with authentic acid.

Detection and Estimation of Quinones

The formation of quinones from ortho-phenols as a result of oxidation by phenolase is an important reaction in disease resistance. Quinones are highly reactive and possess inhibitory activity against microorganisms. Two methods are used to detect quinones in plant extracts.

Starch Reduction Test

The reduction of starch in presence of acid and iodine is used as an indication of quinone in an extract.

Materials

Buffer extract of plant tissues (used for enzyme assays)

Test tubes

Pipettes, 1 ml

Reagents

KI aqueous solution, 30%

H_2SO_4, 30%

Starch indicator solution : Suspend 1 g of soluble starch in 10 ml of water and mix with 100 ml of boiling water. Stir well and continue to heat for 2 min. Cool and decand the supernatant. Add a few ml of 5% NaOH or 10 ml of 1% salicylic acid per *l* to preserve the indicator.

Method

To 1 ml of test solution, add 1 ml of Kl reagent and 0.5 ml H_2SO_4. Shake once, add 1 drop of starch solution; shake once again and rank the colour immediately from 1 to 6. Maintain a water control.

Quinone Reduction Assay

Another method involves the measurement of decrease in extinction value at 400 nm following the addition of a reducing agent (ascorbic acid) which is an indication of reduction of low molecular weight quinone.

The method is sensitive and reliable.

Materials

Buffer extract of plant tissues (used for enzyme assays)

Test tubes

Pipettes, 1 ml, 5 ml

Water bath at 30°C

Buchner funnel

Filter paper, Whatman No. 42

Spectrophotometer

Reagents

Sodium phosphate buffer, pH 6.6

Chlorogenic acid or caffeic acid, 5 x 1^{-3} M

TCA, 0.5 N dissolved in 60% ethanol containing 0.05 M ascorbic acid

Method

Pipette 3 ml buffer, 3 ml substrate and 1.5 ml enzyme extract into a test tube, shake gently and incubate in the water bath. At different invervals, withdraw 2 ml sample. Add 4 ml of TCA reagent to one sample and to the other a corresponding solution containing in addition, ascorbic acid. Filter off the precipitate, record the extinction value at 400 nm in a spectrophotometer. Calculate the amount of quinone produced by the enzyme extract from caffeic acid/chlorogenic acid.

Estimation of Flavanols

Participation of flavanols in disease resistance reaction has been demonstrated. *Fusarium* resistant terminal leaves of cotton plants possess higher concentration of flavanols and synthesis more flavanols than the susceptible leaves in responses to infection. The use of colour reaction for the detection of phoroglucinol employing vanillin and conc HC1 has been reported by Swain and Hills (1959) and suitable modifications are suggested to get reproductible results.

Materials

Alcohol extract of cotton leaves infected by *F. o. F. vasinfectum.*

Erlenmeyer flasks, 25 ml

Pipettes 1 ml, 5 ml

Burette, 10 ml

Water bath

Colorimeter

Reagents

Vanillin reagent : Dissolve 1 g of recrystallized vanillin in 100 ml of 70% conc H_2SO_4; prepare fresh.

70% H_2SO_4 (Add 4 ml of 70% conc H_2SO_4 to 2 ml of dist water)

Method

Place 1 ml of the plant extract containing not more than 0.1 ml of ethanol or methanol in 25 ml flask and dilute to 2 ml with dist water. Add 4 ml of vanillin reagent from a burette rapidly (within 10-15 sec) to flask A and 4 ml of sulphuric acid to flask B. Prepare a blank in flask C containing 4 ml of vanillin reagent and 2 ml of water. Shake the contents of flasks A and B in a water bath to keep its temperature below 35°C. Keep the flasks at room temperature for exactly 15 min. Measure the absorbance of the contents of flasks A, B and C in 1 cm cells at 500 nm against 47% sulphuric acid (flask D). Subtract the absorbancies of the contents of flasks B and C from that of flask A. Alternatively, arrange the cells in the holder and read the absorbance of the contents from flasks A + D against B + C. Calculate the flavanol content using a standard curve prepared with phloroglucinol or express the results as optical densities (A_{500}).

Comment

Up to 100 ug/ml of flavonoids in the aliquots may be estimated. Measure the absorbance of the solution at a fixed time after the addition of the reagent, since the colour fades slowly.

Estimation of Anthocyanins

The method described by Swain and Hills (1959) to measure anthocyanins is sensitive and avoid the interference by other pigments such as chlorophylls.

Materials

Alcohol extract of plant tissues

Pipettes 1 ml, 5 ml

Test tubes

Colorimeter

Reagents

HCl, 0.5 N, in 80-85% methanol

Anthocyanin reagent : Mix 1 ml of 30%, H_2O_2, with 9 ml of HCl-methanol.

Method

Pipette 1 ml of the sample (alcohol extract) into test tubes and add 3 ml of HCl-methanol reagent. Add 1 ml of peroxide reagent to the samples and 1 ml of methanol-HCl to blank tubes. After 1 min in the dark, measure the absorbance of solutions in a 1 cm cell at 525 nm against the blank. Calculate the anthocyanins present in the samples from a standard curve prepared with cyanin HCl or express the results as A_{525}. Cyanin HCl, 10 ug/ml in methanol-HCl gives optical density of 0.405 in a 1 cm cell at 525 nm.

Estimation of Leuco-anthocyanins

Lignification in plants is associated with the presence of leuco-anthocyanins. Lignification occurring after infection is involved in resistance reactions. Nevertheless, little has been understood about the participation of leuco-anthocyanins in host parasite interactions.

Materials

Alcohol extract of leaf spot infected groundnut leaves

Glass stoppered test tubes, 18 x 150 cm

Pipettes, 1 ml, 10 ml

Water bath, with heater

Colorimeter

Reagent

Leuco-anthocyanin reagent : Dilute 25 ml of 36% HCl to 500 ml with n-butanol

Method

Pipette 1 ml of the alcohol extract into a glass stoppered test tube. Reduce the volume to 0.5 ml on a hot water bath so that the sample does not contain more than 0.5 ml of methanol or ethanol. Add 0.5 ml of dist water followed by 10 ml of leuco-anthocyanin reagent. Mix the contents thoroughly. Heat the tubes in a water bath at 97°C or 3 min without covering the tubes. Then cover the tubes with the glass stopper and continue to heat the tubes for a total of 40 min. Cool the tubes under a running tap, similarly maintain a blank with the extract, but without heating. Measure the absorbance of the solution in a 1 cm cell at 550 nm in a colorimeter. If chlorophyll is present, measure the absorbance at 650 nm. Express the results as A_{550} or A_{650}.

Estimation of Lignin

Lignification of host cell walls forms a mechanical barrier for pathogen penetration and effectively interferes with pathogenesis (Mahadevan, 1991). It increases the resistance of cell walls for dissolution by fungal enzymes, prevents the diffusion of enzymes and toxins of the pathogen to the host cells and water and nutrients from the host cells to the pathogen and creates a toxic environment through the low molecular weight phenolic precursors. Fungi degrade lignin (Muthukumar and Mahadevan, 1983; William *et al.,* 1986). In rice, nitrogen fertilization known to predispose plants to blast, reduces the lignin content of leaves. Lignin can be measured in a spectrophotometer.

Spectrophotometric Method

Materials

Dried cotton plant tissues

Pestle and mortar

Beakers, 50 ml

Test tubes

Pipettes, 1 ml, 2 ml
Centrifuges
Water bath at 70-80°C
Spectrophotometer

Reagents

Diethyl ether
Sodium phosphate buffer, 0.1 M, pH 7.0
NaOH, 0.5, N, 0.1 N
HCl, 2 N

Method

Moisten 100 mg of oven dried tissue placed in a mortar with dist water. Grind the tissue with ether until it is free from chlorophyll pigments. Centrifuge at 2,000 g for 5 min and decant the supernatant. Wash the sediment with dist water, recentrifuge and discard the supernatant. Repeat the washing twice. Add 2 ml of NaOH to the residue and extract at 70-80°C for 12-16 h. Cool the suspension, add 0.45 ml of HCl and adjust the pH to 7 or 8 with NaOH. Make up the volume to 3 ml with dist water. Centrifuge the suspension at 2,000 g for 5 min. To 0.8 ml of the hot alkali extract, add 0.8 ml of sodium phosphate buffer. To another aliquot of 0.8 ml of the alkali extract, add 0.8 ml of 0.1 N NaOH (pH 12.3). Measure the optical densities at 245 and 350 nm. Derive the lignin concentration from the difference between optical densities (E) at pH 7.0 and 12.3 (samples diluted with buffer and NaOH, respectively). Express the amount of lignin as E_{350} per sample.

Comments

Non-conjugated phenols give peaks at 250 and 300 nm while phenols with large conjugated side chain such as hydroxycinnamic acid derivatives give longer peaks at wave length higher than 300 nm.

The conc of lignin may be calculated from the phenol values of hot alkali extract. Pipette 0.55 ml of the extract in dist water containing 1 to 3 ug of phenol into test tubes. Add 0.4 ml of 0.5 m Tris buffer pH 9.0 and 0.05 ml of freshly prepared alcohol solution containing 25 ug of 2, 6 dichloroquinonechlorimide. After 1 h at room temperature, measure optical density of the solution at 610 nm using guaiacol as standard. Use a conversion factor of 32 (32 x mg phenol = mg of lignin, calculated from bagasse sample) to calculate the lignin content.

The hot alkali extract can be used to measure the intensity of browning formed during host-parasite interaction. The optical density of pH 7 sample measured at 420 nm will give the measure of the brown pigment content of the tissue. Express the results as E_{350} per sample.

Phenylalanine and Tyrosine Ammonia-lyases

Aromatic compounds arise from phenylalanine and tyrosine by the action of phenylalanine ammonia-lyase (PAL) and tyrosine ammonia-lyase (TAL), respectively. The mechanism of conversion is

$$R\text{-}CH_2\text{-}CH\ (NH_2)\text{ - }COOH \rightarrow RCH_2\text{ - }CO\text{-}COOH \rightarrow RCH_2\text{ - }CH\ (OH)\text{ - }COOH \rightarrow R\text{-}CH = CH\text{ - }COOH$$

A direct determination of the side chain is the mechanism of enzyme action.

In disease resistance mechanism, these enzymes play an important role in the conversion of phenylalanine and tyrosine to coumaric acids. These provide the phenyl-propane carbon skeleton for the synthesis of flavonoids, phenolic phenylpropanes and lignin (Purushothman, 1974). Resistant varieties are characteristics of rapid conversion of phenylalanine and tyrosine to coumaric acids. The significance of PAL in phytoalexin synthesis has been stressed (Mahadevan, 1979).

Preparation of Enzyme Extracts

Plant tissues can be homogenized in chilled acetone and the resulting acetone powder used as enzyme source. Aqueous enzyme extracts often exhibit inconsistent activities due to the presence of enzyme inhibitor. It is possible to obtain reproducible results by including either sodium bisulphate or mercaptoethanol in the extraction medium. The specific activities of the enzyme extracted and assayed from acetone powders and aqueous extracts do not differ much. However, aqueous extraction is preferred to preparation of acetone powder because, it is rapid and decreases variation between samples introduced by powdering the tissues (Mahadevan and Rangamannar, 1980).

Buffer Extraction

Materials

Fresh groundnut leaves infected by rust

Glass homogeniser or pestle and mortar

Pipettes, 1 ml, 5 ml

Refrigerated centrifuge

Reagents

25 mM Borate-HCl buffer, pH 8.8

5 mM Mercaptoethanol (or 2 mM sodium bisulphite).

Method

Homogenize the plant tissue weighing 150-200 mg in a glass homogenizer with 5 ml of cold borate HCl buffer, containing either 2 mM sodium bisulphite or 0.4 ml of mercaptoethanol per litre (ca 5 mM). Centrifuge the homogenate at 12,000 g for 10 min. The clear yellow green supernatant is used as the enzyme source. The clarified extract can be stored frozen overnight without loss of activity.

Assay of Phenylalanine Ammonia Lyase

The amount of trans-cinnamic acid synthesized from L-phenylalanine is measured in a UV spectrophotometer.

Assay with Acetone Powder Preparation

Cinnamic acid formed during the reaction is extracted with ether and its concentration is estimated.

Materials

Enzyme preparation

Cheese cloth

Test tubes

Pipettes, 1 ml, 10 ml

Separatory funnel

Beakers, 50 ml

Water bath at 35°C

Ice bath

Hot plate

Spectrophotometer

Reagents

Boric-acid borax buffer, 0.1 M

L-Phenylalanine, 0.05 M in boric acid-borax buffer, pH 8.8

HCl, 5 N

NaOH, 0.05 N (Prepare fresh from 1 N NaOH)

Diethyl ether

Method

Mix the acetone powder with about 10 times its weight of boric acid-borax buffer. Cool the mixture in an ice water bath, stir manually from time to time for 30 min and filter through two layers of cheese cloth. Centrifuge the filtrate at 2000 g at 2-4°C for 15 min. Use the supernatant for assay. The extract is not stable and must be used within a few hours of preparation.

To 1.5 ml of the enzyme extract, add 1 ml of L-phenylalanine and 2.5 ml of borate buffer. Incubate the mixture on a water bath at 40°C for 1 h. Similarly maintain a zero time control. In addition, maintain controls with boiled enzyme and without substrate. Stop the reaction by adding 0.1 ml of 5 N HCl. Extract twice the acidified mixture with 7.5 ml of ether and pool the ether phases in a 50 ml beaker. Evaporate the ether to dryness under a stream of air. Dissolve the residue in 5 ml of NaOH and measure the absorbance at 268 nm. Express the results as ug of cinnamic acid formed per g fresh weight of tissue per hour.

If no UV spectrophotometer is available, the residue may be dissolved in a suitable volume of ethanol and treated with Folin-phenol reagent. Measure the colour at 560 nm in a colorimeter.

Comment

The recovery of trans-cinnamic acid in the ether layer is about 70-80% while that of phenylalanine is nil. The acetone powder may be directly assayed for PAL activity by suspending 100 mg of the powder into a test tube containing the substrate dissolved in the buffer or using enzyme preparations extracted from acetone powder prior to the assay.

Assay with Aqueous Buffer Extract

The extinction of trans-cinnamic acid dissolved in 0.1 M borate buffer at pH 8.8 is slightly more than 10,000 at 290 nm and phenylanine has almost no absorption at this wave length. Therefore, the enzyme catalyzed deamination of phenylalanine trans-cinnamic acid may be conveniently measured in a spectrophotometer at 290 nm.

Materials

Enzyme extract

Pipettes, 1 ml, 2 ml

Test tubes

Water bath at 30°C

Spectrophotometer

Reagents

Boric acid-borax buffer 0.1 M, pH 8.8

Phenylalanine, 0.1 M

Method

To 2 ml of 0.1 M borate buffer kept in a test tube, add 1 ml of enzyme extract (200-300 ug of protein), 0.6 ml phenylalanine and 2.4 ml dist water. The reaction mixture contains 200 u moles of borate buffer and 60 u moles of L-phenylalanine. Preincubate the mixture at 30°C for 15 min to allow for an initial non-enzymatic decrease in absorbance, which is usually detected in control reactions containing boiled enzymed or no substrate. Incubate the tubes on a water bath at 30°C. Measure the absorbance of the reaction mixture at 290 nm at 20 min interval for an hour. Use a reaction mixture without phenylalanine or enzyme extract as the control.

Comment

The period of assay can be shortened and the sensitivity increased by incubating the reaction mixture at 40°C on a water bath. Record the measurement of A_{350} at 5 or 10 min interval by returning the reaction mixture to the water bath between two readings. Calculate the rate of reaction. One unit of enzyme activity is that amount of enzyme producing an increase in A_{290} of 0.01 per hour (3.37 mu of cinnamic acid). Express the results as milli units (mU) of enzyme per 100 mg original fresh weight of plant tissues.

Assay of Tyrosine Ammonia-lyase (tyrase) (TAL)

In spite of its participation in phenol biosynthesis, TAL activity has not been studied in diseased tissue. Narayanaswamy (1980) examined it in some detail in groundnut leaves infected by *Alternaria alternata* and Mohanty *et al.* (1982) in rice leaves infected by *Xanthomonas oryzae* pv. *oryzae*.

All the methods used to assay PAL can be used to measure TAL except that the substrate is tyrosine. Isolate p-coumaric acid from the reaction mixture by partitioning with ether and measure its conc at 333 nm. Alternatively, measure the colour intensity in a colorimeter at 650 nm employing Folin-phenol reagent.

Phenol Oxidase of Plant Tissues

Phenol oxidase is also known as polyphenol oxidase, tyrosinase, DOPA oxidase, catechol oxidase, potato oxidase, etc. The enzyme catalyses the oxidation of monophenols and ortho-diphenols. Monophenols particularly tyrosine and p-cresol and ortho-phenols like caffeic acid, pyrogallol and substituted catechols are important substrates of the enzyme. It is present in almost all the plant species either as active or latent (`latent phenolase') and the enzyme has four copper atoms/mole (Mukherjee and Ghosh, 1975).

One of the foremost changes noted immediately after inoculation with a parasite is the increase in phenol oxidase and this observation has prompted a few pathologists to suggest that the first reaction which a pathogen brings in the host plants is to alter its oxidation-reduction potential.

Measurement of Phenol Oxidase

Several methods are employed to measure phenol oxidase; these include (i) colorimetric method, (ii) oxidation of catechol in a manometer, (iii) oxidation of catechol in a colorimeter and (iv) measuring the rate of oxidation of catechol in an oxygen electrode.

Colorimetric Method

Materials

Enzyme extract prepared from cotton plants infected by *F. oxysporum f. vasinfectum*

Pipettes, 1 ml, 5 ml

Colorimeter

Reagents

Phosphate buffer, 0.1 M, pH 6

Catechol 0.01 M dissolved in 9.1 M phosphate buffer, pH 6

Method

Pipette 2 ml of the extract and 3 ml of phosphate buffer into cuvettes, mix by inverting. Place in a colorimeter set at 495 nm and adjust the absorbance to zero. Remove and add 1 ml of catechol in phosphate buffer and mix again. Place the tube immediately in the colorimeter and record changes in absorbance at every 30 sec up to 3 min. Plot the changes in absorbance between 30 sec to 150 sec of incubation and calculate the enzyme activity. Maintain control with heated enzyme.

Comments

Compare the relative enzyme activity of healthy and diseased plants. With monophenols as substrates there is an induction period for activity, hence monophenols are not commonly used as substrate. Laccase type of enzyme catalyses the oxidation of p-phenylene diamine, hydroquinone and o-diphenols and does not oxidize mono-phenols. The reaction may be performed with 0.02 M citrate-phosphate buffer, pH 6.0 and addition of 0.5 ml of proline (5 mg/ml) to the reaction mixture gives consistent results.

Catalase

Catalase catalyses the breakdown of hydrogen peroxide to water and molecular oxygen (catalytic) or peroxidatively, H_2O_2 is converted to water in presence of hydrogen donor.

$$H_2O_2 \rightarrow H_2O + 1/2\ O_2 \text{ (catalytic)}$$

$$H_2O_2 + A\ H_2 \rightarrow A + 2H_2O \text{ (peroxidative)}$$

The importance of the enzyme in metabolism is still uncertain but peroxidative removal of H_2O_2 is important as it is toxic to cells. Each catalase molecule has four haemin groups as active centres and the iron in normally in the trivalent form. Increase in catalase activity have been recorded in several host-parasite interactions (Sarkar and Joshi, 1977).

Measurement of Catalase

The enzyme activity is measured by estimating the residual H_2O_2 in the reaction mixture. Determination of H_2O_3 by either spectrophotometric method or by permanganate titration is simple and dependable.

Spectrophotometric Method

Materials

Enzyme extract

Pipettes 1 ml, 5 ml

U.V. Spectrophotometer

Reagents

Phosphate buffer, 0.1 M, pH 6.4

H_2O_2 1% (use reagent grade)

Method

Pipette 2.5 ml of phosphate buffer into a cuvette and add 0.1 ml of H_2O_3 and 0.2 ml of tissue extract. Follow the changes in absorbance at 230 nm for 75 sec at 15 sec interval taking the first reading 15 sec after the addition of peroxide. Use cuvette containing tissue extract and buffer to adjust the absorbance to zero. Maintain controls with boiled tissue extract.

Comment

Compare the catalase activity of diseased and healthy plants at different stages of development.

Permanganate Titration Method

Kar and Mishra (1976) proposed a method to estimate the residual H_2O_2 in the reaction mixture as a measure of catalase activity. This is easy and convenient.

Materials

Enzyme extract

Pipettes, 1 ml, 5 ml

Porcelain crucible

Burette, 25 ml

Stop watch

Reagents

Phosphate buffer, 0.1 M, pH 6.8

H_2O_2, 0.1 M

Sulphuric acid, 2%

Potassium permanganate, 0.01 N

Method

Pipette 3 ml of phosphate buffer into a porcelain crucible. Add 1 ml of H_2O_2 and 1 ml of enzyme extract. Incubate the reaction mixture at 25°C for 1 min. Stop the reaction by the addition of 10 ml of H_2SO_4. Titrate the reaction mixture against 0.01 N $KMnO_2$ to estimate the residual H_2O_2 until a faint purple colour persists for at least 15 sec. Similarly, maintain a control in which the enzyme activity has been stopped by adding H_2SO_4 prior to the addition of enzyme extract. Express the activity as enzyme units. One unit of catalase is defined as that amount of enzyme which breaks down 1 μ mol of H_2O_2/min under the assay condition.

Peroxidase

Peroxidase catalyses the oxidation of various `hydrogen donors" like p-cresol, benzidine, guaiacol, ascorbic acid, nitrite and cytochrome C in presence of H_2O_2. The reaction may be represented as :

$$H_2O_2 + 2AH \rightarrow 2A + 2H_2O$$

Peroxidases are widely distributed in the plant kingdom. The prosthetic group of the enzyme is iron and can be split off readily. Peroxidase changes in diseased plants have been reported in many host-parasite combinations.

Measurement of Peroxidase

To assay peroxidase, the tissue extract is added to pyrogallol or guaicol which in presence of H_2O_2 is oxidized to a coloured derivative. The amount of purpurogallin or tetraguaiacol formed during the reaction can be followed in a colorimeter.

Materials

Enzyme extract

Pipettes, 1 ml, 5 ml

Stop watch

Colorimeter

Reagents

Pyrogallol, 0.05 M dissolved in 0.1 M phosphate buffer, pH 6

H_2O_2, 1%

Method

Pipette 3 ml of pyrogallol in phosphate buffer and 0.1 ml of the enzyme extract into a cuvette and adjust the absorbance to zero at 420 nm. Add 0.5 ml of H_2O_2 and invert the cuvette immediately to mix the contents and replace in the colorimeter. Measure changes in absorbance at 20 sec interval for a period of 3 min. Use the average change in absorbance per 20 sec between 40 and 160 sec to plot peroxidase activity. Maintain suitable controls with boiled enzyme extract.

Comment

Compare the enzyme activity of healthy and diseased tissues at different stages of growth. For assays with other hydrogen donors, guaiacol, o-dianisidine or phenylene diamine, read the absorbance at 470, 460 or 486 nm, respectively.

Ascorbic Acid Oxidase

Ascorbic acid oxidase catalyses the direct oxidation of ascorbic acid by molecular oxygen according to the equation :

$$\text{Ascorbic acid} + 1/2\ O_2 \rightarrow \text{Dehydroascorbic acid} + H_2O$$

This enzyme is widely distributed in the plant kingdom. The prosthetic group of the enzyme contains 6 copper atoms per molecule. The enzyme has been located in the soluble parts of the cytoplasm and in some instances, in cell wall materials. Marked increases in ascorbic acid oxidase have been reported in several host-parasite interactions.

Measurement of Ascorbic Acid Oxidase

The enzyme is assayed either in a manometer, by measuring the amount of oxygen consumed during the reaction or in a spectrophotometre by determining the residual ascorbic acid in the reaction mixture.

Manometric Method

A simple method based on the principle that within a certain range of enzyme concentrations the rate of oxygen consumption during the oxidation of L-ascorbic acid is proportional to the amount of enzyme present, is used to assay the enzyme.

Materials

Enzyme extract

Pipettes, 1 ml, 2 ml

Filter paper

Warburg respirometer

Reagents

Phosphate buffer, 0.1 M, pH 6

L-Ascorbic acid, 0.01 M dissolved, in 0.1 M phosphate buffer, pH 6

Aqueous, KOH, 20%

Method

Pipette 0.2 ml of KOH into the centre well of the Warburg flask and place a small, folded filter paper. Pipette 1.5 ml of tissue extract into the main compartment and 1 ml of phosphate buffer. Transfer 0.5 ml of ascorbic acid into the side arm, close the manometers and take readings at 10 min interval for three times beginning at zero time. Tip the side arm contents into the main chamber and take readings at 10 min interval for another 40 min. Maintain appropriate controls with (i) boiled enzyme extract and (ii) side arm without the substrate but with the buffer.

Spectrophotometric Method

The rate of oxidation of ascorbic acid is measured in a spectrophotometer. This method is relatively simpler than the manometric method.

Materials

Enzyme extract

Pipettes, 1 ml, 2 ml

Spectrophotometer

Reagents

Phosphate buffer, 0.1 M, pH 6

L-Ascorbic acid, 0.1 M dissolved in phosphate buffer, pH 6

Method

Pipette 0.1 ml of the enzyme extract into a cuvette and add 1.0 ml of phosphate buffer, 1.8 ml of dist water and 0.1 ml of ascorbic acid in buffer. Follow the reaction by plotting the changes in absorbance at 265 nm at 30 secondars interval up to 3 min. Cuvette containing tissue extract and buffer is used to adjust the absorbance to zero. Use the average change in absorbance between 30 and 150 seconds to plot oxidase activity. Maintain suitable controls with boiled enzyme extract.

REFERENCES

Agarwal, P. and Mahadevan, A. 1984. Lectins in germinating seeds and their effect on fungi. *Acta Phytopathology* 19 : 219-235.

Anwar, M.V., Majumdar, S.K. and Shetty, H.S. 1995. Changes in phenolic acids in sorghum and maize leaves infected with *Peronosclerospora sorghi*. *Indian Phytopathology* 48 : 21-26.

Chattopadhyay, A.K. and Samaddar, K.R. 1980. Effects of *Helminthosporium oryzae* infection and ophiobolin on the phenol metabolism of host tissues. *Phytopathologiscic Zeitschrift* 98 : 193-202.

Chattopadhyay, S.B. and Bera, A.K. 1980. Phenols and polyphenol oxidase activity in rice leaves infected with *Helminthosporium oryzae*. *Phytopathologiscic Zeitschrift* 98 : 59-63.

Kalaichelvan, P.T. and Mahadevan, A. 1979. Prohibitins in groundnut (*Arachis hypogea*). In : A. Mahadevan (ed.) *Physiology of Host Pathogen Interaction*, pp. 297-300. Today Tomorrow's New Delhi.

Kalaichelvan, P.T. and Mahadevan, A. 1987. Prohibitins from groundnut roots *Current Science* 56 : 1291-1292.

Kalaichelvan, P.T. and Mahadevan, A. 1988. Effects of phenolic prohibitins from groundnut in fungi. *Current Science* 57 : 193-195.

Kar, M. and Mishra, D. 1976. Catalase, peroxidase, and polyphenol-oxidase activities during rice leaf senescence. *Plant Physiology* 57 : 315-319.

Mahadevan, A. 1969. A theory concerning the specificity of plant pathogens. 65 : 358-363.

Mahadevan, A. 1973. Theoretical concepts of disease resistance. *Acta Phytopathology* 8: 391-423.

Mahadevan, A. 1979a. Biochemical aspects of plant disease resistance. *Biochemical Reviews* 69: 51-66.

Mahadevan, A. 1979b. Biochemical aspects of plant disease resistance. *Journal of Science and Industrial Research* 38 : 156-171.

Mahadevan, A. 1982. *Biochemical Aspects of Plant Disease Resistance. Part I. Preformed Inhibitory Substances ``Prohibitins''*. Today & Tomorrow's New Delhi.

Mahadevan, A. 1991. *Part II Post Infectionally Formed Defense Mechanisms.* Today & Tomorrow, New Delhi.

Mahadevan, A. and Rangamannar, S. 1980. Phenylalanine ammonia-lyase in groundnut (*Arachis hypogaea*). *Indian Journal of Plant Physiology* 23 : 288-291.

Mahadevan, A., Muthukumar, G. and Arunakumari, A. 1982. Lectins : Are they involved in recognition of parasites by plants. *Current Science* 51: 263-268.

Mall, S. and Suresh, K. 1989. Phenolic compounds associated with *Rhizoctonia solani* to potato. *Indian Phytopathology* 42 : 183-184.

Mohanty, S.K., Reddy, P.R. and Sridhar, R. 1982. Phenylalanine and tyrosine ammonia-lyases in bacteria leaf blight syndrome of rice. *Journal of Plant Disease Protection* 89 : 422-426.

Mukherjee, P.K. and Ghosh, J.J. 1975. Phenol oxidase activity in relation to resistance of rice to infection by *Helminthosporium oryzae*. *Science and Culture* 41 : 433-435.

Muthukumar, G. and Mahadevan, A. 1983. Microbial degradation of lignin. *Journal of Science and Industrial Research* 42 : 518-528.

Narayanaswamy, P. 1980. Phenylalanine ammonia lyase and tyrosine ammonia-lyase in *Alternaria alternata* infected groundnut leaves. *Indian Phytopathology* 33 : 507-508.

Purushothoman, D. 1974. Phenylalanine ammonia lyase and aromatic amino acids in rice varieties infected with *Xanthomonas oryzae. Phytopathology Z.* 80: 171-175.

Raychaudhari, R. 1980. Effect of extracts of certain Solanaceaous plants on virus infection. *Acta Botanica India* 8 : 91-94.

Raychaudhari, R. 1984. Changes in solanine content in *Solanum khasinum* during the course of its development, virus infection and glycoalkaloid treatment. *Indian Journal of Plant Pathology* 2 : 16-20.

Raychaudhuri, S.P. and Prasad, H.C. 1965. Effect of plant extracts and microbial growth products on the infectivity of radish mosaic virus. *Indian Journal of Microbiology* 5: 13-16.

Reddy, K.K. and Prasad, N.N. 1979. Changes in phenolics in the shoots of resistant and susceptible rice varieties infected by *Pyricularia oryzae* In : A Mahadevan (ed.) *Physiology of Host-Pathogen Interaction* : pp. 417-426. Today & Tomorrow's New Delhi.

Sarkar, K.R. and Joshi, R.D. 1977. Peroxidase and catalase activity in MLO infected brinjal. *Indian Phytopathology* 30 : 518-521.

Sridhar, R. and Mahadevan, A. 1979. Physiology and biochemistry of rice plants infected by *Pyricularia oryzae, Helminthosporium oryzae, Xanthomonas oryzae* and *Xanthomonas translucens* f. sp. *oryzicola. Acta Phytopathology* 14 : 49-82.

Swain, T. and Hills, W.E. 1959. The phenolic constituents of *Prunus domestica* l. The quantitative analysis of phenolic constituents. *Journal of Food Science and Agriculture* 10 : 63-68.

Verma, H.N. and Prasad, V. 1987. Systemic induced antiviral resistance by plant extract alters physiology of susceptible test host. *Indian Journal of Plant Pathology* 5 : 69-72.

William, F., Vasudevan, N., Boominathan, K. and Mahadevan, A. 1986. Microbial degradation of lignin and tannin. *Journal of Science and Industrial Research* 45 : 232-243.

CHAPTER 12

GROWTH INDICES AND LEAF AREA

INTRODUCTION

Growth is irreversible increase in size, weight and number. It is basic characteristics of plants due to various physiological and biochemical processes. However, plant growth and development is quantified by measuring the growth rates of individual organ or whole plant. The effect of various treatments can be compared with each other. Morphological parameters are evaluated by measuring the same plant at different time intervals adopting indirect methods. But, to determine dry matter production rate, the plants are sampled at various time intervals by using direct methods. Valuable inferences are drawn by calculating the different growth indices (Evans, 1972; Coombes *et al.,* 1995).

Principle

Plant growth and development indicate quantitative and qualitative changes during ontogeny. These changes are correlated and coordinated and can be determined by counting the number of organs, time required for the appearance of specific phonological stage. At each stage, rate of growth and development shows extent of functional aspects of the system (Evans, 1972). The ratio between various parameters gives correlation and allometric coordination between organs. Observations can be recorded after giving response period after the treatment application. In stress physiology experiments, observation can be determined before and during the stress treatment. Then recovery potential of plants is judged by giving a specific revival period and observations are determined.

METHODS TO MEASURE PLANT GROWTH

To determine quantitative growth of plants, we have to sample the plants at suitable time after the response period. It is important to make random samplings to minimize experimental error. Sample size should be sufficiently large to have uniformity in the data (Hunt, 1978; Hall *et al.,* 1993).

Indirect Methods

Indirect methods of growth analysis involves non-destructive sampling of the materials and give number of particular organs or plant stand in a community (Evans, 1972).

Materials

Extracts/Leachates, field or screen house grown plants, scale, jewel tags, hot air oven, scales and leaf area meter.

Select uniform plants, count and tag and the leaves present on the plants. Depending upon the objectives of experimentation, apply any of the following treatments like soil or plant leachates or fertilizers or salinity or waterlogging or withholding of watering for a suitable time interval (wait till you get some visible response which depends on treatment intensity and duration). Keep one set of the plants as control (without any treatment). Measure leaf area, length of elongating internodes and stem height with the help of scale before the treatment. Repeat all above observations after a suitable period of treatment. The values will indicate plant response to the treatment. Continue with soil or plant leachates or fertilizer or salinity or terminate waterlogging or drought by removing excess water under field conditions or remove pots from containers having water or by rewartering the plants. Give 15-20 days recovery period, so that new growth can occur. Again repeat all above observations. Pool the values from elongating (upper 3-4 internodes/leaf) and rest of the plant positions. Count the number of shed leaves. The values will indicate plant potential to recover from the treatment effects. Record all observations in tabulated form for individual parameter.

Likewise one can tabulate data of the internodes elongation, stem height, leaf area, number of shed leaf, duration of growth phases i.e. vegetative and reproductive stages, days required for bud, flower and fruit setting and maturation.

Direct Methods

Direct methods of growth analysis involve sampling of plant, then counting the organ number and their fresh/dry weight. These are destructive methods.

Materials

Extracts/Leachates, field or screen house grown plants, scale, jewel tags, hot air oven, scales and leaf area meter.

Harvest the plants before, during and after 15, 30 and 45 days of recovery. Remove excess sand from the root with water. Wash plants under running tap water and dry them with filter papers to remove excess water. Separate the different plant parts and keep separately in paper bags and dry in an oven at 80±1°C for 48 h. Weigh the dried plant material and record the data.

Calculations and Interpretation

Values of various parameters determined by non-destructive or indirect methods are used for qualitative measurement of growth. If the treatment affects differneitation at the cellular level, then growth will be influenced. We can determine absolute growth rate with direct methods to judge quantitative aspects of growth.

$$Absolute\,growth\ rate = \frac{Weight\ of\ plant\ organ\ or\ total\ plant}{Unit\ time\ (day, week\ or\ month)}$$

Its unit is (g day^{-1} or $week^{-1}$ or $month^{-1}$)

Changes in the duration of specific phase due to some treatment indicate degree of earliness. Alterations in absolute growth rate, indicate effect of the treatment on cell division and elongation, consequently organ or plant dimensions are changed. However, absolute growth rate may not always correlate with specific treatment, for which various growth indices are calculated as explained below.

CALCULATION OF GROWTH INDICES

In field grown crops, land area per plant is calculated from row to row and plant to plant spacing. Make sampling at suitable time interval (fortnightly) after giving treatment of leachate/ allelo-chemicals preparation, fertilizers etc. Separate out plant parts and measure leaf area. Keep the samples for drying in oven and note down the dry weight.

Calculations and Interpretation

Various growth indices viz., biomass duration (BMD), leaf area duration (LAD), crop growth rate (CGR), leaf area index (LAI), leaf area ratio (LAR), leaf weight ratio (LWR), specific leaf area (SLA), relative growth rate (RGR), net assimilation rate (NAR) or unit leaf rate (ULR), unit shoot ratio (USR), production rate (PR) and root shoot ratio (R/S) can be determined instantly or over the time t_2-t_1 by using following formulae :

(1) Biomass Duration (BMD)

BMD indicates duration of total biomass production to support plant growth. It is time for which total biomass is maintained i.e., average product of total DW with time interval. It cannot be determined instantly. BMD is generally correlated with yield. Its unit is weight unit multiplied per time unit.

$$BMD = (W_1 + W_2)\ (t_2 - t_1)/2$$

Generally, crop yield is determined by the interaction between BMD and RGR. Hence, higher BMD values indicate better production potential.

(2) Leaf Area Duration (LAD)

LAD indicates duration of leaf area to support the net photosynthesis of plant. It is similar to BMD and can be calculated either from leaf area or leaf area index. Higher value of LAD contributes towards BMD. Its unit is area unit multiplied by per time unit.

LAD (on LA [Leaf Area] basis) = $(LA_1 + LA_2)\ (t_2 - t_1)/2$

LAD (on LAI basis) = $(LAI_1 + LAI_2)\ (t_2 - t_2)/2$

Generally crop yield = LAD x NAR

(3) Crop Growth Rate (CGR)

This is rate of accumulated dry matter expressed per unit land area. It indicates stand productivity under field conditions. Its unit is weight area^{-1} time^{-1}

$$CGR = \frac{1}{P} \times \frac{dw}{dt} \text{ or } \frac{1}{p} \times \frac{w_2 - w_1}{t_2 - t_1}$$

P = land area plant^{-1}

Indirectly CGR is coefficient of solar energy utilization. Higher CGR is desirable for higher yield of crops.

(4) Leaf Area Index (LAI)

This is leaf area maintained per unit land area. LAI indicates extent of leafiness and time required to achieve it in crop stand. It is calculated as under :

For instant value LAI = LA/P, while over the time -

$$t_2 - t_1 = LA_2 - LA_1 / P$$

LA, LA_1 and LA_2 are leaf area in square meters at any time or at t_1 and t_2 time.

Where,

P = land area in m^2 occupied by one plant.

An early high value of LAI is more desirable, to fully utilize the incident radiation.

(5) Leaf Area Ratio (LAR)

It is ratio between leaf area and total plant dry weight. Higher value of LAR indicates coupling of LA with photosynthesis to produce more dry matter. Its unit is area per weight.

Like LAI it can be instantly calculated by the ratio of LA/W, while over the time t_2–t_1 it is calculated by LAR = $(LA_1/w_1)\ (LA_2/w_2)/2$

It indicates the ratio of photosynthesizing portion to respiratory portion of the plant.

(6) Leaf Weight Ratio (LWR)

LWR indicates the average fraction of total plant dry matter in the form of leaf, or leafiness of the plants on weight basis. It is similar to LAR and here instead of leaf area, leaf weight is used.

LWR = LW/W or $(LW_1/w_1)\ (LW_2/w_2)/2$ dimensions less

(7) Specific Leaf Area (SLR)

It is ratio between LA and its weight. It indicates leaf thickness and can also be used to determine leaf area of bulk sample if its weight is known by multiplying leaf weight with SLA of representative leaf sample.

SLA = LA/LW or $(LA_1/LW_1)\ (LA_2/LW_2)/2$ unit of SLR is area wt^{-1}

(8) Relative Growth Rate (RGR)

This is rate of dry matter accumulation and is expressed per unit of RGR is weight over the previous sampling.

$$RGR = \frac{1}{w} \times \frac{dw}{dt} \; or \; \frac{\log_e w_2}{dt} - \frac{\log_e w_1}{t_2 - t_1}$$

unit of RGR is (wt. wt^{-1} $time^{-1}$)

RGR = NAR x LAR

RGR is fundamental measure of dry matter production and can be safely used to compare the performance of species or effects of different treatments under specific conditions. An error due to initial weight of organ is taken care of, while determining RGR. It can be measured for individual organ or for whole plant. From RGR values, we can't find out the causal factors, which determine plant or crop performance

(9) Net Assimilation Rate (NAR)

It is rate of dry matter accumulation and expressed on leaf area.

$$NAR = \frac{1}{LA} \times \frac{dw}{dt} \text{ or } \frac{w_2 - w_1}{t_2 - t_1} \times \frac{\log_e LA_2 - \log_e LA_1}{LA_2 - LA_1}$$

Unit of NAR is weight leaf area^{-1} time^{-1}

Where, $\log_e$ is natural or Neparian log.

This formula is used when ratio between plant weight and leaf area is one. However, when this ratio is 2 or more than following formula is used.

$$NAR = \frac{2\left(w_2 - w_1\right)}{\left(LA_2 + LA_1\right)\left(t_2 - t_1\right)}$$

Changes in RGR can be better explained in term of NAR and LAR.

(10) Production Rate (G)

It is rate of accumulation of metabolite/mineral expressed per unit of another metabolites i.e. accumulation of protein nitrogen per unit RNA

$$GPN, RNA = \frac{1}{RNA} \frac{dPN}{dt} \text{ or } \frac{PN_2 - PN_1}{t_2 - t_1} \frac{\log_e RNA_2 - \log_e RNA_1}{RNA_2 - RNA_1}$$

Unit of G is weight. weight^{-1} time^{-1}

Where

PN = Protein nitrogen;

PN_2 and PN_1 = Protein nitrogen at time t_2 and t_1.

RNA, RNA_2 and RNA_1 = RNA instantaneous time, at t_2 and t_1 time, respectively.

(11) Unit Shoot Ratio (USR)

This is rate of accumulation of dry weight expressed per unit shoot weight.

$$USR = \frac{1}{sw} \times \frac{dw}{dt} \text{ or } \frac{w_2 - w_1}{t_2 - t_1} \times \frac{\log_e sw_2 - \log_e sw_1}{sw_2 - sw_1}$$

Unit of USR is weight. weight^{-1} time^{-1}

(12) Harvest Index (HI)

This is ratio of harvested product to the total cummulative

$$HI = \frac{Economical\ yield}{Biological\ yield} \times 100$$

Economical yield is actual harvested yield and biological yield is total dry matter including the economical yield. It is a very stable parameter of growth indices.

Precautions

(i) Carefully select uniform plants for tagging.

(ii) Wash plant material carefully without removing fine roots.

(iii) Do not subject plant material to high (100°C and more) or low (60°C and below) temperature.

(iv) On cloudy days, heat the material to remove absorbed excess moisture before weighing.

METHODS TO MEASURE LEAF AREA

Several principles are applied to determine the leaf area. The choice of methods depends on the main purpose of measurements i.e.

(a) Whether the total leaf area or (b) individual leaves (c) it should be destructive or (d) non-destructive. Some important methods are described below (Golovin and Migunov, 1968).

(1) Leaf Outline on Graph Paper

In this method the leaf area of the plant is measured by using the graph paper and grid on the acrylic plastic sheet. The method is very economical and can be used without the electrocity/powder in the field conditions.

Materials

Leaf sample, graph paper, acrylic plastic sheet, etc.

Procedure

(i) Place the leaf on graph paper or grid squares made on trasnparent acrylic plastic sheet for calculating the area blocked by leaf in sq cm and outline its margin.

(ii) Obtain the area the counting the squares within the outline.

(iii) In field conditions, mark a grid on a clear piece of acrylic plastic sheet, against which the leaf is held, and count the squares blocked out by the leaf.

Advantages

This method is fairly accurate.

Disadvantages

(i) The accuracy of this method depends on leaf shape, for highly divided or compound leaves, this type of leaf area measurement is impractical.

(ii) It requires long time and difficult to record leaf area of large number of leaves.

(2) Planimeter Method

Hand planimeter is a standard technique, by which other methods of leaf area measurement are calibrated. Measurements with planimeter are more accurate, when the angle between the two arms of instrument deviates from 90° as little as possible. Its accuracy, however, should not be taken for granted without checking.

Materials

Leaf sample, hand planimeter, photosensitive paper.

Procedure

(i) Measure the leaf directly, or obtain an outline of the leaf by tracing or by printing on a photosensitive paper.

(ii) Measure the area within the outline using a hand planimeter.

(iii) Repeat planimeter readings at leaf twice (with the arms of instrument in different position) and record the mean of the values indicated by the counter.

Advantage

This method is best suited to measure leaves in which width are similar to length (orbicular, elliptic, cordate, ovate leaves, etc.). This method is used only for small detached leaves.

(3) Gravimetric Method

This method is based on weight of photosensitive paper on which shape of the leaf is traced. The leaf area is calculated by weight to area relationship of paper.

Materials

Leaf sample, balance, good quality heavy paper.

Procedure

(i) Trace on outline of leaf, or make a print on a photosenstive paper, which has a uniform weight distribution with area.

(ii) Cut the leaf shape from paper and weigh.

(iii) Calculate the leaf area from the weight to area relationship of the paper by weighing pieces of the paper of known area.

Disadvantages

Cutting out the leaf copies is very tedious. This can be obviated by tracing of leaf outlines on paper with 15% sulphuric acid using a glass pen. After slight heating, the sulphuric acid becomes concentrated, attacks the paper and the copy falls out. The shape of the copies is satisfactorily preserved if thin but heavy or hard paper is used.

(4) Dot Counting Method

This is the modification of the graph paper or grid method, which is suitable for field use. Some electronic devices have also been developed to record the number of holes (dots) covered by the leaves. The accuracy of the dot counting method is subject to limitations, similar to the method used for estimating vegetation cover in the field.

Materials

Dotted sheets of paper or acrylic plastic.

Procedure

(i) This is a modification of graph paper method and is used in field conditions by making a grid of regularly spaced dots.

(ii) Count the number of dots obscured by the leaf.

(iii) Repeat the measurement with the leaf in different position and record the mean value.

Precautions

The most suitable distance between the dots depends on leaf size. One cm x one cm spacing is most common. The increased accuracy obtained by closure spacing (0.5 cm) do not commensurate with extra time spent on counting the dots. This method is suitable for broad and small leaves.

(5) Light Interception Method

This method based on the light interception, is commonly used for indirect measurement of leaf area. The instruments employed are planimeter. There are various established features of the method, including requirement for a light proof housing, an even illumination at the level of the inserted leaves, cooling of light source in some instances, protection of photocells from strong illumination and the desirability of knowing details of the light source emission spectrum and that of photocell sensitivity.

Principle

Intensity of light emitted from a constant source and reaching a detector is proportional to the area of leaves placed between light source and detector. The detector is usually a photovoltaic cell, whose output is directly measured against the constant output from a reference photocell under standard illumination.

Materials

Photoelectric planimeter, leaf sample.

Procedure

The leaf area is measured, when leaf passes through the light source and amount of light intercepted is determined by photo tube. Photoelectric planimeter can be of any required size. Large planimeters can be adjusted to accommodate small samples by inserting black paper masks with openings of suitable size. The resulting reduction in luminance may require compensation by increasing the sensitivity of photocells in some way. Even illumination over the area on which the leaves are to be positioned (active area) is essential. It is achieved if light source is a bank of fluorescent tubes or incandescent lamps. If the light is evenly scattered before reaching the leaf surfaces then even luminance may also be obtained by inserting an opaque glass sheath between the light source and the leaves. Photoelectric planimeter can be classified as being either `selective' or `non-selective' with respect to the differences in light transmission by the leaves. Most of the modern photoelectric planimeters are non-selective, the error due to differential light transmission by the leaves being less than 3% or even 1%. With scattered light, the error due to light transmission can be reduced by using very weak light 162 lux at the active area level. This is direct and easy method.

Precautions

(i) Calibration of the instrument should be checked during measurements.

(ii) The elaborate models are suited only for laboratory measurements of relatively large leaves.

(iii) Loading the instrument with very small leaves is very time consuming.

(iv) Leaf wilting can become a problem.

(v) Condensation of water evaporating from the leaves on the glass sheets (between which they are pressed) cause problem.

(6) Air Flow Interception Method

This method is best suited to measure area of evenly distributed detached small leaves of clover, flax and grasses etc or pieces of them, rather than for larger leaves. Its modified version of airflow planimeter show less errors against hand planimeter. The accuracy of the airflow technique is greater than gravimetric method.

Materials

Air flow planimeter, leaf sample.

Procedure

In this method leaves are interposed between a bulk source of air and receiver (vacuum line). The resistance to air flow offered by the leaves is measured by balancing the flow in second flow line. The balancing device is calibrated in leaf area units. The leaves are pressed on to the `active area' (a perforated plate with densely and evenly spaced small holes) automatically by air suction, which makes loading the instrument and measurement less time-consuming than with a photoelectric planimeter.

Disadvantages

(i) In forage plants, in which leaf area is mostly measured with air flow planimeter, separation of leaves from the stem is very tedious.

(ii) The retardation of the flow of very fine sand instead of air is caused by leaves.

(7) Indirect Methods of Leaf Area Estimation

These methods are based on the statistically defined mathematical relationships between some leaf characteristics and the leaf area.

(A) Relationship Between Linear Measurements and Leaf Area

Estimation of leaf area from linear measurement is possible with both leaves and their images without destroying the leaves. Different authors have proposed different relationships but the general form of relationship was proposed by Golovin and Migunov (1969) and has been found satisfactory.

$$A = b \times LB$$

Where,

A is leaf area.

b is a coefficient that requires checking from time to time with leaf shape and age.

LB is length and breadth of leaf.

Lal and Shubba Rao (1951) proposed similar relationship for cereal as under :

$$\log A = \log L + \log B - \log b$$

Where,

L is leaf length.

B is leaf breadth measured at L/2

b is a constant LB/A.

These values of b i.e. constant or coefficient used for converting L and B to leaf area for different plant species has been provided by Kvet and Marshell (1971).

These relationships are simple and sufficiently accurate. The full form of linear equation, $y = bx + a$ is frequently used, in which y represents leaf area and a and b are constants. Relationships have been examined in which x has been leaf length (L), leaf breadth (B), their product LB and L + B and also determine the smallest number of leaves required to obtain a reliable relationship.

(B) Leaf Weight Method

This method depends upon the leaf area : leaf weight ratio and is used in growth analysis of plant stands. The leaf area : leaf weight is estimated in a representative sub sample, using any methods of leaf area measurement. The relationship between leaf area and leaf fresh weight is used in determining the leaf area (Turell, 1961). The variations in leaf area : leaf weight ratio, taking leaf fresh weight, air dry weight or oven dry weight was presented by Alenkseenko (1965). Kvet (1971) showed that leaves of different species differ in their adaptability of their specific leaf area to environmental conditions. Hence, he presented a range of leaf area : leaf weight ratio (specific leaf area) in different crop species in tabular form (Kvet and Marshall, 1971).

Relationship Between Leaf Area and Leaf Water Content

Leaf area is linearly related to absolute leaf water content (leaf fresh weight - leaf dry weight) when

$$A = a + bw$$

Where,

A is total leaf area.

w is the absolute leaf water content.

a and b are the constent.

(C) Leaf Counting Method

The mean area of one leaf or a set of known number of leaves is assessed in a representative sub sample and the total area is obtained by multiplying this mean value by total number of leaves in the sample. This method gives unbiased results, only when the area of individual leaves fluctuates within a relatively narrow range and the leaves are easy to count.

MEASUREMENT OF LEAF AREA BY LEAF AREA METER

Portable leaf area meter provides non-destructive, precise method for monitoring crop canopy development or individual plant growth in the field or growth chamber. Leaf area reduction and recovery associated with condition such as insect infestation, air pollution, drought and disease can be evaluated by measuring the same plant throughout the season (Sestak *et al.*, 1971; Hall *et al.*, 1993).

Principle

Leaf Area Meter uses electronic methods to stimulate a grid pattern on the leaf. The scanning head uses a row of 128 narrow bands, red light emitting diodes (LEDs), spaced on one mm centers, to examine 128 grid cells across the width of leaf. The LEDs are sequentially pulsed (only one LED is lit at a time) to examine a particular grid cell in the row. The LEDs are located along a line 0.62 cm from the edge of the upper section of the scanning head. The base of the scanning head contains a Lense-Photodiode system, respond only to collimated, pulsed LED light. This design makes the measurment insensitive to other light sources.

Materials

Portable leaf area meter have two components, the scanning head and the readout consol. The consol can also be used with transparent belt conveyor, to determine leaf area of detached leaf sample.

Procedure

(i) Connect the scanning head to the readout consol.

(ii) Turn the ON - OFF switch to ON.

(iii) Open the scanning and position it over a leaf. The leaf doesn't need to be detached from the plant.

(iv) Close the scanning over the petiole, so that the leaf base is in the scanning head but not between the LED and photodiode windows.

(v) Draw the length encoding cord to some location, where it can be held steady. The knob can be held against a stem or in broad leaf plants. The knob can be hold firm, by clasping it together with the petiole between the second and third fingers.

(vi) Press the scanning head button, to clear any reading registered before starting the measurements.

(vii) Draw the closed scanning head over the leaf, while the length encoding knob is held stationary. Drawing speed needs to be constant.

(viii) After the scanning head has passed over the leaf, let the length encoding cord slowly rewind itself back into the scanning head.

(ix) Note down the reading (sq. cm) on the console unit.

(A) Using a Transparent Sheath

A transparent sheath can be used to measure the area of detached leaves and that require sample support using the following steps :

(i) The sheath should be clean and the empty sheath to ensure that no spurious area is measured.

(ii) Place an object in the sheath about six cm from one end.

(iii) Open the scanning head and place the sheath in the head with about four cm protruding from the leaf edge and allow the head to close.

(iv) Grasp the length encoding knob with your left hand thumb and forefinger and hold it stationary against the sheath.

(v) Press the scanning head button once to clear the value on the display. Draw the sheath through the scanning head, simultaneously drawing out the leaf encoding cord.

This method is used to measure area of small objects and detached leaves. Many types of leaves attached to plants, cannot be measured directly through scanning head. Area of several leaf types can be measured with this method :

(i) Compound leaves can be measured by either scanning each leaflet or by enclosing the entire intact leaf through transparent belt.

(ii) Area of tender leaves can be measured.

(iii) Insect damaged or natural dissected leaves having fine lobes usually require support of transparent belt.

(iv) Small leaves like alfalfa and some grasses can be measured.

Limitations

If width of leaf is more than 12.8 cm, leaf area meter can't be used, because of its scanning head width. In such cases, PC-coupled with scanners are used and special software (LAMZ 101) or image analysis systems are available to measure area of diseased or senescence portion of leaf.

(B) Using a Transparent Conveyor Belt

Procedure

Fix the scanner head in the conveyor belt unit, tighten the knob. See that conveyor belt moves freely through the scanner head.

(i) Place and spread the leaves on the supporting platform.

(ii) Switch ON the ON/OFF switch. Conveyor belt will start moving and thus leaves samples passes through the scanner head.

(iii) Go on putting more leaves till all the leaves of a plant are measured.

(iv) Note down the reading from display panel, press memory clearing button on the conveyor belt unit and leaf area meter is ready for another leaf sample.

(v) Repeat step (ii) to (v) omitting step (iii).

Precautions

(i) The conveyor belt should be thoroughly cleaned and it should not be too loose or tight in the scanner head

(ii) Leaf should be turgid and well spread on the supporting platform. Folded leaves will give inaccurate values.

REFERENCES

Alekseenko, L.N. 1965. Weight method of leaf area determination of meadow plant and communities. *Botanical Journal* 50 : 205-208.

Coombes, J., Hall, D.D., Long, S.P. and Scurlock, J.M.O. 1995. *Techniques in Bioproductivity and Photosynthesis.* pp. 41-49. Pergamon Press. New York.

Evans, G.C. 1972. *Quantitative Analysis of Plant Growth.* Blackwell Scientific Publications, Oxford, 734 p.

Golovin, V.V. and Migunov, V.S. 1968. Leaf area determination in soybeans by means of parameters. *Vestn. Sel'skokhoz. Nauki* 13 : 90-91.

Hall, D.O., Scurlock, J.M.O., Nordenkampp, H.R.B., Leegood, R.C. and Long, S.P. (eds.) 1993. *Photosynthesis and Production in a Changing Environment: A Field and Laboratory Guide.* Chapman and Hall, London. 147 p.

Hunt, R. 1978. *Plant Growth Analysis.* Edward Arnold, New York. 65 p.

Kvet, J. 1971. Reaction of woodland herbaceous plants to the removal of tree canopy. *Folia Geobot Phytotax* 6 : 36-45.

Kvet, J. and Marshall, J.K. 1971. Assessment of leaf area and other assimilating plant surface. In: *Plant Photosynthetic Production : Manual of Methods,* (Eds. Z. Sestak, J. Catsky and P.G. Jarvis), pp. 517-546. Dr. Junk Publications, The Hague.

Sestak, Z., Catsky, J. and Jarvis, P.G. (Eds.) 1971. *Plant Photosynthetic Production : Manual of Methods.* Dr. Junk Publications, The Hague. 818 p..

Turrell, F.M. 1961. Growth of photosynthetic area of citrus. *Botanical Gazette* 122 : 284-298.

CHAPTER 13

MEASUREMENT OF PHOTOSYNTHESIS, RESPIRATION AND TRANSPIRATION

INTRODUCTION

Growth is irreversible increase in plant size or weight, thus result of integration of many physiological and biochemical processes. Crop production and productivity is the integration of the crop growth occurred during the various growth phases. The quantum of growth is determined by the genetic potential of the crop and their integration with the growing environment. Growth is the net result of gross photosynthesis minus the energy lost through respiration. Physiological processes like photosynthesis, respiration, transpiration and canopy temperature are important growth and plant health indices as well. Therefore, methodologies to measure these important growth processes under field conditions are described in this chapter.

PHOTOSYNTHESIS

Increase in biomass of a plant per unit time is called growth and is dynamic process dependent on time. Growth can be measured in terms of quantitative increase per unit time (dry matter accumulation, plant height etc.), in terms of physiological processes like photosynthesis and respiration. Photosynthesis is a biochemical reaction, which combine the inorganic compounds like water and CO_2 and form organic compounds like sucrose and starch. Many aspects of photosynthesis are better studied in the laboratory than in field. However, single point measurement on small part of leaf is not conclusive. Seasonal measurements (Mahon and Domey, 1979) on whole crop canopy, gave better relation with yield (Zelitch, 1982).

Photosynthesis can be described by the chemical equation :

$$6CO_2 + 12\ H_2O + 12\ NADH + 18ATP\ (112.3\ Kcal) \rightarrow$$
$$C_6H_{12}O_6 + 12\ NAD + 18\ ADP + 6\ O_2 \uparrow + 6H_2O + 18\ Pi$$

The photosynthetic assimilation of carbon is partly counter balanced by the loss of carbon by respiration.

$$C_6H_{12}O_6 + 6O_2 + 12H_2O + 32\ ADP + 32Pi \rightarrow 6CO_2 \uparrow + 6H_2O + 32ATP\ (672\ Kcal)$$

Therefore in photosynthesis, CO_2 is consumed and in respiration CO_2 is evolved. Hence, the decrease or increase in the conc. of CO_2 in the chamber enclosing crop canopy will give the estimation of photosynthesis and respiration rates, respectively. Different workers have used different methods to measure growth and photosynthesis but presently gasometric methods are most widely used to determine gas exchange rates i.e. photosynthesis and respiration of plants.

The Photosynthesis Measurement

There are many different ways to measure photosynthesis. Techniques like growth analysis or sophisticated *in vitro* biochemical and biophysical techniques have been applied. But especially the gas exchange method provides the information to study the macroscopic plant response to the environment (McDermitt *et al.*, 2000).

Therefore, agronomists, crop physiologists, plant ecophysiologists and horticulturists have applied gas exchange techniques to study the plant response to experimental treatments or environments. Plant physiologists and biochemists also used this technique to get knowledge about the exact chemical and physical mechanisms, which are causal for the plant response. But, how does it work?

As mentioned above, there is no discrete photosynthesis sensor. And photosynthesis is a calculated parameter based on CO_2 concentration measurements, gas flow, and others parameters. Therefore, it seems understandable that usually a system is used to determine photosynthesis - a gas-exchange system (Field *et al.*, 1989).

In the nineties the calculation of photosynthesis was based on the measurement of the CO_2 rate changes of a leaf enclosed in a largest or shortest chamber. This system is called `closed system' (Fig. 1). Air circulates a closed circuit through cuvettes. When a leaf is enclosed in the chamber, the CO_2 concentration in this chamber will decline until the experiment is finished or until the CO_2 compensation point is reached (Davis *et al.*, 1987; McDermitt *et al.*, 1989).

If the enclosed plant leaf makes photosynthesis, the CO_2 concentration will decrease. The decrease of CO_2 concentration per unit time is the base of calculation of net photosynthesis :

$$A = \frac{(c_b - c_s) \times V}{L \times \Delta t}$$

where A is the net CO_2 assimilation rate, expressed in mol CO_2 m^{-2} s^{-1}

c_b is the CO_2 concentration at the beginning of the measurement, expressed in mol mol^{-1}

c_s is the CO_2 concentration at the end of the measurement, expressed in mol mol^{-1}

L is the leaf area, expressed in m^{-2}

V is air volume inside of the system, expressed in mol

Δt is the duration of the measurement, expressed in s

An analyser (closed system) is represented schematically in Fig. 1. It consists of four parts :

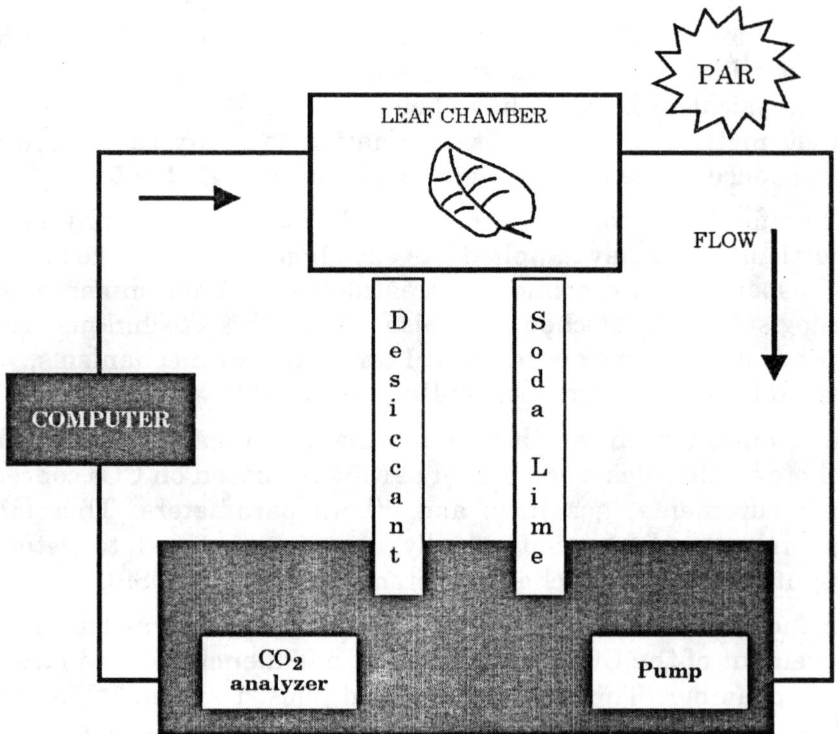

Figure 1. Scheme of closed system to measure photosynthesis (Li-Cor, 1990)

1. Leaf Chamber : Photosynthesis depends on many factors. The most important factors are PAR (Photosynthetically Active Radiation), the concentration of CO_2 and the water vapour pressure in the air. That's the reason by what the leaf, whose photosynthesis will be measured, must be normally enclosed in a chamber and as many as possible external factors must be controlled. The leaf chamber can be made by several materials, the chamber of a LI-6200, for example, is made of acrylic glass. Inside the chamber are a lot of more or less utilitarian sensors and ventilators. Generally, the ventilators are to homogenise the CO_2-concentration. The sensor heads check for several ambient parameters, for example leaf and air temperatures, relative air humdity and also the PAR.

2. Pump : Its function is to provide the system with a continue airflow along the closed circuit between the leaf chamber and the gas analyser. The air stream could be detoured to pass through a desiccant to avoid excess of humidity and/or through a soda lime to catch any CO_2, before its entry into the gas analyser. This is only required for calibration.

3. Infrared Gas Analyser (IRGA) : It makes exact measurments of the whole CO_2 and performs every second one measurement of the air stream, which comes from the leaf chamber. In general accuracy and resolution of 1 mmol mol^{-1} (1 ppm) is required, especially in a differential system. The heart of any gas exchange analyser is the infrared gas analyser (IRGA). The basic principle is the following : CO_2 is a strong absorbent of infrared radiation. So, an infrared source shines through a sample chamber and on to a detector. The energy at the detector is the total energy entering in the system minus that absorbed in the sample chamber (Field *et al.,* 1989).

The main problem is the discrimination between CO_2 and other gases that absorb in the IR. Therefore, most modem IR uses a broadband (non-disperse) source and a detector sensitive only to the CO_2 absorption bands. Recently, practically all CO_2 IRGA incorporated either a parallel or series Luft detector, which consists of two CO_2 filled chambers, separated by a flexible membrane. Different IR absorption in the chambers can be measured by connecting the membrane to a plane. So, one of the two detector chambers is exposed to IR passing through the sample cell, while the other one is exposed to IR coming through a reference cell.

Another problem must be taken under consideration: Both CO_2 and H_2O vapour absorb IR, even at different wavelength. So, different strategies have been applied. Either the incoming air stream have been completely dried before reaching the IRGA, or the IRGA has been provided with filters to eliminate the wavelength band for H_2O, or the effect of water vapour has been compensated by calculation. So, the best way seems to be to dry the incoming air stream (Field *et al.,* 1989).

4. Computer : It is provided with a program to store and handle the data. It permits the modification of inner parameters of the system in function to the use of the equipment. It disposes of storing data memory necessary for the measurement, and permits the introduction of codes in any measurement for a later and correct data treatment.

Another kind of system for photosynthesis measurement was also developed. It is called open system (Fig. 2).

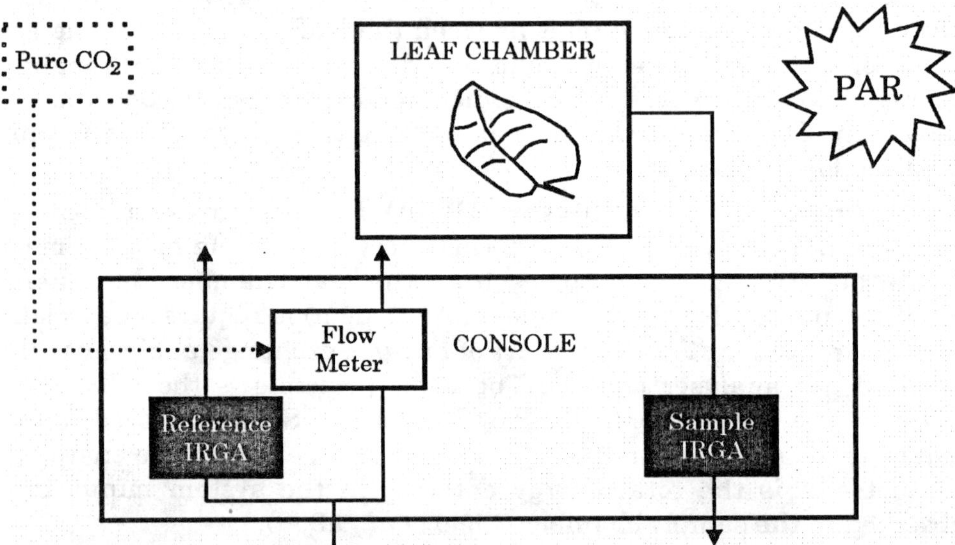

Figure 2. Scheme of a traditional, differential open system. The broken line presents a compensated open system (Li-Cor, 1998).

The measurement of photosynthesis in open system is based on the differences of CO_2 and H_2O in an air stream, which flows through a chamber where a leaf is placed in.

In an open system, especially a differential system, fresh air from outside of the console will be aspirated and introduced into the system during the whole measurement. Photosynthesis will be calculated from the CO_2 depletion, which occurs when a known airflow rate passes the leaf chamber where a photosynthesising leaf is enclosed (Field *et al.*, 1989; Steubing and Fangmeier, 1992).

$$A = \frac{u_c \times c_e - u_a \times c_a}{L}$$

Where, A is the net CO_2 assimilation rate, expressed in mol CO_2 m^{-2} s^{-1}

u_e is the airflow rate in the entry of the system, expressed in mol s^{-1}

c_e is the CO_2 concentration at the entry of the system, expressed in mol mol^{-1}

u_a is the airflow rate at the exit of the system, expressed in mol s^{-1}

c_a is the CO_2 concentration at the exit of the system, expressed in mol mol^{-1}

L is the leaf area, expressed in m^{-2}

Some gas exchange analysers can control microclimatic conditions in the leaf chamber (humidity, temperature, PAR, CO_2 concentration). In the case of compensating systems, CO_2 depletion by photosynthesis is compensated by CO_2 injection, and so the CO_2 concentration in the air exiting from the chamber is the same as that in the air stream that enter in the chamber. Therefore, the CO_2 addition means exactly the net photosynthesis rate (Field *et al.*, 1989; Steubing and Fangmeier, 1992).

In an open system measurement, even more as in a closed system, all depends on an accurate CO_2 measurement. Under normal outdoor conditions $78\%_{vol}$ of the air is nitrogen, $21\%_{vol}$ is oxygen, $0.95\%_{vol}$ are some inert gases and only about $0.035\%_{vol}$ or 350 µmol mol^{-1} is CO_2. But the potential for fluctuations is high. So your breath could reach a CO_2 concentration up to 50,000 µmol mol^{-1} (Li-Cor, 1998).

Light Measurement

Radiation seems to be the most important environmental factor in photosynthesis, but also the most difficult to measure (Li-Cor, 1998).

To measure the photosynthetically active radiation normally two types of sensors are used (Pearcy, 1989).

(a) **Quantum sensor**: Its relatively low cost has made it standard for measurement of PPFD. It is a filtered silicon-blue photocell covered by a 400-700 nm broadband interference filter, some piece of heat absorbing glass and a cloured gelatine filter. On the top of the housing an opaque white plastic disk is mounted, which has the function of a diffuser. The heat absorbing glass provides further blockage beyond 800 nm. The coloured gelatine filter corrects the spectral sensitivity so the sensor will respond more closely to an ideal photon response. Different firms offer models based on these characteristics.

(b) **GaAsP photocells** : They are very cheap, and so where a large amount of sensors is required the gallium arsenide phosphide (GaAsP) photocells will be very attractive. This sensor uses the photovoltaic effect to generate a voltage when it is exposed to light. It consists in a bare plate coated with a layer of GaAsP covered by a thin layer. The absorption of a photon with sufficient energy excites an electron, creating a positively charged hole. It begins a drift towards the junction of the layers, and the result is a negatively charged GaAsP layer and another, positively charged layer. This kind of sensor has the peak sensitivity at 610 nm, with a spectral response from 300 to 680 nm.

Sunlight is an excellent light source, but only on sunny settings. In understories or at cloudy days frequent measurments must be taken to minimise light differences. The slow response of the gas exchange equipments in relation to the changes in light and the responses of the light sensor present a special problem if PAR measurements are made in the conjunction of photosynthesis measurements (Pearcy, 1989).

So, sunlight is also unpredictable. It varies during the day and can be obscured by clouds. Therefore, artificial light sources are typically used for controlled environmental experiments (Field *et al.,* 1989).

DIFFERENT GAS EXCHANGE SYSTEMS

The folliwng two apparatus are able to measure net carbon photosynthesis in plant leaves as well in the laboratory as in the field. Li-Cor 6200 and Li-Cor 6400 are two different systems to obtain the `same' measurement in two diferent ways. Both can be applied in a broad range of different fields, as in :

- Ecology: with the possibility to know the photosynthetic behaviour of a same species in different ecosystems or of different species in different ecosystems.
- Agronomy: environmental stress, productivity, the influence of the soil conditions in the crop status, the study of the herbicide effects.
- Entomology: the effect of the pest as of the insecticides on the crop evolution.
- Biotechnology: comparisons between different genotypes, etc.

CLOSED SYSTEM (LI - 6200)

LI-6200 (Li-Cor model 6200, Lincoln, NE) was one of the first CO_2 analyser, based in a closed system for CO_2, which can measure in a short time interval (10-20 sec) the net interchange between a leaf enclosed in a leaf chamber and the atmosphere inside it.

Other measurements that can be determining the final value of photosynthesis, like the photosynthetically active radiation (PAR), relative humidity, leaf temperature, etc., can be also obtained here. An improvement of this apparatus according to previous gas exchange systems is the possibility to get transpiration measurements. A decrease in the water potential of leaves can be related with a photosynthetic reduction (Boyer, 1996). The water loss by the leaf enclosed in the chamber is determined measuring the necessary dry airflow to maintain a constant relative humidity inside the chamber. So, it is possible to obtain an idea about the stomatal aperture / closure seeing the transpiration values (mol m^{-2} s^{-1}).

LI-6200 is a compact, small size and portable instrument, which allows measuring directly in the field (Davis *et al.,* 1987; McDermitt *et al.,* 1989). The leaf chambers are interchangeable and available in different sizes, which permits measuring whole leaves from very different plants. These chambers hold a light sensor in the outside and a fan (for circulating air) in the inside, and they are made from acrylic glass and Teflon to minimise CO_2 and H_2O exchange by chamber materials (Li-Cor, 1990).

Measurements

Before starting the measurement it is also necessary to select the desired program in the software of the apparatus. That is so because this gas exchange system allows also the soil respiration measurment with a soil chamber (LI-6000-09).

However, some things are necessary to take in account when we work with this apparatus (LI-6200). For example, there is an important influence of handling, because the person who takes the measurments is who decides the appropriate moment to start this measurment. By other hand, there is no control on the light, the temperature, the CO_2 concentration or the PAR into the chamber used by the system to calculate the photosynthetic value. So, if the measure is made in the field along several days, the user must be careful to select those days with similar environmental conditions or to keep in mind in the data interpretation, the environmental differences.

In the same way, altered conditions into the chamber (an increase of the temperature or a shadow by the complicate angle necessary to measure the leaf) make that the microenvironmental conditions around the leaf surface could change along the measurement, and so the finishing conditions will be different from the starting conditions into the chamber.

OPEN SYSTEM (LI-6400)

The photosynthetic measures can be influenced by several external and internal factors (Kirschbaum and Pearcy, 1988; Al-Khatib and Paulsen, 1990; Allard *et al.,* 1991). Controlling total conditions required for a true and stable photosynthesis measurment is very difficult. This inconvenience will be even worst if the measure is done in the field, where it is impossible to have control over environmental parameters such as temperature, high light, sun fleck, humidity, etc. Therefore, it is important that the selected apparatus be able to measure, monitor and control all these conditions. That is one of the most prominent advantages of the LI-6400 on the LI-6200, the capacity to fix a desired value of light, CO_2, humidity, temperature, or other parameter.

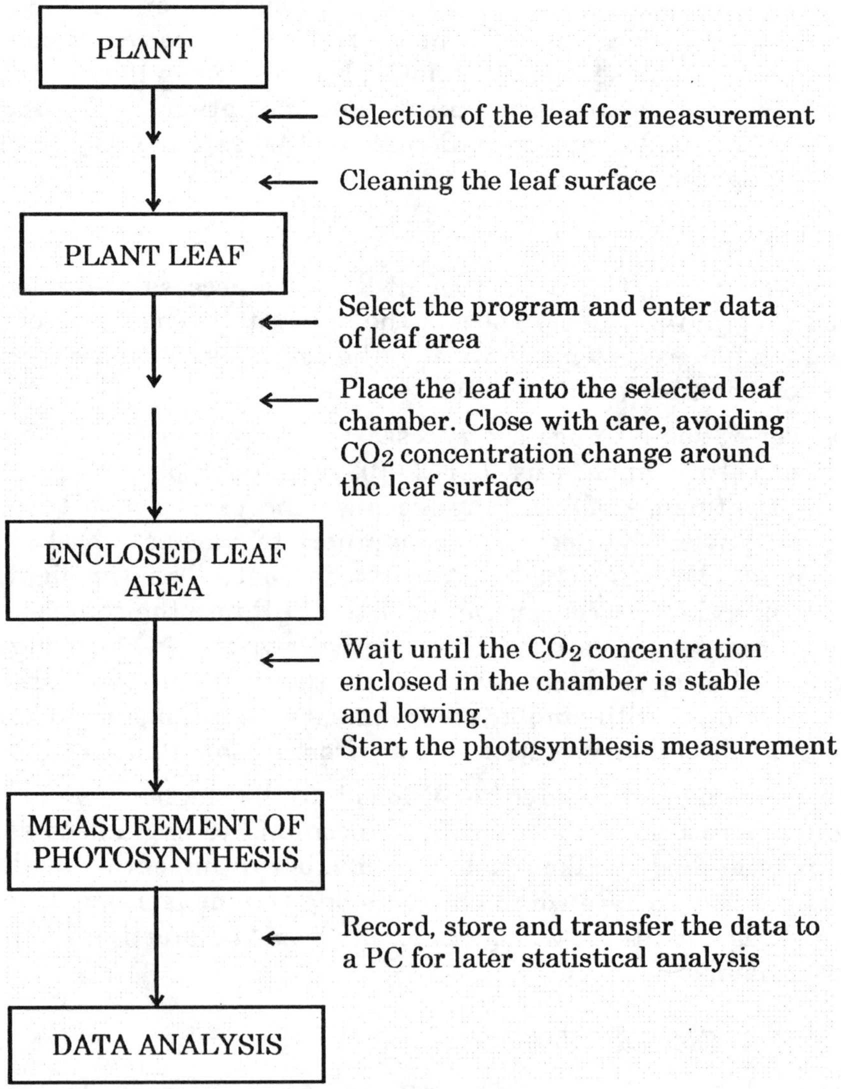

Figure 3. Scheme of the photosynthetic measurement procedure in plant leaves.

Carbon dioxide response curves give us important information about the physiology and biochemistry of the photosynthesis system (Davis *et al.,* 1987; McDermitt *et al.,* 1989). Response curves can be made and displayed by the LI-6400 software, and their knowledge can help to the person that is recording the measurements to correct default preset measuring conditions.

LI-6400 (Li-Cor model 6400) is an open system for laboratory or field photosynthesis measurement of whole leaves (Fig. 4). It has the step up of having the gas analysers in the sensor head, eliminating the connection tube between the chamber where the leaf is placed and the console where the value is calculated.

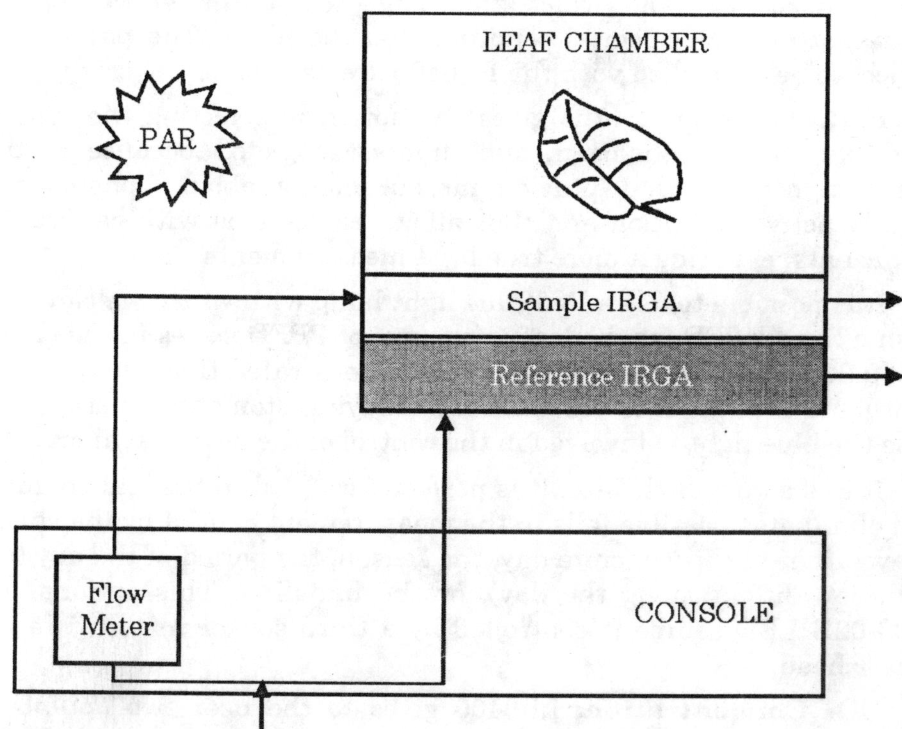

Figure 4. Scheme of the developed open system used by the Li-Cor 6400 (Li-Cor, 1998).

This instrument has two gas analysers in the sensor head, which allow obtaining absolute CO_2 and H_2O values as well from the reference as from the sample. With this progress, gas losses, water condensation or contamination in the return tube are eliminated and the measurement gives a dynamic knowledge about what is happening in the leaf in every moment, in real time and in real microenvironmental conditions (Li-Cor, 1995; 1998).

Measurements in the sensor head come from `the difference between intensity of absorbing wavebands and nearby non-absorbent wavebands in the same optical path' (Li-Cor, 1995).

Control on the Measurement

Likewise, this system has the possibility to fix all desired parameters, and so the user can control the environmental conditions in the leaf chamber.

Until twelve parameters can be recorded and displayed by LI-6400 at the same time while the measurement of photosynthesis is obtained. It allows possible corrections in the measuring conditions giving more real and exact knowledge about the effects of the treatment or the ecological conditions on the plant. The parameters subject to be controlled with the LI-6400 are as below :

Light : The apparatus gives the option of selecting the type of used light source (halogen, sun, fluorescent, etc.) because of two quantum sensors that provide measurements about photosynthetically active radiation and that allows corrections with calibrating coefficients, ensuring a more true light measurements.

It is possible to use a Red/Blue light lamp with which we can establish a fixed PPFD for the leaf or a range of PPFD series in the chamber (0-2000 $\mu mol\ m^{-2}\ s^{-1}$). It has been demonstrated that the use of an additional blue light is essential for studying stomatal kinetics, because the blue light is involved in the control of the stomatal aperture.

It has a very little size, it is perfectly coupled to the leaf chamber and eliminates possible falls in the measurement caused by the shade, the weather of this concrete day, the season, the period of the day (the radiation differs along the day), or the handling. This optional LI-6400-02B LED source is controlled by a third sensor installed in the sensor head.

CO_2 Concentration: LI-6400 gives to the user two options to control the CO_2 concentration. One of them is the manual control with the help of a tube of soda lime (that catches the CO_2), which is limited to the capacity of this substance to catch the CO_2 and to the incoming CO_2 concentration. The second one is the use of the optional CO_2 mixer (LI-6400-01 CO_2 mixer). Mixing scrubbed air and pure CO_2 (in 12 g cylinders) can control the CO_2. This option allows constant CO_2 concentration in the leaf chamber (known as `sample $CO_2$') or maintains constant incoming $CO_2$ (known as `reference CO_2') close to the environmental value. With this injector, CO_2 and H_2O absolute values can be separately controlled.

Temperature : Block or leaf temperatures (by way an indirect measurement of the air temperature) can be maintained constant by measn of internal peltier coolers. So, leaf temperature increases can be eliminated without uses of additional instruments that complicate the management.

Humidity : The humidity can be also maintained constant along the measurements. Two mechanisms are needed for it : the control of the incoming humidity (known as `reference $H_2O$') drying the incoming air with the desiccant or moistening it, and the air flow that goes through the chamber (known as `sample H_2O'). When the humidity value is fixed, the flow value varies and must be manually adjusted.

Measuring Time : A predetermined time to provide measurements can be fixed before starting measurement. So, data will be recorded and curves will be made in function to this preset time period.

Measurements

For starting the work with the LI-6400 a previous warming (10 min), as with LI-6200 (30 min), is necessary. At this time, checking of leaf temperature, CO_2 flow, light sensors, leaf fan, etc. should be done (Fig. 5).

Once finished the warming and after calibrating the LED lamp (optional), the CO_2 mixer (optional), and the zero of sample and reference H_2O and CO_2 flow, one can start to establish the fixed parameter values for the experiment.

So, it is possible to fix the block and leaf temperature value. As the leaf temperature is an indirect measure it is better to fix the block temperature in a value near to the starting temperature. If we have an LED light source we can fix also the desired PPFD. In any case the light source type must be selected at this moment (sun + sky, quartz halogen, LED lamp, fluorescent, etc). It is necessary to enter also the value of the leaf area enclosed in the chamber and the stomatal ratio. The stomatal ratio is the stomatal ratio in one leaf side with respect to the other side (0-1).

Before fixing the flow it is important to decide if we prefer to maintain constant humidity or the CO_2 flow. If we want to have a stable humidity for the measurement the apparatus will give us the humidity value at that moment and then we can enter the value (mmol mol^{-1}) that we want to have along the experiment. When the humidity is controlled we must manually adjust the flow, because it changes.

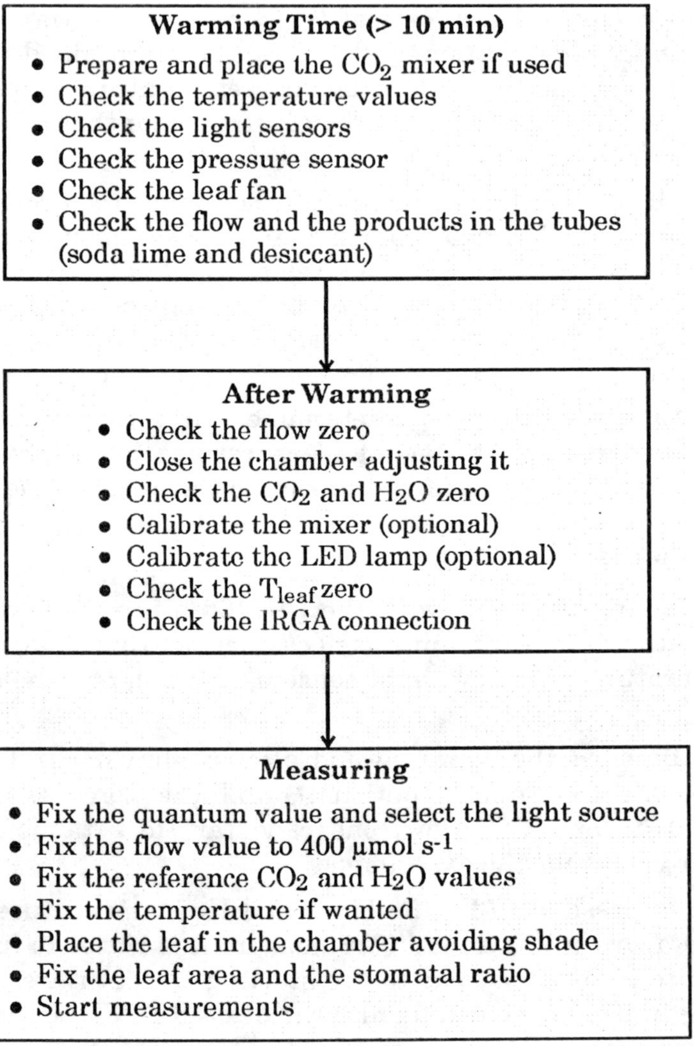

Warming Time (> 10 min)

- Prepare and place the CO_2 mixer if used
- Check the temperature values
- Check the light sensors
- Check the pressure sensor
- Check the leaf fan
- Check the flow and the products in the tubes (soda lime and desiccant)

↓

After Warming

- Check the flow zero
- Close the chamber adjusting it
- Check the CO_2 and H_2O zero
- Calibrate the mixer (optional)
- Calibrate the LED lamp (optional)
- Check the T_{leaf} zero
- Check the IRGA connection

↓

Measuring

- Fix the quantum value and select the light source
- Fix the flow value to 400 µmol s^{-1}
- Fix the reference CO_2 and H_2O values
- Fix the temperature if wanted
- Place the leaf in the chamber avoiding shade
- Fix the leaf area and the stomatal ratio
- Start measurements

Figure 5. Scheme of the photosynthetic measurement procedure in plant leaves with a LI-6400 Portable Photosynthesis System.

For fixing the CO_2 reference value the apparatus gives use two options. If a CO_2 mixer is installed and we are going to use it, the reference value must be close to the ambient conditions (±400 mmol mol^{-1}). But if no CO_2 mixer is connected, then we must to use the Soda lime until obtaining the desired CO_2 concentration.

Finally, we can fix the time interval to obtain the photosynthesis measurments and the response curves that we want to record.

Once the leaf is placed into the selected leaf chamber and the CO_2 flow is stable the data recording can start. During this time we can display the parameter values of our interest and correct them if necessary.

Data can be stored in a data memory and transferred to a PC with a data transfer program for later analysis.

MEASUREMENTS WITH TYPE 225 MARK II IRGA

Materials

IRGA (Type 225 Mark II) with inbuilt air pumps, Plexiglass chamber with inbuilt fan of required size. Photometer with quantum sensor to measure PAR, Tygon tubings, Leaf area meter and stopwatch.

Installation of IRGA

(1) Exclude all effects of the local environment, which would interfere with its operation. The instrument can be unfavourably affected by vibration, temperature or pressure changes and unsuitable composition of the atmopshere. The place of analyser must be chosen in such a way that it is not subjected to direct effects of source of heat i.e. radiators, steam pipes, sun shine etc especially when it is not equipped with a thermostat.

(2) IRGA should be switched on at a reduced amplifier output and allowed to operate for 2 hours to stabilize its temperature. The amplification is then increased up to the degree which approximately corresponds to normal working operation of the analyser. The sample tube is filled with dry neutral gas or both tubes are filled with the ambient air in differentially measuring IRGA, the deflection of the meter is adjusted to the center of the scale by means of the optical shutter and the output is followed until it is steady.

(3) **Zero setting :** With the tubes filled with dry air the amplification is set so that the measuring instrument deflection attains 3/4 of the scale. This deflection is first reduced by the optical shutter. On attaining zero deflection the amplification is increased and the procedure be repeated. The instrument records the absolute value of the energy difference between the two radiation paths. It is important to carry out correct phase adjustment of IRGA, so that the counter current equalizes the detector signal brought about by reducing the radiant energy in the sample tube path by

means of the optical shutter. The correctness of the phase adjustment can be checked by slightly screning off the sample tube path at a set zero value. When the adjustment is correct the indicator immediately exhibits a positive movement.

(4) **Sensitivity adjustment :** After setting zero, the sensitivity of the IRGA must be adjusted in relation to the required scale. With IRGA of moderate sensitivity, the highest concentration indicated on the scale is adjusted after filling the sample tube with a standard gas containing this concentration.

(5) **Gas handling system :** A sample of the analysed air must be trasnsferred into the IRGA in steady state i.e. suitable amount, temperature, pressure, humidity and purity. A gas handling or sampling system for measuring CO_2 exchange in plants contains tubings connected to a probe, a pump, drying device with a filter, flow meter, pressure regulator and a sequential valve. After connecting all elements the entire system must be checked precisely for leaks.

(6) **Probe** : Taking air samples from big assimilation chambers containing plant communities. To obtain spatially integrated samples in the field, probes with orifices spaced at 25 or 50 cm have been used both in and above crop canopies. Such orifices should be covered with fine mesh nylon to prevent insects or dirt from entering.

(7) **Gas sample line :** Metal and glass tubings of an internal diameter of 4 to 6 mm is to be preferred, high-density nylon tubings may also be used. Rubber tubings or some brands of plastic tubing especially polyethylene and tygon absorb gases and permeable to CO_2 and so are not suitable. Short sections of these thick walled plastic tubes may be used to provide flexibility of the line. Hopwever considerable lengths of such tubes should not be used.

(8) **Pump** : The pump is usually inserted between the assimilation chamber and the IRGA because of greater accuracy and stability of measurement. Pumps, which neither leak or contaminate the air sample be used.

(9) **Flow meter :** A rotameter with a range of 3 to 40-60 litre/hour is used to measure the airflow. These are designed to be connected immediately before the IRGA. It has the advantage of slight pressure drop.

(10) **Pressure and flow regulators** : IRGA output depends on the gas pressure. The inlet pressure must be 20% higher than

that in the outlet. The pressure regulators are often combined with flow regulators. A check against a systematic error arising from pressure differences between the reference and sample tubes of the IRGA, it is convenient to place a simple differential u-tube manometer across the reference and sample tube outlets.

(11) **Drying device :** Air must be well dried because of overlap between absorption bands for CO_2 and H_2O in 2.7 μm wave band. Chemical desiccants should not absorb or emit CO_2 and must not be conducive to growth of microorganisms. Magnesium per chlorate is most useful. However, for drying an air stream $ZnCl_2$ is more suitable, because it absorbs water faster than magnesium per chlorate.

Procedure for Measurement

(1) The measurement of rate of CO_2 exchange, the crop plants are enclosed in an air tight transparent plexiglass or iron frame covered with transparent polythene sheet chamber of 50 x 50 x 50 cm^3 size or as per the crop height.

(2) A micro fan is fixed in the chamber for mixing the air in the crop canopy. The rate of CO_2 depletion/addition is measured with the help of IRGA table model (Type 225 Mark II manufactured by Analystic Development Corporation Ltd. Hoddesdon, England, operated in differential mode.

(3) The time required for certain value of depletion / addition of CO_2 (normally 5-10 ppm) in the chamber is recorded by a stopwatch.

The gas exchange rate is calculated using the following formula.

$$CER = \frac{D \times DRC \times VC}{AC}$$

Where,

CER = CO_2 exchange rate ($mg/dm^2/h$) on ground area basis,

D = Density of CO_2 at atmospheric temperature and pressure,

DRC = Depletion/addition rate of CO_2 in the chamber,

VC = Volume of chamber (cm^3) and AC-Area of chamber (cm^2).

The value of CER measured on ground area basis are divided by LAI of enclosed plants in the chamber will determine the gas exchange rate on leaf area basis.

RESPIRATION

Respiration is the process of oxidation of carbohydrate as substrate provided by the photosynthesis to release stored energy. This free energy released during the respiration is incorporated into ATP form. This energy is essential for various active metabolic processes like absorption and translocation of mineral nutrients and photosynthates, which are needed for growth and development of different plant parts. Respiration from plants can be divided into growth and maintenance based on its function. Respiration is just reverse to process of photosynthesis, because here CO_2 is released. Hence, the estimation of respiration rate is essential with the rate of photosynthesis for the calculation of gross photosynthesis and quantum efficiency (Peters *et al.,* 1974).

Rate of CO_2 release in CO_2 free air is monitored using absolute mode of IRGA.

(1) Make the connections as shown in panel 1 and set zero to left hand side of meter with zero control knob after one hour of warm up.

(2) A standard gas (atmospheric gas in the present case) is used to standardize the analyzer. The fine gain control is adjusted to set the correct readings on meter. The first and second steps are repreated until no further adjustment is necessary.

(3) The reference air stream to be measured is connected to the analysis cell inlet. While plant or parts of plant is placed in a transparent Perspex chamber and placed in light and enclosed gas system is used for measurement.

(4) Additional gas circuit is made by using a pump and a CO_2 absorbing tube to reduce the CO_2 concentration in chamber to zero. When CO_2 concentration decreased to zero, pump is put off and rate of CO_2 released is reocrded. Light intensity is measured with quantum photometer. Rate of photorespiration is then calculated as

$$F = (AC/T) \times (V/A)$$

By using portable photosynthesis system like ADC or PP Systems Model CIRAS-1 or 2 we can simultaneously determine stomatal conductance, transpiration, leaf temperature, PAR and internal CO_2 concentration in addition to assimilation rate. These values can be used to calculate various derived parameters like water use efficiency, carboxylation efficiency, quantum yield, mesophyll conductance and transpiration cooling.

Precautions

(1) Gas exchange systems are designed to measure small differences in gas composition. If parts of system absorb or release water vapour or CO_2, the overall accuracy of the device will be reduced. Materials used to avoid absorption of water vapour include aluminium, plexiglass and most plastic tubings. Gas tubings in the field photosynthesis system must be flexible. The best plastic tubings, especially for gases containing water vapour, include high density polyethylene, some of teflons and polypropylene.

(2) IRGA is sensitive to mains voltage fluctuations. A voltage stabilizer should be used.

(3) It is important to calibrate field instruments often and whenever possible in the field, as instruments are sensitive to variation in temperature. Maintain the temperature and humidity inside the chamber at the same levels as outside the chamber.

CANOPY TEMPERATURE

The canopy temperature gives the estimate of crop health and growth. Its measurements do not disturb the physiological properties and energy exchanges with the environment. Infra red thermometry is an indirect method, which evaluates the reflected radiation. It is ideal for many situations, where contact thermometer cannot be used (Bhosale *et al.,* 1996). Infra red thermometer is a hand held instrument that measures temperature without coming in contact with plant and is very easy to use.

Materials

Infra red thermometer, Model AG 42.

Principle

Infrared thermometer operates by measuring the infrared energy emitted from all objects. It works on the principle of Stephen's Law of black body radiation. According to this law, radiation emitted by a surface at any temperature is given by

$$E = \varepsilon \sigma T^4$$

Where, E is energy radiated per unit time per unit area (emittance) W/m^2

ε is emissivity i.e. ratio of emissive power of the given surface to that of black body (dimension less quantity), σ is Stepen Boltzman constant (5.67×10^{-8} W/m/K^4), T is temperature of the surface (°K).

Procedure

(1) When the handle of the Infrared Thermometer (AG 42) is held, there will be an indication on the digital display.

(2) Two gold plated electrodes embedded in the handle grip cause the power to turn `on' when the handle is grasped in the hand. The reading on the digital display will indicate temperature plus °C.

(3) When the handle grip is released, power to the Infrared Thermometer (AG 42) is cut off.

(4) To make the temperature measurements, aim the Infrared Thermometer (AG 42) at the object, whose temperature is to be measured, using the straight line on the top of the case and the reading will be displayed.

(5) Operate the trigger on the handle, the Infrared Thermometer (AG 42) will display the temperature minus the ambient temperature.

(6) Release the trigger. The measured temperature will again be displayed.

If the Infrared Thermometer (AG 42) is to be used on a tripod for long-term measurements, or if the operator wears gloves, conductive cuff may be used to maintain power to the unit.

(1) Position the thermometer on a tripod or other support, wrap the cuff around the handle of the thermometer with the black side facing inward, overlapping the end with the hooks on the blue side.

(2) When the cuff has been secured around the lower portion of the handle, power will be applied to the unit and continuous reading on the digital display will be activated.

Precautions

(1) The observations should be taken on calm and sunny day with a non-variable sky.

(2) When measuing the temperature of an object from 1.0 m away, the object must be larger than 10 cm to get accurate measurements.

(3) The field should be fully covered by crop canopy.

(4) Charge the battery regularly.

(5) Do not point thermometer at sun.

TRANSPIRATION RATE

Crop growth is not possible without the water loss from the crop canopy through transpiration. It helps in maintaining the canopy temperature and diffusion of CO_2 for the process of photosynthesis. During transpiration, the crop canopy loses water and the water lost from the soil profile is called evaporation. The combined loss of moisture from soil and crop canopy is called evapotranspiration. The increase in concentration of water vapour in the chamber per unit time is called the evapotranspiration rate. The canopy temperature depends on the rate of transpiration, which is dependent on the growth and vice-vera.

TRANSPIRATION MEASUREMENT

Transpiration measurement is one of the most determinant measures in the knowledge of the leaf energy balance and plant water status (Pearcy *et al.*, 1989). This importance and its relation with the CO_2 exchange made that several researchers developed different techniques for measuring and calculating it. Many techniques and several humidity sensors were developed for industry but the reliability and repeatability of their measurements was not enough for research measurements. So, psychrometric systems, infrared gas analysers and other known sensors are used for this measurement.

In plant ecophysiology, transpiration measurement is very important, but a basic condition must be taken in account : the plant must not suffer damage during the transpiration measurement, and so the number of techniques available for measuring transpiration is limited. If additionally, the instrument must be able to take the measurement as well in laboratory as in the field, the most widely used instruments for measuring stomatal conductance `the diffusion porometers' are the only choice. Three main diffusion porometers are available to obtain transpiration and stomatal conductance values (Pearcy *et al.*, 1989).

Transient Porometers: These porometers, primarily developed by Wallihan in 1964, are based on the diffusion of water vapour from a leaf placed in a small chamber to the humidity sensor inside the chamber. The poromter will measure the time required to reach a determined humidity into the chamber or the humidity value that is detected by the sensor in a prefixed time.

This chamber has the inconvience that for a good measurement the chamber must totally enclose the leaf. Therefore thin leaves (as those from conifers) can not be measured by this method. Furthermore, we obtain a transpiration value, but to calculate the stomatal conductance is necessary an empirical calibration by comparing the measurement of the leaf into the chamber with that of a plate of pores in the place of the leaf.

Constant Flow Porometers: Parkinson and Legg (1972) and Day (1977) developed a diffusion porometer based in a constant flow of dry air through the chamber where the leaf is placed. So, the humidity value, at which a steady state is reached, is recorded.

This method is exposed to several differences between the environmental conditions outside the chamber and the microenvironmental conditions at which the leaf is exposed into the chamber. Furthermore, if a change in the vapour pressure occurs, the stomata can react in front of it and the actual plant status will be different from the starting plant status.

Instruments based on a constant flow system able to measure CO_2 exchange are also available (Schulze *et al.,* 1982). This type of porometer has inside an infrared analyser that allows the measurement of the differences in the CO_2 decrease and H_2O increase. These apparatus are interesting when simultaneous measurements of photosynthesis and transpiration are needed in the experiment, or when the environmental humidity is really high. In any case, they are more precise than the transient porometer but also more expensive and more difficult to use.

Null Balance Porometers: The first null balance porometer appeared in 1972. Beardsell *et al.* (1972) developed a diffusion porometer based in a dry airflow through the chamber to balance the humidity generated into the chamber by a transpiring leaf reaching a steady-state humidity. With this method it is necessary taking care when measuring the air and leaf temperatures, because differences can occur with the dry airflow. That is one of the advantages of this porometer, that is does not assume as equal the leaf and air temperatures, avoiding important errors in the measurement.

NULL-BALANCE POROMETER (LI-1600)

LI-1600 (Li-Cor model 1600, Steady State Porometer) is a diffusion porometer operating in the null-balance principle.

The null-balance porometers are based in the measurement of the dry air flow rate necessary to maintain a constant relative humidity inside a chamber where a transpiring leaf is placed. So, we can know the water loss by this leaf. According to this principle, the apparatus gives direct values of stomatal resistance (s cm^{-1}) calculated from the relative humidity, leaf and air temperature and flow rate measured values. Stomatal resistance values are also corrected for boundary layer conductance (Li-Cor, 1989).

The equation used by LI-1600 to obtain the leaf transpiration rate (mg cm^{-2} s^{-1}) is as follows :

$$E = (p_c - p_a)\frac{F}{A}$$

Where, E is the leaf transpiration rate, expressed in mg cm^{-2} s^{-1}

P_c is the water vapour density in the chamber, expressed in mg cm^{-3}

P_a is the water vapour density in the dry air flow entering in the chamber, assumed as a 2% of relative humidity

F is the volumetric flow rate in the chamber, expressed in cm^{-3} s^{-1}

A is the leaf area, expressed in cm^2

As shown in this equation, the calculated leaf transpiration rate directly depends on the volumetric flow rate into the chamber. This flow rate depends also on the chamber temperature and the barometric environmental pressure.

Considering above statements, it is necessary to use an instrument able to obtain these measurements and calculations implying on it, as many environmental parameters as possible. In this way, LI-1600 allows inclusion of atmospheric pressure data, measurments of chamber and leaf temperatures or relative humidity. Light intensity is also measured although it is not used in the calculations.

A LI-1600 porometer is represented schematically in Fig. 6. It consists of two main parts :

1. **Console** : It has some control elements that make possible the transpiration measurement. The pump draws air from inside the console and pumps it to the air tank where the airflow is dispersed in two directions. A part escapes via the Internal Flow Controller (that controls and adjusts the relative humidity in the chamber to the set point), and the other part goes to the Null Adjust Valve (that limits the maximum flow rate entering in the chamber and therefore, forces a major air flow to escape from the system).

This air dispersion is essential in the system because leaves with very different transpiration rates must be measured and control on the flow rate entering in the chamber will be necessary to maintain the relative humidity in the set point. For example, if the leaf placed in the chamber has a very low transpiration rate, only a small dry air flow rate will be necessary into the chamber to balance the leaf transpiration rate; if the air entering in the system is too high the relative humidity can drop below the setpoint and the measurement will be not real. Therefore, the system has the portion of venting off excess flow.

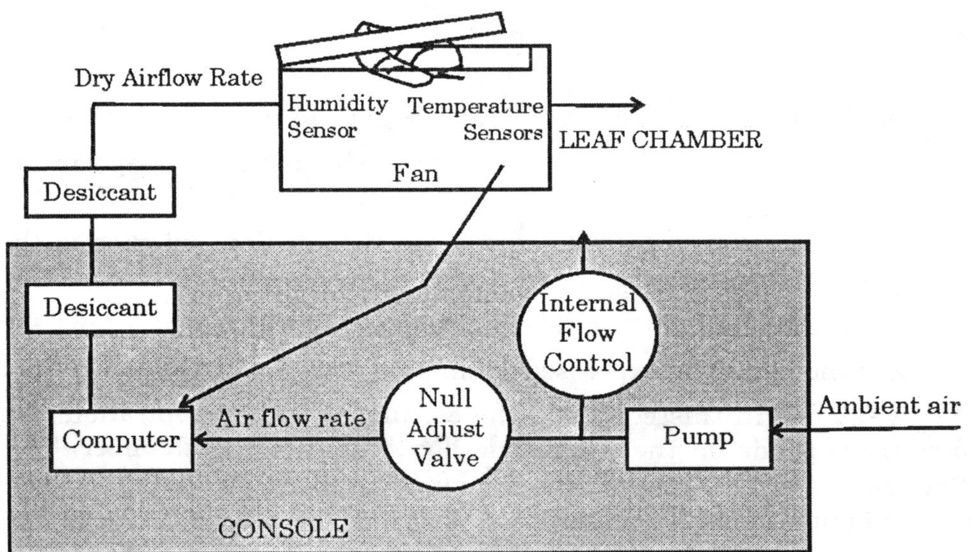

Figure 6. Scheme of null-balance porometer to measure transpiration (used by the LI-1600 Steady State Porometer).

Once that the airflow passess through the null adjust valve, it will pass through two desiccants, one in the console and the other one in the sensor head. So, we ensure that the airflow entering in the chamber is dry.

2. **Leaf Chamber** : Transpiration depends on many factors. Therefore, the leaf which will be measured, must be normally enclosed in a chamber where the major number of possible external factors can be controlled. In the same way the size of the selected chamber must permit the measurement of very different leaves, with very different sizes but also with very different stomatal resistances. Therefore, a high variety of chambers are available for measuring (with different shapes and sizes). The cylindrical chamber of a LI-1600, for example, is made by nickel plated aluminium to minimise water absorption. The sensors indise the chamber check for several environmental parameters (Fig. 6), for example leaf and air temperatures, relative air humidity and also the PAR (photosynthetically active radiation).

Once that the just dry air flow enters in the chamber it is mixed with the water vapour loss by the transpiring leaf, and a dry air flow rate will be necessary to balance the relative humidity into the chamber. This value is the key for the later stomatal resistance and transpiration calculations.

Measurements

The first step to use the LI-1600 is to check that the desiccant from the console is completely dry, changing it if necessary. Every week the desiccant from sensor head must be changed too.

The chamber must be installed at this time. The selected leaf chamber and the aperture (size and form) will depend on the plant species that will be measured.

Once that the sensor head is cleaned without presence of pollen or dust (it can increase the water vapour absorption with unusually increased values), the apparatus is ready to work.

For starting the work with the LI-1600 a previous habitation to the environmental parameters is necessary (30 min). Therefore, we must be careful and to install the instrument in the definitive measuring place. Direct sun must be avoided to protect the chamber against abrupt temperature changes. At this time, checking of chamber and leaf temperatures, flow, relative humidity, PAR etc. can be done.

Once finished the acclimatisation and after calibrating relative humidity sensor if necessary, and setting the null point humidity to ambient relative humidity (null points different from ambient values can also be used), and the dry air flow rate entering in the chamber, data recording can start. Along this time, we can display the parameter value of our interest.

The LI-1600 needs generally a mean time of 10 to 20 seconds to reach a Steady-state situation on a leaf with diffusive resistance of 5 s cm^{-1}. With higher values the instrument needs a little longer time.

Once that the instrument is at or near the null point relative humidity and the diffuse resistance value is stable, data can be stored in a data memory and transferred to a PC with a data transfer program for later analysis.

REFERENCES

Al-Khatib, K. and Paulsen, G.M. 1990. Photosynthesis and productivity during high temperature stress of wheat genotypes from major world regions. *Crop Science* 30 : 1127-1132.

Allard, G., Nelson, C.J. and Pallardy, S.G. 1991. Shade effects on growth of tall fescue. II. Leaf gas exchange characteristics. *Crop Science* 31 : 167-172.

Beardsell, M.F., Jarvis, P.G. and Davidson, B. 1972. A null balance diffusion porometer suitable for use with leaves of many shapes. *Journal of Applied Ecology* 9 : 677-690.

Bhosale, A.M., Jadhav, A.S., Bote, N.L. and Varshney, M.C. 1996. Canopy temperature as an indicator for scheduling irrigation in wheat. *Journal of Maharashtra Agricultural Universities* 21 : 106-109.

Boyer, J.S. 1996. Advances in drought tolerance in plants. *Advances of Agronomy* 56 : 187-217.

Davis, J.E., Arkebauer, T.J., Norman, J.M. and Brandle, J.R. 1987. Rapid measurement of the assimilation rate versus internal CO_2 concentration relationship in green ash (*Fraxinus pennsylvanica* Marsh) : the influence of light intensity. *Tree Physiology* 3 : 387-392.

Day, W.A. 1977. Direct reading continuous flow porometer. *Agricultural Meterology* 18 : 81-89.

Field, C.B., Ball, J.T., Berry, J.A. 1989. ``Photosynthesis : Principle and Field Techniques" In: *Plant Physiological Ecology*. R.W. Pearcy, J.R. Ehleringer, H.A. Mooney, P.W. Rundel (eds.) London : Chapman and Hall.

Kirschbaum, M.U.F. and Pearcy, R.W. 1988. Concurrence measurements of oxygen and CO_2 exchange during light flecks in *Allocacia macrorrhiza*. *Planta* 174 : 527-533.

Li-Cor Inc. 1989. *Steady State Porometer.* Lincoln : Li-Cor.

Li-Cor, Inc. 1990. Using the Li-6200. *Portable Photosynthesis System.* Lincoln : Li-Cor.

Li-Cor, Inc. 1995. *Using the Li-6400 Portable Photosynthesis System. Feature and Spec's.* Lincoln : Li-Cor.

Li-Cor, Inc. 1998. *Using the Li-6400. Portable Photosyntheisis System.* Lincoln : Li-Cor.

Mahon, J.D. and Domey, J. 1979. A light weight battery operated Infra-red gas analyser for field measurement of photosynthetic CO_2 exchange. *Photosynthetica* 13 : 459-466.

McDermitt D.K., Garcia, R.L., Welles, J.M. and Demetriades-Shah, T.H. 2000. ``Common Errors in Gas Exchange Measurements." In: *Probing Photosynthesis.* M. Yunus, U. Pathre, P. Mohanty (eds.) London and New York : Taylor and Francis.

McDermitt, D.K., Norman, J.M., Davis, J.E., Arkebauer, T.J., Ball, J.T., Welles, J.M. and Roemer, S.R. 1989. CO_2 Response curves can be measured with a Field Portable Closed Leaf Photosynthesis System." In *International Symposium on Forest Tree Physiology*. Nancy.

Parkinson, K.J. and Legg, B.J. 1972. A continuous flow porometer. *Journal of Applied Ecology* 9 : 669-675.

Pearcy, R.W. 1989. ``Radiation and Light Measurements" In: *Plant Physiological Ecology*. R.W. Pearcy, J.R. Ehleringer, H.A. Mooney, P.W. Rundel, (eds.) London : Chapman and Hall.

Pearcy, R.W., Schultz, E-D. and Zimmermann, R. 1989. Measurement of transpiration and leaf conductance". In: *Plant Physiological Ecology*. R.W. Pearcy, J.R. Ehleringer, H.A. Mooney, P.W. Rundel (eds.) London : Chapman and Hall.

Peters, D.B., Clough, B.F., Garves, R.A. and Stahl, G.R. 1974. Measurement of dark respiration, evaporation and photosynthesis in field plots. *Agronomy Journal* 66 : 460-462.

Schulze, E.D., Hall, A.E. and Lange, O.L. 1982. A portable steady state porometer for measuring the carbon dioxide and water vapour exchanges of leaves under natural conditions. *Oecologia* 52 : 141-145.

Steubing, L. and Fangmeier A. 1992. *Pflanzenokologischen Prakitikum:* Gelande- und Laborprakilkum der terrestrischen Pflanzenokologie. L. Steubing, ed. Stuttgart : Ulmer.

Wallihan, E.F. 1964. Modification and use of an electric hygrometer for estimating relative stomatal apertures. *Plant Physiology* 39 : 86-90.

Zelitch, I. 1982. The close relationship between net photosynthesis and crop yield. *Bioscience* 32 : 796-800.

CHAPTER 14

STRESS INDICATORS IN PLANTS

INTRODUCTION

Plant species inhabit a wide range of environments with diverse combinations of abiotic conditions and biotic interactions. Physical attributes of environment can affect phyisological patterns of a species in a positive or negative manner. The response of a particular plant species is not always equal. At different times during its life cycle, a plant will be differentially sensitive to a particular environmental stress. Stress is defined as an external factor that exerts a disadvantageous influence on the plant. In most cases stress is measured in relation to plant survival, crop yield, growth or the primary asimilation processes (CO_2, mineral uptake), which are related to overall growth. Environmental stresses can be of diverse kinds like drought, temperature, radiation, flooding, heavy metals soil pH, salinity etc. In general, crops are limited to about 25% of their potential yield by the impacts of environmental stresses.

Plants subjected to abiotic stresses show decreased levels of various leaf metabolites and enzymes on the one hand and increased concentrations of free proline, amino acids, sugars and activities of hydrolysing enzymes on the other hand. Tolerance to environmental stresses occurs when a plant under stress is able to maintain high metabolic activity similar to that attained under optimum natural conditions. However, plants differ in their capacity to accumulate certain metabolites and maintain enzyme activity which confers resistance to a particular stress. Some of such increases and decreases in physiological responses and biochemical paremeters are of special significance for imparting stress tolerance. In other words these parameters or indices may be termed as stress indicators which are of varying nature under different stresses. Few of these stress indicator are particularly important for which methodology is described in this chapter for their measurement in plant tissues.

(1) CHLOROPHYLLS AND CAROTENOIDS

Rationale

The pigments of photosynthesis, the chlorophylls and carotenoids, are by far the most abundant biological pigments on the planet. The net annual production of the chlorophylls, aquatic and terrestrial, exceeds 1 x 10^9 tonnes (1) and much of this is turned over several times each year. The events associated with pigment destruction are numerous and have been discussed elsewhere (1, 2). Apart from net destruction at critical points in the plant's life cycle, the loss of pigments during environmental stresses or on premature death is a highly visible indicator of such events as :

Disease

Herbicide application

Industrial pollution

Mineral deficiencies

Mineral toxicity

Water deficiency

Temperature extremes

UV-B exposure

In addition, the ratio of chlorophyll a to b in higher plants has been used widely as an indicator of response to shade and as an early indicator of senescence. The ratio between chlorophyll and carotenoids has been much less widely used diagnostically, though we have found this ratio to be a sensitive marker distinguishing natural full-term senecence and senescence due to environmental stresses such as desiccation in mosses (3) and drought in flowering plants (4). Chlorophylls are particularly sensitive to oxidative attack and photo-damage, while carotenoids function naturally as anti-oxidants and in quenching photo-induced excitations (5). Changes in chlorophyll : carotenoid ratios are then potentially sensitive indicators of oxidative damage.

Method

Chlorophylls and carotenoids are extracted rapidly from leaf (or shoot) material and their concentration determined spectrophotometrically. By use of simultaneous equations it is possible to determine the concentration of chlorophylls a and b, total chlorophylls and an estimate of total carotenoids on one and the same extract. Greater accuracy in the determination of the several carotenoids present in most plant tissues can of course be obtained but only by separation of the individual pigments, a procedure which does not lend itself to bulk screening. The use of ammoniacal acetone rather than the more usual aqueous acetone containing a carbonate salt may be preferred, as the former avoids the haze sometimes found in acetone extracts with carbonates (6, 7).

Reagent

Ammoniacal acetone (81.8 $(CH_3)_2$ CO/18 H_2O/0.2 NH_4OH v/v) (prepare daily).

Procedure

Homogenize leaf or shoot tissue twice in ammoniacal acetone.

Centrifuge at 3000 g for 5 min.

Place all extracts in the dark immediately following centrifugation.

Immediately before recording the spectrum, dilute extract to a convenient volume.

Assay

Record the spectrum from 470 nm to 710 nm against ammoniacal acetone.

Alternatively record the absorbances at 480, 645, 663 and 710 nm.

Calculations

The absorbance at 710 nm should be used as an isobestic point or theoretical base line and should be deducted from all other absorbance readings from the same solution. This is an important point, particularly when comparing pigment values in different tissues or species.

Chlorophylls

The extinction coefficients provided by MacKinney in 1941 are used to calculate chlorophyll data (5).

However, newer and probably more accurate extinction values, are also available. Restated, MacKinney's equations for the determination of chlorophyll in an 80% acetone extract are :

Chlorophyll a mg/l

$$= 12.7 \times A663 - 2.69 \times A645$$

Chlorophyll b mg/l

= 22.9 x A645 – 4.68 x A663

Total chlorophyll in mg/l

= 8.02 x A663 + 20.2 x A645

Conventionally, it is now more usual to express the chlorophyll concentrations in μmol per l rather than in mg. The conversion factors are :

Chlorophyll a mg to μmol ... multiply by 1.119

Chlorophyll b mg to μmol ... multiply by 1.102

Ratios of chlorophyll a to b should always be expressed as the mole ratio (A mole ratio of 3.00 is equivalent to a mass ratio of only 2.95, a source of error in comparisons of data).

Carotenoids

The absorbance of the carotenoids at 480 nm is determined using the equations provided by Kirk and Allen (8) which compensate for interference at this wave-length from chlorophyll a and particularly b. To quantify the carotenoids, It is assumed that

(a) Plant leaves contain 45% carotenes and 55% xanthophylls

(b) Extinction coefficients based on beta-carotene of 1940 and lutein of 2160

(c) The average molecular weight of carotenoids is 545.

Based on these assumptions and approximations, an estimate of total carotenoids can be made in the presence of chlorophylls from the following spectrophotometric observations :

Carotenoids μmol per unit area or weight =

$$\frac{(A480 + 0.114 \times A663 \times A645) \times V \times 10^3}{112.5 \times unit\ area\ or\ weight}$$

where, V = extract volume

REFERENCES

1. Hendry, G.A.F., Houghton, J.D. and Brown, S.B. 1987. The degradation of chlorophyll - a biological enigma. *New Phytologist* 107 : 255-302.

2. Brown, S.B., Houghton, J.D. and Hendry, G.A.F. 1991. Chlorophyll breakdown. In: *Chlorophylls* (ed. H. Scheer) CRC PRess, Boca Raton, pp. 465-489.

3. Buckland, S.M., Price, A.H. and Hendry, G.A.F. 1991. The role of ascorbate in drought-treated *Cochlearia atlantica* Pobed. and *Armeria maritima* (Mill.) Willd. *New Phytologist* 119 : 155-160.

4. Seel, W.E., Hendry, G.A.F. and Lee, J.A. 1992. The combined effect of desiccation and irradiance on mosses from xeric and hydric habitats. *Journal of Experimental Botany* 43 : 1023-1030.

5. MacKinney, G. 1941. Absorption of light by chlorophyll solutions, *Journal of Biological Chemistry* 144 : 315-323.

6. Arnon, D.I. 1949. Copper enzymes in isolated chloroplasts. Polyphenol oxidases in *Beta vulgaris*. *Plant Physiology* 24 : 1-14.

7. Lichtenthaler, H.K. and Wellburn, A.R. 1983. Determination of total carotenoids and chlorophylls a and b of leaf extracts in different solvents. *Biochemical Society Transactions* 603 : 591-596.

8. Kirk, J.T.O. and Allen, R.L. 1965. Dependence of chloroplast pigment synthesis on protein synthesis : effects of actidone. *Biochemical and Biophysical Research Communications* 21 : 523-530.

(2) CHLOROPHYLL FLUORESCENCE

Rationale

Measurement of chlorophyll fluorescence is now established as an important technique in the study of physiological plant ecology and stress physiology (1). The yield of chlorophyll fluorescence reveals the level of excitation energy in the pigment system driving photosynthesis. This level depends upon the balance between irradiance and the sum of the rate of photosynthetic electron transport and thermal dissipation. This can be measured, as described below, as qP (Photochemical) and qN (non-photochemical) quenching, respectively (2,3). qN is induced in response to the approach of light saturation and indicates the operation of photoprotective regulation. Using the technique of chlorophyll fluorescence, it is also possible to estimate both the actual photosynthetic efficiency of a leaf, under any condition at any time (ΦPS2) and also the maximum potential quantum efficiency (F_v/F_m). The former indic-ates the effectiveness of light utilization by the leaf and reveals the extent of other limitations to photosynthesis (besides light) and the presence of adaptation or acclimation. The second parameter has been shown to be a reliable indicator of stress.

Fluorescence Parameters and Plant Stress

F_o : It is affected by any environmental stress, which causes alterations at the pigment level of PS II. For example, thermal damage of the PS II is characterised by a drastic increase in F_o. But freezing damage of the thylakoids does not affect F_o.

F_v : The difference $F_M - F_o$ is denominated as variable fluorescence. Normally this value is lowered by environmental stresses, which cause damage on the thylakoids. Examples for this kind of stress are heat, freezing and photoinibition.

F_M : The maximal fluorescence decreases after exposure of the leaf to high but not injurious temperatures. More severe heat treatment causes an increase of F_o and a decrease of F_M. The PS II activity will be inhibited.

F_M/F_o : This ratio depends on the leaf water potential. Under drought conditions it will drop to 1. Under severe drought conditions no F_v was produced, but after watering F_M/F_o were restored up to 3.0 (2,3).

F_v/F_M : This relation is typically in the range of 0.832 ± 0.004 (Krause and Weis, 1991) and it is proportional to the quantum yield of the photochemical phase of the photosynthesis (4,5). A decrease of this relation is a good indicator for damage of photoinhibition caused by the light when the plants undergo diverse types of environmental stresses as drought, cold, freezing, excessive illumination or salinity. This ratio is used thoroughly as an indicator of stress induced on the photosynthetic apparatus.

Measuring Fluorescence on a Fluorometer

A fluorometer of continue excitement, such as the Plant Efficiency Analyser (PEA) of Hansatech Instruments Ltd. is adequate to obtain the Kautsky curve.

The Plant Efficiency Analyser (PEA) measures the chlorophyll fluorescence, which has been emitted from green plants. This equipment has been designed to be easy to use, and its functioning is equally good in either laboratory or in the field (6).

The measurements of the fluorescence induction are made using dark-adapted intact plant leaves, needles or other samples including micro and macro algae. For the last sample types, special accessories are necessary. The PEA systems consists of the following items.

Leafclips

Before any measurement the leaf or needle, which will be analysed, must be covered with a small, lightweight leafclip. The leafclip is a kind of pincer. It is made of white plastic to reduce the development of heat during the time where the leafclip is placed on the leaf. It consists of a locating ring for the sensor unit, a shutter plate to close the exposed leaf area in the centre of the locating ring, and a foam pad in order to minimise possible damages of the leaf structure.

When the leafclip is placed at the chosen leaf, the shutter plate must be closed so that light is excluded and dark adaptation starts. During dark adaptation any fluorescence yield is quenched. This process takes time. This period is variable and depends on the analysed plant species, the light levels prior to the dark adaptation, and if the plant is stressed. Normal time periods are about 30 minutes. Two strategies are offered to reduce waiting time : We can measure during a non-light period, or several leafclips can be simultaneously used to ensure sufficient dark-adapted leaves.

Sensor Unit

It houses an optical assembly, which provides illumination of the exposed leaf area and the detection of the consequent fluorescence signal. Therefore, the sensor unit consists of an array of 6 ultra bright LED, which are focused onto the leaf surface to provide even illumination over the exposed leaf area (4 mm diameter). They provide red light with a peak wavelength of 640 nm and a maximum intensity of 3000 μmol m^{-2} s^{-1}. The LEDs dissipate low heat radiation and they rise fast to full light intensity (microseconds) after switched on. So inaccuracies of F_o measurements are eliminated.

The detector is a high performance Pin photodiode and associated amplifier circuit. An infrared long pass filter (50% transmission at 720 nm) ensures that it responds maximally to the longer wavelenght of the chlorophyll fluorescence signal and blocks the reflected shorter wavelength of the light source (LED light). The sensor unit plugs in the main control box via a 1.5 m cable. The sensor unit is located over the leafclip at the location ring. So, daylight will be excluded. During the measurement the leaf with the sensor unit connected should be held in its correct position with the hand.

Control Box

It consists of three components : The LCD display, a waterproof membrane keypad and a side panel of connectors. PEA is totally solid-state, no moving parts or chemicals are needed. A rechargable battery and a microprocess control all instrument functions. Data and measurement setting could be stored for several days by backup device. But it is recommended to download data almost every day and don't store critical data for longer time periods.

Light level and recording interval are selectable from the keypad. During the initial fast rise the fluorescence signal is measured by a reading frequency of 100.000 readings s^{-1}, first of all to provide a good resolution of the F_o value. After 2 milliseconds and unitl 1 s a slower frequency is taken (about 1.000 readings s^{-1}). After 1 s a rate of 1 reading s^{-1} is adopted. At the end of the recording time of data points are expressed automatically the fluorescence parameters F_o, F_M, F_V and the ratio F_v/F_M are calculated.

Experimental Procedure

Two settings of crucial importance must be determined before starting the measurments. The first is the time that the plant leaf needs to be dark adapted. The other one is the determination of the saturating light level, or in other word how much light must be provided to get all reaction centres closed.

Determination of the Dark Adaptation Time

When the measurments are taken in actinic light absence, the determination of the minimum dark adaptation period is important. During dark adaptation all reaction centres have to oxidise or to get `open'. The time period may vary with the light level prior to the dark adaptation. So it is recommended that the determination be carried out on plants which have been in high light conditions. So, the dark adaptation period will be enough for all light conditions.

Hansatech (1996) recommends to place 20 leafclips on the leaves or needles and to close the shutter plates. Measure in intervals of 2 minutes, using the 100% light level. Build a table of dark time period and F_v/F_m. The ratio should plateau at the minimum satisfactory dark adaptation period' (Hansatech, 1996).

Determining the Saturating Light Level

After the determination of the required time for dark adaptation place 10 leafclips on the leaves or needles and close the shutter plates for at least the determined dark adaptation time period.

Then make measurements at each clip using increasing light levels from 10 to 100%. The F_v/F_M should plateau at the saturation point. Generally, the 100% light level will not cause over scaling problems.

Sample Preparation and Measurement

Attach a leafclip to a leaf and close the shutter plate to start the dark adaptation. Ensure that the base of the leafclip is not directly exposed to the light source, otherwise light will penetrate the clip and spoil the dark adaptation.

When the leaf is dark adapted fit carefully the sensor head over the location ring of the leafclip in order to seal out the light. Hold the clip and the sensor unit in your hand, open the shutter plate and press either the button on the sensor head or ENTER on the keypad of the console. The measurement starts. With the PEA full data files (FULL DATA - 75 full data sets) or only the fluorescence parameters (PARAMS ONLY -8500 parameter sets) could be recorded. To measure only the fluorescence parameters a recording time of 3 s is enough. To achieve a complete OIDPSMT curve a recording time of 3-5 min should be used. After the measurement data could be stored by pressing ENTER or the sensor switch.

Data Analysis

For the data evaluation, software is supplied as well to download the data points as for graphical analysis of full data files or parameter files. These files could be converted to ASCII format and than imported into other software packages for further analysis.

REFERENCES

1. Ball, M.C., Butterworth, J.A., Roden, J.S., Christian, R. and Egerton, J.J.G. 1994. Applications of chlorophyll fluorescence to forest ecology. *Australian Journal of Plant Physiology* 1994 : 22 : 311-19.
2. Hansatech Instruments Ltd. 1996. *An Introduction to Fluorescence Measurements with the Plant Efficiency Analyser* (PEA). Pentney, Norfolk, England : Hansatech.
3. Hansatech Instruments Ltd. 1995. *Operating Instructions for Plant Efficiency Analyser* - Special JIP Firmware Version. Pentney, Norfolk, England : Hansatech.
4. Krause, G.H. and Somersalo, S. 1989. Fluorescence as a tool in photosynthesis research : application in studies of photoinhibition, cold acclimination and freezing stress. *Philos T. Royal Society* B : 323 : 281-93.
5. Krause, G.H. and Weis, E. 1991. Chlorophyll fluorescence and photosynthesis : the basics. *Annual Review of Plant Physiology and Molecular Biology* 42 : 313-49.
6. Krause, G.H. and Weis, E. 1984. Chlorophyll fluorescence as a tool in plant physiology. II. Interpretation of fluorescence signals. *Photosynthetic Research* 5 : 139-57.

(3) STORAGE CARBOHYDRATES

Rationale

The major non-structural storage carbohydrate reserves of vascular plants are starch, fructan and sucrose. As a broad generalization, starch (a polymer of glucose) is widespread throughout the plant kingdom where it is the major storage carbohydrate. Sucrose (a dimer of glucose and fructose) is probably present in all plants but only in a minority does it accumulate to relatively high concentrations. Fructan (an oligomer of fructose) is the dominant reserve carbohydrate in six major orders of angiosperms (the Asterales and Campanulales, and in certain families or sub-tribes of the Poales, Liliales, Dipsacales and Polemoniales), all of relatively recent evolution (1). Changes in the concentration of carbohydrates or in the relative abundance of one form over another have been linked to changes in environment, particularly to declining temperature and water availability (2). The association of these carbohydrates with particular habitat types, with early season and low temperature growth, with genome size and with availability to herbivores has been well known (2, 3).

Method

The following methods have been specifically adapted to the bulk screening of samples. Determination of the concentration of all three carbohydrates in 50 or more samples can be carried out to completion by one person in one full working day. In practice it is often more efficient to determine, say, starch alone in several hundred samples before turning to another carbohydrate. The method also allows storage of samples in freezers (at –14 to –18°C) for several months.

Carbohydrate concentrations may directly reflect the rates of photosynthesis ruling at the time the carbohydrates where formed. Rates of photosynthesis vary both with the external environment and with a diurnal pattern. It is essential that in making comparisons between species, between tissues or treatments, that the plant material be harvested and extracted within the same hour and at the same time each day. Considerable variations in otherwise identical samples will be experienced if this is not followed.

Reagents

50 mM KPi buffer, pH 7.6

1 N NaOH

(a) Starch determination :

50 mM Na acetate buffer, pH 4.6

1,4-α-D glucan glucohydrolase (amyloglucosidase) (from *Aspergillus niger* or *Rhizopus* sp.)

Solution A : 26 mM o-anisidine-HCl

Solution B : 20 mM KPi buffer, pH 7.0 containing

Peroxidase (from *horse radish*) 1 U/ml

Glucose oxidase (from *A. niger*) 5 U/ml

Solution C : 1.6 ml of soln. A per 100 ml soln. B

β-D-glucose standard 1 mg/ml in phosphate buffer

Potato starch internal standard 5 mg/ml in H_2O

(b) Fructan determination :

Solution D 3 mg $CuSO_4$ in 50 ml concentrated HCl added to

75 ml glycerol and 25 ml H_2O

Solution E 0.45% weight by volume aqueous resorcinol

β-D-fructose standard 1 mg/ml in phosphate buffer

1 mg/ml inulin in KPi buffer

(c) Sucrose determination :

50 mM Na acetate buffer, pH 4.6

β-D-Fructofuranoside fructohydrolase (Bakers yeast invertase)

100 U/ml in 50 mM acetate buffer

Solution C as for starch assay

Sucrose internal standard 1 mg/ml in acetate buffer

Procedure

Homogenize 50-100 mg tissue in 1.5 ml

KPi buffer, pH 7.6.

Centrifuge at 13000 g for 60 s.

The supernatant contains fructan and/or sucrose and may be conveniently stored, or processed as below.

The pellet contains starch and should be immediately resuspended in 1 ml 1 M NaOH and gelatinized at 80°C for 90 min.

Gelatinize potato starch standard (prepared as 5 mg/ml in 1 M NaOH).

On cooling, the starch extracts may be stored at –14°C or processed further as below.

Further procedure for starch determination

Adjust pH of gelatinized samples to between 4.3 to 4.8 with 50 mM Na acetate/5 M HCl.

Add 10 U amyloglucosidase to each 1 ml of sample and standard. incubate sample and standards for 10 min at 50°C (or longer at a lower temperature)

Add 1 ml of reagent C to :

100 µl of sample and

50 µl potato starch internal standard and

0–50 µg glucose as the product standard

Mix and incubate for 30 min at 37°C

On cooling read absorbance of samples and standards at 460 nm.

Further procedure for fructan determination

Add 1 ml solution D to :

200 µl soln. E

20 µl sample and

50 µl inulin standard and

0–50 µg fructose as the product standard Mix. Incubate for 12 min at 100°C Cool. Determine absorbance at 470 nm.

Further procedure for sucrose determination

Add 0.8 ml Na acetate buffer, pH 4.6 and 100 µl invertase to :

100 µl sample and

100 µl sucrose as internal standard and

0 –50 µg glucose as the product standard Mix. Incubate at 50°C for 20 min

Cool. Add 1 ml solution C. Mix. Incubate at 37°C for 30 min

Determine absorbance at 460 nm.

Calculations

For starch and sucrose, the absorbances can be converted to glucose equivalents from the glucose standards. Sucrose can be calculated by multiplying by 2 (1 mol sucrose = 1 mol glucose plus 1 mol fructose).

For fructan, after converting absorbance values to fructose equivalents, deduct half of the value of any sucrose present in the sample (half of the sucrose molecule is fructose which would otherwise overstate the value of fructan).

For precision the concentration of any free glucose and free fructose should also be determined and deducted from the calculations above.

In all three determinations if the calculated concentration of the internal standards varies significantly from the actual concentration steps should be taken to check that all protocols have been followed and that added enzyme preparations are active.

REFERENCES

1. Hendry, G.A.F. and Wallace, R. 1992. The origin, distribution and evolutionary significance of fructans. In: *Science and Technology of Fructans* (eds. M. Suzuki and J.N. Chatterton), CRC Press, Boca Raton.
2. Brocklebank, K.J. and Hendry, G.A.F. 1989. Characteristics of plant species which store different types of reserve carbohydrates. *New Phytologist* 112 : 255-260.
3. Hendry, G.A.F. 1987. The ecological significance of fructan in a contemporary flora. *New Phytologist* 106 (Suppl.) 201-216.

(4) ELECTROLYTE LEAKAGE

Rationale

Visual symptoms of stress-induced injury are difficult to quantify when plant species or genotypes are screened for differential susceptibility to a specific stress. Undoubtedly, the most reliable indicator is to measure relative growth rate but this is time consuming and requires extensive provision of controlled environment facilities (1, 2). Alternatively, a number of physiological parameters, including electrolyte leakage, have been used as more rapid indicators. Stresses such as chilling, freezing, heat or desiccation lead to an increase in passive ion efflux due to a decrease in membrane integrity. In particular, electrolyte leakage has been successfully utilized to determine different susceptibilities to chilling in closely related species (3, 5).

Method

Two different experimental approaches can be employed :

A. Intact plants or detached organs such as leaves, are subjected to a known degree of stress for a fixed period of time (6). The organ in question is then suspended in an aqueous medium with no stress imposed and the resultant electrolyte leakage measured. In practice, diffusion problems created by the cuticle can be largely overcome by the use of tissue strips (or discs).

B. Tissue strips in aqueous medium are subjected to stress generally for the full duration of the electrolyte leakage period (3, 5). This approach lends itself to work under field conditions where strips can be incubated in boiling tubes within a temperature controlled facility (5).

Procedure

Tissue strips are cut with a razor blade, avoiding major veins, leaf edges and senescent tissue. Useful dimensions are 1-5 mm in width and 15-30 mm in length depending on organ dimensions, the bulk of tissue required and the amount of labour available! The strips are washed for about 2 hours and rinsed in a further three changes of aqueous medium in order to remove electrolytes from damaged cells. Suitable volumes of deionized water are generally used but it may be sensible to use a standard nutrient solution in order to avoid any effects due to hypotonic shock (6). If desired, the tissue batches can be blotted dry and weighed at this stage so that electrolye leakage can be expressed on a fresh weight basis. The washed strips are then suspended in a fixed aqueous volume for electrolyte leakage determination. It is desirable to gently agitate the strips using a mechanical shaker as well as providing controlled conditions of temperatrure and illumination. However, it may be impractical to satisfy all these criteria under field conditons (5).

Leakage is often expressed on the basis of total tissue electrolyte content. At the end of the incubation, each tissue batch is boiled for 10 minutes, left for 24 hours on ice and electrolyte content of the solution measured.

Measurements

Electrolyte content can be measured at suitable intervals using a conductivity meter. Alternatively, potassium as the major cation lost from the tissue (3) can be determined using a flame photometer.

REFERENCES

1. Hardacre, A.K. and Eagles, H.A. 1980. Comparison among populations of maize for growth at 13°C. *Crop Science* 20 : 780-784.

2. Hetherington, S.E. and Oquist, G. 1988. Monitoring chilling injury - a comparison of chlorophyll fluorescence measurements, post chilling growth and visible symptoms of injury in *Zea mays. Physiologia Plantarum* 72 : 241-247.

3. Patterson, B.D., Murata, T. and Graham, D. 1976. Electrolyte leakage induced by chilling in *Passiflora* species tolerant to different climates. *Australian Journal of Plant Physiology* 3 : 435-442.

4. MacRae, E.A., Hardacre, A.K. and Ferguson, I.B. 1986. Comparison of chlorophyll fluorescence with several other techniques used to assess chilling sensitivity in plants. *Physiologia Plantarum* 67 : 659-664.

5. Earnshaw, M.J., Carver, K.A., Gunn, T.C., Kerenga, K., Harvey, V., Griffiths, H. and Broadmeadow, M.S.J. 1990. Photosynthetic pathway, chilling tolerance and cell sap osmotic potential values of grasses along an altitudinal gradient in Papua New Guinea. *Oecologia* 84 : 280-288.

6. Larcher, W. and Bodner, M. 1987. Criteria for chilling stress in *Saintpaulia ionantha. Agnew Botanik* 61 : 309-323.

(5) MEMBRANE STABILITY INDEX

Rationale

Stress induced damage of plant membrane can be determined using membrane thermo-stability (MTS) assay, which measures electrolyte leakage from leaf tissue after exposure to high temperature. This is the fastest method to screen plants for temperature tolerance and measures the increased electrolyte diffusion resulting from heat induced cell membrane permeability (Sullivan, 1972). Membrane thermo stability of plant is expressed in terms of electrical conductance. The electrolyte leakage is measured by bathing the heat stressed tissue in deionised water and reading electrical conductance.

Materials

Conductivity bridge with conductivity cell, thermo-statically controlled water bath, autoclave, long test tube to accommodate conductivity cell for measurements, aluminium foil, measuring cylinder and de-ionised water.

Procedure

(i) Collect the upper most fully expanded leaf for assaying MTS.

(ii) Remove the mid rib from the middle portion of the leaf and take approximately 7 cm^2 of leaf and cut it into 1 cm pieces.

(iii) Put leaf pieces into 16 x 150 mm test tube and wash thrice with deionized water to remove unwanted ions from leaf pieces. After final wash, drain water from test tube.

(iv) Cover the tubes with aluminum foil and incubate in a water bath at 49°C for 30 min.

(v) After the heat treatment period, the tubes are kept over night at room temperature.

(vi) Conductance of solution is measured with an electrical conductivity meter after calibration with a 1 N KCl solution.

(vii) Autoclave the test tubes at 120°C and 0.10 MPa for 10 min. Cool the tubes at 25°C and measure the conductance again.

(viii) Membrane thermal stability (MTS) is expressed in % as the reciprocal of relative leakage and can be calculated using the following formula :

$$MTS (\%) = (1 - T_1 / T_2)$$

Where T_1 is the initial conductivity reading after heat treatment and T_2 is the final conductivity reading after autoclave. Whitlow *et al.* (1992) suggested improvement in expressing results of membrane stability by using two formula : 1st is index of injury (Id) = 100 (Rt RO/(1-RO)

Where Rt = Ec initial / Total Ec for stressed plant, RO = same ratio of EC for control plant. Initial Ec is Ec after a given period of incubation at room temperature and total Ec is Ec after heat killing of the tissues.

REFERENCES

1. Sullivan, C.Y. 1972. Mechanisms of heat and drought resistance in grain sorghum and methods of measurement. In: *Sorghum in the Seventies.* (Eds. N.G.P. Rao and N.R. House), pp. 247-267. Oxford & IBH Publishing Co., New Delhi, India.
2. Whitlow, T.H., Bassuk, N.L., Ranney, T.G. and Reichest, D.I. 1992. An improved method for using electrolyte leakage to assess membrane competence in plant tissues. *Plant Physiology* 98 : 198-205.

(6) LIPID PEROXIDATION

Rationale

Free radical reactions and lipid peroxidation are believed to be a universal and fundamental mechanism of membrane and cell damage. The formation of lipid peroxides, mediated by production of reactive forms of oxygen as well as by activation of certain oxygenases, has been shown to take place in plants exposed to herbicides, pollutants, heat, chilling and drought, and as a result of a hypersensitive response to invasion by pathogenic microorganisms as well as during senescence (1-3). Due to its simplicity, the most widely used method for the quantification of lipid peroxidation in isolated membranes and tissues is a photometric determination of a complex (absorption maximum at 532 nm) produced in the reaction between malondialdehyde (MDA) and the 2-thiobarbituric acid (TBA) reagent (4). TBA is known to form coloured complexes not only with MDA (as a principal product of lipid peroxidation) but also with carbohydrates, pyrimidines, amino acids, aliphatic aldehydes, haem-containing substances and particularly in plants, various phenolic compounds including anthocyanins (2,5).

Method

Two simple modifications to the original method are presented here which significantly improve the sensitivity and accuracy by including an antioxidant in the assay to reduce artefactual lipid peroxidation and by using derivative absorption spectroscopy to overcome problems of contamination with plant pigments.

Reagents

400 mg thiobarbituric acid (TBA) dissolved in

100 ml 0.25 M HCl then add

15 g trichloracetic acid then add

1 ml, 2% butylated hydroxy toluene in EtOH and mix

Procedure

Plant tissue of about 100 mg is homogenized in a minimum volume of extraction medium (typically KPi buffer, pH 7-7.6). The homogenate is centrifuged at 13000 g for 60 s.

Assay

n ml assay supernatant (typically 100-200 µl for shoots or leaves and 500 µl for root tissue) added to

3.0 ml minus n ml reagent

Heat to 100°C for 25 min using glass stoppers.

Cool. Centrifuge at 4000 g 3 min

Read the absorbance at 532 nm minus non-specific absorbance at 600 nm or preferably the second derivative. The points of interception of the derivative spectra with the base line are at 520 and 542 nm.

Calculation against 1,1,3,3, tetraethoxy-propane standards.

REFERENCES

1. Thompson, J.E., Legge, R.L. and Barber, F.R. 1987. The role of free radicals in senescence and wounding. *New Phytologist* 105: 317-344.
2. Merzlyak, M.N. 1991. Syndrome of lipid peroxidation in plants, in *Plant Lipid Bio-chemistry, Structure and Utilization* (eds. P.J. Quinn and J.L. Harwood), Portland Press, London, pp. 281-288.
3. Seel, W., Hendry, G., Atherton, N. and Lee, J. 1991. Radical formation and accumulation *in vivo*, in desiccation tolerant and intolerant mosses. *Free Radical Research Communication* 15: 133-141.
4. Heath, R.H. and Packer, L. 1968. Photoperoxidation in isolated chloroplasts. I. Kinetics and stoichiometry of fatty acid peroxidation. *Archives of Biochemistry and Biophysics* 125: 189-198.
5. Gutteridge, J.M.C. and Halliwell, B. 1990. The measurement and mechanism of lipid peroxidation in biological systems. *Trends in Biochemical Sciences* 15: 129-135.

(7) NITRATE REDUCTASE ACTIVITY

Rationale

Nitrate reductase is the first enzyme of the nitrate assimilatory pathway and its activity in different plant species has been used as a measure of nitrogen nutrition and nitrate availability. It is a substrate inducible enzyme, and despite differences between species in levels and inducibility of nitrate reductase activity many studies have demonstrated a close relationship between soil nitrate availability and leaf nitrate reductase activity (1-4). Comparison of *in situ* nitrate reductase activity with that present after induction provides a measure of nitrate availability and the potential of different species to assimilate nitrate. Induction can be carried out by direct fertilizer addition or by supplying cut shoots with nitrate via the transpiration stream.

Method

For the processing of large numbers of samples or the comparison of different species in the field or laboratory, an *in vivo* method of determining nitrate reductase activity is to be preferred. The following method, based on (5, 6) has been employed for many years and has been applied to hundreds of different species.

Shoots/Leaves

Whole shoots of small herbaceous species or terminal leafy shoot material of other species are sliced into small fragments; approximately 0.1-0.5 g fresh materials is placed in a small reaction vial and vacuum infiltrated with assay solution. This contains 100 mM KPi buffer, pH 7.5; 100 mM KNO_3; 1.5% (v/v) propan-1-ol. After infiltration the samples are incubated for at least 1 hour, in darkness at 25°C. 1 ml of the reaction mixture is removed and assayed colorimetrically for nitrite using a modified Griess-Ilosvay reaction. To 1 ml sample add 1 ml sulphanilic acid (1% in 3 M HCl), followed by 1 ml a-napthyl ethylenediamine dihydrochloride (0.02% w/v). The colour is allowed to develop for 20 minutes and absorption at 540 nm is then read. A standard curve is prepared using 0-100 nmol per ml $NaNO_2$. Tissue blanks should be prepared using 1 ml reaction mixture (after incubation), 1 ml sulphanilic acid and 1 ml water. This corrects for leakage of pigments, etc. from the tissue. This assay gives a measure of relative levels of nitrate reductase activity (NRA) of the enzyme present in leaf tissue when assayed with a non-limiting amount of substrate. A similar assay can be carried out using a reaction mixture from which nitrate has been omitted. This assay measures nitrate reductase activity dependent solely on endogenous pools of nitrate available to be enzyme system during the assay. This provides an easy identification of species storing nitrate ions in the shoot system.

Roots

Samples (0.25-0.5 g fresh weight) of freshly excavated root material are assayed anaerobically in the presence of added substrate (75 mM KNO_3 and 1% v/v propan 1-ol) as in the Standard Assay for shoots. Wherever possible only the finer lateral feeding roots of a species are utilized.

REFERENCES

1. Taylor, A.A., De-Felice, J. and Havill, D.C. 1982. Seasonal variation in nitrogen availability and utilization in an acidic and calcareous soil. *New Physiologist* 92 : 141-152.
2. Adams, M.A. and Attiwill, P.M 1982. Nitrogen mineralization and nitrate reduction in forests. *Soil Biology and Biochemistry* 14 : 197-202.
3. Hogberg, P., Granstrom, A., Johanson, T., Lundmark-Thelin, A. and Hasholm, T. 1986. Plant nitrate reductase activity as an indicator of ava-ilability of nitrate in forest soils. *Canadian Journal of Forest Research* 16 : 1165-1169.
4. Stewart, G.R., Hegarty, E.E. and Specht, R.L. 1988. Inorganic nitrogen metabolism in plants of Australian rainforest communities. *Physiologia Plantanum* 74 : 26-33.

5. Stewart, G.R., Popp, M., Holzapfel, I., Stewart J.A. and Dickie-Eskew, A. 1986. Localization of nitrate reduction in ferns and its relationship to environment and physiological characteristics. *New Phytologist* 104: 373-384.
6. Jaworski, E. 1971. Nitrate reductase assay in intact plant tissues. *Biochemical and Biophysical Research Communications* 43: 1274-1279.

(8) HIGH TEMPERATURE AND PROTEIN SYNTHESIS

Rationale

A wide range exists in temperature tolerance both between species and amongst genotypes within a species. The thermal environment can thus greatly affect plant distribution. Protein synthesis is a very thermosensitive metabolic process. Analysis of the response of protein synthesis to temperature provides information on the adaptation of a given species and its ability to acclimate to changes in temperature.

Method

Protein synthesis is monitored by the addition of a radioactive amino acid to the incubation medium of the plant, or plant part (at a given temperature), and the determination of its incorporation into newly synthesized protein. The tissue is homogenized and the proteins fractionated on the basis of their molecular size by sodium dodecyl sulphate polyacrylamide gel electrophoresis (SDS PAGE). The radiolabelled proteins can then be detected by fluorography using X-ray film.

Reagents

Incubation buffer 20 mol m^{-3} Tris, pH 7.5, 5 mol m^{-3} $MgCl_2$, 30 mol m^{-3} sucrose, 50 µg cm^{-3} chloramphenicol

[^{35}S]-methionine (specific activity > 1000 Ci $mmol^{-1}$)

Homogenization buffer. 50 mol m^{-3} Tris, pH 7.5, 6 mol m^{-3} β mercaptoethanol, 1% w/v Triton X-100

SDS PAGE sample, buffer. 250 mol m^{-3} Tris, pH 6.8, 8% w/v sodium dodecyl sulphate, 20% w/v glycerol, 20% v/v β-mercaptoethanol, 0.04% bromophenol blue

Coomassie Blue stain : 0.5% w/v Coomassie Brilliant Blue R, 10% v/v HOAC, 45% v/v MeOH

Destain : 7% v/v HOAC, 25% v/v MeOH

Amplify (Amersham)

D19 developer and Rapid Fixer (Kodak)

Procedure

Pre-heat incubation buffer to the treatment temperatures before use.

Approximately 250 mg (FW) plant tissue (e.g. seedlings, leaves, plant parts) are placed in vials containing 250 ml preheated incubation buffer and 20 μCi [^{35}S]-methionine is added to each sample. Vials are placed in incubators at appropriate temperatures for 2 hours.

After incubation with radiolabel, the tissue is washed thoroughly with distilled water and homogenized in an ice cold mortar in 4 μl homogenization buffer per mg FW. The homogenate is then centrifuged for 10 minutes (12000 g, 4°C). A 100 ml aliquot of supernatant is added to 33 μl SDS sample buffer, the mixture heated at 95°C for 5 min and stored at -20°C until analysed by SDS PAGE. The remaider of the supernatant can be used to determine the protein concentration [1] and the incorporation of radioactivity into protein by precipitation with trichloro-acetic acid (TCA) (2). SDS PAGE is carried out using a 12.5% running gel and a 4% stacking gel (3, 4). Equal quantities of protein are applied to each lane of the gel with molecular weight standards applied to the outermost lanes.

Once electrophoresis is complete, the gel is stained to detect unlabelled marker proteins by covering the gel with stain solution for 15 minutes. The stain is then removed and the gel placed in destain solution until the background is sufficiently clear to see the protein bands. The gel is then prepared for the detection of radioactively labelled polypeptides by fluorography by removing the destain solution and replacing it with Amplify (or other commercially available fluor such as EnHance) for 15 minutes. The gel is dried down on to filter paper under vacuum at 60°C, exposed to X-ray film (e.g. X-OMAT) in complete darkness and kept at -80°C until ready to develop.

Exposure time is dependent on the incorporation of radioactivity into protein but usually 4 days are sufficient to detect protein synthesis. X-OMAT film is developed according to the manufacturer's specifications (5 minutes in D19, followed by 30 seconds in stopbath and 5 minutes in fixer).

REFERENCES

1. Lowry, O.H., Rosebrough, N.J., Farr, A.L. and Randall, R.J. 1951. Protein measurements using the Folin phenol reagent. *Journal of Biological Chemistry* 193 : 265-275.

2. Mans, R.J. and Novelli, G.D. 1960. A convenient, rapid and sensitive method for measuring the incorporation of radioactive aminoacids into protein. *Biochemical and Biophysical Research Communications* 3 : 540-543.
3. Laemmli, U.K. 1970. Cleavage of structural proteins during the assembly of the head of bacteriophage T4. *Nature* 224 : 680-685.
4. Hames, B.D. and Rickwood, D. 1989. *Gel Electrophoresis of Proteins. A Practical Approach*. IRL Press, Oxford.

(9) CARBON ISOTOPE DISCRIMINATION AND WATER USE EFFICIENCY

Rationale

Two naturally occurring stable isotopes of carbon exist, ^{12}C and ^{13}C. The majority of the carbon is ^{12}C (98.9%), with 1.1% being ^{13}C (1). The ratio of these two stable isotopes varies in plant material as a result of isotope fractionation during photosynthesis and respiration. The ratio of $^{13}C/^{12}C$ is thus reduced compared to the atmospheric ratio because of biochemical discrimination against the heavier isotope (2), (3). Recently, carbon isotope discrimination (Δ) has been found to be affiliated with the extent to which plants preferentially fix $^{12}CO_2$ rather than $^{13}CO_2$ and with the ratio of intercellular (C_i) and atmospheric (C_a) carbon dioxide concentrations (4). Consequently, a strong correlation exist between carbon discrimination and the ratio of carbon assimilation and the rate of transpiration (A/E), i.e., instantaneous water use efficiency (5).

Because of these instantaneous physiological relations, water-use efficiency is negatively related to Δ (4). That is, when Δ is relatively large, water-use efficiency is low due primarily to high stomatal conductance and high intercellular CO_2 concentrations (C_i). However, when stomatal conductance is low, C_i decreases, resulting in less ^{13}C discrimination in plants and relatively higher water-use efficiencies. Under constant environmental conditions (similar leaf-to-air vapour pressure differences and similar p_a), Δ will be negatively related to whole plant water use efficiency if little variation in respired CO_2 exists and little difference in cuticular transpiration occurs (6). Carbon discrimination (Δ) can therefore, be used as a tool to examine integrative measures of physiological plant performance including estimates of C_i and of estimates of water use efficiency provided a series of assumptions are followed and a suite of environmental conditions are considered.

Methods

Carbon discrimination can be accurately measured on bulk plant material using sampling precautions and an isotope ratio mass spectrometer[3]. In general, recently synthesized plant tissue can be removed from laboratory experiments or field studies with care being given to collecting samples at the same time of day; or as is often the case, at the end or near the end of the growing season to take advantage of the integrative nature of carbon discrimination (7, 8). After harvesting, plant material should be dried immediately in air or at 70-80°C for 48 hours. Once dry, samples can be stored for long periods without affecting the isotope composition.

Measurements of isotropic ratios on plant material are conducted on dried biomass which has been ground to pass through at least a 40-mesh screen. Subsamples of larger bulk samples are often used because only 3 mg or less of dried organic materials is needed for $^{13}C/^{12}C$ determinations. Consequently, it is critical that the ground samples are homogeneous.

Differences in carbon isotope ratios must be measured with very high precision because small but significant variations in stable carbon isotope ratios ($^{13}C/^{12}C$) exist in plants. A dual-inlet gas isotope ratio mass spectrometer (GIRMS) equipped with two or more ion beam collectors is capable of such precision. In general, the equipment converts the ground material to CO_2 gas which is then pulled into the system by vacuum. CO_2 molecules are then ionized by an electron beam and an accelerating potential propels the positive ions down an analyzer tube into the curved sector located in a magnetic field. The trajectory of the ions passing through the magnetic field then becomes a function of the energy and mass of the ions. Because of their different atomic weights (44, 45 and 46) the three major species of CO_2 are separated into discrete ion beams and strike separate Faraday cup collectors. The isotope ratios of Carbon-13 and Carbon-12 are then computed based on electrical currents.

Carbon discrimination in C_3 plants can be expressed as :

$$\Delta = \frac{-7.8\times10^{-3} - \delta_p}{1+\delta_p}$$

where, -7.8×10^{-3} is the carbon isotope composition of CO_2 in the air relative to the PDB standard, and δ_p is the $^{13}C/^{12}C$ ratio of the plant material expressed as $\delta^{13}C$ (4).

REFERENCES

1. Craig, H. 1954. Carbon-13 in plants and the relationship between carbon-13 and carbon-14 variations in nature. *Journal of Geology* 62 : 115-149.
2. O'Leary, M.H. 1988. Carbon isotopes in photosynthesis. *Bioscience* 38 : 325-336.
3. Boutton, T.W. 1991. Stable carbon isotope ratios of natural materials : sample preparation and mass spectrometric analysis, in *Carbon Isotope Techniques.* (eds. D.C. Coleman and B. Fry) pp. 155-172. Academic Press, Inc., New York.
4. Farquhar, G.D., Ehlenringer, J.H. and Hubick, K.T. 1989. Carbon isotope discrimination and photosynthesis. *Annual Review of Plant Physiology* 40 : 503-537.
5. Farquhar, G.D. and Richards, R.A. 1984. Isotopic composition of plant carbon correlates with water-use efficiency of wheat genotypes. *Australian Journal of Plant Physiology* 11 : 539-552.
6. Read, J. and Farquhar, G. 1991. Comparative studies in Nothogagus (Fagaceae). Leaf carbon isotope discrimination. *Functional Ecology* 5 : 684-695.
7. Ehleringer, J. 1991. $^{13}C/^{12}C$ fractionation and its utility in terrestrial plant studies, In: *Carbon Isotope Techniques* (eds. D.C. Coleman and B. Fry). pp. 187-200. Academic Press, Inc., New York

(10) FREE PROLINE ACCUMULATION

Rationale

Free proline contents increases have been found under water and osmotic stress (Singh *et al.,* 1972; Bhaskaran *et al.,* 1985) and also when plants were subjected to salt, cold or freezing stress (Chu *et al.,* 1974, 1976). Proline can also accumutate in some non-stressed plant structures that dehydrate, like pollen and seeds (Hua *et al.,* 1997). This fact avails its role in the tolerance to water stress acting as osmoprotectant. So, its measurement would be an excellent stress detector (Aspinall and Paleg, 1981; Bellinger *et al.,* 1991).

Many evidences relate proline metabolism with stress tolerance, and many papers have studied proline accumulation as a parameter whose genetical transmission has been used in selection for tolerance to wilting or salinity in a number of plant species (Lutts *et al.,* 1999; Gibon *et al.,* 2000).

PROTOCOL FOR PROLINE QUANTIFICATION (Bates *et al.*, 1973)

Reagents

Sulphosalicylic acid 3% (w/v)

Phosphoric acid 6 M

Glacial acetic acid

Acid-ninhydrin prepared by warming 1.25 g ninhydrin in 30 ml glacial acetic acid, and adding 20 ml phosphoric acid 6 M (Acid ninhydrin will keep stable only for 24 hours, at 4°C).

Purified proline

Toluene

Preparation of the Standard Curve

Sample absorbances can be converted to proline concentration by comparison to a proline standard curve obtained using known proline concentrations. Every time proline is calculated, a standard curve must be prepared.

A trial-and-error assay will be often necessary to select adequate standard curve points. Several proline solutions must cover be prepared at different concentrations; they must the expected range of unknown sample concentrations. Different solutions are prepared by adding distilled water to an `stock' solution of commercial purified proline.

Method

1. Homogenisation and Filtration

250 to 500 mg of fresh leaf tissue (obtained from the youngest fully expanded leaves) are powdered in a cold mortar with liquid nitrogen until pulverisation and homogenised with 5 ml of sulphosalicylic acid 3% (to get protein precipitation). Then, the homogenate is filtered through a n-2 filter paper (e.g., Whatman@ n2) or centrifuged.

The required amount of fresh sample will depend on the tissue's water content.

2. Reaction with Ninhydrin

2 ml of the filtrated sample are mixed with 2 ml of glacial acetic acid and 2 ml of acid-ninhydrin in test tubes. After agitation, the sample is incubated at 100°C for 1 hour, and this produces the

coloured complex formation. After one hour, reaction is stopped by putting the tubes in an ice bath/cold water.

3. Extraction

4 ml of toluene are added to each tube, and vortexed for 15-20 s. Then, organic and inorganic phases are separated, obtaining the cromophore dissolved in toluene.

4. Spectrophotometric Measurement

The maximum absorbance of the chromophore when dissolved in toluene is attained at a wavelength of 520 nm. Thus, the organic phase of each sample and standard containing the chromophore is collected, and absorbance at 520 nm is measured spectrophotometrically.

Calculations and Data Analysis

Using the data obtained from the standard curve prepared at known concentrations, a linear regression is done (comparing absorbance vs. proline concentration). Thus, we obtain the equation that will allow sample concentration calculations.

Finally, it is advisable to express free proline concentration relative to dry weight (μmol/g of dry weight of plant material).

REFERENCES

1. Aspinall, D. and Paleg, L.G. 1981. ``Proline Accumulation : Physiological Aspects." In: *The Physiology and Biochemistry of Drought Resistance in Plants.* L.G. Paleg, D. Aspinal (eds.) New York, USA : Academic Press.
2. Bates, L., Waldren, R.P. and Teare I.D. 1973. Rapid determination of free proline for water stress studies. *Plant and Soil* 39 : 205-207.
3. Bellinger, Y., Bensaoud, A. and Larher, F. 1991. ``Physiological significance of proline accumulation - a trait of use of breeding for stress tolerance." In: *Physiology and Breeding of Winter Cereals for Stressed Mediterranean Evnironments.* Paris : INRA Les Colloques.
4. Bhaskaran, S., Smith, R.H. and Newton, R.J. 1985. Physiological changes in cultured sorghum cells in response to induced water stress. I. Free proline. *Plant Physiology* 79 : 266-269.
5. Chu, T.M., Aspinall, D. and Paleg, L.G. 1974. Stress metabolism. VI. Temperature stress and the accumulation of proline in barley and radish. *Australian Journal of Plant Physiology* 1 : 87-97.
6. Chu, T.M., Aspinall, D. and Paleg, L.G. 1976. Stress metabolism. VII. Salinity and proline accumulation in barley. *Australian Journal of Plant Physiology.* 3 : 219-228.

7. Gibon, Y., Ronan, S. and Larher, F. 2000. Proline accumulation in canola leaf discs subjected to osmotic stress is related to the loss of chlorophylls and to the decrease of mitochondrial activity. *Physiologia Plantarum* 110 : 469-476.
8. Hua, X.J., Van de Cotte, B., Vann Montagu M. and Verbruggen N. 1997. Developmental regulation of pyrroline-5-carboxylate reductase, gene expression in Arabidopsis. *Plant Physiology* 114 : 1215-1224.
9. Lutts, S., Majerus, V. and Kinet, J.M. 1999. NaCl effects on proline metabolism in rice (*Oryza sativa*) seedlings. *Physiologia Plantarum* 105 : 450-458.
10. Singh, T.N., Aspinall, D. and Paleg, L.G. 1972. Proline accumulation and varietal adaptability to drought in barley : a potential metabolic measure of drought resistance. *Nature New Biol.* 236 : 188-190.

(11) GLYCINE-BETAINE

Rationale

Quarternary amies (QAMs), in particular Glycine-betaine are accumulated during salinity stress in some slightly salt-tolerant glycophytes and halophytes (1). They serve multiple purposes related to both osmotic adjustment and osmo-protection damage (2). Glycine-betaine can protect enzymes from heat damage also which indicate its capability to protect enzymes from stress damage. The absence of glycine-betaine in some important crop species (rice, tobacco etc) have led to speculate that genetic engineering of this trait may be a successful avenue for increasing stress tolerance in crops.

Estimation of Glycine-betaine

The plant samples can be extracted and estimated for glycine betaine following the method of Grieve and Grattan (1983) described below.

Reagent

(1) Toluene (0.05%) in 2 N HCl

(2) Potssium tri-Iodide solution : Dissolve 7.5 g Iodine and 10 g potassium iodide in 100 ml of 1 N HCl.

(3) 1, 2 Dichroethene (Chilled at 10°C)

(4) Glycine-betaine for standard curve preparation

(5) 1 N H_2SO_4

Extraction

The samples are extracted and estimated following the method of Grieve and Grattan (1983).

Five hundred milligrams of finely ground dry plant sample are mechanically shaken with 20 ml of distilled water for 24 hours at 25°C. Time required for this step may be determined by extracting the plant samples for 4, 8, 16, 24 and 48 hours. The samples are then filtered through Whatmann No. 1 filter paper and the filtrates are made upto 20 ml with deionized water and used for estimation immediately.

Estimation

One millilitre of the extract is diluted with one millilitre of 2 N H_2SO_4 and 0.5 ml of this acidified extract is cooled in ice water for 1 hour. Later 0.2 ml of cold potassium tri iodide solution is added and mixed gently with a vortex mixture and the tubes are stored at 4°C for 15 minutes and then centrifuged at 10,000 g for 15 minutes at 0°C. The supernatant is aspirated with a fine tipped glass tube. The per iodide crystals are dissolved in 9 ml of 1,2-dichloroethane with vigorous vortexing. After 2.5 hours the absorbance is measured at 365 nm in a Spectrophotometer. Reference standard of glycine-betaine is prepared in 1 N H_2SO_4 and used for estimating the glycine-betaine content and the results are expressed in µg/g dry weight.

REFERENCES

1. Storey, R. and Wyn Jones, R.G. 1977. Quartary ammonium compounds in plants in relation to salt resistance. *Phytochemistry* 16 : 447-453.
2. Pollard, A. and Wyn Jones, R.G. 1979. Enzyme activities in concentrated solutions of glycine-betaine and other solutes. *Planta* 144 : 291-298.
3. Grieve, C.M. and Grattan, S.R. 1983. Rapid assay for the determination of water soluble quartery ammonium compounds. *Plant and Soil* 70 : 303-307.

CHAPTER 15

DETERMINATION OF PLANT WATER STATUS

INTRODUCTION

The two basic parameters which describe the degree of unsaturation, i.e. the plant water deficit are: (i) the water content, and (ii) the energy status of the water in the cell. The water content is usually expressed as relative to that at full saturation, i.e. the relative water content (RWC) or water saturation deficit (WSD). The energy status of the water is usually expressed as the total plant water potential (Ψ_w).

However, metabolic status of tissues is related with its water status. Water content of the plants is always in a state of continuous flux,with the water status in surrounding soil and air. Therefore, the knowledge of the subtle changes in the plant water status is imperative. Depending upon the tissue, scale of operation and the objectives of observation, several methods are available to monitor plant water status. These are discussed in this chapter under appropriate heads.

METHODS TO DETERMINE WATER STATUS

Several methods to determine the water potential (Ψ_w) of cells, tissues and organs are available and are of three kinds: (1) Compensation methods (2) Psychrometric methods or direct vapour pressure measurement methods and (3) Pressure equilibration methods (Slavik, 1974).

Compensation Methods

In these methods, a solution of known osmotic potential (Ψ_s), which is equal (isotonic) to the Ψ_w of the tissue or individual cells, is used. The net transfer of water between graded test solutions and samples of sets of tissues dipped in test solutions is determined. This water transfer is measured by (i) change in concentration of test solution by refractometer or smear methods, (ii) change in size (volume/weight) of the tissue sample and (iii) change in the volume of the test solution by potometric method.

Measurement of Relative Water Content of Leaf

Principle

Plant leaves generally have lower (more negative) water potential than pure water; hence, they osmotically absorb water and become turgid. A measure of this property is the Relative Water Content (RWC) which expresses the leaf water content (%) of the turgid leaf water content. It is calculated by the formula :

$$RWC\% = \frac{Fresh\ weight - Dry\ weight}{Turgid\ weight - Dry\ weight} \times 100$$

A similar parameter is the Water Saturation Deficit (WSD) that is computed by the formula :

$$WSD\% = \frac{Turgid\ weight - Fresh\ weight}{Turgid\ weight - Dry\ weight} \times 100$$

The water is absorbed till the existence of a Water Saturation Deficit (i.e. till the tissue becomes fully turgid and stops absorbing water) so that RWC+WSD=100% or WSD = 100–RWC (Kramer, 1983).

Materials

Leaf tissue, deionised water, Petri dishes, balance, etc.

Procedure

Excise the leaf tissue (or any other tissue) whose RWC is to be determined. Weigh the tissue and now dip the tissue in deionised water in covered Petri dishes or any other suitable container for 4-5 h at room temperature till the tissue is fully saturated in a perfectly humid environment. Remove the tissue, blot it to remove surface water and weigh to get the turgid weight. Now dry the tissue to a constant weight in an oven at 65-70°C for about 72 h and weigh it to record the dry weight.

Precautions

(i) For fresh weight, weigh the tissue immediately.

(ii) For turgid weight, tissue should be kept dipped in water for sufficient time (4-5 h) till a constant weight is obtained.

(iii) If leaf tissue is too large, then leaf discs can be taken.

Measurement of Water Potential (Ψ_w) of Plant Tissues

In terms of thermodynamics, work done by a system is dependent upon its free energy. Free energy is the sum total of all types of energies i.e. kinetic, potential, chemical etc. For a system, in solution the free energy of one mole of a substance is defined as its chemical potential. As water is the solvent for all activities in the living systems it is conventional to take into account the chemical potential of the water which is abbreviated as Ψ_w. Thus, free energy of one mole of water can be defined as the water potential. The concept of water potential has caught up with the physiologists as a convenient mode as a unit to express the flux of water in living systems. To simplify, water potential of pure water is regarded equal to zero. Water moves from a region of higher water potential to a region of low water potential or a more negative water potential. The presence of solute molecules in the solvent water decreases its water potential. This decrease is proportional to the amount (moles) of solute dissolved in water. Therefore, like freezing point and boiling point etc. water potential is also a colligative property of solution.

Chardakov's Method (Dye Method)

This method developed by Chardakov a Russian scientist, can be used in the laboratory as well as the field and does not require any sophisticated instrumentation. Further, it is reasonably sensitive and can generate reliable experimental data.

Principle

Concentration of the solution in which a plant tissue is immersed changes (increases or decreases), depending upon the water potential of the tissue and the solution. If the solution is hypertonic (with lower Ψ_w) the tissue loses water and the solution becomes dilute. If the solution is hypotonic (with higher Ψ_w) the tissue absorbs water and the solution becomes concentrated. If the solution is isotonic (with Ψ_w equal to the tissue sap), the tissue neither loses nor absorbs water and the solution concentration does not change. Thus the Ψ_w of this solution gives the Ψ_w of the tissue (Salisbury and Ross, 1969).

Materials

Leaves or any other tissues of the plant to be tested, test tubes, sucrose or polyethylene glycol (PEG) 6000 solution, methylene blue, dropper.

Procedure

Prepare graded concentrations of sucrose or PEG 6000 of different molalities i.e. 0.05, 0.1, 0.15, 0.2, 0.25, 0.3. Take these solutions in two test tubes each. Colour one set of each solution by dissolving a small crystal of methylene blue dye in it (control solution). This does not change the Ψ_w of the control solution compared to its non-coloured counterpart (test solution). Immerse tissue samples of approximately of equal size in each colourless test solution test tube (Fig. 1). Allow the tissue to come into equilibrium with the solution for about 15-20 min. at room temperature. Remove the tissue from the test tubes. Put a drop of the corresponding coloured control solution in the colourless test solution in the centre. Record the movement of the drop i.e., whether it rises up, diffuses or sinks.

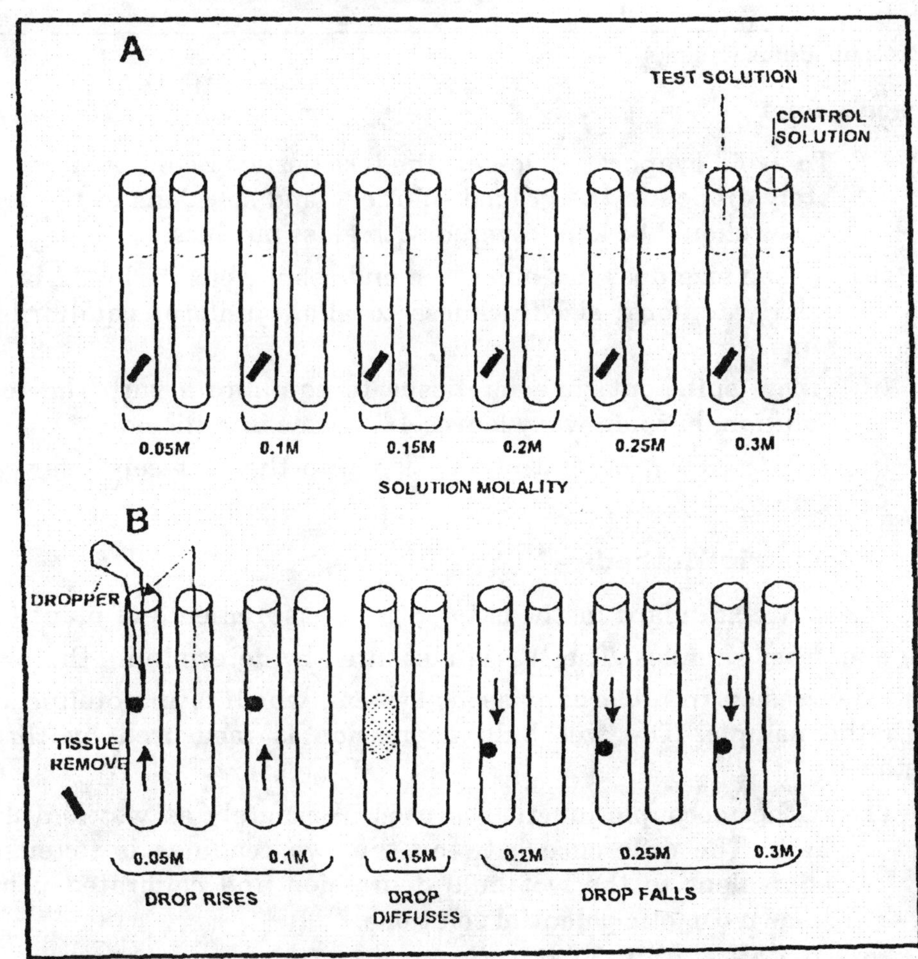

Figure 1. Measurement of water potential of plant tissues by Chardakov's method.

The movement of control solution drop in test solution should be recorded in following Table 1.

Table 1. The movement of control solution drops in test solution.

S.No.	Molality of solution	Direction of drop movement*
1.	0.05	
2.	0.10	
3.	0.15	
4.	0.20	
5.	0.25	
6.	0.30	
7.	0.35	

* Rises up, diffuses, sinks.

Precautions

(i) To avoid evaporative losses, the test tissue should be immediately dipped in the graded solutions and solutions in the test tubes should be kept covered with glass marbles.

(ii) Tissue size does not affect the end result, but it should be of nearly uniform size/thickness to allow uniform equilibrium time.

(iii) Allow sufficient time for tissue to equilibrate with the test solution. For non-waxy leaves 4-5 h time is sufficient.

(iv) Put the drop of control solution into the test solution very gently.

Psychrometric Methods

These are standard methods for Ψ_w measurements of plant tissue and other samples. The Ψ_w is measured by determining the `wet bulb' depression in a closed gaseous system, which is in equilibrium with the sample. The wet bulb depression is measured by three methods :

(1) A thermocouple junction is used alternately as wet and dry bulb. The difference between the two readings is taken as equivalent to the wet bulb depression and calibrated using known osmotic potential solutions.

(2) Wet loop or droplet psychrometer, the output of thermocouple with the thermal junction permanently wetted by a small drop of pure water is maintained. This psychrometer needs a long term thermal stability of the vessel or an additional similar thermocouple without a water droplet included in every sample vessel and measured as a reference. Instead of

pure water, sucrose solution of different known osmotic potentials can be used on wet junction of the thermocouple by the `Isopiestic' technique.

(3) A thermistor with attached miniature chamber, which can be filled with water, is used to measure wet and dry bulb temperature.

Thermocouple Psychrometer Method

Thermocouple psychrometers are excellent instruments based on sensitive electric and electronic gadgetry and are well accepted for their speed, ease and accuracy. These can measure both the water potential of the plant tissues as well as the solute potential of solutions or expressed plant sap.

Principle

Water vapour at equilibrium with a plant tissue in a closed chamber will have the same Ψ_w as that of the tissue. The Ψ_w of tissue can thus be measured by knowing the vapour pressure of the chamber (Hopkins, 1995).

Psychrometery *per se* refers to the measurements of difference in temperature between an atmosphere and a freely evaporating moist surface in that atmosphere. The rate at which the water molecules evaporate and therefore, the rate of cooling depends primarily on the vapour pressure of water in atmosphee. Lower the vapour pressure, steeper is the vapour pressure gradient and more rapid is the evaporation and cooling. The psychrometric technique is thus used to determine small differences in vapour pressure and this is the principle behind the use of a dry and wet-bulb thermometer to measure relative humidity of the atmosphere. In a thermocouple psychrometer, thermocouples are used instead. A thermocouple is formed when wires of two dissimilar metals such as copper and constantan (a copper nickel alloy) are joined. If both ends of the wires are joined to form a close circuit with two thermocouples and the two thermocouple junctions are held at different temperatures an electric current will flow through the circuit. This is known as the Seebeck effect. The magnitude of the current is a measure of the difference in temperature between two junctions.

Another property of the thermocouple called the Peltier effect is also used. This pertains to the cooling down of the junction, when a suitable current is passed through the circuit. Detailed relevance of these effects in the instrument is explained below.

Instrumentation

Various types of thermocouple psychrometers are available but a typical instrument (Fig. 2). has a small chamber in which the plant sample can be introduced. The chamber is maintained at a constant temperature bath that acts as a heat sink. The chamber also contains a thermocouple junction with a silver ring (silver has high thermal conductivity) on which a drop of water is produced by Peltier current effect. The chamber is very small so that the atmosphere quickly saturastes with water vapour.

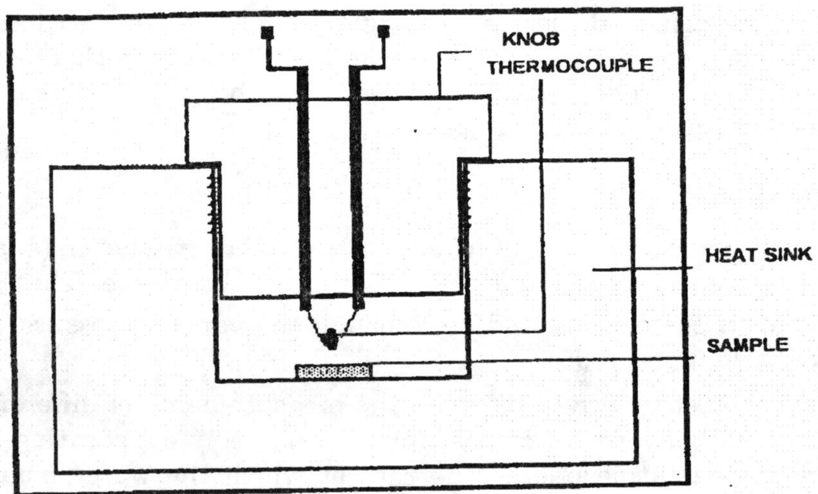

Figure 2. Diagram of thermocouple psychrometer sample chamber.

As the tissue Ψ_w is more negative than the water drop on the thermocouple, water from the chamber atmosphere will diffuse into the tissue. As the water continues to evaporate from the drop to replace the vapours taken up by the tissue, the resulting evaporating cooling depresses the temperature of the thermocouple (wet bulb depression) which can be measured due to Seebeck effect against a standard reference.

Procedure

To measure Ψ_w, plant sample is placed on the sample holder. For Ψ_w measurement filter paper discs saturated in cell sap expressed by freeze-thaw [Freeze-thaw is a technique to express the plant tissue sap. The tissue is first frozen (up to -10°C). The cell sap forms ice/solution cystals. The frozen tissue is thawed i.e. allowed to come to room temperature. The ice melts and the sap is expressed by the gentle pressure]. Plant tissues are placed on the sample holder, The holder is then inserted in the chamber. The chamber is sealed by tightening the knob. When the tissue/sap attains equilibrium with the chamber atmosphere, the voltage is measured on the display panel. The Ψ_w or Ψ_s values are read from the calibration curve. Calibration of the instrument is done so that the voltage measured is converted into Ψ_w units. This is done by measuring voltage out put, in response to standard KCl or PEG solutions of known Ψ_w. Then a Ψ_w versus microvolt calibration curve is drawn. Such a curve is usually a straight line.

Precautions

(i) The Ψ_w of vapours is very sensitive to temperature. Therefore, the chamber should not be subjected to temperature variations.

(ii) The sample holder, the chamber and the thermocouple must be kept clean.

(iii) Sufficient time (about 5 min) should be allowed to lapse to allow equilibration of the sample with chamber atmosphere.

Pressure Equilibration Methods (Pressure Chamber Method)

It is necessary to know the water potential of the leaves or whole shoots rather than bits of tissues. Pressure chamber method gives exact values of the cumulative water potential (negative hydrostatic pressure) of the sap in such cases.

Principle

A negative hydrostatic pressure (tension) exists in the xylem column of actively transpiring shoots or leaves. This hydrostatic pressure is fairly equal to the average Ψ_w of the whole shoot because the osmotic potential of the xylem water *per se* is negligible and xylem water is in intimate contact with most of the cells of the plant shoot. When a plant shoot is cut, water is pulled up in the xylem capillaries due to the negative hydrostatic pressure. An applied balance pressure that counteracts this hydrostatic pressure in therefore, equal to the shoot Ψ_w (Scholander *et al.,* 1965).

Materials

Pressure chamber apparatus, plant shoot, compressed nitrogen gas cylinder.

Procedure

The pressure chamber apparatus (Fig. 3) comprises of a steel chamber with a connection to compressed nitrogen gas cylinder to pressurize the chamber and a pressure gauge to measure the chamber pressure. The chamber has an air tight lid with a hole and seal. The chamber also has a gas release stopcock.

Excise the shoot of the plant, whose Ψ_w is to be determined and fit it air tightly in the seal hole of the chamber with the inverted shoot inside the chamber and a part of the cut stem protruding out. At this point cut surface of the stem appears dull and dry. Now release the gas into the chamber, gradually increasing its pressure. At a point the cut surface shall just appear wet and shiny due to reverse flow of xylem sap (Fig. 3). Record this pressure of the chamber, which represents the balance pressure. This balance pressure is equal in magnitude (but negative in sign) to the Ψ_w of the shoot. A single leaf can also be used in the pressure chamber. Because the pressure chamber method is rapid and does not require delicate instrumentation or temperature control, it can be used extensively under field conditions.

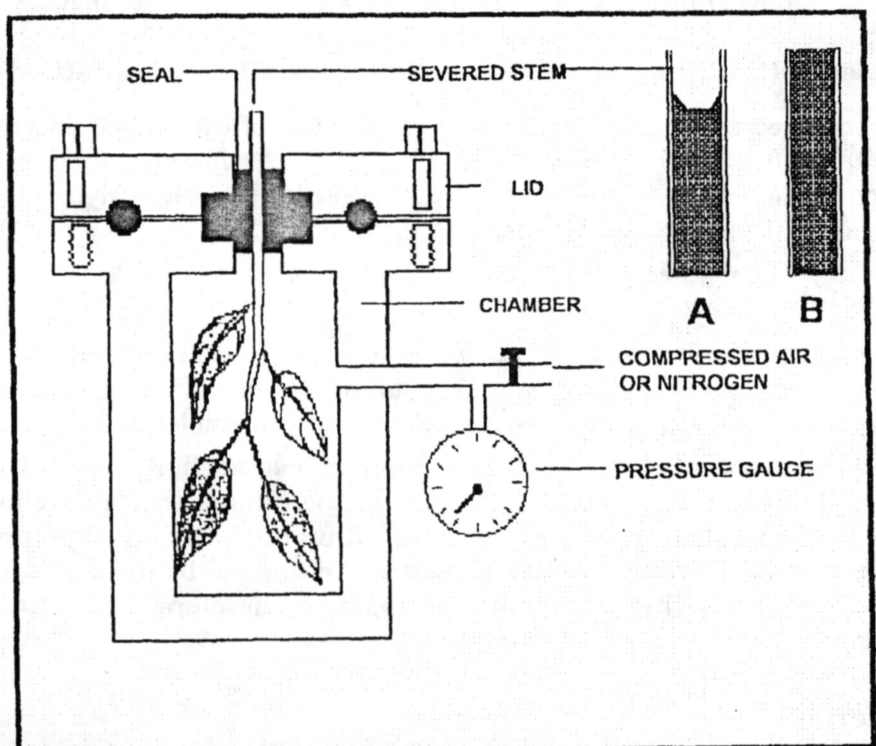

Figure 3. Pressure chamber apparatus for the measurement of shoot water potential.

Precautions

(i) The seal of the apparatus and plant stem in the lid of chamber should be air tight.

(ii) Water loss from the shoot be prevented by reducing time between the cutting and enclosing in the pressure chamber. Shoot should ber enclosed in humid polyethylene bag before cutting.

(iii) It is essential to prevent the trasnpiration loss from the shoot during measurement. This can be done by humidifying the air or nitrogen entering the pressure chamber or by covering the inner walls of the pressure chamber with wet filter paper.

(iv) The stem should not be recut as this shall give wrong reading of the balance pressure.

(v) Pressure in the chamber should be gradually increased.

(vi) Observations at the cut end should be made using hand lens or binocular microscope.

Measurement of Water Potential and its Components

In plant tissues the water pontential has two components i.e. osmotic (Ψ_s), and pressure potential (Ψ_p) which need to be measured to study the water relations of plants.

$$\psi_w = \psi_s + \psi_p$$

Water Potential

Water potential can be measured by the following methods :

- Liquid equilibration
- Pressure chamber
- Psychrometric

1. Liquid Equilibration

The sample is equilibrated in a series of solutions of known concentrations, and the solution in which no change in volume or weight of the sample occurs is identified. This indicates no net gain or loss of water by the sample which means that water potential of the tissue is equal to that of the solution. Usually sucrose, sorbitol, mannitol or polyethylene glycol (PEG) is used, PEG being the best. The best solute for such measurements is one that does not easily penetrate through membranes or harm the tissues. Measuring change in tissue volume is dificult and weight change may introduce error. An alternative method was developed by a Russian scientist, V.S. Chardakov. This method is simple, efficient and can be used effectively in field conditions also (Slavik, 1974).

A graded concentrations or sucrose of PEG solutions of known osmolarity are placed in test tubes (one pair of test tubes for each concentration). In one test tube a small pinch of methylene blue dye is added to make it coloured (test solution). Tissue samples are placed in he other test tubes that contain equivalent concentration but no dye, and allowed to equilibrate for 30-60 minutes. During this period tissue samples exchange water with solution. At the end of the incubation period, the tissue samples are removed from the solution using clean glass rod or long forcep.

Then with the help of a lambda or bent pasture pipette a small drop of the equivalent coloured solution is placed gently in the middle of the test tube and the movement of the drop is recorded. If the coloured drop rises the solution in which tissue was placed has become concentrated (dense) due to absorption of water by the sample tissues. The tissue in this case has lower water potential than the origian solution. However, the drop will sink if the tissue has lost water to the external solution during equilibration. The tissue water potential in this case is higher than the external solution. If the coloured drop diffuses evenly into solution the water potential of tissue equals to the external solution. Record the concentration and compute the water potential from the known values of the solution.

2. Pressure Chamber Method

Principle

The amount of pressure that must be applied to force water out of the leaf cell into the xylem until it is refilled, is equal to the tension originally existing in the xylem sap. Thus the applied pressure is equal to the water potential of the leaf cells.

Procedure

A leafy shoot or single leaf is sealed in the vessel of the pressure chamber apparatus in such a way that the cut end of the shoot or leaf is protruding out of the chamber. Close the exhaust value and gradually increase the pressure by forcing compressed nitrogen gas from a cylinder until xylem sap just appears to oose out at the cut end of the leaf/shoot protruding outside the vessel. The pressure applied is recorded on the pressure gauge fitted with the vessel, which is equal to the water potential of the leaf cells. A magnifying lens or microscope may be used to ascertain oosing of xylem sap. Slowly release the pressure by turning exhaust valve open.

3. Psychrometric Method

Principle

A thermocouple psychrometer is used to measure the vapour pressure of the air in a closed volume of chamber. A thermocouple is a pair of junction made of copper and constantan wires. One of these thermocouples function has relatively larger mass and second one is small. When a tissue is placed in a small air tight chamber the vapour pressure of the air equilibrates with the water potential of the tissue at constant temperature and pressure and is measured. The measurement is based on the principles of ``Seebeck" and ``Peltier" effects. The RH is linearly relative vapour pressure is related to water potential by the following equation :

$$\psi_w = \frac{\mu_w - \mu_w{}^o}{\overline{V}_w} = \frac{RT\ In\ e/e^o}{\overline{V}_w}$$

Where, ψ_w is water potential.

μ_w = chemical potential of water in solution.

$u_w{}^0$ = Chemical potential of pure free water.

$\overline{V}_w$ = Partial molal volume of water (cm^3/mole)

e/e^0 = Vapour pressure of the solution relative to pure free water (relative humidity)

R = Gas constant (0.0820 litre atm degree^{-1} mole).

T = Absolute temperature (273°C)

Ln = Natural log (2.303 log 10)

Seebeck effect

When a drop of water is placed on the wet junction of the thermocouple while the other junction is dry, i.e. a temperature difference exists between the two junctions, an electric current will flow between them in such a direction as to tend to equalize the difference. This is Seebeck effect. The strength of the current is measured when a steady state of evaporation is attained.

Peltier effect

If both the junctions of a thermocouple are initially at the same temperature, then by passing an electric current through them, one junction will cool and the other will heat. This is Peltier effect.

Procedure

The instrument is calibrated using filter paper discs soaked in standard solutions of known vapour pressure. After calibration the leaf sample is placed in small psychrometer chambers and left as such for equilibration for few hours. The sample will attain equilibrium vapour pressure with the air inside the chamber. The vapour pressure of this air inside the chamber is then measured by the wet-bulb hygrometry.

Thermocouple psychrometer can also be used to measure osmotic potential of plant samples. After measuring the water potential the sample chamber is immersed in freezing bath to kill the tissue and then return to the water bath of the psychrometer for a second measurement. Since turgor pressure (pressure potential) is eliminated by tissue death, the new measurment gives the osmotic potential of the tissue.

The difference between the first (water potential) and the second (osmotic potential) measurements gives an indication of the pressure potential of the plant sample.

Osmotic Potential

Osmotic potential is the measure of potential generated by the presence of solutes in the system. Solutes dissolved in pure solvent change the following solvent properties :

(i) The freezing point in depressed.

(ii) The boiling point is raised.

(iii) The osmotic potential is decreased.

(iv) The vapour pressure is lowered.

Measurement of osmolality of a solution can be made by comparing one of the above mentioned colligative properties of the solution viz-boiling point, vapour pressure or freezing point.

Measurement

(A) Vapour Pressure Osmometer

Principle

It is based upon the measurement of vapour pressure depression which is made possible by thermocouple hygrometry. A thermocouple hygrometer, a sensitive temperature sensor operates on the basis of an unique thermal energy balancing principle to measure the dew point temperature depression within the chamber.

Calibration

(i) Place the filter paper disc in the sample well of the osmometer and load 10 uL of a 290 mmol/kg standard. Set the display to read 290 using CALIBRATE 290 control.

(ii) Then calibrate with 1000 mol/kg standard. Locate the instrument reading on the left hand (READ) side of the calibration Nomograph. Then using the calibrate 1000 control, adjust the display to read corresponding ``SET'' value

(iii) Repeat the step one to correct the offset. Calibration of the osmometer is now complete.

Sample Measurement

(i) Keep the sample in -70°C for the required duration and express the sap.

(ii) Open the sample chamber and withdraw sample slide.

(iii) Place the filter paper disc in sample holder.

(iv) Load about 10 u L of sap on the paper disc.

(v) Gently push the sample slide fully until it stops lock the chamber.

(vi) The ``IN PROCESS'' indicator will go out and a peep sound appears when the measurement is completed. The osmolality of the plant sap is displayed on the front panel.

(vii) Unlock the chamber by rotating locking lever to open (vertical) position, withdraw the sample slide, and clean the sample holder with smooth tissue paper.

(B) Freezing Point Osmometer (Advanced Digimatic Model 3D$_2$)

Principle

When solute is dissolved in a pure solvent the freezing point of the solvent decreases. The proportionality constant is characteristics of the particular solvent. For water, one gram mole (Avogadro's number) of non-ionizing solute dissolved in one kilogram of water will depress the freezing point by 1.86°C. If the solutes dissociate, then all ionized and indissociated species must be counted. For example, one mole of sodium chloride (58.4 g), i.e., one osmol would really be 2 osmols (Avogadro's number) on complete dissocation (1 mole of Na^+ and 1 mole of Cl^-).

The quickest and most precise way to measure the freezing point of a solution is to super cool a sample several degrees below its freezing point and mechanically induce it to freeze. The heat of fusion suddenly liberated causes the sample temperature to rise towards a plateau where in an ice/water equilibrium is maintained. This equilibrium temperature is, by definition, the freezing point of the solution.

Procedure

Calibration

(i) Turn the osmometer `ON' and provide 30 minutes warm up time.

(ii) Check that the refrigeration bath is full and also, check for bath temperature when BATH cycle light beings to cycle off and on.

(iii) Pipette 0.2 ml of one of the standards with freezing point close to that of expected of unknown in a sample tube and place the tube in elevator well (NaCl solutions are best standards).

(iv) Lower the head by pressing the `HEAD CONTROL' button and observe the following action :

(a) The read out resets from the previous reading to 8888.

(b) The 8888 reading holds for about 30 seconds and then changes to 0 before beginning to count up to the 3000 freeze point.

(c) At 3000 there is a buzz as a one-second freeze pulse vibrates the stir wire violentily to freeze the sample.

(d) The read out begins counting down rapidly as the heat of fusion is released and the sample warms to its actual freezing point. The instrument is now sensing and following the plateau of the sample.

(e) When the plateau is constant, the read out is automatically locked into the display and the head rises.

The readout remains locked onto the display until the next test is begun (or the instrument is turned `OFF").

If `PLEASE REPEAT TEST' light comes on, the test is automatically discontinued because the sample did not freeze correctly. Now check for :

(i) Bath filling with 3 LA 811 Bath liquid.

(ii) Appropriate bath temperature (-6.2 ± 0.2°C for 0-2000 m osm range in 3D2).

(iii) Volume of solution in the sample tube (0.2 - 0.25 ml).

(iv) Repeat steps 3 and 4 above four more times on aliquots of the same standard and record the results. Calibration checks are made by running determination of enough different standards to bracket the expected range of unknowns.

Sample Measurement

Once the standard check is completed go for sample measurement as under :

(1) Clean the sample tube and wipe dry with soft tissue paper (lint free).

(2) Gently wipe the probe, stir/freeze wire, mandrel and top of the elevator well.

Osmotic Adjustment (Osmoregulation)

Osmotic adjustment refers to the maintenance of turgor by lowering of the osmotic potential by accumulating solutes in response to water deficits or salinity (Turner and Jones, 1980). This term is derived from the changes in solute concentrations of the cells in response to the osmotic potential of external environment, i.e. plants try to adjust their osmotic potential vis a vis external environment enabling them to maintain the turgor and growth. Solutes which help in osmotic adjustment are sugars, polyuols, amino acids, organic acids and ions such as nitrate, chloride, potassium and glycine-betaine. Higher the osmotic adjustment better is the adaptation to adverse environment (drought and salinity).

Estimation of Osmotic Adjustment

Osmotic adjustment is usually determined by the differences between osmotic potentials in stressed and non-stressed plants. The degree of osmotic adjustment is measured as the change in osmotic potential at a particular level of water potential or relative water content. Osmotic adjustment could conveniently be measured either at full turgor or zero turgor.

Procedure

(A) Measurement at Full Turgor

The leaves from non-stressed and stressed plants are detached and are rehydrated upto 100% turgidity by floating on water by enclosing in a humid chamber. The osmotic potentials of these rehydrated leaves are measured using the standard procedures mentioned earlier. The difference in osmotic potential of non-stressed and stressed tissues is the measure of osmotic adjustment.

(B) Measurement of Zero Turgor

The leaves of non-stressed and stressed plants are placed in a freezing bath to kill the tissue which eliminates the turgor pressure. Measure the osmotic potential of both the samples. The difference between stressed and non-stressed sample osmotic potential is the extent of osmotic adjustment.

Source of Error

A possible means of error could be the apoplastic water (water present in cell wall and xylem element). Tissue killing results in mixing of the symplastic water (chiefly vacuolar sap) with the apoa-plastic water because freezing of tissue causes cell wall rupture, and thus osmotic potential could be underestimated.

Pressure Potential (Turgor Pressure)

Pressure potential (ψ_p) is calculated from the measured values of water potential (ψ_w) and osmotic potential (ψ_s) using the following equation :

$$\psi_p = (\psi_w - \psi_s)$$

Pressure Volume Curve

Pressure volume curve can be used to determine different components of water potential. It also measures the bound water in plant tissues.

Procedure

1. Select the sample (Leaf or branch) and equilibrate in pure water overnight (or any desired period), cover in a plastic bag to ensure 100% RH.
2. Fix the equilibrated sample in pressure chamber with cut end protruding out and seal it properly with Suba Seal.
3. Attach a thin capillary filled with water to the cut end and make this junction air tight. The other end of capillary is placed in a beaker half-filled with water. The beaker is placed on a sensitive balance capable of micro weighing.

4. Apply a small pressure by pushing N_2 gas from the cylinder and note the amount of water exuded. Record the pressure applied on the gauge of the pressure chamber.
5. Repeat step-4 by gradually increasing pressure and note the amount of sap exuded with every increase in pressure. Continue this process until oosing of sap ceases.
6. Discontinue the operation and take the weight of the plant sample.
7. Dry the sample in an oven at 70 ± 1°C till constant dry weight.
8. Plot a curve using volume of sap exuded on `X' axis and the reciprocal of the pressure (1/P) applied on `Y' axis.
9. Draw the curve and extrapolate the linear portion of the curve on both sides to intersect the X and Y axis (Figure 4).

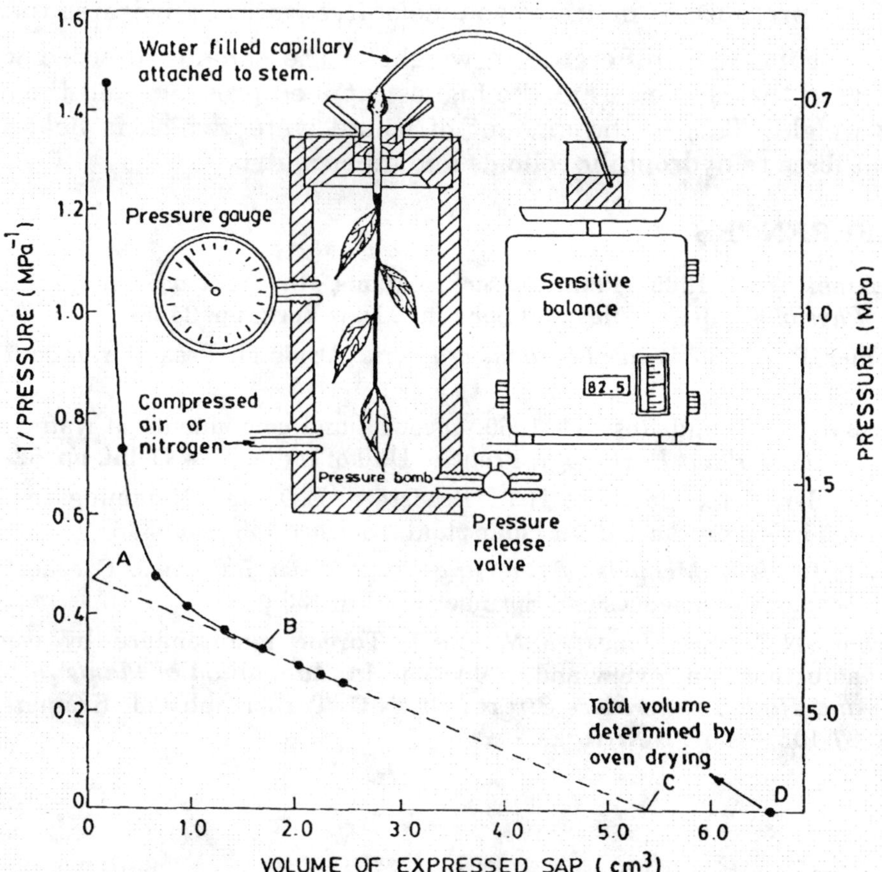

Figure 4. Plotting the pressure volume curve by pressure chamber technique.

Calculation

1. For cell protoplast the product of pressure (P) and volume (V) is a constant (PV = K). So, 1/P against V is linear. Initially with a small applied pressure bulk of sap is exuded but later on this ratio goes on decreasing. Thus the line is curved in the beginning but subsequently becomes linear. The point at which the curve becomes linear is the turgor lost point of the cell i.e. the internal pressure is zero. At this point, therefore, the negative value of the applied pressure is equal to osmotic potential.
2. On extrapolating linear part of the line intersects Y axis on point `A' and X axis on point `C'. At point `A' the cell is fully turgid and the water potential is zero (ψ =0). Thus;
3. The intersect of X axis at point `C' gives the total volume of liquid originally present in the cell and could be exuded under pressure. It indicates volume of free water in hydrated tissue.

Calculate the difference in weight of the sample recorded when oosing of the sap ceases at the highest applied pressure and the oven dry weight. This is the amount of bound water which is held with great force to hydrophilic colloids and other matrix.

REFERENCES

Hopkins, W.G. 1995. Plant water relations. In : *Introduction to Plant Physiology.* Jonh Wiley and Sons Inc., New York. pp. 34-38.

Kramer, P.J. 1983. *Water Relations of Plants.* Academic Press, Orlando, USA. 665 p.

Salisbury, F.B. and Ross, C. 1969. Osmosis and components of water potential. In: *Plant Physiology.* Prentice Hall of India, New Delhi. pp. 52-74.

Scholander, B.F., Hammel, H.T., Brodsteer, E.D. and Hemmingsen, E.A. 1965. Sap pressure in vascular plants. *Science* 148 : 339-346.

Slavik, B. 1974. *Methods of Studying Plant Water Relations.* Czechoslovak Academy of Sciences and Springer, Berlin. 433 p.

Turner, N.C. and Jones, M.M. 1980. Turgor maintenance by osmotic adjustment: a review and evaluation. In: *Adaptation of Plants to Water and High Temperature Stress* (eds. N.C. Turner and P.J. Kramer). pp. 87-103, Wiley, New York.

CHAPTER-16

AERENCHYMA : FORMATION, FUNCTION AND MEASUREMENT

INTRODUCTION

Aerenchyma is an air channel in the roots of some plants which allows exchange of gases between the shoot and the root. The channel of large air-filled cavities provides a low-resistance internal pathway for the exchange of gases such as oxygen and ethylene between the plant above the water and submerged tissues.

Aerenchyma is the continuous longitudinal system of gas filled cavities between shoots and roots. Internal oxygen transport through these gas filled cavities (aerenchyma) allow many species to survive and grow in flooded soils. Poor aeration induce aerenchyma formation in many plants. Flooding tolerant plants have consitutive as well as inducible (induced in response to hypoxia or anoxia) aerenchyma. Aerenchyma is formed schizogenously by cell division in the apical meristem or lysigenously by programmed lysis of cells in cortex region (Kozlowski, 1984). In rice, aerenchyma occupies a significant volume of the roots, internodes and leaf sheaths. However, aerenchyma is poorly developed in the lower half of the leaf blade and is completely absent in the upper half of the blade (Blom *et al.,* 1994).

When studying aerenchyma it is important to distinguish between measuring its formation and function. The formation of aerenchyma can be measured by :

(i) Cross sections which describe the proportion of gas space at any given point on the root and shoot systems and,

(ii) Porosity by Archimede's principle which describes the porosity of the total root system.

In some cases this is adequate, however, these measurements do not account for the factors which affects the function of the aerenchyma in the role to adequately aerate the root system such as :

- Tortuosity.
- Consumption of oxygen by respiration.

- Radial loss of oxygen to the rhizosphere.
- Total diffusion path length.
- Blockage of aerenchyma, or areas of the roots where aerenchyma formation has been restricted.

MEASUREMENT OF AERENCHYMA

- Cross sections
- Fractional root porosity.

(i) Cross Section Method

Fix root sample between split pith and cut fine sections, place in Petridish having distilled water. Select nice sections and place them on watch glass for staining. Put few drops of eosin dye solution (0.02% w/v) onto the sections and after few minutes wash thoroughly with distilled water. Mount on the slide and place a cover slip avoiding entrapping of air bubbles. Observe the section under microscope fitted with micro photographic attachment.

On clear focusing, take the photograph and develop the prints. Calculate the proportion of aerenchyma present in plant samples. Per cent aerenchyma can be calculated by:

(a) Cut and Weigh Method

Take the weight of the photoimpression of the total cross section. Then cut the portions of crosssection under aerenchyma and weigh on a sensitive balance, express as per cent of the total weight of the paper.

(b) Per cent Area

Remove area under aerenchyma by cutting and measure using Leaf Area Meter. Alternatively the area could be computed by measuring length and breadth of the aerenchyma cavities with the help of occular micrometer attached to the eye piece of the microscope. Express the proportion of aerenchyma on per cent area basis. Error may be introduced due to irregular size and shape of aerenchyma, when microscopic measurement is made.

(ii) Root Porosity

Many physical parameters of roots, leaf and other plant organs viz-porosity (internal air spaces), volume, density and volume of air layer surrounding plant organs can be easily determined by applying Arechimede's principle (Raskin and Kende, 1983). According to

Archimede's principle, the buoyant force acting upon a body immersed in fluid is equal to the weight of the fluid displaced by the body :

$$B = pV \tag{1}$$

Where, B, is the buoyant force, p is the density of the fluid (weight unit volume) and V, is the volume of the immersed body.

Thus, a close approximation of B acting on the body completely submerged in water can be obtained from the differences in the weight of the body in air (W_1) and its weight in water (W_2).

$$B_{root} = W_{air} - W_{water} \tag{2}$$

Combining equations (1) and (2)

$$B_{root} = W_{air} - W_{water} = pV \tag{3}$$

Procedure

Gently blot plant sample (root) to remove adhering water and weigh in air on a balance (W_1). Attach a metal clamp with a wire from the pan hook beneath an analytical balance. Place a large beaker beneath the balance, filled with distilled water to reference point. It is important to ensure that the water level in the beaker remains constant so that the same proportion of the holder is under water for each measurement. Attach plant sample with the metal clamp and record weight of sample and clamp submerged in water (W_2). The weight of the submerged clamp (W_3) is also measured.

The weight of the root in water can be calculated by subtractng the weight of the submerged holder with the sample (W_2) from the weight of the submerged metal clamp (W_3). The weight of the root in water is usually negative because of the positive buoyancy of the roots.

Treat the root sample with Triton X-100 or Tween-20 to eliminate the air layers, and measure the weight of the submerged clamp with root (W_4). Elimination of air layers lead to a decrease in buoyant force (B), where,

$$B_{air} = W_4 - W_2 \tag{4}$$

According to Archimede's principle, the buoyant force exerted on the air layers is equal to the weight of water displaced by the air layers. The volume of air layers (V_{air}) can then be calculated from :

$$V_{air} = \frac{B_{air}}{pH_2O} \tag{5}$$

Where, pH_2O is the density of water. The volume of the plant sample (root/leaf) can also be calculated by applying Archimede's principle:

$$V_{sample} = \frac{B_{sample}}{pH_2O} \tag{6}$$

Where, B_{sample} is the buoyant force acting on the submerged plant sample without air layers and equals :

$$B_{sample} = W_1 - (W_4 - W_3) \tag{7}$$

The density of the plant sample (p sample) is determined as :

$$p_{sample} = \frac{W_1}{V_{sample}} \tag{8}$$

Determination of the Volume of the Internal Air Spaces (Porosity)

Gases in the internal spaces in the leaf can be replaced by vacuum infiltration with a 0.05% Triton X-100 solution. The plant material fixed to the metal clamp is placed in the bottom of the desiccator or vacuum flask partially filled with surfactant solution so that the tissue is submerged totally. Agitation of the desiccator and sectioning of the plant material assist the infiltration process (Raskin and Kende, 1985).

The vacuum about 500 mm Hg, should be applied and released at least 4 times, with every evacuation lasting about 40 seconds. The infiltrated tissue attached to the clamp is then weighed in water (W_5) as described above. The buoyant force exserted on the intercellular air space ($B_{air\ space}$) is calculated.

$$B_{air\ space} = W_5 - W_4 \tag{9}$$

For all practical purposes, the density of the 0.05% Triton X-100 solution inside the plant material can be considered equal to the density of water. Therefore, the internal gas volume of the plant material (V_{plant}) is given b :

$$V_{plant} = \frac{B_{air}}{pH_2O} \tag{10}$$

$$V_{plant} = \frac{W_5 - W_4}{pH_2O} \tag{11}$$

No difference in electrical conductivity of Triton X-100 solution before and after infiltration of rice leaves were detected, indicating that infiltration does not cause any detectable leakage of ionic solutes from the cells (Raskin, 1983).

Fractional root/leaf porosity can be calculated as :

$$Porosity\ (\%) = \frac{Volume\ of\ air\ space}{Volume\ of\ root/leaf} \times 100 \tag{12}$$

Staining for Suberin

Stain preparation

Put about 0.5 g of sudan black powder in 20-30 ml of 70% (v/v) ethanol. Mix well and filter the solids. Store in glass bottle.

Staining

Place fine cut sections in a watch glass and put stain solution on it. Cover the watch glass to prevent evaporation and leave for 2 hours. Wash the sections successively in 50, 25 and 5% ethanol, then in distilled water. Mount the section on the slide and observe in microscope and take photographs.

AERENCHYMA FUNCTION – RADIAL OXYGEN LOSS

(a) Methylene blue-qualitative study

Methylene blue is bright blue in the oxidised state and colourless in the reduced state. Hence it can be used to obtain qualitative indications of regions of the roots which lose O_2.

The following method is used

- Prepare a 0.1% agar solution by the addition of 1 g of agar powder to 1 litre of water and then boil until dissolved, continuously stir agar solution as it cools to prevent flocculation.
- Add methylene blue at 12 mg l^{-1}, the dissolving can be left to the bubbling with N_2 gas.
- Bubble with N_2 gas until solution is anoxic, 1-3 hours depending upon volume.
- Now dissolve 130 mg l^{-1} sodium dithionite, continue to bubble with N_2. It is best if solution is used as soon as possible after the addition of sodium dithionite.
- Decant solution to perspex containers, add plants and seal lid.
- Colour change can occur within a couple of minutes but it is best to watch development over 2-4 hours.

Detail of reactions

The reduction of methylene blue

$$MeB + 2H_2O + S_2O_4^{2} \rightarrow MeBH + Cl^- + 3H^+ + 2SO_3^{2}.$$

And the oxidation of methylene blue :

$$MeBH + 0.5\,O_2 + H^+ \text{ --- } MeB^+ + H_2O$$

(b) O_2 transport and root elongation

In your planning remember that the experiment could either be short over a few hours or run for long periods. These studies on aerenchyma function may be backed up by descriptions of aerenchyma (cross sectional area and fractional root porosity). The radial oxygen loss (ROL) from root can be measured following the method described by Armstrong and Wright (1975).

Cylindrical Platinum Electrode Technique

The polarographic determination of oxygen flux from roots is based on the characteristics of the current-voltage curve obtained when oxygen in aqueous solution is electrolytically reduced in a cell in which one electrode, the cathode, consists of a sleeve insulated thermo-pure platinum cylinder (Figure 1) while the other of some standard half cell.

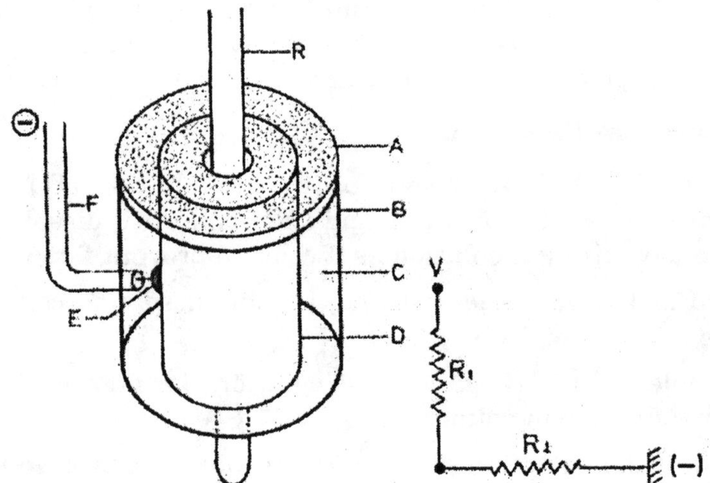

Figure 1. A Cylindrical platinum electrode and inserted root A : Celluloid guide (one at each end) to position, R, in a central path through the electrode. B : Perspex tube. C : Eponepoxy resin, D : Platinum cylinder, E : Solder joint, F : Copper wire (Sleeved) to polarographic circuit (B) Simple circuit analogue of root/electrode system, R_1 = Plant resistance, R_2 = Liquid cell resistance, V = oxygen concentration difference between the atmosphere and electrode surface.

The reduction of oxygen at a platinum surface is thought to proceed in two stages. At pH 3.5 or above the overall reaction follows the equation :

$$O_2 + 2\,H_2O + 4e^- \rightarrow 4\,OH^-$$

For each molecule of oxygen reduced there is a current transfer of $4e^-$. A current plateau placed somewhere between -0.2 V and -0.8 V applied E.M.F.

Armstrong (1967) indicates that the rate of oxygen reduction is independent of voltage and dependent on the rate of oxygen diffusion to the electrode surface. When a voltage setting in the plateau region is sustained, the current equilibrates to a value which is related to the oxygen flux to the electrode according to the equation :

$$ir = n\ FAfx - 0.T \qquad (1)$$

Where, ir = the diffusion current in amperes at the time of equilibration, T

N = number of electrons required for the reduction of one molecule of $O_2 = 4e^-$

F = Faraday, 96,500 coulombs

Fx-0.T = Oxygen flux at zero distance (x) from the platinum surface at equilibrium time, T ($mol^{-2}\ S^{-1}$).

A = Area of electrode surface (cm^2).

Oxygen flux from roots is measured in de-oxygenated liquid medium (+supporting electrolyte). Roots are inserted through the Pt electrode and a shell of liquid lies between the root surface and the inner electrode surface; the above ground parts of the plant remain above the liquid surface. Under the appropriate polarizing voltage the platinum acts as a sink for oxygen (substituting for a soil sink), and a diffusion gradient is set up between root and electrode. The rate of oxygen loss from the root at equilibrium can be calculated from the equation (1) by substituting root surface area within the electrode, in place of electrode surface area (Armstrong, 1968, 1979).

Equation (1) simplifies the expression :

$$ROL = \frac{4.9741}{A_1} \qquad (2)$$

Where, ROL = Radial oxygen loss or flux, (O_2 ng cm^{-2} root surface min^{-1})

I = [Diffusion current (mA) with the root in the electrode] [diffusion current (mA) with a glass rod through the electrode] (residual current)

A_1 = Surface area of root within the electrode (cm^2).

Assay of radial oxygen loss can be divided into four distinctve stages (Armstrong, 1979).

(1) Set Up

Equipment Components

(a) Reaction chamber

This is a small perspex box of about 300 ml volume and about 150 mm high. The chamber is filled with deoxygenated agar/nutrient solution with supplemented KCl to increase the total ion content to 5-10 mM, to ensure adequate electrical conductivity, and no metallic micro-nutrients. The preparation of the anoxic media is of vital importance. The agar solution should be autoclaved or boiled, and continuously stirred as it cools to prevent flocculation and maintain a high optical clarity. Once cooled agar should be placed in a one litre measuring cylinder, plugged with a non-absorbent cotton wool plug and bubbled slowly with high purity nitrogen for at least 24 hours. When transferring the agar solution to the perspex boxes care should be taken to ensure minimum exposure to air. The perspex box is fitted with a dense rubber lid and all holes are sealed with terostat to reduce the chance of oxygen entering the solution.

(b) Reference electrode-Anode

The reference electrode is an Ag/AgCl half cell, it consists of a sleeve of silver inserted into 100 mmol KCl. The silver sleeve is rejuvenated by soaking in saturated KCl when not in use. A dowel bridge connects the reference electrode to the reaction chamber.

(c) Cylindrical Pt electrode-Cathode

The Pt electrode should be cleaned prior to use and fitted with guides. Guides are made out of a small piece of celluloid film with a hole of a similar diameter as the roots. A guide is glued (with a small drop `Araldite') to the top and bottom of the electrode so that their holes are lined up and centred in the middle of the hole in the electrode. This ensures that the roots are equidistant from the surface of the electrode.

(d) Polarogram box and chart recorder

Connect the reference electrode to the (+) cell terminal, connect the cylindrical Pt electrode to the (–) cell terminal.

(2) Pre-start

Insert plant into medium, hold in place with terostat. It is desirable to allow the plant a couple of hours to equilibrate with its environment.

Insert reference electrode, it is important to ensure that the reference electrode has no gas bubbles trapped inside.

Insert Pt electrode. Air bubbles often get trapped within the electrode and under the guides, it may be necessary to shake the electrode vigorously to remove bubbles, care should be taken to ensure that oxygen is not mixed into the solution.

An initial polarogram is run to identify a point where the rate of oxygen reduction is independent of the voltage, this point will be identified as a plateau on the polarograph.

Procedure

Switch box to voltage, ensure that voltage reads zero, otherwise reset. Briefly flick back to current and ensure that reading is close to zero, otherwise allow to equilibrate.

- Switch back to voltage.
- Turn on chart recorder, set zero and speed to 1 cm/min
- Press start, voltage will increase by 0.02 V approximately every 20 seconds.
- Allow voltage to increase upto above 0.8 V
- Press stop and reset.

Look at the chart recorder out put and identify plateau. Out puts A and B are polarogram outputs. If the polarogram looks like output A with a well defined plateau then the system is working well and the medium is devoid of oxygen. If the polarograph looks like output B then there are problems. Usually this is the result of oxygen in the system, either there is a bubble on the electrode or the agar medium was not de-oxygenated. The plateau on the polarograph will determine the active range for the electrode function. Over this range oxygen reduction is independent of voltage and current will be dependent on oxygen consumption. Insert the root and repeat the standard polarogram, in this case there will be a small amount of oxygen present and another slightly different plateau will be formed.

Select a voltage which coincides with the plateau on both standard polarograms (the volgate is determined by counting the number of steps and multiplying by 0.02 V). This will be the voltage used for measurements (approx-0.46 V).

(3) Measurements

Insertion of Roots

It is important to avoid damaging roots when inserting them to the electrode. Most problems occur when roots are bent or squashing them against the side of the perpex container blocking of the aerenchyma. The use of a bicycle spoke to guide the root into the electrode can make this easier.

Once electrode is in place seal all holes with terostat. Conduct the second standard polarogram as described above.

Apply the prescribed voltage using the step button on the box, switch to current. Wait for current to equilibrate. Time to equilibrate will depend greatly on the conditions and the plant tissue, it can be from 15 minutes to 2 hours. Record the current on the graph paper every 10 minutes, especially record the equilibrium value (values from 0-2 μA).

Apply any manipulation such as altering oxygen concentration around the shoot and observe effect on current, or shift electrode to new positions etc.

(4) Residuals and Calculations

After completion of measurements, remove the root and insert a glass rod of a similar diameter, allow current to equilibrate and record. This current is the residual current (approx. 0.1 μA).

Advantages of Aerenchyma

The large air-filled cavities provide a low-resistance internal pathway for the exchange of gases between the plant organs above the water and the submerged tissues. Some of the oxygen transported through the aerenchyma leaks through root pores into the surrounding soil. The resulting small rhizosphere of oxygenated soil around individual roots support microorganisms that prevent the influx of potentially toxic soil components, such as sulfide, iron, and manganese. Nitrifying bacteria provide the roots with a favourable nitrogen source (Blom *et al.,* 1994).

During drought, aerenchymas allow plant roots to grow deeper for water, even through compacted layers, thick and tough roots are formed. As the roots dieback and decay, the resulting voids are paths in which new roots can grow and elongate when resources are available.

Disadvantages of Aerenchyma

Water and nutrient uptake may be less efficient; large intercellular spaces decrease the tissue available to transport water and nutrients from the root surface to the root xylem (Visser *et al.,* 2000).

Large root diameters reduce biomass-to-surface ratio, resulting in less uptake of water and nutrients and the reduced opportunity to explore all microzones for nutrients.

Some roots with aerenchymas are less likely to resist the physical strain of compacted soils. Those roots that penetrate and survive dense and compact drained soils have a higher bulk density and a strongly lignified layer of cells surrounding the aerenchyma, which strengthens the root. This dense, lignified layer prevents radial leakage of oxygen from the aerenchyma and may block some water and nutrient uptake (Visser *et al.,* 2000).

During drought, roots with aerenchyma may be less tolerant to water stress as the open structure of the cortex is probably a low-resistance pathway for water vapor, as well as for air, thereby increasing the susceptibility of the root to water loss.

REFERENCES

Armstrong, W. 1967. The relationship between oxidation-reduction potentials and oxygen diffusion levels in some waterlogged organic soils. *Journal of Soil Science* 18: 27-34.

Armstrong, W. 1968. Oxygen diffusion from the roots of woody species. *Physiologia Plantarum* 21: 539-543.

Armstrong, W. 1979. Aeration in higher plants. *Advances in Botanical Research* 7: 225-332.

Armstrong, W. and Wright, E.J. 1975. Radial oxygen loss from roots: The theoretical basis for manipulation of flux data obtained by the cylindrical platinum electrode technique. *Physiologia Plantarum* 35: 21-26.

Blom, C.W.P.M., Vosenek, L.A.C., Banga, M. Ringders, J.H.G. and Visser, E.J.W. 1994. Physiological ecology of Riverside species: Adaptive response of plants to submergence. *Annals of Botany* 74: 253-263.

Kozlowski, T.T. 1984. *Flooding and Plant Growth.* Academic Press, Inc. New York. 356 p.

Raskin, I. 1983. A method of measuring leaf volume, density, thickness and internal gas volume. *HortScience* 18: 698-699.

Raskin, I. and Kende, H. 1983. Regulation of growth in rice seedlings. *Journal of Plant Growth Regulation* 2: 193-203.

Raskin, I. and Kende, H. 1985. Mechanism of aeration in rice. *Science* 228: 327-328.

Visser, E.J.W., Varde Strag, H.M, Pierik, R. and Blom, C.W.P.M. 2000. Flooding tolerance in *Carex* species in relation to yield distribution and aerenchyma formation. *New Phytologist* 148: 93-103.

CHAPTER - 17

ELECTROPHORESIS

INTRODUCTION

The term electrophoresis describes the migration of charged particles in an electric field. The earliest technique, pioneered by A Tiselius in 1937, was a moving boundary electrophoresis. It has a drawback of simple diffusion and mixing of a sample components due to convection currents. These limitation have been overcome in zone electrophoresis in which a buffer saturated solid matrix is used as an electrophoresis medium in place of a buffer solution. The technique is widely used for separation and analysis of a large number of biomolecules such as amino acids, peptides, proteins, nucleotides and nucleic acids etc. which posses ionisable groups and therefore, at any given pH, exist in solution as electrically charged species either as cations (+) or anion (–) (Andrews, 1993; Berg *et al.,* 2002). Under the influence of an electric field these charged particles will migrate either to the cathode or to the anode, depending on the nature of their net charge. The sample is applied, depending upon the nature of support medium. Electrophoresis is an analytical method frequently used in molecular biology and medicine. It is applied for the separation and characterization of proteins, nucleic acids and subcellular-sized paticles like viruses and small organelles. Its principle is that the charged particles of a sample migrate in an applied electrical field. An electric current is passed through a medium containing the molecules, and each molecule travels at a different rate depending on its electrical charge, size and shape. Separation by electrophoresis is based on these differences. In electrophoresis, agarose and acrylamide gels are used for electro-phoresis of nucleic acid and proteins (Dhillon, 2000).

Principle of Electrophoresis

Any charged ion or molecules migrates when a potential difference (voltage) is applied across the electrodes, its generates a potential gradient (E). The force on a molecule bearing a charge of q coulombs is

E.q newtons. This force drives a charged molecule towards an electrode. However, there is also a frictional resistance that retard the move-ment of the charged molecules. This frictional force is measure of hydrodynamic size, shape and net charge of molecules, the pore size of medium and the applied current. Therefore, v, of charged molecules can be represented by the following equation :

$$v = \frac{E.q}{f}$$

where, v = velocity of migration of the molecules

E = electric field in volt/cm

q = net electric charge on the molecule

f = the frictional coefficient

The movement of charged molecules in an electric field is often expressed in terms of electriphoretic unit (μ) which is defined as the velocity per unit of electric field.

$$\mu = \frac{V}{E} = \frac{E.q}{f.E}\left(Since\ v = \frac{E.q}{f} = \frac{q}{f}\right)$$

For molecules with similar conformation, f varies with size but not with the shape. Thus electrophoretic mobility (μ) of a molecule is directly proportional to the charge density (charge/mass ratio).

Classification

Depending upon the nature of support medium, zone electroph-oresis is of different types such as paper, starch, polyacrylmide and agarose gel electrophoresis (Andrews, 1993; Dhillon, 2000).

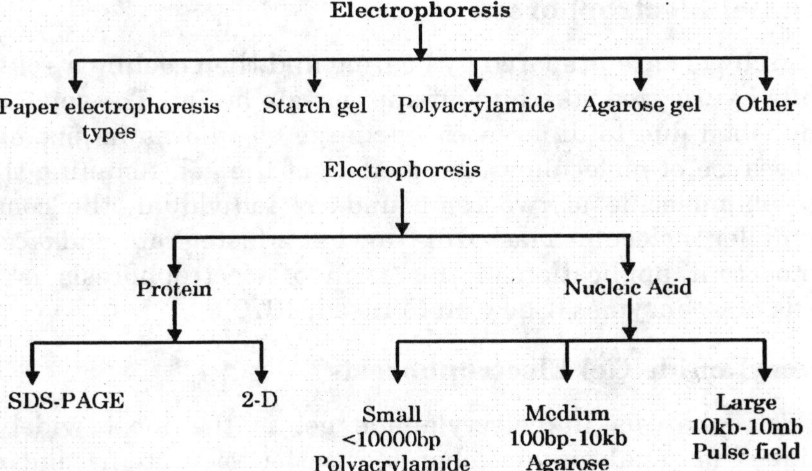

Electrophoresis can be classified based on:

1. Molecules to be separated
   a. Nucleic acids (DNA, RNA)
   b. Protein
      i. 1D
      ii. 2D
2. Kind of gel matrix used
   a. Agarose gel electrophoresis
   b. Polyacrylamide gel electrophoresis
      i. Native PAGE
      ii. Denaturing PAGE
   c. SDS-PAGE
3. Size of the molecules: Pulse Field Gel Electrophoresis.

Paper Electrophoresis

In this form of electrophoresis, the sample is applied as a circular point or spot on a strip of whatman filter paper or cellulose acetate paper moistened with the buffer solution. The ends of the paper are emerged in separate reservoirs containing buffer and in which the electrodes are fitted. Upon passing electric current the ions in the sample migrate towards oppositely charged electrodes at characteristics rates. This method is suitable for separation of low molecular weight compounds such as amino acids, small peptides and nucleotides.

Starch Gel Electrophoresis

Starch gels are prepared by heating and then cooling a solution of partially hydrolyzed starch in an appropriate buffer. The separation is accomplished due to differences in charge as well as in size of molecules because of molecular sieving effect of the gel. Asuming that the charge on molecule of two compounds is individual, the compound with smaller molecular mass will travel at a faster rate and *vice versa*. An important application of this form of electrophoresis is in the analysis of isoenzymes (Show and Prasad, 1970).

Polyacrylamide Gel Electrophoresis

Electrophoresis in polycrylamide gel is the most widely used techniques for analysis and characterization of proteins and nucleic acids. Use of polyacrylamide gel has several advantages such as : it is chemically inert, gives superior resolution, is amenable to preparation of gels with wide range of pore sizes and that the gel are stable over a wide range of pH, temperature and ionic strength (Rickwood and Hames, 1982; Dhillon, 2000).

Agarose Gel Electrophoresis

Electrophoresis through agarose or polyacrilamide gels is the standard method to separate, identify, quantify and purify DNA fragments. Although agarose gels have a lower resolving power than polyacrilamide gels, they have a greater range of separation and are much simpler and easier to handle than the polyacrilamide gels, and have been more widely used in DNA electrophoresis. Agarose is a polysaccharide extracted from seaweed. It is typically used at concentrations of 0.5 to 2%. The higher the agarose concentration the "stiffer" will be the gel. Agarose gels are extremely easy to prepare you simply mix agarose powder with buffer solution, melt it by heating, and pour the gel. It is also non-toxic (Dadlani and Verier, 1993; Berg, 2002).

Other Types of Electrophoresis

An isoelectric focussing (IEF) method for seperation of proteins which differ in their isoelectric points. Similarly, the technique was developed by O' Farrell (1975) and combines the resolving power of isoelectric focussing (IEF) and SDS-polyacrylamide gel electrophoresis (SDS-PAGE). Proteins in the first dimension are separated according to their charge by IEF and in second dimension SDS-PAGE facilities their separation on basis for their size. It is standard method for judging purity of protein preparation.

Study of Protein Profile

There is no doubt that of the biochemical technique available for traits identification, the analysis of seed and seedling proteins and enzymes is by far the most widely used and successful to data. Because they are direct gene products, proteins can be used as markers for the genes that encode them (Laemmli, 1970). During gene expression the information encoded within the genome is converted by a process of transcription to corresponding primary RNA transcript (mRNA). The mRNA is transported to cytoplasm where it is bound to ribosomes to act as template for translation in to corresponding protein. The resulting proteins are the primary working molecules within the cell and the protein, which it expresses, determines a cell's phenotype (Thomas *et al.,* 2009). The diversity of the functional properties of proteins is truly remarkable for example :

1. Acting as enzymes
2. Providing mechanical support
3. Controlling permeability, transport and storage process
4. Acting as hormones, receptors
5. Controlling gene expression etc.

Protein and enzyme electrophoresis is undoubtly the most widely applied for a wide range of species of agricultural and horticultural importance. The precise methods employed both for protein analysis and for interpretation of data, vary according to species in question and are governed to large extent by the reproductive behaviour and the consequent genetic structure of varietis. However, in general electrophoresis provides a rapid and relatively straight forward information. Proteins are composed of subunits called amino acids. The amino acids are arranged in a linear array, to form polypeptides. The protein further folds in a unique 3-D structure and certain proteins exhibit quaternary structure consisting of several sub units (Laemmli, 1970; Show and Prasad, 1970).

The side chains of four of the amino acids are highly ionised and are therefore charged at the near neutral pH values. Protein's charge at a particular pH independent of the content of acidic and basic amino acids. For every protein there is particular pH, known as the isoelectric point (PI) where the net charge is zero. These electrical properties of proteins are exploited in the various techniques of electrophoresis to achieve the effective separation and resolution of mixtures of proteins and enzymes required for their analysis. The versality and resolving capacity of these techniques has resulted in their becoming the most popular tool in the armory of the cell and molecular biologist.

Tissue Source and Protein Extraction

Seeds provide some of the easiest tissue for analysis because storage tends to be similar than for other tissue types and newly germinated embroys are relatively free of the secondary plant products that can inhibit enzyme activity.

Plants require generally more complex buffer for extraction that prevent undesirable interaction between proteins and phenol. The common ingredients include addition of polyvinyl pyrrolidione (PVP), ascorbic acid, dithio threitol (DTT) in extraction buffers.

Various continuous and discontinuous, anionic and neutral pH buffers are being employed to enhance the resolution of enzyme bound in genetic characters as possible. The enzyme or isozyme data can be analysed statistically using genotypic and phenotypic scoring. The isoenzyme pattern could be used as finger print of a variety.

Polyacrylamide Gel Electrophoresis Technique for Separation and Characterization of Protein

Electrophoresis in polyacrylamide gel is the most widely used technique for analysis and characterization (Hames and Rickwood, 1990). Polyacrylamide gel is prepared by polymerization of acrylamide monomers into land chains and crosslinking these by bifunctional reagents such as N, N' - methylene bisacrylamide. Polymerization is initiated by ammonium per sulphate or riboflavin, N'-tetramethyl ethylenediamine (TEMED) catalyses formation of free radicals from persulfate which in turn initiate polymerization. In the case of riboflavin-TEMED system, light causes photodecomposition of riboflavin and production of necessary free radicals. Oxygen inhibits polimerization. Gel solution, therefore, should be degassed prior to the polymerization step (Weber and Osborne, 1969).

The pore size of polyacrylamide gel can be varied by changing total concentration of acrylamide and bisacrylamide monomers in a fixed volume of gel solution. Gel ranging from 3 to 30% acrylamide concentration can be made and these can be used for separation of molecules size upto 1 x 10^6 daltons. Polycrylamide gels can be formed in cylindrical glass tube (rod gels) or in flat slab (slab gels). Separation of protein molecules on PAGE depends upon a number of factors including pH and types of buffers.

Buffer Systems

PAGE can be carried out in buffers ranging from pH 3-10. Based upon the types of buffer system used, electrophoresis can be of following types :

(a) Native continuous PAGE

(b) Native discontinuous PAGE

(c) Dissociating or denaturating PAGE (SDS-PAGE)

(a) Native Continuous PAGE

In native continuous PAGE same buffer ions are present throughout the same, gel and electrode vessel reservoirs at constant pH such that proteins retain their native conformation and biological activity. In this system, protein is loaded directly on to the gel, referred to as the resolving gel, in which the separation of the sample component occurs.

(b) Native Discontinuous PAGE (or native disc gel electrophoresis)

In contrast to continuous system, it has following features :

1. Contain different buffer ions in the gel than those in the electrode reservoir.
2. Consist of two types of gels-a large pore stocking gel which is polymerized on top of small pore resolving gel. The buffer used to prepare these two gels are of different ionic strength and pH.
3. Most discontinuous buffer systems have discontinuous or both buffer composition and pH.

Discontinuous buffer systems have an advantage over continuous system as relatively large volumes of dilute solutions of protein samples can be applied into the gels and yet good resolution of sample components can be obtained.

(c) Dissociating or Denaturating PAGE (SDS-PAGE)

In this system, proteins are dissociated into their constituent subunits by anionic detergent such as sodium-dedecyl sulphate (SDS) or sodium lauryl sulphate. The protein mixture is denatured by heating the sample at 100°C in the presence of excess SDS. SDS-polypeptide complexes have identical charge densities (charge/mass ratio) and these migrate in polyacrylamide gel according to their molecular size or mass. Under reducing conditions, in addtion to SDS, β-mercaptoethanol is included in the sample buffer to cleave disulfide bond between polypeptides.

The choice of buffer system depends on the particular proteins of interest. There is no universal buffer system ideal for electrophoresis of all proteins. In native PAGE both protein stability and resolution are important. Some native proteins may aggregate and precipitate at very high concentrations reached during stacking in discontinuous electrophoresis. Consequently either they may not enter the gel or might cause streaking. In such cases it is best to use to a continuous buffer system.

Sample Preparation

For best results with continuous system the concentration of protein in the sample should be at least 1 mg/ml to keep the sample volume at a minimum. The sample should be loaded in a buffer with ionic strength about 1/2 to 1/5 that of gel and electrode buffer. Tracking dye and glycerol or sucrose are added to sample to monitor progress of run and make samples heavy, respectively. Generally 50-200 μg protein in a minimal volume of sample buffer is loaded per well. For SDS-PAGE, in addition to glycerol and tracking dye, sample buffer also contains SDS and or β-mercaptoethanol.

Voltage

For electrophoresis to run, direct current (DC) power supplied designed for electrophoresis are used. Generally, constant voltage of 10-20 V/cm or current in the range of 15-25 mA/mm of gel thickness is applied.

Detection

Detection of protein bonds in gels is commonly done by staining with coomassive brilliant blue R-250 dye which can detect mg protein/band. Dye such as amino black or silver stain may also be used. Very low amount (0.1 mg protein/band) can be detected by silver strain. For detection of isozymes, gels are immersed in solutions containing enzyme specific stains.

Methodologies

Disc-gel (Ornstein, 1974)

(a) Glass tubes (10 x 0.6 cm) parafilm, disc-gel electrophoresis set.

(b) Stabilized power pack, capable of supplying 500 V/100 mA unit constant voltage or current.

(c) Stock solution: These solution are prepared using distilled water and are filtered and stored in brown glass bottle in refrigerator.

A		B	
Medium	**Reagents quantity**	**Medium**	**Reagents quantity**
In HCl	48 ml	In HCl	48 ml
Tris	36.6	Tris	5.98

Temed	0.23	Temed	0.46
Water to	100 ml (pH 8.9)	Water to	100 ml (pH 6.7)

C

Medium	Reagents quantity
Acrylamede	23.0 g
Bisacrylamide	0.735
Water to	100 ml

D

Medium	Reagents quantity
Acrylamede	10.0 g
Bisacrylamide	2.5 ml
Water to	100 ml

E

Medium	Reagents quantity
Ribofluvin	4 mg
Water to	100 ml

F

Medium	Reagents quantity
Sucrose	40 g
Water to	200 ml

G

Ammonium per sulphate 0.14 g/100 ml H_2O (freshly prepared)

Working Solutions

These solutions should be used the day they are prepared.

Running gel[a]	Spacer gel	Stock buffer solution[b]
1 Part A	1 Part B	Tris 6.0 g
2 Part C	2 Part D	Glucine 28.9 g
3 Part Water	1 Part E	Water to 1 litre
4 Part G	4 Part F	pH 8.3

a - Immediately after mixing, the gel solution should be used for filling.

b - Dilute the stock buffer 10 times with dist. water before using it for electrophoresis and this is termed diluted buffer.

Experimental Method

(A) Reagents

(i) Tris-HCl buffer (pH 7.2, 0.05 M)

A = 0.2 M solution of tris hydroxymethyl aminomethane (24.2 g in 1000 ml with distilled water).

B = 0.2 M HCl

Mix 50 ml of A and 44.2 ml of B and dilute it to 200 ml with distilled water, adjust pH to 7.2 with HCl.

(ii) Add 1 mM DTT solution to the buffer in the ratio of 1 : 5 (W/V).

(iii) **Reagent I : 2% Na_2 CO_3 in 1 N NaOH :** Dissolve 40 g NaOH and 20g Na_2CO_3 to make 1 liter in distilled water.

Reagent II : 2% Sod. Pot. Tartrate : Dissolve 2 g Sod. Pot. Tartrate in distilled water to make 100 ml.

Final reagent (C) is prepared by mixing 50 ml of 1 + 0.5 ml of 11 ±0.5 ml of **Reagent III : 1% $CuSO_4$ $5H_2O$:** in distilled water to make 100 ml.

(iv) **Protein standard :** 100 ml of Bovine serum albumin (BSA) in 100 ml of distilled water, to give 1.0 mg protein/ml

(v) **Folin-ciocalten reagent :** 2 N solution is commercially available. Dilute it to 1 N using distilled water.

Extraction

Grind the seed materials in 0.05 M Tris-HCl buffer pH 7.2, Containing 2 mM DTT. The homogenate is centrifuged at 1200 x g for 15 min and supernatant is immediately used for gel electrophoresis.

(B) Gel Electrophoresis of Protein Sample (Davis, 1964; Hames and Rickwood, 1990)

I. Preparation of Gel

Rod (disc) gel : (i) Clean the gel tubes and dry them (ii) Cut small pieces of parafilm and lay a piece on the top of the grommets located in rack and insert the tube down to the grommets pushing the parafilm to form a leaf proof seal (iii) Fill the gel tubes up to deserved height (generally leaving 0.5 to 1.0 m space on the top with the running gel using a posteur piette (avoid air bubbles). Layer a small amount of water (0.1 ml) on the top of the gel slowly with a syringe to prevent a meniscus formation. Keep the rack near a 15 W day light fluorescent lamp for 30 min. Till a visible interface develops between the gel and water layer (iv) When polymerization is complete invert the rock and blot the ends of the tubes to remove any moving water droplets. Return it to upright position (v) Layer 0.2 ml (0.3 cm ht) of the stacking gel (spacer gel) solution on top of the polymerizedgel (in case of discontinuous system). Allow it to polymerize in front of a fluorescent lamp.

II. Sample Loading and Electrophoresis

(i) Gel farmers/parafilm, etc. are removed from the base of the gels and they are mounted in place on the electrophoresis apparatus. Rod gels are placed in rubber grommet located in the upper reservoir of the apparatus.

(ii) Reservoir buffer is added to the lower trough taking care that no air bubble remains in it. Next, add buffer in the upper reservoir.

(iii) Using a microsyringe or micropipette with fine tip, load the sample (approx. 250 μg protein) carefully on to the gel surface (in sample wells in case of slab gel). The sample liquid being densor (due to addition of source or glycerot) settles down in the wells or on the gel surface. Add a few drop of romo phenol dye solution to the upper buffer reservoir.

(iv) Connect the electrophoresis apparatus to the power pack with the anode (+) connected to the lower reservoir and cathode) -+ connected to the upper reservoir.

(v) Connect the power pack to the main and switch it on. Electrophoresis conditions are quoted as either constant voltage or constant current, which is adjusted accordingly. Generally, slab gels is electrophoresed at 25-30 mA constant current on 120 V followed by 200 V constant voltage. For rod gels 2-4 mA constant current sod or 100 V constant voltage is applied. However, power out put can be increased or decreased with adjustment of running time depending upon the gel concentration ensuring that it is not too high, which way cause excessive heating, or too low, which may which increase electrophoresis time considerably and thus may reduce the resolution as a result of band diffusion. To ensure that the sample has traveled the complete length of the gel check the brocomophenol stain. When it reaches a distance about 0.5 - 1.0 cm above the bottom level, switch off the mains and disconnect the electrophoresis unit. The whole process is preformed at 5-10°C either in a cold room or in a chamber. In case of Isozymes the extraction as well as the electrophoresis is performed at 5°C or below.

III. Recovery and Staining of Gels

(i) Lower the slabs in a staining tray (preferably glass or solvent resistant and plastic trays). Containing diluted buffer solution or distilled water. Remove the side spaces and gently remove the glass plates away. Rod gels can be removed by rimming with gently steam of water using a syringe.

(ii) To mark the position of sample and direction of electrophoresis either cut a small piece at one corner or use a little India Ink.

(iii) Stain the gel in 0.1% Amido black dye [(in 7% acetic acid (1 g) dye/1000 ml of 7% acetic acid for about 30 minutes 1 hr.)]. Alternatively, gels can also be stained with 1% coomassic brilliant blue (R_2SO), prepared in acetic acid : methanol, water, 1 : 1 : 8

(iv) Destain the gels b decenting excess stain and washing the gels with 7% acetic acid in case of amido black or in acetic acid : methonal water, 1 : 1 : 8 solution in case of coomassic blue for overnight. In case of isozymes the gels are incubated with the specific reaction mixture till the bands are developed on the gels and then the reaction is by stopped the gels in and % acetic acids.

Table 1. Some electrophoresis buffers and their pH range.

S.No.	Buffer System	pH
1	Aluminium-lactate buffer	3.1 - 3.2
2	Sodium-lactate buffer	3.1
3	Sodium-acetate buffer	4.5
4	B-alanine-lactic acid buffer	4.5
5	Histidine-citrate buffer	5.0
6	Sodium-phosphate buffer	7.2
7	Borate/Tris-citrate buffer	7.5
8	Tris-glycine/Tris HCl buffer	8.0
9	Borate/Tris-citrate borate buffer	8.3
10	Lithium borate/Tris-cittrate buffer	8.5
11	Tris-citrate buffer	8.6
12	Tris-borate buffer	9.0

(indicates discontinuous system).

Electrophoresis of denatured seed proteins is one of the simplest and routine laboratory techniques used for identification of traits SDS-PAGE of tris soluble proteins.

(A) Methodologies Involved in Seed Protein Extractions for Gel Electrophoresis (Laemmli, 1970; Hames and Rickwood, 1990)

Protocol - 1 : Acid extraction (pH 2.41, Sodium Ascorbate) of storage seed proteins.

Mature seed

Soaked in distilled water (2-4 h) for removal of testa ad embryo.

Air dried and ground in micromill.

Flour is passed through 40 mesh screen. 1 gram sample is soaked in 20, 20, 10 ml for the period of 24/5/5 h (three time extract 0.25 M L-Ascorbic acid Centrifuge tube.

Centrifuge 10,000 x g for 5 min Collection of supernatant (contains albumin, globulins and free amino acid).

Separation of G1 fraction

A five fold volume of prechilled distilled

Water is added. An cloudy of precipitate is formed

Separated by centrifugation

Precipitate is used as G1

Protein (soluble in 0.5 M NaCl)

Total protein

(Contain G1, G2 and albumin fraction)

Protocol -2 : Crude protein extraction at neutral pH (7)

Soaking of seed flour in 0.5 M NaCl (for 25 min).

Centrifuged at 5000 x g for 10 min Equal volume of supernatant and cracking buffer is used and boiled for 3 min., before being applied to gel.

Protocol-3 : Finely ground seed meal (at alkaline pH)

Extracted in 0.125 m Tris-HCl (pH 8.6) upto 6 hour. (1 ml/seed)

Clear supernatant is obtained after centrifugation at 15,000 rpm (15.600 g)

Diluted with cracking buffer (0.5 ml sample and 1.5 ml cracking buffer)

Boiled for 3 minutes and loaded.

(B) Electrophoretic Analysis of Seed Protein (Cotyledonery) : SDS - PAGE System

Protein source : Seed meal (1-2 gm.)

---

(1) Extraction procedure*

1. 0.5 M NaCl & 1 M NaCl (pH 7.0)
2. 0.5 M NaCl & 0.25 M Ascorbic acid (pH 7.25-7.5)
3. 0.125 M Tris-HCl (pH 7.25-7.5)
4. 1 M NaOH (pH 7.0)
5. 0.125 M Tris-HCl (pH 6.9), 10% SDS (0.5 ml)
6. 0.5 M NaCl, 0.5 Tris-HCl (pH 6.8-7.0)
7. 0.5 M NaCl, 0.25 M Ascorbic acid (pH 2.4), 10% SDS (1 ml).
8. 0.5 M NaCl, 0.25 M Ascorbic acid, 0.5 M Tris-HCl (pH 6.8)

* Extraction period = 24-36 hrs.

(2) Stock sample buffer (SDS-reducing buffer): 0.06 M Tris-HCl (pH 6.8), 2% SDS, 10% Glycerol, 0.025% Bromophenol Blue.

Water = 4.8 ml
0.5 M Tris-HCl = 1.2 ml
10% SDS = 2.0 ml
Glycerol = 1.0 ml
0.5% Bromophenol Blue = 0.5 ml

(3)

Seed protein Sample		Sample buffer
1	:	1
1	:	2
1	:	3
1	:	4

(4) 50 μ 1 75 μ 1, (16-walled comb), 100 μ 1, 125 μ l (8-walled comb)

(5) Electrode buffer: Tris-glycine pH 8.6

(6) Acrylamide-Bis Acrylamide concentration: Stacking gel : 4%, 4.5%, 6% (40% T).

Running gel: 7.5%, 10%, 12.5% & 15% (single layered)

Running gel : (7%, 10%) (10%, 12.5%), (10%, 14%), (12%, 14%) two layered

(7) Current flow: Disc gel 1. : 20 mA
: 40 mA
2. : 10 mA

: 20 mA

3. : 10 mA

: 30 mA

Slab gel

Current	**Constant Voltage**
20 & 40 mA	175 V
10 & 30 mA	200 V
30 & 50 mA	200 V
30 & 60 mA	200 V
40 & 60 mA	200 V
30 & 40 mA	200 V
30 & 50 mA	175 V
30 & 40 mA	150 V

Table 2. Artifacts and troubleshooting.

(A) Polycrylamide gel electrophoresis (Dadlani and Varier, 1993)

S.No.	Problem	Cause	Remedy
1.	Slow or failure of gel to polymerize	(a) Catalysts absent or solution old (b) Presence of oxygen	Check conc. of catalyst (APS). Use freshly prepared solution. Deaerate gel solution
2	Failue of sample to enter and sterile into wells	Low density of sample	Increase conc. of glycerol or source in the sample
3	Buffers become mouldy	Improper storage	Prepare fresh buffer, store in refrigerator
4	Buffer pH changes during storage	Buffer not stabilized while preparation	While preparation of buffer take time to let pH stabilize before storing the buffer
5	Gel polymerizes very fast	Catalyst conc. too high	Use less conc. of catalyst
6	Gel cracking (a) during polymerization	(a) Excessive heat of reaction	Use cold solution
	(b) during electrophoresis	(b) Excessive current input	Lower the current
7	Poor resolution of bands	(a) pH of buffers not correct	Check pH of all solutions and buffers
		(b) Resolving get not proper	Change conc. of resolving gel or use gradient gel

		(c) Buffer system not proper	Try a different buffer system
		(d) High voltage	Check power settings
8	Band streaking	(a) In native PAGE, precipitation of proteins	Load a less conc. sample Use conditions system
		(b) In SDS-PAGE, protein remain aggregated	Use additional SDS and boil samples properly
		(c) Insoluble material in the sample	Remove insoluble material by centrifugation
9	Protein bands in all tracts of gel (bands found in lanes originally not loaded with samples)	(a) contamination of sample buffer	Use fresh sample buffer
		(b) Upper buffer contaminated during loading of samples	Load gels carefully
10	Bands broaden at the bottom of the gel	Sample density	Load equal volume of samples in all wells
11	Wavy bands	Excess APS	Use proper conc. of APS
12	Detachment of gel from glass plates during electrophoresis	Glass plates not clean	Use glass plates cleaned with swab of acetone or ethanol and air dried
13	Gel sticks to the plate or tube	(a) Dirty tube or plate	Clean plates very thoroughly before use with a swab of acetone or ethanol, then air dry
		(b) Low concentration of gel	
14	Gel gets torn during removal of plate	Improper handling	Squirt water along sides of the gel with the help of a syringe. Handle the gel gently.
15	Samples migrate in opposite direction (up into the buffer)	Improper electric connection	Check electrode connection and properly connect cords (reverse the connections)
16	Dye migration not even	(a) insufficient cooling	Improve cooling or run at low current

		(b) air bubbles	Remove bubbles before electrophoresis
17	Poor staining	(a) Protein concentration low	Load protein
		(b) Amount of dye low or dye is old	Use more concentrated stock of dye or use fresh stock
		(c) For isoenzyme staining, pH of buffer, time and temperature of incubation	Check pH of buffer use proper time and temperature in incubation
18	Patchy staining	Insoluble material in dye	Dissolve dye properly and filter the solution
19	High background of staining	(a) Sample proteolysis	Use protease inhibitors in protein extract
		(b) Impure grade SDS	Use pure grade SDS and boil samples properly
		(c) Exposure of light sensitive stains to light during staining process	Incubate in the dark unless otherwise indicated. Store all stains in dark
20	Long duration required for electrophoretic run	Air bubbles	Remove air bubbles from gels and use deaerated solutions
21	Variations in staining intensity along width of stained band	Uneven gel surface	Overlay water gently
22	Non-reproductibility of protein bands or loss of protein bands after staining	(a) Sample proteolysis	Use protease inhibitors during sample preparation
		(b) Inappropriate fixative	Try another fixative
23	Gel is blank (no bands where normally there are bands)	(a) Samples may have deteriorated or solutions old	Use fresh samples and fresh solutions
		(b) Staining solution may be missing some component or very old	Prepare staining solution carefully
		(c) Improper grade of chemicals	Use proper grades of chemicals especially substrates and cofactors in enzyme staining

24	Dye front is curved (sampling or sad face of zymogram)	(a) Non-uniform cooling	Check cooling
		(b) Loading unequal	Load equal amount of protein in all wells
25	Dye front straight but slanted	(a) Combs not placed properly	Check that comb is placed straight
		(b) Loading unequal	Load equal amount of protein in all wells
26	Bands in adjacent lanes merged	(a) Wells not proper	(a) Check that comb fits tightly and wells
		(b) Overloading	(b) Load less protein
		(c) Delay in electrophoresis after loading	(c) Start electrophoresis immediately after loading the samples
27	No current registers on the meter of power supply unit	Loose connections or break in cables	Inspect power supply cables and clean contacts regularly
28	Gel heats up	(a) Current high	Check the power supply setting
		(b) Ionic strength of buffer too high	Provide supplementary cooling while using buffers in high ionic strength
		(c) Improper cooling	Check power supply settings
29	Marker dye moves too slowly	(a) Low current	Use more current for thicker gels compared to thin gels. If two gels are connected to same power supply, use double current.
		(b) Gel or buffers not properly prepared	Use freshly prepared buffers and gel solutions
30	Voltage too high	(a) Insufficient electrode buffer	Check level of buffers in buffer reservoirs
		(b) Ionic strength of buffer not proper	Prepare fresh buffer

(B) Starch Gel Electrophoresis

Many of the problems discussed above for polyacrylamide gels also apply to starch gel electrophoresis (SGE). However, some are peculiar to SGE and are as given below (Show and Prasad, 1970; Andrews, 1993) :

S.No.	Problem	Cause	Remedy
1.	Wick imbibes extract too slowly	Protein extract too thick	Use proper ratio of buffer tissue for extraction
2	Lumpy, specky gel	(a) Starch powder not properly dissolved	Suspend starch powder properly before adding the hot buffer Do not degas for too long
		(b) Got drought for a long time	Use clean glassware
		(c) Dirty glassware	Use clean glassware
3	Starch gel is cloudy	Insufficient cooling of gel	Cook the gel for correct time
4	Gel is too chick and viscous	(a) Insufficient buffer	Use proper proportion of buffer
		(b) Prolonged degassing	Do not degas for too long
5	Rough gel surface after cooling	Gel wrapped before cooled properly	Always warp the gel only after it has cooled properly and make sure wrapper is put smoothly over the gel
6	Cracking noise when gel is poured into mold	Gel is too hot	Let the gel cool just a little before pouring into mold
7	Gel is too hard to slice	(a) Starch concentration high	Reduce starch concentration
		(b) Excessive cooling during electrophoresis	Adjust cooling temperature to about 4°C
8	Dye front is wavy	(a) Cloth wick front uneven	Make a clean and straight cut at the origin for sample wicks.
		(b) Slits not proper	Wicks. Wicks should be placed straight and parallel to origin.
		(c) Presence of air in slits	Close up the slit after dewicking.

REFERENCES

Andrews, A.T. 1993. *Electrophoresis Theory, Technique and Biochemical and Clinical Applications.* Claredon Press, Oxford.

Berg, J.M., Tymoczko, J.L. and Stryes, L. 2002. *Biochemistry* (5th Ed.) W.H. Freeman.

Dadlani, M. and Varier, A. 1993. *Electrophoresis for Variety Identification.* Technical bulletin, Division of Seed Science & Technology, IARI, New Delhi.

Davis, B.J. 1964. Disc electrophoresis. II Method and application to human serum protein. *Annals of the New York Academy of Sciences* 121 : 404-447.

Dhillon, S. 2000. Electrophoresis In: *Introductory Practical Biochemistry* (Eds.) S.K. Sawhney and Randhir Singh. Narosa Publishing House, New Delhi.

Hames, B.D. and Rickwood, D. 1990. *Gel Electrophoresis of Proteins- A Practical Approaches.* 2nd Edition, IRL Press, Oxford, London.

Laemmli, U.K. 1970. Cleavage of structural proteins during the assembly of the head of bacterophage T4. *Nature* 277 : 680-685.

O'Farrel, P.H. 1975. High resolution two-dimensional electrophoresis of proteins. *Journal of Biological Chemistry* 250: 4007-4021.

Ornstein, L. 1974. Disc Electrophoresis. I. background and theory. *Annals of the New York Academy of Sciences* 121 : 321-349.

Rickwood, D. and Hames, B.D. 1982. *Gel Electrophoresis of Nucleic Acids : A Practical Approach.* IPL Press, Oxford, England.

Thomas, S., Thirmalpura, N., Crossley, E.C. and Walker, D.H. 2009. Antigenic protein modifications in *Ehriczia parasita. Immunology* 31: 296-303.

Tiselius, A. 1937. A new apparatus for electrophoretic analysis of collectical mixtures. *Transaction of the Forestry Society* 33: 524-531.

Show, C.R. and Prasad, R. 1970. Starch gel electrophoresis of enzymes - A compilation of recipies. *Biochemical Genetics* 4 : 297-320.

Weber, K. and Osborne, M. 1969. The reliability of molecular weight determination by docecyl sulfate polyacrylamide gel elephophoresis. *Journal of Biological Chemistry* 244 : 4406-4412.

CHAPTER - 18

MOLECULAR BIOLOGY TECHNIQUES

INTRODUCTION

Molecular biology is the branch of biology that deals with the molecular basis of biological activity. This field overlaps with other areas of biology and chemistry, particularly genetics and biochemistry (Karp *et al.,* 1996). Molecular biology chiefly concerns itself with understanding and the interactions between the various systems of a cell, including the interactions between the different types of DNA, RNA and protein biosynthesis as well as learning how these interactions are regulated (Alberts *et al.,* 2008).

The shape and functions of cells are determined by their proteins. Enzymes catalyze the reactions, which synthesize membranes, cell walls and pigments, and they are needed to extract energy from substrates. Proteins in membranes are responsible for the transport of molecules from one cellular compartment to another, and between the inside the outside of the cell. Synthesis of proteins is itself catalyzed by proteins, by DNA, which carries all the information needed to specify the structure of every protein the cell can make, ultimately directs this process. The realization that DNA lies behind all the cell's activities led to the development of molecular biology, which aims to explain biological processes in terms of the structures and interactions between nucleic acids and proteins (Cashmor, 1982; Cherry, 1973).

Molecular biology has led directly to the immensely powerful, and potentially profitable, techniques of genetic engineering (Sambrok *et al.,* 1987). A great deal of effort (and money) is being directed into the genetic manipulation, or genetic engineering, of microorganisms, to make then produce a range of valuable polypeptides such as insulin, blood clotting factor VIII, growth hormone, or interferons, which they would not normally produce and which are expensive to prepare by conventional biochemical means (Alberts *et al.,* 2008; Tomar *et al.,* 2010). Since, it is easy and cheap to grow microorganisms on a large scale, they are very attractive as potential source of these polypeptides. Genetic manipulation of plant and livestock should also permit the introduction of beneficial characteristics, such as resistance to herbicides or disease that could not readily be obtained by conventional breeding.

Few of the basic molecular biology techniques have been discussed in this chapter in detail.

TECHNIQUES OF MOLECUALR BIOLOGY

Since the late 1950s and early 1960s, molecular biologists have learned to characterize, isolate, and manipulate the molecular components of cells and organisms. These components include DNA, the repository of genetic information; RNA, a close relative of DNA whose functions range from serving as temporary working copy of DNA to actual structural and enzymatic functions as well as a functional and structural part of the translational apparatus; and proteins, the major structural and enzymatic type of molecule in cells (Karp *et al.,* 1996). Some of these are discussed below:

ISOLATION OF GENOMIC DNA

Genomic DNA is the blueprint of life, which is inherited and passed to generations directing the development of cells organelles and organisms.

Isolation of high molecular weight genomic DNA free of protein and RNA contamination is the prerequisite for molecular cloning techniques. The isolated DNA should also be amenable for the action of restriction endonucleases. Plant genomic DNA is usually isolated by lysing the nuclei in buffer containing detergents like Sodium Dodecyl Sulphate (SDS) or Cetyl Trimethyl Ammonium Bromide (CTAB) in the presence of EDTA, followed by phenol and chloroform extractions to remove proteins and other contaminations. DNase free RNase is used to remove RNA and subsequently proteinase K is used to digest DNA bound proteins and RNase.

The resulting DNA is of the order of about 100 kb which is suitable for most molecular cloning techniques.

In the two procedures described here, the first one uses a modified procedure of Dellaporta *et al.* (1983) and second method uses the salient features of Murray and Thompson (1980) and Sanghai-Maroof *et al.* (1984).

SOLUTIONS

* SDS extraction buffer (For SDS method) :

1 M Tris HCl (pH 8.0) - 100 ml

0.5 M EDTA (pH 8.0)	-	40 ml
3 M NaCl	-	167 ml
Polyvinyl pyrrolidone (PVP)	-	10 gm
10% SDS	-	170 ml

Make up to 1000 ml with distilled water and autoclave

* Nuclear extraction buffer (for CTAB method)

Sorbitol	-	56 g
1 M Tris HCl (pH 8.0)	-	50 ml
0.5 M EDTA (pH 8.0)	-	10 ml

Make up to 1000 ml and autoclave

* Nuclear lysis buffer (For CTAB method)

Cetyl Trimethyl Ammonium Bromide (CTAB)	- 20 g
1 M Tris HCl (pH 8.0)	- 200 ml
0.5 M EDTA (pH 8.0)	- 400 ml

Adjust pH to 7.5 with concentrated HCl. Make up to 1000 ml and autoclave

* 5 M Potassium acetate (pH 5.8)

* Distilled phenol - Melt phenol at 68°C. Add hydroxyquinoline (0.1%) : Add equal volume of 0.1 M Tris HCl (pH 8.0) and stir for 15 minutes. In a separating funnel collect phenolic phase. Repeat the procedure till pH is >7.8 (Test with pH paper). Store in 100 ml aliquots, layered with 1 M Tris HCl at 4°C.

* Phenol : chloroform (1 : 1)

* Chloroform : isoamyl alcohol (24 : 1)

* RNase (10 mg/ml)

RNase	-	10 mg
1 M Tris HCl	-	10 µl
3 N NaCl	-	5 µl

Make up to 1 ml with sterile water. Boil for 15 minutes (this is to remove nucleus). Cool and store at 4°C.

* Proteinase K solution : 20 mg/ml in sterile water

* Absolute Ethanol, *70% Ethanol*

* T_{10} E_1 solution

1 M Tris (pH 8.0)	-	1 ml
0.5 EDTA (pH 8.0)	-	0.2 ml

Make up to 100 ml. Autoclave.

(A) SDS METHOD

- Collect young leaves and weigh up to 3-5 g.
- Freeze tissue in a sterile mortar-pestle using liquid nitrogen and grind to a fine powder.
- Collect the powdered tissue in a centrifuge tubes and add 20 ml SDS extraction buffer preheated to 65°C. Dispense the powder evenly with a sterile glass rod.
- Incubates tubes in a 65°C water bath for 15 minutes.
- Add 5 ml of 5 M potassium acetate and mix for 5 minutes with gentle inversion (A rotary shaker set to 80 rpm may also be used).
- Add equal volume of chloroform : isoamyl alcohol and mix for 5 minutes.
- Centrifuge at 3000 rpm for 20 minutes.
- Collect the aqueous supernatant and repeat the extraction step.
- Collect supernatant and add equal volume of isopropanol. Mix gently by inverting the tube several times.
- Hook out the precipitated DNA fibers by a glass hook and dry on a paper towel.
- Dissolve the dried DNA in 4 ml T_{10} E_1 solution.
- Proceed to purification protocol.

(B) CTAB METHOD

- Collect about 5 g of young leaves from field and freeze them in a sterile mortal- pestle using liquid nitrogen.
- Grind to a fine powder and transfer to 50 ml chilled extraction buffer in a sterile beaker. Homogenize to slurry using a magnetic stirrer.
- Squeeze the homogenate through two-three layers of cheese cloth and distribute evenly in tubes.
- Centrifuge at 700 x g for 15 minutes at 4°C to pellet nuclei. Discard the supernatant, which contains plastids and cellular debris.
- Resuspend pellet in 10 ml nuclear lysis buffer preheated to 65°C and incubate in 65°C water bath for 90 minutes.

- Add equal volume of chloroform isoamyl alcohol solution and mix for 5 minutes.
- Centrifuge for 5000 rpm for 15 minutes and collect aqueous supernatant. Repeat the step twice.
- Collect supernatant and equal volume of isopropanol. Mix gently by inverting the tube several times.
- Hook out the precipitated DNA fibers using a glass hook. Dry on a paper towel and dissolve in 4 ml T_{10} E_1 solution.
- Proceed for purification protocol.

PURIFICATION OF DNA

- To the DNA solution, add 4 ul of 10 mg/ml RNase and incubate for 1 hour at room temperature.
- Add 10 ul of proteinase K and incubate at room temperature for 30 minutes.
- Add equal volume of equilibrated phenol and mix for 5 minutes.
- Centrifuge at 5000 rpm for 10 minutes and collect supernatant.
- Add equal volume of phenol : chloroform : isoamylalcohol and mix for 5 minutes (25 : 24 : 1).
- Centrifuge at 5000 rpm for 10 minutes.
- Add equal volume of chloroform : isoamyl alcohol (24 : 1), mix gently and collect supernatant at 5000 rpm for 10 minutes.
- To the supernatant add 1/10th volume of 3 M sodium acetate (pH 5.8) and 2.5 volumes chilled absolute ethanol and mix by inverting the tubes several times.
- Hook out the precipitated DNA fibers, dry them on a paper towel and dissolve in minimum amount of T_{10} E_1.

The final product usually gives a yield of 0.5 mg/g tissue DNA which is protein and RNA free.

ISOLATION OF RNA

In 1909, Aaron Theodor Levene was the first to determine that the nucleic acid contains a ribose sugar. Twenty years later, he found that certain nucleic acids contain a different sugar deoxyribose. It was proposed that RNA (Ribonucleic acid) has ribose sugar different from DNA. Levene also worked out the chemical nature of RNA and the chemistry was explored in detail in 1930's by Alexander Todd.

The isolation of pure RNA to be approached in a sterile RNase free atmosphere in the lab. All the glassware and solution should be kept separate from other experiment (Davies *et al.,* 1972). Glassware should be usually baked at 180°C in an oven a day before the experiment. Wearing of gloves is a must to avoid nuclease in the form of sweat. All glassware and solutions should be treated with 0.1% DEPC (Diethyl Pyrocarbonate).

The key step involved in the isolation is lysis of cells in a high pH buffer and inactivation of RNase using phenol which also removes protein and partitions nucleic acid into the aqueous phase. Subsequent chloroform extraction removes out the RNase cellular debris from homogenate and in the final setp pure, high molecular weight RNA is selectively precipitated using lithium chloride (Schuber and Zielinsia, 1989; Tomar *et al.,* 2010).

SOLUTIONS

* RNA extraction buffer :

1 M Tris HCl (pH 8.5)	-	5 ml (Final solution pH 8.5)
0.5 M EDTA (pH 8.0)	-	2 ml
NaCl	-	1.2 g

Make up to 100 ml and autoclave

* T_{10} E_1 solution

1 M Tris (pH 8.5)	-	1 ml
0.5 M EDTA (pH 8.0)	-	0.2 ml

Make up to 100 ml Autoclave.

* Water Distilled Phenol

Melt phenol crystals in a 61°C water bath. Add water and stir to a milky emulsion for 20 minutes. Allow phases to separate. Discard the upper water phase by pipetting out. Store at 40°C in an amber colored bottle with cap wrapped tightly in paraffin.

* Phenol : chloroform : isoamyl alcohol (25 : 24 :1)

* Chloroform : isoamyl alcohol (24 :1)

* 8 m urea : (Do not autoclave, Urea breaks down when heated above 550°C).

* 10 m LiCl

* 5 m NaCl

* 3 m Potassium acetate.

PROTOCOL

- Weigh about 2 g of fresh leaves and freeze in liquid nitrogen.
- Grind to a fine powder and transfer into a blender jar containing 20 ml RNA extraction buffer. In a fine hood, add 20 ml phenol : chloroform : isoamyl alcohol. Cover the jar with plastic wrap and blend at top speed for 2 minutes.
- Decant the homogenate into 50 ml centrifuge tubes and shake for 10 minutes.
- Centrifuge at 5000 rpm for 20 minutes and collect the aqueous supernatant. Add equal volume of chloroform : isoamyl alcohol and shake for 5 minutes.
- Centrifuge at 5000 rpm for 20 minutes and collect the aqueous supernatant.
- Repeat the chloroform extraction step twice.
- Add equal volume of isopropanol and mix by inverting the tubes several times. Incubate at -70°C for 20 minutes.
- Thaw the tube and collect the precipitated nucleic acid by centrifuging at 5000 rpm for 20 minutes at 4°C.
- Resuspend pellet in 5 ml T_{10} E_1 solution. Add 200 ul of 5 M NaCl and 12.5 ml chilled ethanol to reprecipitate nucleic acid by incubating tube in ice for 20 minutes.
- Resuspend pellet in 2 ml T_{10} E_1. Hold on ice.
- Distribute the solution in eppendorf tubes. Add 0.25 volume of 8 M urea, mix and add 0.25 volumes of 10 M LiCl. Incubate overnight at 200°C. This treatment causes all high molecular weight RNA to precipitate while most of tRNA and DNA remains soluble.
- Collect the precipitate RNA by centrifuging at 5000 rpm for 20 minutes at 40°C. Discard supernatant and resuspend pellet in 2 ml T_{10} E_1 buffer. Store at -70°C.

ELECTROPHORESIS OF NUCLEIC ACIDS

The standard method used to separate, spot and purify DNA fragments (As little as 1-10 μg) is, electrophoresis through agarose gels. Agarose gels can be used for isolation of DNA from 200 bp - 50 kb in length. The gels are usually run in a horizontal configuration in an electric field of constant strength and direction. The DNA can be detected by direct examination of the gel in ultraviolet light after staining with fluorescence intercalating dye ethidium bromide (10 mg/l). If necessary, the DNA bands can be recovered by electro elution from the gel, which can be used for cloning purpose. Agarose gels are casted by melting the agarose in a desired buffer and allowing them to gel in a mould. The hardened matrix when placed in an electric field, allows the negatively charged DNA to migrate towards anode at a neutral pH. Molecules of linear double stranded DNA which tend to become oriented in electric field in an end on position migrate through the gel matrices at rates inversely proportional to log 10 of the number of base pairs i.e. larger molecules migrate more slowly. Super helical circular form of DNA migrates faster than nicked circular and linear DNA.

SOLUTIONS

* TAE buffer (50 x)

Tris base - 24.2 (Make up to 100 ml and autoclave)

Glacial Acetic Acid - 5.71 ml

0.5 EDTA (pH 8.0) - 10 ml

working conc = 1 x

* Agarose : Molecular Biology Grade

* Blue Juice : in water

Bromophenol blue - 0.25%

Xylene cyanol - 0.25%

Sucrose - 40%

* Ethidium Bromide : 10 mg/ml in water. (note : EtBr is carcinogenic - avoid contact with skin. Always use gloves). Store in a bottle wrapped with aluminium foil.

PROTOCOL

- Seal the open ends of the gel tray supplied with electrophoresis apparatus with plastic tape to form a mould. Set the mould on a horizontal section checked with a spirit level.
- Prepare 1 x TAE buffer to fill the electrophoresis tank and also to prepare gel.
- Add 0.8% agarose to desired amount of 1 x buffer and melt in a water bath till agarose dissolve. The smallest gel mould holds 30 ml of solution.

- Cool the solution and add ethidium bromide to a final concentration of 0.5 mg/ml and mix thoroughly.
- Pour the solution in the mould and position the comb. Allow to harden at room temperature. Gel should be abotu 5 mm thick. Air bubbles should be avoided.
- After the gel is set (about 20-30 minutes) carefully remove the comb and tape and mount the gel in electrophoresis tank. See the buffer level up to 1 mm above the gel.
- Mix around 10 ul DNA solution with 5 ul blue juice and load into the slots. Add the Lambda.
- Hind III digest which acts as the molecular weight marker.

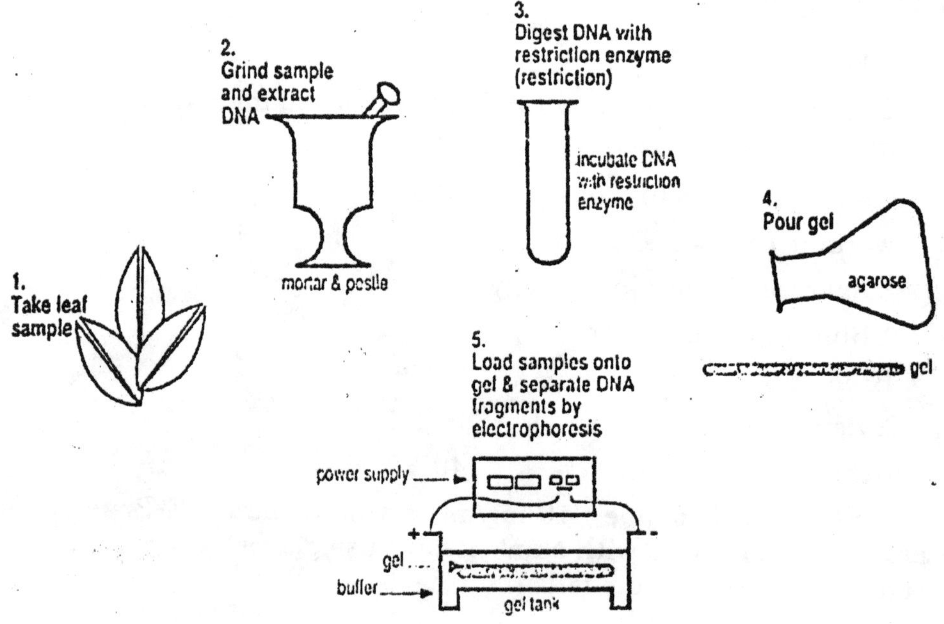

Figure 1. Restriction and electrophoretic separation of DNA.

Note

- Close the lid of gel tank and attach the electrical leads so that DNA migrates towards anode (red lead). Apply a voltage of 5 v/cm (measured as a distance between electrodes).
- Run the gel until bromophenol blue dye has migrated up to more than half of the gel.
- Turn off the electric current and remove leads and lid. View the gel under UV light to detect DNA bands.

- Lambda DNA digested with Hind III on migration shows DNA bands corresponding to 23.13, 9.416, 6.557, 4,361, 3.322, 2.027, 0.564, 0.125. The last two bands are not seen if very little sample is loaded.

SOUTHERN BLOTTING

Analysis of localized genomic sequences is usually accomplished after transferring them onto a solid support like nitrocellulose (Southern, 1975). After digestion with a restriction enzyme and electrophoresis, the DNA is denatured *in situ* and trasnsferred onto a solid support. The relative position of DNA fragments is preseved during their transfer to the filter. DNA attached to the filter is then hybridized to radiolabel led DNA probes and autoradiography is used to locate the position of bands complementary to the probe.

In the original capillary transfer method of Southern, DNA fragments are carried from the gel in a flow of liquid and deposited on the surface of solid support. The liquid is drawn through the gel by capillary action that is established and maintained by a stack of dry absorbent paper towels.

In the present protocol, possiblot apparatus (STRATAGENE) is used which does not require the paper towel stack for capillary action. Here, the gel is placed in contact with the nitrocellulose membrane supported on a porous screen in the pressure chamber. Buffer absorbed sponge placed on the gel maintains fluid through the gel onto the membrane by the pressure, which elutes the nucleic acid onto the support.

DNA is invariably linked to the membrane using UV stratalinker at 1200 joules for 45 seconds.

SOLUTIONS

- Depurination solution : 0.25 N HCl.
- Denaturation solution : 5 M NaCl + 0.5 M NaOH, dissolve 29.22 g of NaCl and 20 g of NaOH in sterile water. Make up to 1000 ml and autoclave.
- Neutralising solution : 0.5 Tris (pH 7.0) and 1.0 M NaCl. Dissolve 60.55 g Tris in water and adjust the pH to 7.4. Dissolve 58.44 g NaCl adjust pH to 7.4 make up to 1000 ml and autoclave.
- Transfer solution : 10 x SSC. Dissolve 87.66 g of NaCl and 44.1 g of sodium citrate in 800 ml of water. Adjust pH to 7.0 with NaOH. Adjust to 1000 ml and autoclave.

- 0.4 N NaOH : Dissolve 6.40 g of NaOH in sterile water and make up to 400 ml.
- 0.2 M Tris-HCl (pH 7.5) - 2 x SSC : Add 80 ml of 1 M Tris-HCl (pH 7.5) to 80 ml of 10 x SSC and make upto 400 ml.

PROTOCOL

- Treat the electrophoresed gel for 10 minutes in 0.25 N HCl with gentle agitation in a plastic tray.
- Pour of the HCl, rinse with distilled water and add the denaturation solution and continue treatment for 30 minutes.
- Pour of the denaturation solution, rinse with distilled water and add the neutralizing solution and treat for 30 minutes.
- Wash the gel in sterile water.
- Wearing gloves cut out the nitrocellulose membrane and two sheets of Whatman 3 mm paper about 1.5 times greater the gel size.
- Prewet the membrane and Whatman paper for 5 minutes in water and 5 minutes in the transfer solution.
- Expand the cellulose sponge in water and wring out the water. Saturate the sponge with the transfer buffer.
- Cut off a rectangular whole slightly smaller thn gel in the plastic mask of the apparatus (about 0.3 cm smaller).
- Now place Whatman paper on the porous screen followed by the membrane and on the top of this place the plastic mask.
- Place the treated gel over the mask avoiding air bubbles.
- Lay the sponge over the gel and close the lid of the apparatus with clamps.
- Apply 75 mm pressure and carry out rapid trasnfer for about 45 minutes.
- After transfer, check the gel under UV light to ensure complete transfer of DNA.
- Remove the membrane, air dry for about 10 minutes to evaporate buffer and take onto a whatman filter paper.
- Place in the UV crosslinker and autocross link the DNA to the membrane.
- Store the membrane in a sealed cellophane bag for hybridization.

CONVENTIONAL PROTOCOL

Here the first 6 steps remain the same

- Take a plastic tray and fit in a sponge saturated with transfer buffer.
- Overlay the sponge with a Whatman paper saturated with the transfer buffer.
- Place the gel on the paper and cover the rest of the area with parafilm.
- Place the nitrocellulose membrane on top of the gel and on top; place Whatman paper exactly cut to size of the membrane.
- Place a stack of filter paper on top of the gel assembly and put a weight as support.
- Allow transfer to occur by capillary action overnight.
- The next day, check the gel under UV to ensure complete transfer.

ANALYSIS OF RNA AND NORTHERN TRANSFER (NORTHERN BLOTTING)

Immediately following the isolation of RNA, it is subjected to gel electrophoresis under denaturing conditions through formaldehyde gels (Sharp *et al.,* 1973).

RNA is then transferred onto a nitrocellulose membrane. Care should be taken that all apparatus are treated with 0.1% DEPC.

SOLUTIONS

- 5 x formaldehyde gel - running buffer :

  0.1 M MOPS (pH 7.0) (20.6 g MOPS/lit.)

  40 mM Sodium acetate

  5 mM EDTA
- Agarose
- Formaldehyde
- Formamide
- 0.5 M EDTA (pH 8.0)
- Loading dye

  50% Glycerol

1 mM EDTA (pH 8.0)

0.25% bromophenol blue

- DEPC treated autoclaved water and glassware.

PROTOCOL

- Make agarose (1.5%) gel solution in water, cool to 600°C and mix the following

  1 part of formaline (37% solution)

  3.5 part agarose (1.5%) in water

  1.1 part 5 x formaldehyde gel bufer
- Cast the gel in chemical hood, add I μ 1 of 1 mg/ml ethidium bromide solution to the gel before solidification.
- Prepare the samples by mixing in a sterile microfuge tube

RNA	-	4.5 μl (20-30 μl)
5 x formaldehyde buffer	-	2.0 μl
Formaldehyde	-	3.5 μl
Formamide	-	10.0 μl

  Water to final volume of 20 ul
- Incubate the samples for 15 minutes at 65°C, chill over ice. Centrifuge at high speed for 5 second to remove the particles.
- Prerun the agarose gel for 5 minutes 5 V/cm.
- Load the entire sample along with 2 ul of sterile loading dye.
- Run the gel at 3-4 V/cm in 1 x formal dehyde gel buffer for around 2 hour.
- Run RNA marker (28S, 18S) along with the samples in the same treatment.
- If ethidium bromide has not been added in the gel itself, stain the gel with ethidium bromide (0.5 ug/ml) solution in 0.1 M ammonium acetate for 30 minutes.
- Observe under transilluminator.

NORTHERN TRANSFER OF RNA TO NITROCELLULOSE

- Handle the gel carefully with sterile gloves. Place the gel in 0.05 n NaOH solution for 20 minutes.
- Transfer to DEPC treated 20 x SSC.

- Cut off the bottom of the gel to facilitate the orientation of the gel.
- Place two-piece of Whatman 31 mm (1 cm larger than gel) on the matrix.
- Cut out the nitrocellulose (Schleicher and Schuell, B.A. 85) 1 mm larger than gel.
- Float nitrocellulose on the surface of a dish of deionised water until it wets completely from beneath, and then immerse the filter in 20 x SSC for atleast 5 minutes.
- Place the gel over the support in an inverted position so that it is centred on the wet 3 mm paper. No air bubbles between 3 mm paper and the gel. Arrange the transfer by capillary means as detailed in the southern blotting head.
- After overnight capillary transfer as done in the case of Southern transfer, soak the filter in 6 x SSC for 5 minutes at room temperature.
- Remove the excess of SSC and UV cross link the membrane for 20 seconds at 0.12 KJ in a cross linker.
- Proceed for the hybridization.

WESTERN BLOTTING

The term ``Western Blotting'' (Burnette, 1981) refers to a procedure which does not directly involve nucleic acids, but which is of importance in gene manipulation. It involves the transfer of electrophoresed protein bands from a polacrylamide gel onto a membrane of nitrocellulose or nylon which they bind strongly (Gershoni and Palade, 1982; Renart and Sandoval, 1984). The bound proteins are then available for analysis by variety of specific protein ligand interactions.

ALTERNATIVE BLOTTING TECHNIQUE

The original blotting technique employed capillary blotting but now days the blotting is usually accomplished by electrophoretic transfer of polypeptides from and SDS-polyacrylamide gel on to the membrane (Towbin *et al.*, 1979). Electrophoretic transfer also is the method of choice for trasnsferring DNA or RNA from low pore-size polacrylamide gels. It can also be used with agarose gels. However, in this case, the rapid electrophoretic transfer process requires high currents, which can lead to extensive heating effects resulting in distortion of agarose gel. The use of an external cooling system is necessary to prevent this.

Another alternative to capillary blotting is vaccum-driven blotting for which several devices are commercially available. Vacuum blotting has several advantages over capillary or electrophoretic transfer method : transfer is very rapid and gel treatment can be performed *in situ* on the vacuum apparatus. This ensures minimal gel handling and, together with the rapid transfer, prevents significant DNA diffusion.

CONCLUSIONS

In terms of pure research, all the techniques of molecular biology, from cloning to sequencing, have revolutionized our understanding of gene structure and function. In long term it is anticipated that plant and animal breeders shall introduce genes for disease resistance, improved yield etc., into crops and livestock, by using genetic engineering.

REFERENCES

Alberts, B., Johnsen, A., Lewis, J., Raff, M., Roberts, K. and Walker, P. 2008. *Molecular Biology of the Cell.* pp. 538-539, 5th Ed. Garlard Science, Taylor and Francis Group, NY.

Burnette, W.N. 1981. "Western flotting": electrophoretic transfor of proteins from sodium dotecyl sulfate-polyacrylamide gels to unmodified nitro cellulose and radiographic detection with antiboly and radioiodinated protein A. *Analytical Biochemistry.* 112: 195-203.

Cashmor, A.K. 1982. *Methods in Plant Molecular Biology.* Academic Press Inc., New York.

Cherry, J.H. 1973. *Molecular Biology of Plants: A Text Manual.* Columbia University Press. New York and London.

Davies, E., Larkins, B.A. and Knight, R.H. 1972. Polyibosomes from plants: An improved method for their isolation in the absence of ribonuclease inhibitors. *Plant Physiology* 50: 581-584.

Dellaporta, S.L., Wood, J. and Hicks, J.B. 1983. A plant DNA mini-preparation: Version II. *Plant Molecular Biology Reporter* 1: 19-21.

Gershoni, J.M. and Palade, G.E. 1982. Electrophoretic transfer of proteins from sodium dotecyl sulfate polyacrylamide gels to a positively changed membrane filter. *Analytical Biochemistry* 124: 396-405.

Karp, A., Seber, O. and Buiatti, M. 1996. Molecular techniques in the assessment of botanical diversity. *Annals of Botany* 78: 143-149.

Murray, M.G. and Thompson, W.F. 1980. Rapid isolation of high molecular weight plant DNA. *Nucleic Acid Research* 8: 4321-4325.

Remalt, J. and Sandoval, I.V. 1984. Western blots. *Methods in Enzymology* 104: 455-460.

Sanghi-Maroof, M.A., Biyashov, R.M., Yang, G.P., Zhang, C. and Allard, R.W. 1994. *Proceeding of National Academy of Sciences,* USA 91: 5466-5470.

Sambrok, J., Fritsch, E.F.A dn Maniatis, T. 1987. *Molecular Cloning: A Laboratory Manual.* Cold Spring Harbor Laboratory.

Schuber, S. and Zielinski, K. 1989. *Methods in Plant Molecular Biology.* Academic Press, Inc. New York and London.

Southern, E.M. 1975. Nucleic acid blotting: Southern and nothern. *Molecular Biology* 98: 503-515.

Tomar, R.S., Parakhia, M.V., Patel, S.V. and Golakya, B.A. 2010. *Molecular Markers and Plant Biotechnology.* New India Publishing Agency, New Delhi, 577 p.

Sharp, P.A., Sugden, B. and Sambrook, J. 1973. Detection of 2 restriction endo nuclease activities in *Haemophilal paraifluenze* using analytical agarose - ethidicum bromide electrophorens. *Biochemistry* 12: 3055-3067.

Towbin, H., Stachelin, T. and Gordon, J. 1979. Electrophoretic transfer of proteins from polyaccytamidegels to nitrocellulose sheets: procedure and some applications. *Proceedings of Natural Academy of Science, USA.* 76: 4350-4354.

CHAPTER - 19

DETERMINATION OF POLYAMINES AND PHENOLICS BY TLC AND HPLC TECHNIQUES

INTRODUCTION

Polyamines are positively charged aliphatic nitrogen-containing compounds of low molecular weight that are widely distributed in living organisms. The major forms are (1) *putrescine, (2) spemidine* and (3) *spermine*.

(1) $H_2N - (CH_2)_4-NH_2$

(2) $H_2N-(CH_2)_4-NH-(CH_2)_3-NH_2$

(3) $H_2N-(CH_2)_3-NH-(CH_2)_4-NH-(CH_2)_3-NH_2$

Polyamnes occur in nature not only as free molecular bases but also as conjugates. In plants, the hydroxycinnamic acid amides are the most common polyamine conjugates, but polyamines also occur linked to proteins or as part of macrocyclic alkaloids. Due to their affinity to acidic components (phospholipids, nucleic acids), polyamines are also bound to DNA, RNA, cell walls and membranes (Tiburcio *et al.,* 1985; Martin-Tanguy, 1997). The exact localisation of polyamines is still controversial, but it seems that they are mainly localised in the cytoplasm and in some specific compartments like chloroplast and mitochondria (Bagni, 1991; Rajam, 1997).

Phenolics have been considered as allelopathic substances which are widely distributed in the vascular plant kingdom (Harborne, 1989). The term phenolic or polyphenol can be defined chemically as a substance which posseses an aromatic ring bearing a hydroxyl substituent, including functional derivatives as esters, methyl ethers, glycosides, etc. Most of them have two or more hydroxyl groups and are derived from one of the common dihydric or trihydric phenols. Examples of some phenolics include:

(1) p- hydroxy benzoic acid (2) p-coumaric acid (3) Ferilic acid (4) Quercetin (5) Ellagic acid and (6) Salicyclic acid.

Phenolics can affect membrane permeability of roots and alter the uptake and flux of nutrients. They can be absorbed by the roots and influence physiological processes.

The determination of polyamincs and phenolics in plants by TLC and HPLC techniques has been presented in this chapter.

THIN-LAYER CHROMATOGRAPHY (TLC)

Thin-layer chromatography (TLC) is a chromatographic technique that is useful for separating organic compounds. Together with paper chromatography comprise `planar' or `flat-bed' chromatography. TLC is much simpler and less automated than High Performance Liquid Chromatography (HPLC) with regards to methodology and equipment requirements; a suitable closed vessel containing solvent and a coated plate (the thin layer) are all that are required to carry out separations.

TLC is widely used in experimental biology to analyse lipids, amino acids and vitamins. Its usefulness has been also proven in pharmaceutical, food and environmental analysis of multiple organic compounds (drugs, pesticides, herbicides, etc.). Because of its simplicity and rapidity, it is often used in analytical chemistry to monitor the progress of organic reactions, and to check the purity of products (Sherma, 1990; Fried and Sherma, 1999; Hahn-Deinstrop, 2000).

Sherma (1990) summarises the basic TLC proceuure as follows : The sample (either liquid or dissolved in a volatile solvent) containing a mixture of components is deposited as a spot near one end of the *stationary phase,* a thin layer immobilised on a plate. Then the sample is dried; and the end of the stationary phase with the initial zone is placed into a *mobile phase,* usually a mixture of pure solvents, inside a closed chamber. The components of the mixture migrate at different rates during movement of the mobile phase through the stationary phase (by capillary action), which is termed the *development of the chromatogram.* When the mobile phase has moved an appropriate distance, the stationary phase is removed from the chamber, the mobile phase is rapidly dried, and the separated *zones* or spots are detected by application of a suitable visualisation reagent.

Differential migration is the result of varying degrees of affinity of the mixture components for the solvents and the thin layer, depending on the nature of both phases and solutes. The interactions determ-ining chromatographic retention and selectivity include hydrogen bonding, charge transfer, ion-ion, ion-dipole and van der Waals interactions (Sherma, 1990).

Thin-layers and Solvents

The thin layer consists of an adsorbent material made of cellulose, polyamide, silica gel, alumina or a mixture of these components, usually installed on a glass, plastic or aluminium foil support. There are a great variety of commercial precoated layers for TLC which can be used as received, without preparation, leading to an enormous time saving (Silica gel adsorbent is by far the most frequently used layer material for HPTLC (High Performance TLC) plates. Compared to TLC, HP plates are more efficient, leading to tighter zones, better resolution, and more sensitive detection, due to their physical and chemical properties (narrower pore, smaller particle diameter, etc.)

The solvents for TLC are chosen in relation to the nature of the layer and mixture to be separated. The strength (polarity) of the mobile phase influences the distances moved by the solutes, while the chemical classification of the solvent components determines the interactions and selectivity of the system.

The versatility and flexibility of TLC relative to HPLC is enhanced by the greater choice of solvents available for preparing TLC mobile phases. The choice of solvents for HPLC is limited regarding the exigent nature of the method : a close system operated under high pressure, with on-line detection, besides a column that is continually reused. Solvent components with high vapour pressure (e.g., ethyl ether) or UV absorbance (e.g., benzene) or those that might degrade the column (e.g., NaOH) are difficult to use in HPLC but are readily applicable to TLC.

Advanced TLC

TLC can be automated using forced solvent flow, running the plate in a vacuum-capable chamber to dry the plate. The ability to program the solvent delivery makes it convenient to do multiple developments in which the solvent flows for a short period of time; the TLC plate is dried, and the process is repeated. This method refocuses the spots to achieve higher resolution than in a single run (Poole and Poole, 1994). Robotics and automation either speed up the operation of TLC or allow for more samples to be processed at one time.

Two-dimensional TLC uses the TLC method twice to separate spots that are unresolved by only one solvent. After running a sample in one solvent, the TLC plate is removed, dried, rotated 90°, and run in another solvent. Any of the spots from the first run that contain mixtures can now be separated. The finished chromatogram is a two-dimensional array of spots (Sherma, 1990; Hahn-Deinstrop, 2000).

HPLC TECHNIQUES

In general, chromatography, and HPLC in particular, is a method of separating a mixture of compounds into its components. Depending on the precision of the equipment and the response of the technique it enables to separate trace impurities or major concentrations from each other. From the information provided for that separation (time required for elution) we can identify each component. However, final identification of an unknown chromatographic peak needs other analytical procedures, such as infrared spectroscopy (IR), nuclear magnetic resonance (NMR), or mass spectroscopy (MS). Usually in plant ecophysiology, we need to know the variation in concentration of some known compounds, and quantitative analysis can be carried out by measuring the area of the chromatographic peaks. This makes HPLC techniques very useful to plant ecophysiologists.

Liquid chromatography is important because most natural compouns are not sufficiently volatile for gas chromatography (Robinson, 1995). High performance liquid chromatography uses high pressure to force solvent to pass through closed columns containing very fine particles that give high-resolution separations from liquid natural samples. The HPLC system consists in a reservoir for the solvent delivery, a pump system, a sample injection valve, a column, a detector and a computer to control the system and display results.

The principle of HPLC technique is based on the passage of the constituents to be separated between two non-miscible phases. For this, the sample is dissolved in the mobile phase, which can be composed by different liquids, and moved across a stationary phase, which is inside column. Due to the interactions of the constituents with the stationary phase they separate after sufficient running time and they appear at the end of the column at different times, where they can be detected externally by means of a detector properly attached to a registering device.

ANALYZING POLYAMINES IN PLANT TISSUES

Separation and quantification of each amine in a given extract requires the use of chromatographic techniques. The polyamines contents of various tissues have been analysed using several ion exchange chromatography, thin-layer chromatography (TLC) and high-performance liquid chromatography (HPLC).

Smith and Davies (1987) offered a descriptive review of methods to separate and quantify polyamines by HPLC. Most common HPLC methods require pre-derivations of polyamines. Dansyl derivatised amines-dansylamines - are monitored by fluorescence detectors (Smith and Davies, 1987; Waldren and Genus, 1987; Corbin *et al.*, 1989; Marce *et al.*, 1995). Benzoyl derivatives - benzoylamines - are UV monitored (Flores and Galston, 1982; Slocum and Flores, 1991). Other methods such as ion-pair reverse phase HPLC or ion-exchange chromatography operate with underivatised polyamines; fluorogenic detection is performed following post-column derivatisation (Smith and Davies, 1987). HPLC offer the advantage over TLC of a superior resolution and easier quantification of polyamines. Many plant scientists utilise this sensitive method to follow the changes in the levels of polyamines during plant growth and development (Flores, 1990, 1991). Besides the most common polyamines (putrescine, spermidine and spermine) HPLC has been extended to the analysis of a large number of other naturally occurring amines in plant tissues, including polyamine homologues, heterocyclic and aromatic amines (Smith and Davies, 1987).

The basic steps for the analysis of polyamines in different plant tissues by means of both TLC and HPLC are briefly described below. In these methods polyamines are analysed as their fluorescent dansylated drivatives, dansylpolyamines. Fluorescence monitoring of polyamines through their dansyl derivatives is a rapid and highly sensitive method with limits of detection at the femtomolar level (Smith and Davies, 1987). Both chromatographic procedures allow high resolutive separations of the polyamines diaminopropane, putrescine, cadaverine- another diamine of limited occurrence in higher plants, formed from lysine decarboxylation, spermidine and spermine.

Extraction of Polyamines

Samples of plant material are powdered with liquid nitrogen, using precooled mortars and pestles. Polyamines are extracted by homogenising in 5% (V/V) cold perchloric acid (PCA) at a ratio of 100 to 300 mg fresh mass ml^{-1} Homogenates are transferred to cold centrifuge tubes (e.g., Eppendorf tubes) and incubated in an ice bath for 30 min.

The required amount of plant material will depend on the water content of each tissue and species. As example, *calli* from *in vitro* culture are extracted at 600 mg fresh-mass. ml^{-1} PCA. Fresh material must be processed immediately after being harvested or kept in freezer at –80°C until extraction, regarding polyamines are extremely labile.

After incubation, homogenates are centrifuged at 27,000 g_n for 20 min. Slight modifications can be introduced, e.g., the extraction procedure for leaf tissue proposed by Reggiani *et al.* (1989) includes centrifugation at 12,000 g_n for 15 min at 4°C.

Both the supernatant (S) and the pellet (P) fractions are collected.

The pellet is resuspended in the original volume with 1 M NaOH. Aliquots (300 μl) each of the pellet suspension and the original supernatant are mixed 1 : 1 with 12 M HCl into an injectable vial, which is sealed with a flame and heated at 110°C for 16 to 18 h. The resulting mixtures are filtered through glass wool (to eliminate carbonised material) and dried in vacuum. Then, the dried material is redisolved in 300 μl of 5% (V/V) PCA.

In this way we obtain three fractions : the non-hydrolysed PCA supernatant (S fraction : free polyamines), the hydrolysed PCA supernatant (SH fraction : containing the polyamines liberated from the PCA-soluble conjugates) and the hydrolysed pellet (PH fraction : containing the polyamines liberated from the PCA-insoluble conjugates).

These samples may be kept at 4°C for several hours, or stored long-term at -20°C until dansylation.

Preparation of Standard Solutions

Stock solutions of commercial polyamine standards : diaminopropane, pytrescine, cadaverine, spermidine and spermine (in the form of hydrochlorides) are prepared at the concentration of 1 mM in H_2O (or in 0.01 N HCl).

For the HPLC procedure, a stock solution 1 mM of 1, 7-*diaminoheptane* is also prepared. The unnatural amine 1, 7-diaminoheptane is used in HPLC as internal reference compound because it resolves well from derivatives of endogenous amines, elutes near amines of interest, and it is stable under storage conditions.

- **Working standard solution for HPLC** (0.05 mM) are prepared by 1: 20 dilution of the stock [i.e., 50 ml 1 mM stock + 950 ml H_2O (or 0.01 N HCl)].
- **Working standard solutions for TLC** will be the 1 mM stock solutions, undiluted.

Standards are stable for at least 2 months if stored at –20°C in plastic tubes. Plastic containers (e.g., Eppendorf tubes) must be used for storage because polyamines adsorb to the surface of glass.

Derivation by Dansylation

Dansyl chloride is the most widely used reagent for derivation of polyamines prior to separation by TLC or HPLC (Smith and Davies, 1987).

Dansylation of polyamines is carried out after Seiler (1971), with some modifications (Marce *et al.*, 1995), as follows : Aliquots (200 ml) of the 3 fractions : S, SH and PH, are transferred to reactionm tubes. Then, 400 µl of dansyl chloride (5 mg . ml^{-1} in acetone) are added, besides 200 µl of a saturated Na_2CO_3 solution (in order to produce an alkaline medium for labelling, which is optimal at pH of 9.5 to 10.5 (Smith and Davies, 1987).

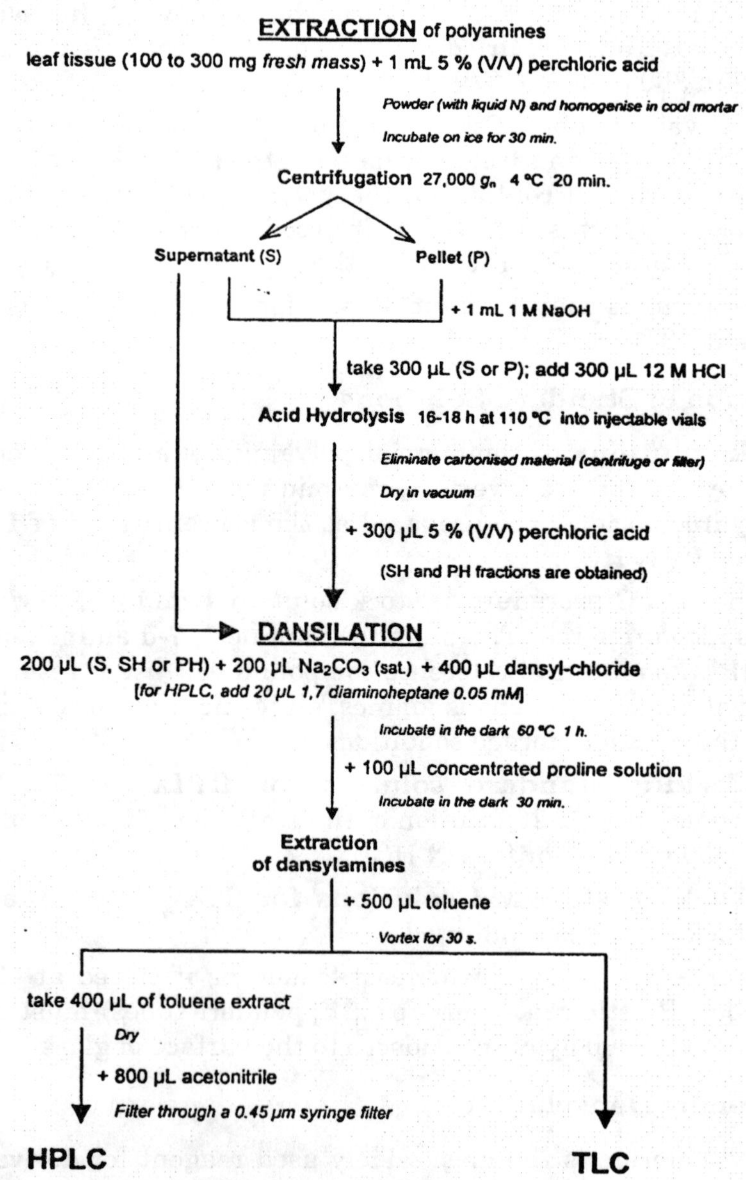

Figure 1. Extraction and dansylation procedures for the analysis of polyamines by TLC and HPLC (Marce *et al.*, 1995).

After vortexing the mixtures for a while, they are incubated in the dark overnight at room temperature (alternatively for 10 min. at 70ºC). The reaction is stopped by the addition of 100 µl of a proline solution (100 mg. ml^{-1} in H_2O) followed by incubation in the dark for 30 min. This way, the excess of dansyl chloride is removed.

Finally, dansylated polyamines are extracted from the alkaline medium with 500 µl toluene by vortexing intensively for 30 s. In this step, dansylated amino acids remain in the aqueous phase, being discarded.

Dansylation of standards is carried out in the same way described for samples considering the following:

- Polyamine standards for HPLC are prepared by mixing 40 µl of each 0.05 mM working solutions (diaminopropane, putrescine, cadaverine, 1, 7-diaminoheptane, spermidine and spermine), having a final volume of 240 µl. Moreover, 40 µl 1, 7-diaminoheptane 0.05 mM are added to each sample before dansylation, as internal reference, thus having equal experimental volumes for samples and standards.
- Polyamine standards for TLC are prepared by mixing 20 µl of each 1 mM stock solutions (diaminopropane, putrescine, cadaverine, spermidine and spermine). Final volume of 200 µl (equal to sample aliquots) is achieved by adding 100 µl H_2O (or HCl 0.1 N). Alternatively, different concentrations of standards can be prepared (e.g., 25, 20, 15, 10 µl of each 1 mM stock + 75, 100, 125, 150 µl H_2O) in order to obtain a standard curve for each amine (i.e., concentration vs. fluorescence).

Dansylation must be performed in the dark since the derivatives are light sensitive, but dim light in the laboratory is tolerable. Solution of dansyl chloride in acetone can be stored for 24 h at 4ºC in the dark. After the dansylation reaction, samples must be kept in the dark or in dark vials.

The TLC Procedure

Separation of the dansylamines is performed on a HPTLC plate of silicagel 60 A, using 4 : 1 (V/V) chloroform/triethylamine as the developing solvent system (Tiburcio *et al.,* 1985). However, a trial and-error approach would be necessary to adequate the proportions of solvents to local conditions; e.g., 5 : 1 (V/V) can be used (Slocum *et al.,* 1989).

Other solvent systems can provide accurate separation of dansylpolyamines. For example, Reggiani *et al.* (1989) and Slocum *et al.*

(1989) develop TLC in 5:4 (V/V) cyclohexane/ethyl acetate solvent system. Nontheless, with the latter solvent the diamines putrescine and cadaverine are poorly separated (Slocum *et al.,* 1989); so its use would be inadequate for cadaverine-rich species (Tiburcio *et al.,* 1985).

Chromatography is developed under penumbra, into a closed TLC chamber. The chamber containing the solvents mixture (usually 90 to 100 ml) must be prepared some hours before chromatography in order to equilibrate the mixture and provide optimal conditions for migration and separation of dansylamines.

Samples and standards are spotted in the application zone (Fig. 2) by using a micropipet. Generally 10-40 μl of each toluene extract are applied. Large volumes must be spotted by repeated applications of small aliquots, with solvent evaporation in between increments. After application, samples are dried in a warm air stream. Then, the plate is placed vertically into the chamber, dipping the immersion zone into the developing solvent. The chamber is then closed hermetically.

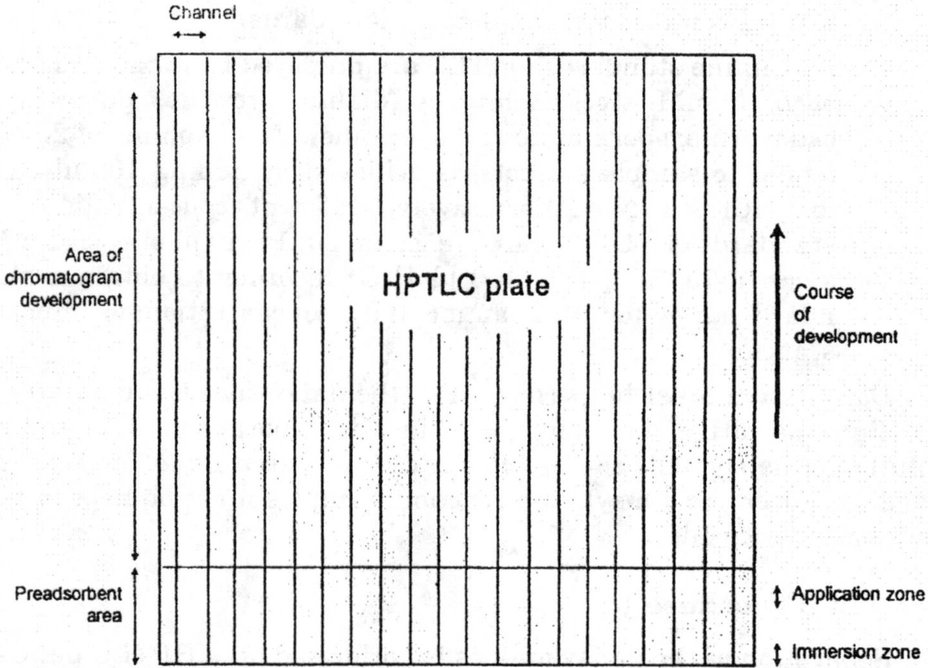

Figure 2. Representation of a HPTLC plate of silicagel with 19 separated channels (to avoid sample mixing), and a preadsorbent or concentration area provided with a thicker layer.

Times of development are different for the different solvent system. Plates are developed for approximately 1 h and 15 min. in 4 :

1 (V/V) chloroform/triethylamine; and 40 min. in 5 : 4 (V/V) cyclohexane/ethyl acetate. Time of development can also vary slightly with local conditions too, specially with room temperature.

Once developed, the plate is removed from the chamber and dried quickly with warm air (e.g., dried for 10 min. at 60º).

Spots corresponding to the different amines are visualised under UV-light, and identified by comparison to standard. When developed in 4 : 1 (V/V) chloroform/triethylamine, the migration order to the different polyamines is diaminopropane firstly separated) followed by putrescine, cadaverine, and finally, spermidine and spermine.

When developed in 5:4 (V/V) cyclohexane/ethyl acetate, the migration order of the different polyamines is spermine (firstly separated) followed by spermidine, and finally, putrescine and cadaverine (if present) poorly separated. Once visualised, the spot bounders are marked with pencil.

Quantification is carried out following the *zone elution* method. The marked silicagel zones corresponding to diaminopropane, putrescine, cadaverine, spermidine and spermine are scraped and transferred carefully to assay tubes. Dansylamines are recovered from the silica gel by eluting in 2 ml ethyl acetate. Analytes and standards are measured with fluorescence spectrophotometer with excitation at 350 nm and emission at 495 nm, using ethylacetate as blank. Sample concentrations of the different amines (expressed as nmol. g^{-1}) are determined by comparison with polyamine standards, taking into account the different dilutions made during the extraction procedure.

As an important final regard, notice that the analysed SH fraction would contain the polyamines liberated from the soluble conjugates but also the free polyamines; thus, for final quantification S fraction value must be subtracted to the SH fraction value, i.e., [SH_{real} = $SH_{measured}$ – S].

The HPLC Procedure

After the extraction of the dansylated polyamines, 400 µl of the organic phase (the toluene extract) is removed, dried (e.g., under a stream of N_2) and re-dissolved in 800 µl of acetonitrile (which is compatible with the HPLC column. Finally, the samples are passed through a 0.45 µm pore size syringe filter before injection in HPLC.

Separation of dansylpolyamines is performed according to the high-performance liguid chrmoatographic method of Marce *et al.,* (1995). The chromatographic system used consists of a reversed phase column and a mobile phase of acetonitrile and water. The separation of 1, 3-diaminopropane, putrescine, cadaverine, spermidine and spermine takes only 9 min. This method provides a good resolution between 1, 3-diaminopropane and putrescine.

The procedure is carried out as follows : Generally 10 - 20 μl of each acetonitrile extract are injected into the HPLC system. The sample is subjected to a gradient elution in acetonitrile and water, both solvents previously sonicated and filtered through a 0.45 μm filter. The initial conditions are 70% acetonitrile and 30% water pumping at a flow rate of 1.5 ml min^{-1}. The mixture is pumped for 4 min; then the concentration of acetonitrile is raised to 100%. This concentration of acetonitrile is kept constant for 4 min. and finally returned to the initial conditions.

Chromatographic separation is such a way developed in a reversed phase mode (in which the eluent strength decreases as the solvent becomes more polar) thus providing optimal separations of low molecular weight organic compounds. The proper described conditions were optimised by Marce *et al.* (1995) for the elution of spermidine and spermine, which are both strongly retained on the column.

Dansylamines are monitored with a fluorescence detector, with excitation at 252 nm and emission at 500 nm. Under the described conditions, spermidine is the last peak eluted from the column (approximately 8.5 min), and the column is then re-equilibrated in the remaining 6.5 min. The elution order of the different dansylamines is diaminopropane, putrescine, cadaverine, diaminoheptane, spermidine and spermine.

Finally, quantification of the different amines is usually performed by means of a software program coupled to the HPLC system. It consists on integrated the areas of each sample peak, compared to those of polyamine standards of known concentration. Final concentrations (expressed as nmol g^{-1}) are calculated regarding the different dilutions made during the extraction procedure.

ANALYZING PHENOLICS IN PLANT TISSUES BY HPLC

The HPLC techniques has been successfully employed in both quantitative and qualitative analysis of low molecular weight phenolics and flavonoids from plants. The polar nature of these substances makes its possible to separate them using a reversed phase chromatographic column. Due to their chemical structure, phenolics absorb electromagnetic radiation in the low ultraviolet range (220 to 183 nm). So they can be analysed by a photodiode array detector.

Extraction of Phenolics

The method used to extract phenolics is very important since all results of the analyses are affected by the extraction step. Sample management does affect phenolic extractability, so all samples should be treated similarly. An efficient optimisation method should be employed during the technique development in order to deal with the optimal process conditions for each plant species.

In the classical studies of phenolis, it was common to separate the phenolic fraction of a plant extract either by precipitation with lead acetate or by extraction into alkali or carbonate, followed by acidification (Harborne, 1989). Alternatively, the powdered plant material might be extracted using several systems in sequence, some to remove lipids and others such as ethyl acetate or ethanol to remove the phenolic fraction. However, it is much more common to make a direct extract of plant material, using boiling methanol or ethanol with fresh or dried tissue.

The extraction of phenolics is commonly accomplished by repeated maceration of crushed or ground material with a small amount of HCl (0.1 - 1.0%) in methanol or ethanol at room temperature. The addition of water (10-50%) might be advisable in some cases to get complete extraction, depending on the plant material.

Several methods can be used for extracting phenolics. The method proposed by Macheix *et al.* (1990), and optimised by Bolano *et al.* (1997) is described here because of its good extraction and repeatability. The scheme of the process can be seen in Figure 3. Start from 1 g dry weight of plant material, air dried, because phenolics are not stable to drying in an oven at high temperature. The fresh plant tissue should be previously cut into small pieces with scissors.

Use a 100 ml Erlenmeyer flask to mix plant material with 50 ml of methanol/HCl (1000 : 1, v/v) and keep it in darkness for 12 h with hand shaking every three hours. Filter the mixture, with a filter paper and a funnel, saving the filtrate in the refrigerator and repeat the process with the residual plant material for another 12 h by adding 50 ml of methanol/HCl. Filter the mixture again and store.

The combined methanolic extract (approximately 100 ml) is then completely dried with the help of a rotary evaporator under vacuum, taking into account that we will never exceed 60°C because phenolics are very sensitive to high temperature. The solid residue is brought to a final volume of 40 ml (20 + 20) of ethanol/water (2:8, v/v) and then filtered.

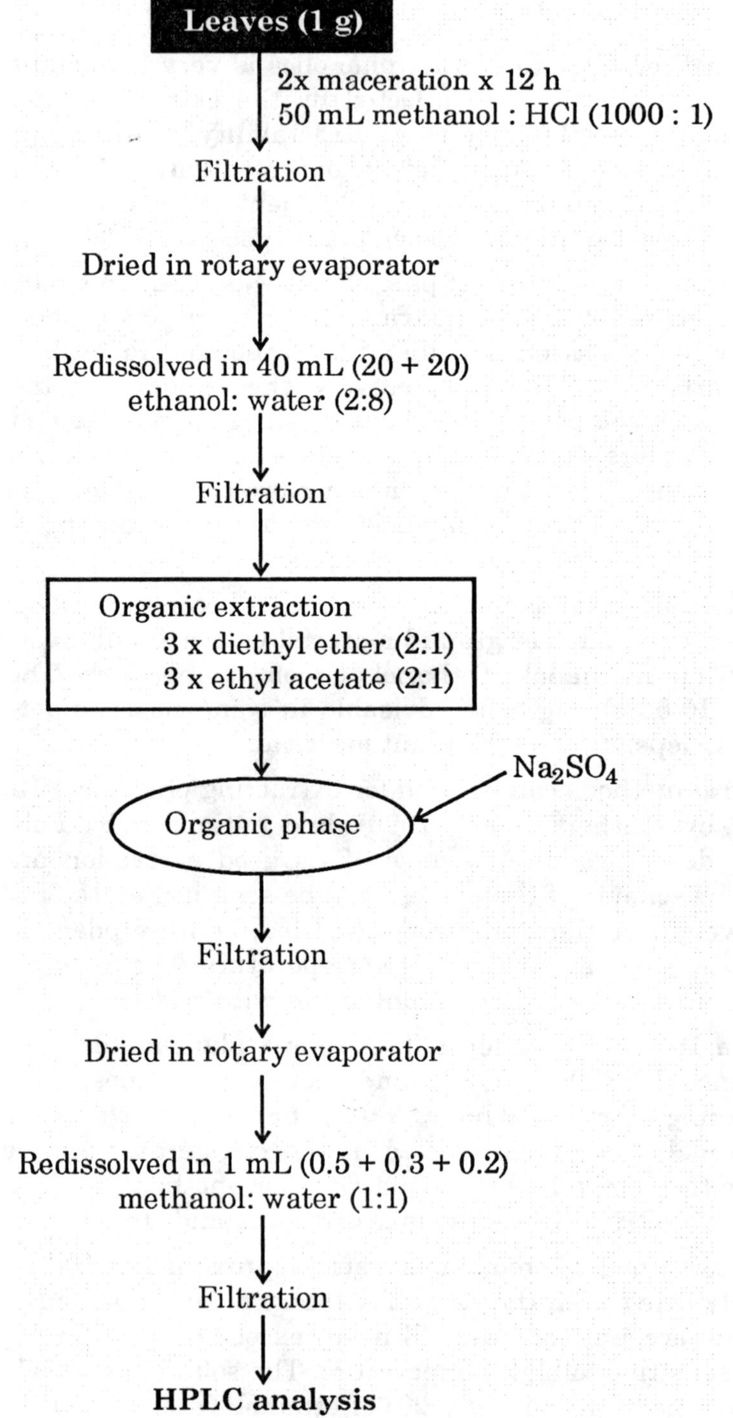

Figure 3. Extraction procedure for phenolic acids (Bolano *et al.*, 1997).

Afterwards, three sequential extractions with 20 ml of diethyl ether are performed by using an extraction funnel and apply vigorous shaking for one minute each time, waiting until the complete separation in two phases; the aqueous one, in the lower part, and the organic one, in the upper part of the funnel, where there will be the ether extracted phenolics. This phase is removed and saved, collecting the three etheral phases (approximately 60 ml) to an Erlenmeyer flask. Now complete another three sequential extractions with 20 ml of ethyl acetate on the aqueous phase, obtaining three new organic phases that will be collected and combined with the ethereal ones.

The total organic fraction obtained by this way is then dehydrated with anhydrous sodium sulphate for 30 minutes to remove the minimum water residues. Subsequently it is filtered to withdraw the sodium sulphate and evaporated to dryness in the rotary evaporator.

The final residue containing the phenolics is re-dissolved in 1 ml (0,5 + 0,3 + 0, 2) of methanol/ water (1 : 1, v/v), then filtered through a 0, 45 μm pore size nylon membrane filter and saved in refrigerator at 4°C unit its use for HPLC anlaysis.

HPLC Analysis of Phenolic Acids

The photodiode array detector has led to great improvements in HPLC analysis of phenolic compounds, because not only the retention time but also the UV spectrum can be used for identification purposes. Quantification also becomes more exact due to the simultaneous recording of different wavelengths. Generally, the most useful wavelengths to get a good resolution are 254, 280 and 320 nm for single phenolic compounds.

The most commonl chromatographic system used for the analysis of the phenolics present in the samples from plant material is built by a pump (which can be binary, ternary or quaternary), and a diode arrary detector with an attached computer. This computer is provided with a data storage system, and all the software needs to process the samples. It is also very common to use of an autoinjector, which makes the work much easier. There are many different columns one can use for analysing phenolics. They are usually reversed phase columns, C-18 with 4 or 5 μm packing, and the length uses to be around 25-30 cm.

Solvents Preparation

Many different solvents are described in the literature for analysing phenolics by means of HPLC. The most common solvents are water, methanol and acetonitrile, in different combinations and amounts depending on the phenolics and columns. It is also very common to add a small amount of a weak acid to one or more solvents. The solvents (HPLC grade) are as follows : solvent A, water/acetic acid (98:2, v/v) and solvent B, water/ methanol/ acetic acid (68 : 30:2, v/v) filtered through 0,45 μ pore size membrane filter, and degrassed for 5 minutes in ultrasonic bath (Bransonic, 1200). Gradient development and elution conditions are given in Table 1.

Table 1. Mobile phase gradient composition and flow-rate.

Time (min)	Flow-rate (ml/min)	A%	B%
0	0.8	100	0
59	0.8	20	80
69	0.8	20	80

Samples Injection

Usually, a trial run is performed on a new sample, to determine the optimal volume to be injected on the HPLC system for the best determination of the phenolic content. An amount of 50 μl is routinely used for the initial run in leaf extract samples following this protocol.

Injections of phenolic acids and flavonoids standards dissolved in 50% aqueous methanol at different concentrations are used to create a spectra library, useful to identify the peaks in the chromatogram.

Identification of Phenols

The total running time for each chromatogram is very variable, and it depends on the length of the column and the solvents used. In the routine described here, each run lasts about 69 minutes, with a flow rate of 0,8 ml/minute. Scanning a range of UV wavelengths from 220 to 400 nm usually performs the detection of phenolic peaks.

Each chromatograph has its own software, needed to identify and quantify the chemical compounds under study. However, most of them allow the user to monitor the chromatogram in different ways : for example in 3 D (all wavelengths at the same time), or selecting one or more fixed wavelengths.

It is also useful to save each chromatogram as a file, which will be managed and analysed later. Many programs allow the user to export the data of the chromatogram as ASCII codes, which is very useful for transferring the data to other computer programs.

The detection of phenolic compounds is based on their UV absorbance in the mobile phase. The photodiode array detector allows saving the complete spectrum of a compound (one peak in the chromatrogram). The computer registers its absorbance in a wide range of wavelengths, which is essential in an ulterior determination.

Different parameters in the UV spectrum from a peak will be measured with the photodiode array detector, in order to determine the identity of phenolics: retention time, position of maximum absorbance ratios and the comparison of the spectrum shape of the unknown peak with those of the pure standards previously injected and analysed.

Peaks from chromatographic runs are then assigned or identified as specific phenolic acids. Examples of chromatograms of phenolic acids and some spectra can be seen in Figure 4.

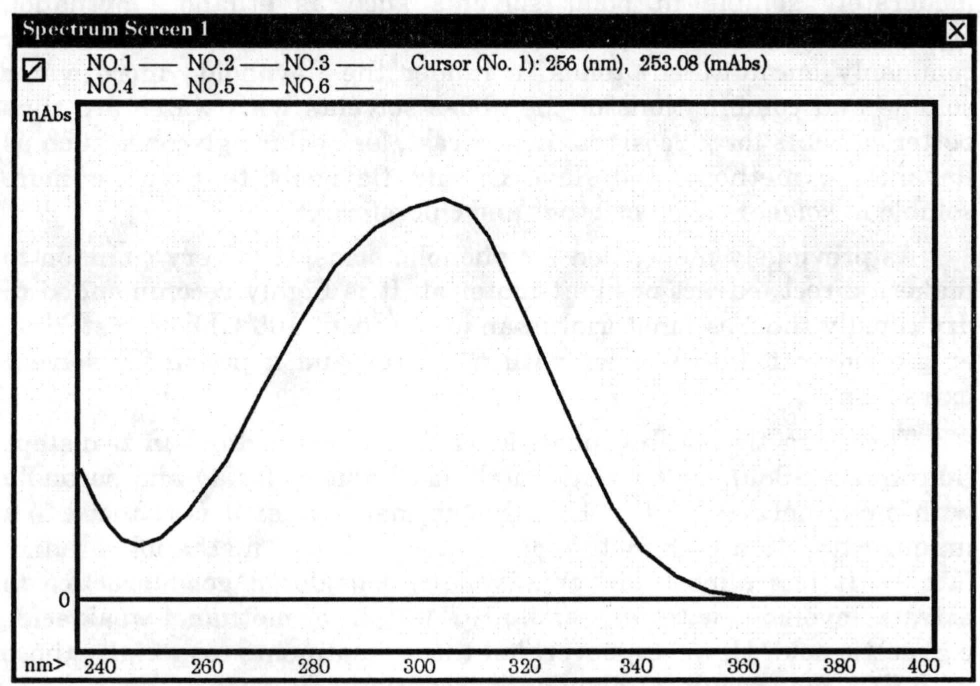

Figure 4. Spectrum of p-coumaric acid, a typical phenolic acid.

Quantification of Phenolics

To quantify the phenolics present in the extracts, each peak area is calculated from chromatograms at 280 nm, with the help of the proper data handling system, and compared with those from the standards injected at different known concentrations. A linear regression gives a slope that allows calculation of the concentrations of each phenolic compound using a single equation.

EXTRACTION OF FLAVONOIDS FROM PLANTS

Flavonoids are compounds chemically related with phenolic acids, and many of their characteristics are similar or identical. It is estimated that about 2% of all carbon photosynthesised by plants is converted into flavonoids or closely related compounds. They constitute one of the largest groups of naturally occurring phenolics, and are virtually ubiquitous in green plants. Many of them have important physiological/ecological effects, including allelopathy.

Flavonoid aglycones posses the chemical properties of phenolics, that is, they are slightly acidic and will thus dissolve in alkali (Markham, 1989). Flavonoids possessing a number of unsubstituted hydroxyl groups, or a sugar, are polar compounds. They are generally moderately soluble in polar solvents such as ethanol, methanol, acetone, water, etc. The presence of an attached sugar (which is very commonly encountered) tends to render the flavonoid more water soluble and combinations of the above solvents with water are thus better solvents for glycosides. In contrast, less polar aglycones such as flavanones, methoxylated flavones and flavonols tend to be more soluble in solvents such as ether and chloroform.

As previously mentioned for phenolic acids, it is very common to make a direct extract of plant material. It is highly recommended to dry rapidly the fresh material in an oven (about 100°C) before storing, or ground to a fine powder with a mortar and a pestle for solvent extraction.

The extraction of flavonoids is commonly carried out in two steps (Markham, 1989), firstly with methanol : water (9 : 1) and secondly with methanol : water (1 : 1), although many times it is reduced to a unique extraction with methanol : water (80:20) or methanol : ethanol (1 : 1). It has almost universally been considered good practice to extract flavonoids with methanol : HCl (99 : 1) or methanol-weak acids e.g. acetic acid. However, several of these treatments (especially those with HCl) may result in pigment degradation.

Souto *et al.* (1995) have selected a method that was used with success in extracting flavonoids and also phenolic acids (Figure 5). The process starts macerating 10 g of plant material (dried and powdered) in 300 ml of methanol : water (80 : 20) for 24 hours at room temperature and constant shaking. Filteration to separate extract from plant material can be carried out rapidly using a Buchner funnel with filter paper (Reigusa Roger, 2001).

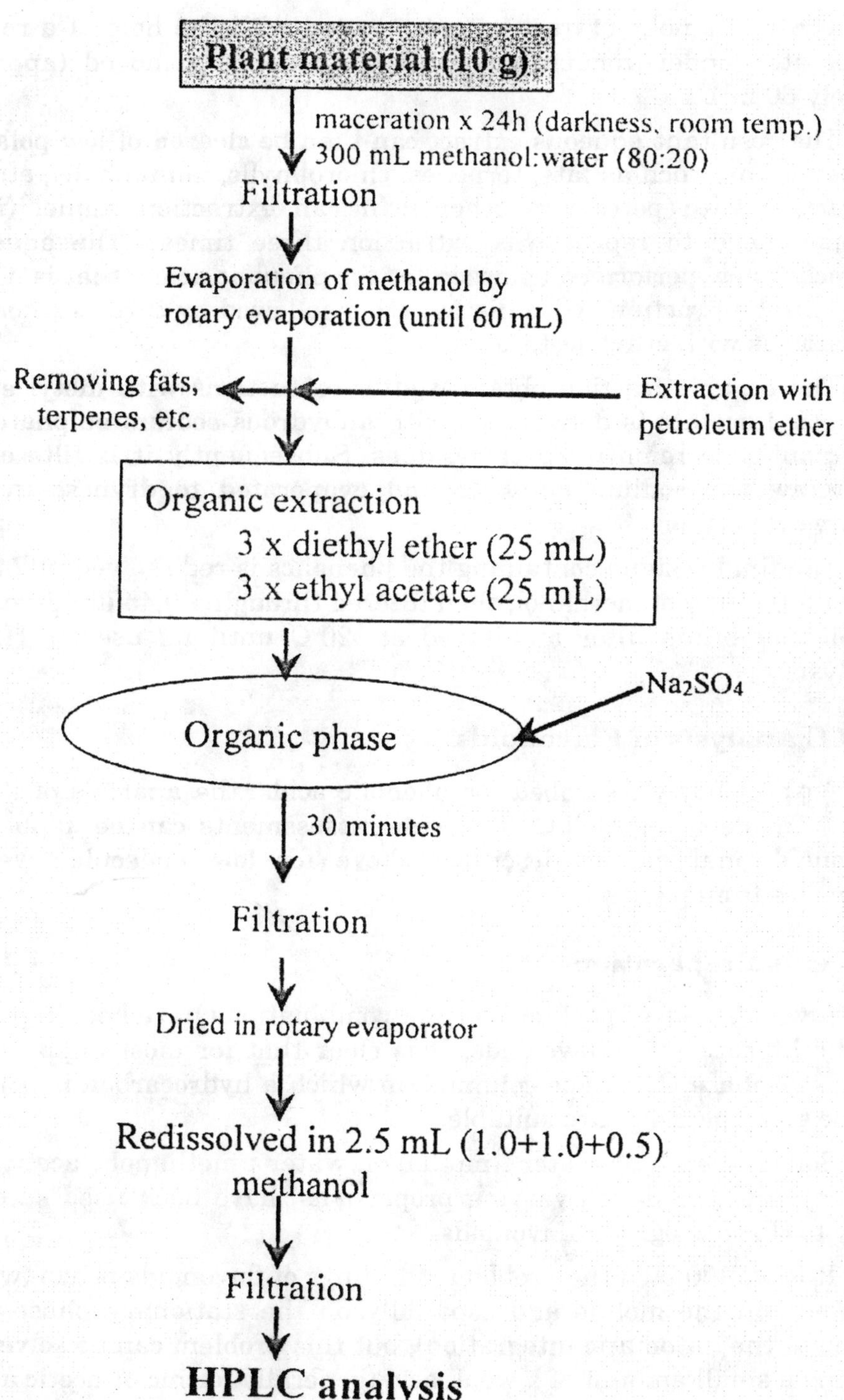

Figure 5. Extraction procedure for flavonoids.

The methanolic extract is concentrated with the help of a rotary evaporator under vacuum unitl all methanol is removed (approximately 60 ml).

The resultant aqueous extract can then be cleared of low polarity contaminants such as fats, terpenes, chlorophylls, xantophylls, etc. by extraction with petroleum ether using an extraction funnel (it is recommended to repeat this extraction three times). The aqueous extract is now performed as described for phenolic acids, that is, three sequential extractions with diethyl ether followed by three sequential extractions with ethyl acetate.

The organic fraction obtained after extractions with dietyl ether and ethyl acetate is dehydrated with anhydrous sodium sulphate for 30 minutes to remove water residues. Subsequently it is filtered to withdraw the sodium sulphate and evaporated to dryness in the rotary evaporator.

The final residue containing the phenolics is redissolved in 2.5 ml (1.0 + 1.0 + 0.5) of methanol, then filtered through a 0,45 μm pore size nylon membrane filter and saved at -20ºC until its use for HPLC analysis.

HPLC Analysis of Flavonoids

As previously described for phenolic acids, the analysis of flavnoids is made using HPLC. The same assessments can be applied to flavonoid analysis, as described above for low molecular weight phenolics from plants.

Solvents Preparation

A wide range of packing/ solvent combinations have been reported in the literature for flavonoids. It is clear that for most applications reversed phase C-18 type columns (in which a hydrocarbon is bonded to the silica packing) are suitable.

Solvents such as water : methanol, water : methanol : acetic and water : acetonitrile (in varying proportions) have been used successfully to chromatograph flavonoids.

It is well known the problem of tailings of flavonoids peaks (which depends on the mobile and especially on the stationary phase, and disturbs the automatic integration), but this problem can be solved by adding a small amount of a weak acid (generally formic or acetic acids, or even orthophosphoric acid) into the solvents.

The solvents (HPLC grade) are as follows : solvent A, methanol : orthophosphoric acid (999 : 1) and solvent B, water : phosphoric acid (999:1) filtered through a 0.45 μm pore size nylon membrane filter and degassed by means of a vacuum filter degassfier.

Samples Injection

This procedure is identical to that of phenolic acids.

Identification of Flavonoids

The flavonoid spectrum typically consists of two absorption maxima in the ranges 240-285 nm (band II) and 300-550 (band I). The precise position and relative intensities of these maximua give valuable information on the nature of the flavonoid and its oxygenation pattern. Characteristic features of these spectra are the low relative intensities of band I in dihydroflavones, and isoflavones, and the long wavelength position of band I in the spectra of chalcones, aurones and anthocyanins. Some examples are presented in Figures 6 and 7.

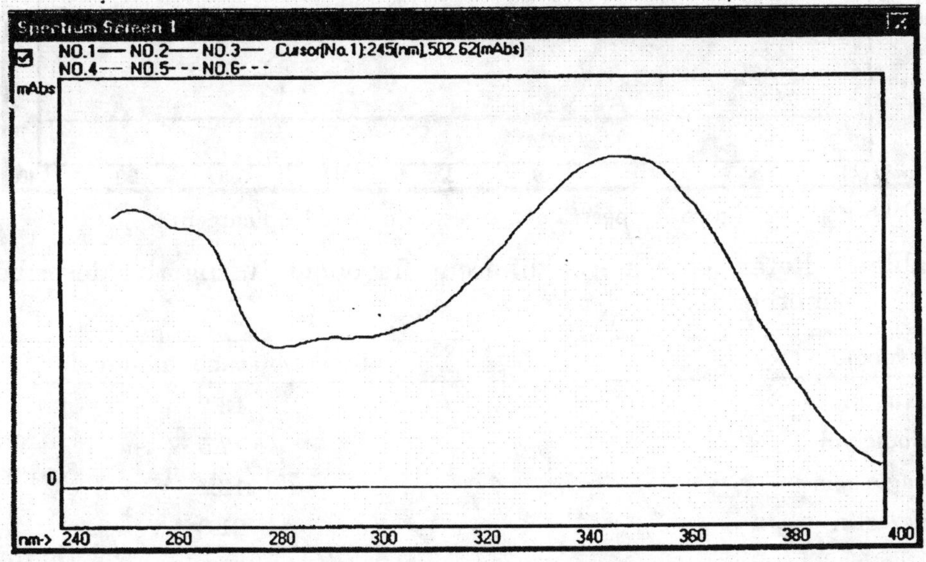

Figure 6. Spectra of a typical flavonoid luteolin.

The detection of flavonoid peaks is usually performed by scanning a range of UV wavelengths from 220 to 400 nm. The total running time for each chromatogram is about 55 minutes. The identification of each flavonoid is based on the comparison of its spectra and retention time with those of authentic standards. The identification of the peaks is made with an automated library search system from the HPLC software. This application includes the peak purity (comparison of the spectra upslope, apex, downslope and correlation between the peak at different wavelength) and compares the spectra of the peaks in the chromatogram with the spectra of the reference compounds stored in the library. This comparison also considers the retention time of the peaks in the chromatogram and in the library. In Table 2 some retention times of flavonoids are presented.

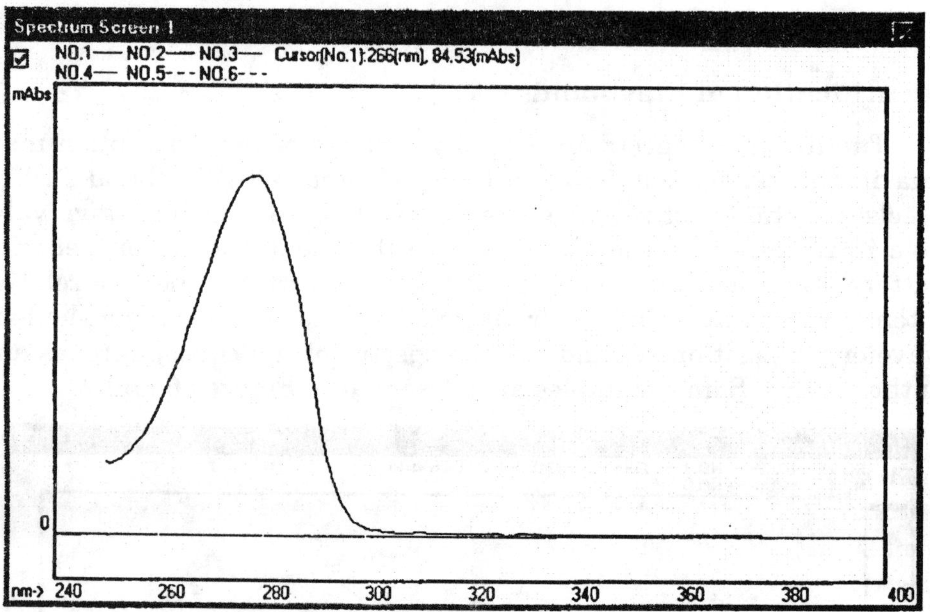

Figure 7. Spectra of a typical flavonoid, epicatechi.

Table 2. Retention time of different flavonoids using the described method.

Flavonoid	Retention time (minutes)
Taxifolin	13.2
Eriodictyol	17.5
Ellagic acid	18.1
Quercetin	21.8
Luteolin	23.7
Kaempferol	25.1
Apigenin	26.7

Quantification of Flavonoids

Quantification of flavonoids follows the same procedure explained for phenolic acids.

REFERENCES

Polyamines

Bagni, N. 1991. Transport and ubcellular compartmentation of polyamines in plants. In: *Polyamines as Modulators of Plant Development.* International Meetings on Biology. Madrid, Espana : Fundacion Juan March.

Corbin, J.L., Marsh B.H. and Peters G.A. 1989. An improved method for analysis of polyamines in plant tissue by precolumn derivatisation with o-phthaladehyde and separation by High Performance Liquid Chromatography, *Plant Physiology* 90 : 434-439.

Flores, H.E. 1990. Polyamines and plant stress. In: *Stress Responses in Plants. Adaptation and Acclimination Mechanisms.* R.G. Alscher, J.R. Cumming, eds. New York, USA : Wiley-Liss Inc.

Flores, H.E. 1991. Changes in polyamine metabolism in response to abiotic stress. In: *Biochemistry and Physiology of Polyamines in Plants.* R.D. Slocum, H.E. Flores, eds. Boca Raton, Florida : CRC Press.

Flores, H.E. and Galston, A.W. 1982. Analysis of polyamines in higher plants by high performance liquid chromatography. *Plant Physiology* 69 : 701-706.

Fried, B. and Sherma, J. 1992. *Thin-Layer Chromatography.* New York, USA: Marcel Dekker Inc.

Hahn-Deinstrop, E. 2000. *Applied Thin-Layer Chromatography. Best Practice and Avoidance of Mistakes.* Weinheim, Germany: Wiley-VCH.

Marce, M., Brown, D.S., Capell, T., Figureas, X. and Tiburcio, A.F. 1995. Rapid high-perfrmance liquid chromatographic method for the quantitation of polyamines as their dansyl derivatives: application to plant and animal tissues. *Journal of Chromatography* 666: 329-335.

Martin-Tanguy, J. 1997. Conjugated polyamines and reproductive development: biochemical, molecular and physiological approaches. *Physiologia Plantarum* 100: 675-688.

Poole, C.F. and Poole, S.K.1994. Instrumental thin layer chromatolography *Analytical Chemistry* 66: 27-37.

Rajam, M.V. 1997. Polyamines. In: *Plant Ecophysiology* M.N.V. Prasad, ed. New York, USA: John Willey.

Reggiani, R., Hochkoeppler, A. and Bertani, A. 1989. Polyamines in rice seedlings under oxygen-deficit stress. *Plant Physiology* 91: 1197-1201.

Robinson, J.W. 1995. *Undergraduate Instrumental Analysis.* New York, USA, Marcel Dekker.

Seiler, N. 1971. Identification and quantitation of amines by thin-layer chromatography. *Journal of Chromatography* 63: 97-112.

Sherma, J. 1990. Basic Techniques: Materials, and Apparatus. In *Handbook of Thin Layer Chromatography.* J. Sherma and B. Fried, eds. New York, USA: Marcel Dekker Inc.

Slocum, R.D. and Flores, H.E. 1991. In : *Biochemistry and Physiology of Polyamines in Plants.* R.D. Slocum and H.E. Flores, eds. Boca Raton, Florida: CRC Press.

Slocum, R.D., Flores, H.E., Galston, A.W. and Weinstein, L.H. 1989. Improved method for HPLC analysis of polyamines, agmatine and aromatic monoamines in plant tissue. *Plant Physiology* 89: 512-517.

Smith. M.A. and Davies, P.J. 1987. Monitoring polyamins in plant tissue in high performance liquid chromatography. In: *High Performance Liquid Chromatography in Plant Sciences* (eds.) H.F. Linskers and J.F. Jackson, Springer-Verlag, Germany.

Tiburico, A.F, Kaur-Sawhney, R. and Galston A.W. 1985. Correlation between polyamines and pyrrolidine alkaloids in developing tobacco cells. *Plant Physiology* 78: 323-326.

Waldren, H.J.P. and Genus, J.M.C. 1987. High speed analysis of polyamines in plant tissues. *Plant Physiology* 83: 232-234.

Phenolics

Bolano, J.C., Gonzalez, L. and Souto, X.C. 1997. Analisis de Compuestos Fenolicos de Bajo Peso Molecular Inducidos por Condiciones de Estres, mediante HPLC. In *Manual de Tecnicas en Ecofisiologia Vegetal.* N. Pedrol, M.J. Reigosa, eds. Vigo, Spain: Gamesal.

Harborne, J.B. 1989. General Procedures and Measurements of Total Phenolics. In *Methods in Plant Biochemistry,* Vol. 1, *Plant Phenolics.* P.M. Dey and J.B. Harborne, eds. London: Academic Press.

Macheix, J.J., Fleuriet, A. and Billot 1990. *Fruit Phenolics.* Boca Raton, Florida, USA: CRC Press.

Markham, K.R. 1989. Flavones, Flavonols and their Glycosides. In: *Methods in Plant Biochemistry.* Vol. 1, *Plant Phenolics.* P.M. Dey and J.B. Harborne, eds. London Academic Press.

Reigusa Roger, M.J. 2001. HPLC techniques for phenolics. In: *Handbook of Plant Ecophysiology Techniques.* (eds.) M.J. Reigosa Roger. pp. 251-282. Kluwer Academic Publishers, Netherlands.

Souto, X.C., Gonzalez, L. and Reigusa, M.J. 1995. Allelopathy in forest environment in Galicia, NW spain. *Allelopathy Journal* 2: 67-78.

CHAPTER - 20

HISTOCHEMISTRY OF ENZYMES AND IMPORTANT METABOLITES

INTRODUCTION

In vivo and *in vitro* enzyme assays are done to determine the enzyme activity, but these methods do not give any information on the inter or intra cellular distribution of enzymes. It is necessary to find the inter and intra cellular location of these enzymes and regulating elements by using histochemical methods. To understand cellular metabolism the knowledge of kinetic and regulatory behaviour of enzymes, the determination of their intra and intercellular localization is important. Recent enzyme localization methods are (a) immunofluorescent labelling, (b) *in situ* hybridization and (c) autoradiography.

These findings must be followed by histochemical localization of their end products or intermediates to elucidate and confirm the intracellular regulation of plant metabolism.

PRELIMINARY REQUIREMENTS

The following preliminary requirements must be fulfilled to study the histochemistry of enzymes :

Fixation

Ideally, enzyme cytochemistry should be performed on unfixed cells obtained from razor sections of fresh tissues. However, razor section can lead to a loss of organelles and soluble enzymes from cut cell surfaces. Under certain circumstances, unfixed frozen sections may be used in optical microscope level, but electron microscope cytochemistry methods require the tissue to be fixed (Hall *et al.*, 1977). When fixatives are employed, the Cytochemist is faced with a situation of compromise between the best cytological preservation possible and the maximum activity of the enzyme under study. Specific fixatives are used for each group of enzymes.

EFFECTS OF ENZYME REAGENTS

For optical microscopy, the product need to be coloured, while in electron microscopy, the product need to be electron dense.

The principles involved in enzyme biochemistry (Dixon and Webb, 1979) are also applicable to enzyme cytochemistry. Thus, the enzyme reaction will be affected by temperature, pH and concen-trations of substrate and trapping agents. The reactants affect the pH of a solution in which they are present, hence, there is need to buffer the solution to maintain optimal pH.

The reaction rate is also important, since longer the incubation period, greater the possibility of the diffusion of the product. Elevation of the temperature from 27°C to 37°C may aid the process by decreasing the time of incubation. Conversely, if the cells react too heavily at room temperature in a very short incubation period, the reaction can be slowed by decreasing the temperature to as low as 4°C.

The problem of the mobility of the end product is a real one, and it is important to know if the product is substantive, i.e., will bind to protein and so stay at its site of origin or is water or lipid soluble.

Controls

If the cytological preservation is good, the reaction is understood and the end product is reliably located, then it is of importance to test the specificity of the reaction. The main controls used are : (a) incubation in the absence of the substrate, (b) treatment of the sections or small blocks of tissue in distilled water at 0°C for 5-10 min to inactivate the enzyme activity and (c) addition of specific inhibitors or known activators to the full incubation medium.

Extrapolation of Methods

Many workers find that a specific method works well with a given plant tissues. However, this does not mean that the method cannot be used for another tissue of the same species (e.g., from root apex to leaf) or to the same tissue of another species (e.g., from root tips of *V. faba* to those of *A. cepa*), and it is essential to check out the methodology on a new tissue.

ENZYME HISTOCHEMISTRY PROTOCOLS

The methods of light microscopic localization, immuno fluorescent labelling and *in situ* hybridization techniques are discussed for important enzymes (Jense, 1962; Malik and Singh, 1980).

(1) Cytochrome Oxidase (E.C.1.9.3.1)

Cytochrome oxidase is the terminal oxidase in the electron transport pathway that results in the final transfer of electrons from the sequence of cytochrome carriers to molecular oxygen. Seligman *et al.* (1968) showed that 3, 3'-diaminobenzidine (DAB) is useful to demonstrate cytochrome oxidase activity in electron microscope studies.

Procedure

Tissue : These may be fixed for 2 h at 4°C in buffered 2-4% depolymerized paraformaldehyde or 0.1-0.2% glutaraldehyde. Tissue should be washed for several hours in the buffers at 4°C.

Incubation Media

Test solution : Nine milliliters of 0.05 M phosphate buffer (pH 7.4) containing 4 mg DAB, 20 μg cytochrome C (type II Sigma) and 750 mg sucrose.

Control solution : Full incubation medium omitting cytochrome C, or including sodium azide or potassium cyanide.

Results : Deposits of brown-blck (electron dense) material found at the sites of cytochrome oxidase activity. Controls should have no deposits.

(2) Peroxidases (E.C. 1.11.1.7)

These enzymes are present in plant tissues and catalyse the reduction of hydrogen peroxide to water by exploiting electron donors other then hydrogen peroxide as under :

$$XH_2 + H_2O_2 - \xrightarrow{\text{Peroxidase}} 2H_2O + X$$

Studies on the peroxidase have shown that a broad range of electron donors are involved and XH_2 could be from phenols and alcohols.

Incubation Media

Test solution: Ten ml of 0.05 M Tris-HCl buffer at pH 7.6 containing 5 mg DAB and 0.2 ml of freshly diluted 1% H_2O_2 (from 30% stock just prior to use). It may be necessary to warm the solution to increase the solubility of the DAB. This should be cooled and filtered prior to the addition of the H_2O_2.

Control solutions : The test solution without H_2O_2 may be used with or without the addition of 0.01% catalase to eliminate any endogenous H_2O_2. Heat treated sections (15 min in boiling water) may react in the test solution.

Results : Brown-black (electron opaque) deponsits are found at sites of peroxidase activity, which should be absent from the controls. Control for the adsorption of DAB may be by incubation of sections in the test solution lacking H_2O_2, washing and treating with the buffer containing 2 x 10^{-3} M potassium ferricyanide to oxidize the DAB.

(3) Polyphenol Oxidases

This group of enzymes is responsible for the oxidation of polyphenols to quinines, e.g. the browning reaction in apples. They are also known as phenolase, catechol oxidase and tyrosinase. Localization in the plants has been done with the dopa test in which dihydro-phenylalanine (dopa) is oxidized to dopa quinine.

Incubation Media

Test solution : Ten milliliters of 0.067 M Sorensen's phosphate buffer at pH 7.0 containing 50 mg D-1,3,4 dihydrophenylalanine.

Control solution : Test solution containing 0.02 M sodium diethyl dithiocarbamate.

Procedure : Incubate sections at 37°C up to 1 h.

Results : Pigments are visible at the sites of polyphenol oxidase activity, but lacking in controls.

(4) Succinate Dehydrogenase (E.C. 1.3.99.1)

Succinate dehydrogenase is the only dehydrogenase that can be directly demonstrated with tetrazolium salts and is not dependent upon NAD(P).

Incubation Media

Test solution : 0.05 M phosphate buffer at pH 7.8 containing 40 mM sodium succinate, 0.5 mg (ml of NBT) and 22% polyvinyl alcohol (BOS/140).

Control solution : (a) As test solution but containing 0.01 M sodium malonate to inhibit succinate dehydrogenase. Check that the solution pH is 7.8. (b) As test solution with the omission of sodium succinate. Oxygen free nitrogen should be bubbled through the solution for 10 min to create the anaerobic conditions.

Incubation Period

This will vary according to the tissue. Sections should be tested for periods of time such as 5, 15, 30, 45, 60 and 90 min. Incubations should be performed at 37°C in micro cells.

Results: Sites of succinate dehydrogenase activity are revealed by the deposition of formazon, whereas in the presence of malonate, there will be no formazon produced. Any formazon produced in the absence of substrate will be due to endogenous deydrogenase activities.

ELECTRON MICROSCOPY

Test solution (after Ekes, 1971) : 0.8 ml of 0.1 M Sorensen's phosphate buffer at pH 7.6 containing 0.7 ml of 1 M aqueous sodium succinate solution, 0.15 ml of 50 mM aqueous potassium ferricyanide solution, 0.35 ml of 0.3 M aqueous copper sulphate solution and 3 ml of 0.5 M sodium potassium tartarate in phosphate buffer at pH 7.6.

Controls : Incubation of the unfixed tissues in the test solution lacking sodium succinate or containing 140 mM sodium malonate.

Procedure

(i) Incubate unfixed tissues for 20-30 min at room temperature.

(ii) Fix tissues in 2% glutaraldehyde in 0.5 M cacodylate buffer pH 7.2 for 2 h at 4°C.

(iii) Post fix in O_sO_4 and embed by standard methods for electron microscopy.

Results

Electron opaque deposits are found in mitochondria at the sites of succinate dehydrogenase activity, which are absent from controls.

(5) Malate Deydrogenase (E.C. 1.1.1.38)

Use fresh hand cut or cryostat cut sections :

Reaction : Incubation for 10-15 min at 35°C in a freshly prepared reaction mixture containing :

Triethanolamine buffer, 0.05 M, pH 7.6	10 ml
EDTA, Disodium salt	16 mg
Sodium malate (Na_3 salt)	36 mg
NAD	10 mg

Nitro-BT	13 mg
Phenazine methosulfate (PMS) (15 mg/ml)	0.10 ml

After incubation wash in water and mount.

Indication : Sites of enzyme activity will turn *blue-violet.*

Control : Incubate in reaction mixture minus sodium malate or alternately use heat inactivated tissues (3 to 5 min in boiling water).

Note : If INT is used as tetrazolium salt then indication will be the appearance of *purple-red* colour.

(6) Glutamate Dehydrogenase (E.C. 1.4.1.2)

Use fresh hand-cut or cryostat cut sections :

Reactions : Incubate for 15-20 mins at 30°C in reaction mixture comprising :

Tris-HCl buffer, 0.05 M, pH 7.6	10 ml
EDTA (Disodium salt)	16 mg
Glutamic acid (monosodium salt)	42 mg
Nicotinamide-dinucleotide (NAD)	10 mg
Nitro-BT	13 mg
Phenazine-methosulphate (PMS) (15 mg/ml)	0.10 ml

Wash section in water and mount.

Indication : Appearance of *blue violet colour* shows the presence of this enzyme.

Control : For control omit the substrate from reaction mixture or use heat-inactivated tissue sections.

Note : INT can be used in place of Nitro-BT.

(7) Glyceraldehyde-3-Phosphate Dehydrogenase (E.C.1.2.1.12)

Use fresh hand cut or cryostat cut sections :

Reaction : Incubate the sections for about 10-15 min at 30°C in following staining mixture (freshly prepared).

Triethanolamine buffer, 0.05 M, pH 7.6	10 ml
EDTA, Disodium salt	16 mg
Sodium arsenate	3 mg

NAD	10 mg
Glyceraldehyde-3-phosphate	17 mg
Nitro-Blue Tetrazolium (or INT)	13 mg
Phenazine methosulphate (PMS) (16 mg/ml)	0.10 ml
Glutathione, reduced	1 mg

After incubation wash in water and mount.

Indication : Sites of enzyme activity will turn *blue-violet* (or *purple-red*).

Control : Omit the substrate in the reaction mixture or use heat inactivated tissue.

(8) Glucose-6-Phosphate Dehydrogenase (E.C.1.1.1.49)

Use fresh unfixed hand cut or cryostat cut sections :

Reaction : Incubate for 10 to 20 min at 35ºC in a reaction mixture comprising :

Triethanolamine buffer, 0.05 M, pH 7.6	10 ml
EDTA, Disodium salt	16 mg
Glucose-6-phosphate (monosdodium salt)	42 mg
NADP	11 mg
Nitro-BT	13 mg
Phenazine-methosulphate (PMS) (15 mg/ml)	0.10 ml

After incubation wash in water and mount.

Indication : Presence of *blue-violet* colour will show the presence of this enzyme.

Control : Omit substrate in the reaction mixture or use heat inactivated tissue (3 to 5 min in boiling water).

Note : Indication will be of *purple-red* colour, if INT is used as tetrazolium salt.

(9) NADP-Isocitrate Dehydrogenase (E.C. 1.1.1.42)

Use fresh cut or cryostat cut sections :

Reaction : Incubate for 20 to 30 min at 35ºC in freshly prepared reaction mixture containing :

Triethanolamine buffer, 0.05 M, pH 7.6	10 ml
EDTA, Disodium salt	16 mg
$MgSO_4 . 7H_2O$	2 mg

DL-Isocitric acid (trisodium salt)	42 mg
NADP	11 mg
Nitro-BT	13 mg
Phenazine methosulphate (PMS) (15 mg/ml)	0.10 ml

Wash the sections in water and mount

Indication : Appearance of *blue-violet* colour shows the presence of this enzyme.

Control : Two types of controls can be maintained :

(a) Omit the substrate in the reaction mixture.

(b) Use heat inactivated tissue (3 to 5 min. in boiling water).

(10) NAD-Isocitrate Dehydrogenase (E.C.1.1.1.42)

(Modified method)

Use fresh hand cut or cryostat cut sections :

Reaction : Incubate for 30 to 40 min at 35°C in a freshly prepared reaction mixture containing :

Triethanolamine buffer, 0.05 M, pH 7.6	1 ml
EDTA, Disodium salt	16 mg
$MgSO_4 . 7 H_2O$	2 mg
DL-Isocitric acid (trisodium salt)	42 mg
Glutathion (reduced)	1 mg
NAD	10 mg
ADP	13 mg
Phenazine methosulfate (PMS) (15 mg/ml)	0.10 ml

After incubation wash in water and mount.

Indication : Apperance of *blue-violet* colour shows the presence of NAD-isocitrate dehydrogenase activity.

Control : Use two types of controls :

(a) Omit the substrate in the reaction mixture.

(b) Use heat inactivated tissue (3 to 5 min in boiling water).

(11) Leucine Aminopeptidase

Use fresh hand cut or cryostat cut sections :

Reaction : Incubate for 20 to 30 min at 35°C in a freshly prepared reaction mixture containing :

Tris maleate buffer, 0.1 M, pH 6.5	10 ml
L-leucyl-β-naphthylamide hydrochloride	4 mg
Fast blue B	5 mg

After incubation, wash the sections in distilled water and mount.

Indication : The reaction product is *reddish brown.*

Control : Use reaction mixture from which substrate is omitted.

Note : In place of fast blue B, other diazonium salts like fast blue BB or fast garnet GBC salt can also be used.

Alanine-β-naphthylamide or other suitable aminoacyl-β-naphthylamides can be used in place of leucyl-β-naphthylamide. The use of these alternative substrates can depict the localization of such aminopeptidases which can preferentially hydrolyse such chromogenic substrates.

(12) Glycosidases (indigogenic method)

A : Indoxyl method

Reaction: Incubate sections for 1-5 h at 35° in the following reaction mixture:

5-Bromo-4-chloro-3-indoxyl-glucoside	3 mg
Dissolve in N, N-dimethylformamide	0.3 ml
0.1 M citric acid phosphate buffer, pH 5.0	7.0 ml
(a) 65% Potassium ferricyanide	0.5 ml
(b) 11% Potassium ferrocyanide	0.5 ml

Mix well, filter it

After incubation, mount in glycerol gelatin.

Indication : Enzyme-active sites are stained *turquoise.*

Control : Use heat-killed control.

Note : Similar method can be used for α-glucosidase, α-and-β-galactosidases by using appropriate indoxyl substrates.

B : Azo indoxyl method

Reaction : Incubate sections for 1-3 h at 35°C in the following reaction mixture :

5-Bromo-4-chloro-3-indoxyl-b-glucoside	4 mg
Dissolve in N, N-dimethyl formamide	0.2 ml

Buffered hexazonium p-rosaniline or fast blue B (made up of 9.9 ml, 0.1 M citrate phosphate buffer, pH 5.5 and 0.1 ml hexazonium p-rosaniline or 8 mg fast blue B; in the case of p-rosaniline adjust pH with NaOH to 4.5). Mix well and filter.

Indication : Enzyme active sites are stained *reddish-brown.*

Note : Because very low concentrations of the coupling agent are employed, the inhibition of the enzyme is negligible. All indoxyl molecules released are converted to the azo-dye, and no loss of indoxyl occurs as is the case in the indigogenic procedure.

(13) Acid-Phosphase (E.C.3.1.3.2)

Use fresh hand cut or cryostat cut section :

Metal Salt Method

Reaction : Incubate for 30-40 min at 35°C in reaction mixture comprising :

Acetate buffer, 0.5 M, pH 5.0	10 ml
Sodium-β-glyceophosphate	30 mg
Lead nitrate	10 mg

After incubation rinse in distilled water twice, for 5 min each. Then transfer to 1% ammonium sulphide solution for 5 min, wash in distilled water and mount in aqueous mounting medium.

Indication : The reaction product is *brown-black.*

Control : The reaction without substrate should be negative.

(14) Alkaline Phosphatase (E.C. 3.1.3.1)

Use fresh hand cut or cryostat sections :

Metal salt method

Reaction : Incubate for about 30 min at 37°C in following reaction mixture :

Distilled water	10.0 ml
Sodium barbiturate	50 mg
Calcium chloride ($CaCl_2$. 6 H_2O)	80 mg
Sodium-β-glycerophosphate	60 mg
Magnesium sulfate ($MgSO_4$. $7H_2O$)	10 mg

Then treat the sections successively with the followings for the time specified.

(i) Distilled water	3 min
(ii) Cobalt nitrate (2%)	5 min
(iii) Distilled water	2 min
(iv) Ammonium sulphide (2%)	5 min

Wash in distilled water twice for 2 min each.

Indication : The reaction product is *brown-black.*

Control : Incubate in reaction mixture omitting substrate (N-β-glycerophosphate).

(15) Glucose-6-Phosphatase (E.C.3.1.3.9)

Use fresh unfixed hand cut or cryostat cut sections :

Reaction : Incubate at 32° to 35°C in the following freshly prepared incubation medium :

Acetate buffer, 0.1 M, pH 6.5	20 ml
Glucose-6-phosphate (monosodium salt)	13 mg
Lead nitrate (0.1 M)	1 ml

Wash in distilled water several times and then immerse in 0.5% (v/v) ammonium sulphide solution for one min. Then again wash in distilled water and mount.

Indication : Sites of glucose-6-phosphatase activity are denoted by *brown black* percipitate of lead sulphide.

Control : Use heat inactivated tissues.

Note : To test the chemical specificity of the enzyme reaction, replace glucose-6-phosphate with 12 mg sodium-β-glycerophosphate.

(16) Adenosine Triphosphatase (E.C.3.6.1.3)

Use fresh or fixed hand cut or cryostat cut sections:

Reaction: Incubate sections for 20 to 40 min at 30°C in reaction mixture containing:

Tris maleate buffer, 0.05 M, pH 7.0	45 ml
Adenosine triphosphate (Disodium salt)	25 mg
Lead nitrate	50 mg
Magnesium nitrate (2.5%)	5 ml

After incubation, rinse the sections in distilled water repeatedly then immerse in 0.5% ammonium sulphide solution for one min. Wash sections in distilled water and mount.

Indication: Site of ATP-ase activity will be indicated by a *brown black* precipitate of lead sulphide.

Control: Use two types of controls.

(a) Incubate sections in medium in which ATP is omitted.

(b) Use normal incubation medium but use sections which had been placed in boiling water for 2-3 minutes.

Note: For checking the specificity of enzyme, substitute ATP in the reaction mixture by the followings:

Adenosine - diphosphate

Adenosine-5' - monophosphate

Sodium - β-glycerophosphate

Inosine - triphosphate

(17) Amylase (E.C. 3.2.1.1)

For the localization of amylase activity only substrate-film method can be used and the recommended method is based on that used by Molnar and Lacroix (1972).

Preparation of substrate film

Prepare 3% suspension of soluble starch containing 1% bacto-agar and heat at 100°C in a water bath for 15 min and while still hot, filter it through sintered funnel. Deposit ten to twelve drops of this preparation on a glass-slide and then spread it over the surface. Place slide vertically on a piece of paper and drain off the excess of suspension. Allow the film to dry at room temperature and fix overnight in methanol : acetic acid : water (5 : 1 : 5 v/v). Finally, wash the film in three baths of distilled water, five min each, and dry at room temperature. Before exposure, keep the film for two hours at room temperature in a moist chamber consisting of a Petri-dish containing a wet filter-paper.

Reaction: Cut the fresh tissue sections into 0.05 M calcium-acetate buffer, pH 5.3, then lift the sections from the buffer solution with brush and drain excess of moisture by placing the tip of the brush on a blotting paper. Then carefully place the section on a humidified starch-agar film and cover with cover-slip to prevent dessication. Incubate at room temperature in the moist chamber for 30 to 45 min. After incubation, flush off tissue sections with a stream of water.

Immerse the film in methanol-acetic acid fixative for 15 min and stain by periodic acid Schiff (PAS) reaction by oxidising the film in 1%

periodic acid for 10 min and washing for 5 min in tap water. Then stain for 10 min in Schiff's reagent, wash directly in three successive baths of 0.05 M sodium metabisulphite for two min each and 10 min in running tap water. Allow to dry at room temperature and observe under the microscope.

Indication: The sites on the substrate-film on which amylase enzyme has leached out will not give *red-stain* wth Schiff's reagent.

Control: Use heat-killed sections as control.

(18) Invertase (E.C. 3.2.1.48)

This enzyme can be localized cytochemically in tissue sections by coupled glucose-oxidation using glucose-oxidase as an auxilluary enzyme.

Reaction: For invertase localization, incubate freshly cut sections in the following incubation medium which comprises:

Citrate-phosphate buffer, 0.1M, pH 6.0	4 ml
Glucose-oxidase	2-3 mg
Nitro-BT	8 mg
Phenazine methosulphate (PMS)	2 mg
Sucrose	2 ml (4%)
1% Agar or Agarose or 10% gelatin	8 ml

Mix thoroughly the reaction mixture and pour it on cover-slip. Allow to set at room temperature. Place sections on slides and gently place coverslip with reaction-mixture on them.

Incubate for several hours at 35°C in the dark.

Indication: The reaction product is *blue-violet.*

Control: For controls omit sucrose from the incubation mixture.

Note: This method does not permit intra-cellular localization because diffusion of glucose produced by the action of invertase cannot be stopped.

IMMUNOFLUORESCENT LABELLING

It is based on the specificity of antigen (enzyme) antibody reaction, combined with a labelling technique to detect the enzyme antibody complex in a tissue section either by light or electron microscopy. For immunoelectron microscopy (Hoyer and Bucana, 1982) ferritin, peroxidase and gold are most widely used markers for direct and indirect labeling.

***In situ* immunofluorescent labeling of ribulose-bisphosphate carboxylase in leaves of C_3 and C_4 plants:** The chloroplastic enzyme Rubisco (EC 4.1.1.39) (RuP_2case) may be localized by *in situ* immunofluorescent labeling in hand cut leaf sections of C_3 and C_4 species by the method of Hattersley *et al.* (1977).

Procedure

Purify RuP_2case from wheat leaves by $(NH_4)_2SO_4$ fractionation and DEAE and Sephadex chromatography. Inject 3 month old female rabbits in the vicinity of axillary and popiteal lymph nodes with 0.5 ml (10 mg RuP_2case) wheat enzyme emulsified with 0.5 ml Freund's complete adjuvant. After two booster injections, bleed the rabbits from marginal ear vein and take the serum in the usual way. Take the pre-immunization control sera from untreated rabbits.

Immerse segments of fresh blade material from young leaves in 20% ethanol for 2 h. Rinse 10 mm thick trans sections in phosphate saline (PBS) and incubate in antiserum for 1 h. Rinse again a couple of times with PBS and incubate in dark with fluorescein isothiocyanate labeled sheep anti rabbit immunoglobulin diluted 1:4 with PBS. Rinse the sections with PBS and mount in aqueous glycerol solution (50% containing 1% thymol). Keep in dark. Observe under a fluorescent microscope.

In situ Hybridization

It is new method and specific nucleic acid sequences within tissue sections or whole mount preparations of single cells, organelles or chromosomes are detected and localized. It works as follows - a labeled single stranded fragment of DNA or RNA (probe) exogenously applied, will hybridize to a complementary sequence on cellular DNA or RNA (target) forming stable hybrids. The hybrids are then located using a detection system of (visible markers), which can be coupled to the label on the probe and visualized in the microscope. If the probe is labelled with a radioisotope, the hybrid is detected by autoradiographic procedures, whereas non-radioactive labels (biotin, bromodeoxyuridine or digoxigenin) are detected by histochemical techniques (Enzyme reactions, affinitity or immunocytochemical methods). We are able to determine the location of particular DNA or RNA sequence with respect to cell morphology.

In situ hybridization methods were initially developed for animal systems, but were transposed as such to plant systems. Plant material presents morphological features which constitute drawbacks for certain technical aspects of ISH in that (a) most tissue is composed of a highly heterogenous cell population which necessitates compromises in fixation/permeabilization, (b) cell walls constitute barriers to probe pernetration adding to those of cell and organelle boundary membranes and (c) vacuoles, fragile membrane bound compartments must remain intact to avoid leakage/diffusion of *in situ* targets.

Fixation

There are two types of fixatives, (i) precipitating fixatives viz. acetic acid-alcohol mixtures and (ii) aldehydes. For plant material, aldehyde (glutaldehyde and paraformaldehye) are the fixative of choice. Pretreatment of samples with a detergent or proteinase digestion is a standard procedure in almost all protocols to increase probe penetration and accessibility. This renders the matrix formed between fixative and cell components porous, permeabilizes cell membranes and digests proteins which may targets.

Embedding

There are two ISH approaches in plant systems. The pre-embedding appraoch starts with small samples (blocks) of prefixed tissue which must be permeabilized by a "pore forming" treatment to facilitate the probe penetrationand its accessibility to tartes throughout the volume of the tissue block. Once inside the cells, the probe drives the reaction and virtually all specific targets would be bound, greatly increasing the signals. The labelled probes are detected either by autoradiography on sections or embedded in a diffusion of available marker into the tissue block. The tissue is then embedded in a resin for sectioning and observation. The post embedding method also starts out with a light fixation procedure; samples are processed, embedded in wax or resin for sectioning. Probe target formation (hybridization) and its detection take place on the sectioned material.

PROTEIN KINASES PSK_4 AND PSK_6

Protein kinases PSK_4 and PSK_6 are known to exhibit differential expression patterns in petunia flower buds throughout their development. The tissue specific expression of their mRNAs in the whole flower and in particular in the ovules and stamens can be studied by *in situ* hybridization of petunia flower buds at different stages of male and female gametogenesis.

Plant Material

Petunia flower buds from stage 1 (less than 0.5 cm long) to mature stage (one day before anthesis).

In situ Hybridization

1. Preparation and Labelling of RNA Probes

PSK_4 and PSK_6 sequences used as probes are amplified using the RT-PCR specific primers to produce fragments of 600 bp and 500 bp, respectively. Probe generated fragments are subcloned into the P Bluescript K S vector and used for the synthesis of RNA probes ^{35}S labelling is carried out by *in vitro* transcription of the linearlized plasmid as follows:

(i) Add the following to a microcentrifuge tube on ice : 1-2 µg of linearized plasmid, 1 x buffer T7 (5 x buffer T7-200 mM, Tris-HCl pH 8, 100 mM $MgCl_2$, 25 mM DTT, 500 µg/ml BSA), 1 mM r NTP, 3 µl $rUTPS^{15}$, 10 U enzyme T7RNA polymerase. Incubate for 1 h at 37°C.

(ii) Stop the reaction by adding 10 U DNase (RNase free), 10 µg of TRNA (carrier) is added prior to purification by phenol/ chloroform. Non-incorporated nucleotides eliminated by column filteration (Sephadex).

(iii) Test labeling in a scintillation counter and dilute probe to 10^5 cpm/µl.

II. Fixation and Preparation of Sections

(i) Flower buds, excised ovaries and ovules are fixed in 4% paraformaldehyde and 0.25% glutaraldeyde in 1 x PBS (130 mM NaCl, 7 mM Na_2HPO_4, 3 mM NaH_2PO_4; 0.27 mM KCl pH 7.4) for 1 h under vacuum. Renew the fixative and fix for 2-3 h at room temperature. Anters may be preembedded in an agarose block to avoid loss of pollen grains during subsequent steps. Rinse in 1 x PBS then in H_2O.

(ii) Dehydrate progressively in a graded ethanol series and then in an ethanol/xylene series 3:1 (3 volumes of ethanol for 1 volume of xylene), 1: 1, 1:3, then into pure xylene.

(iii) Infiltrate progressively with filtered paraffin (Paraclean Klinipath at 42°C infiltration for 4 to 16 h is necessary to ensure complete penetration of tissue. Infiltrate in pure paraffin at 60°C for 48 h changing paraffin baths 4 x.

(iv) Cut routine sections (7-10 µm) and place them on coated (poly L-lysine) slides.

(v) Bake the sections for 35 min (42°C), dewax in xylene and rehydrate them in a graded ethanol series.

(vi) Treat sections with Proteinase K (1 μg/ml) in a buffer of 100 mM Tris-HCl (pH 7.5), 50 mM EDTA for 30 min at 37°C. Rinse is distilled water and then in 2 x SSC for 5 min.

(vii) Dehydrate in a graded ethanol series : sections are air-dried and are ready for ISH or can be stored at -20°C.

III. Hybridization and Post Hybridization Washes

(i) Hybridize using probes diluted in hybridization buffer (0.01 M Tris-HCl, pH 7.5; 12.5% Denhardts solution, 2 x SSC, 50% formamide, 0.5% SDS, 250 mg/ml solution of sperm DNA).

(ii) Cover section with cover slips, seal Klear tak rubber cement and hybridize overnight at 42°C in a sealed box.

(iii) Remove cover slips by immersion in 4 x SSC, renew washing solution 3 x 10 min.

(iv) Incubate with RNase (Sigma) at 37°C for 30 min (50 μg/ml in 500 mM NaCl, 1 mM EDTA, 10 mM Tris-HCl, pH 7.5) followed by 4 washed in same buffer.

(v) Dehydrate sections in a decreasing gradient of ethanol and 0.3 M ammonium acetate to limit RNA loss.

(vi) Air dry sections prior to application of emulsion.

IV. Autoradiography

(i) Dip each slide in emulsion (LM-1 emulsion Amersham), avoiding air bubbles, air dry for 1 h in dark at room temperature.

(ii) Place in slide box, cover and store at 4°C for 2 weeks.

(iii) Develop in D-19 Kodak 5-10 min at 15°C following instruction material.

Protocol for LM ISH on LR-white Embedded Sections using Digoxigenylated-DNA Probes

I. Labelling of DNA Probes

(i) Use 0.5-1 μg linearized DNA in 10 μl sterilized distilled water.

(ii) Denature in a boiling water bath for 10 min and chill on ice.

(iii) Add 2 μl of hexanucleotide mixture (10 x conc.), 2 μl DIG-DNA labeling mixture (10 x conc.) and bring volume upto 19 μl with sterile distilled water.

(iv) Add 1 μl Klenow enzyme, mix centrifuge briefly and incubate either at 37°C or overnight at room temperature.

(v) Stop labeling procedure by heating in boiling water for 10 min.

(vi) Add water and 20 x SSC to final concentration of 100 μl with a final conc. of 5 x SSC, add 1 μl of 10% SDS to give 0.1% SDS.

(vii) Store at -20°C.

II. Fixation and Embedding

Follow fixation and embedding protocol as discussed earlier in above sections.

III. Pre-hybridization and Hybridization

(i) Wet sections with distilled water.

(ii) Digest with Proteinase K (100 μ/ml) in a buffer of 100 mM Tris-HCl, pH 7.5, 50 mM EDTA for 30 min, fix in 4% paraformaldehyde solution for 10 min to stop the protease reaction, wash in 2 x SSC for 5-10 min.

(iii) Prehybridize sections in hybridization buffer of 50% formamide 10% dextran sulphate, 6% Denhydrats 2 x SSC and sonicated Salmon sperm DNA at 250 μl/ml to block non-specific biding for 2 h.

(iv) Prepare probe probe-hybridization mixture 2 μg/ml of digoxigenylated probe in hybridization buffer, denature double stranded probes prior to use by boiling for 2 min, immediately cool to 0°C in ice where kept until use.

(v) Cover sections with probe-mixture; cover slip the sections, seal with rubber cement and hybridize overnight at selected T° (40°C-42°C) in a sealed humid chamber.

(vi) Remove cover slips by washing several times in 2 x SSC.

IV. Immuno-gold Detection

(i) Pre-incubate grids in 0.1 M Tris-HCl, 0.1 M NaCl, 1% BSA and 1% Tween 20, pH 7.2 for 30 min. Drain of excess solution.

(ii) Incubate with sheep anti-dig antibody conjugated to 1 nm gold diluted 1/100 in pre-incubation buffer containing 0.1% BSA and 0.1% Tween 20.

(iii) Rinse in Tris-HCl (as above) without BSA for 4 x 5 min. Drain off excess buffer. Wash with PBS 3 x 5 min.

(iv) Silver-enhance gold particles using silver enhancement kit (Biocell); cover slip sections and monitor under microscope until black deposits appear.

(v) Rinse by dipping girds several times in distilled H_2O.

(vi) Counter stain lightly with blue toluidene.

LOCALIZATION OF METABOLITES

To test the applicability of the biochemical findings to all living cells, it is necessary to do the biochemical measurements within the confines of the living cell, so as to disturb its dynamic state as little as possible. Biochemical methods estimate the level of metabolites in a particular organ ie., leaf, root, stem, fruit and seed etc. To find out the exact location and amount of metabolites at the cellular level, one has to use histochemical localizations. Such approach has been developed through histochemistry. This section discribes the the localization of (a) carbohydrates, (b) nucleic acids and (c) proteins.

(1) Total Carbohydrates (Insoluble Polysaccharides)

Few methods are available for determining insolube polysaccharides and the most reliable is the periodic acid Schiff (PAS) reaction (Hotchkiss, 1948; Mc Manus, 1948). This reaction of an oxidative agent on the polysaccharide, resulting in the formation of two free aldehyde groups. The aldehydes then react with leucofuchsin dye producing highly coloured complexes. The most commonly used oxidant is periodic acid (HIO_4). The reaction of aldehydes with the dye fuchsin produces colour. This reagent is called Schiff's reagent and the carbohydrate test is called the periodic acid-Schiff's (PAS) reaction.

Materials

Basic fuchsin, potassium or sodium metabisulphite, HCl, charcoal, filter paper, periodic acid.

Preparation of Schiff's Reagent

Standard Schiff's reagent can be made according to Longley (1952). Dissolve 0.5 g of basic fuchsin and 0.5 g of sodium metabisulphite in 100 ml of 0.15 N HCl, shake the mixture at intervals of 2-3 h, add 300 mg of fresh decolorizing charcoal and shake for at least 5 min, filter through filter paper. The filtrate should be clear and colorless, if not, repeat the last two steps. Schiff's reagent should be kept in black colour bottle and stored in the refrigerator.

Procedure

(i) Use chemically fixed tissue (F.A.A. fixed tissue), dehydrates, paraffin infiltrate and section at 8-10 µm and mount on slides with any standard adhesive.

(ii) Deparaffinize and pass the slides through a gradual alcohol series to water.

(iii) Place the slides in 0.5% periodic acid solution in distilled water at room temperature for 25 min. Test for the optimum time in periodic acid as per your test material.

(iv) Wash in running tap water for 10 min.

(v) Stain in Schiff's reagent for 10-30 min.

(vi) Rinse the sections in distilled water and place them in 2% sodium metabisulphite for 2 min.

(vii) Dehydrated, cleared in xylol and mounted in D.P.X. The polysaccharides will stain an intense purplish red.

Control

Control test is by leaving out periodic acid treatment.

(2) Nucleic Acids Localization

The nucleic acids play an important role in chromosomes structure and their function in protein synthesis. Both deoxyribonucleic acid (DNA) and ribonucleic acid (RNA) have been involved in many important functions of the cell. Each of the three major parts of the nucleic acid molecule can be used as a means to estimate the amount of nucleic acid present.

DNA-Feulgen Method

The deoxyribose present in DNA is the basis for histochemical procedure - the Feulgen reaction. The tissue is treated with warm HCl, which hydrolyses the purine-deoxyribose linkages, thus exposing free aldehyde groups (Lessler, 1953). The method was modified by Goyal and Pillai (1985) where hydrolysis is done for 50 min in 5 N HCl at 20°C. The aldehyde groups then form a highly colored complex with fuchsin.

Materials

Basic fuchsin, potassium or sodium metabisulphite, HCl, charcoal, filter paper, celloidin, alcohol and ether.

Procedure

(i) Use chemically fixed tissue (Carnoy;s solution, dehydrate, paraffin infiltrate and section the tissue at 7-8 μm and mount on slides with Haupt's adhesive.

(ii) Deparaffinize and pass the slides through a gradual alcohol series to water.

(iii) Before hydrolysis, coat the sections with 0.5% celloidin solution in alcohol-ether (1 : 1).

(iv) Hydrolyze for 15 min in 1 N HCl at 60°C. Some times in modified procedures of Goyal and Pillai (1985) hydrolysis is done for 50 min in 5 N HCl at 20°C.

(v) Wash in running tap water for 10 min.

(vi) Stain in Schiff's reagent for 10-30 min.

(vii) Rinse the sections in distilled water and place them in 2% sodium metabisulphite for 2 min.

(viii) Dehydrate, make clear in xylol and mount in D.P.X.

Structures containing DNA will stain violet. The color produced in proportional to the amount of DNA present.

RNA-[Pyronine-Y Method (Tepper and Gifford, 1962)]

The nucleic acids have marked tendency to bind basic dyes in acidic pH. The dye binding is ionic in nature. The nucleic acid molecule is anionic or positively charged and the dye molecule is cationic or negatively charged. The pyronine alone can be used to demonstrate RNA in fixed plant tissue.

Materials

Pyronine-Y, chloroform, butanol, perchloric acid, celloidin, alcohol, ether, sodium carbonate.

Procedure

(i) Use chemically fixed tissue (F.A.A. fixed tissue), dehydrate in tertiary butanol series, embed, and section the tissue at 7 μm and mount on slides with Haupt's adhesive.

(ii) Deparaffinize and pass the slides through a gradual alcohol series to distilled water.

(iii) Stain at room temperature for 6-7 min in a 2% aqueous solution of pyronin-Y. The dye should be extracted 5 times with equal volume of $CHCl_3$ and its should have pH 3.5.

(iv) Slides are blotted and differentiated in two, 5 min changes of normal butanol.

(v) Clear in two, 5 min changes of xylol and mount in D.P.X.

The RNA containing structures will stain red.

Control (RNA and DNA)

The perchloric acid extraction procedure is based the selective removal of first RNA with cold perchloric acid and then of DNA with warm perchloric acid. The basis of this procedure is the differences in size and probably in structure of the two nucleic acids, whereby, RNA is more easily hydrolyzed than DNA.

Extraction of DNA and RNA Perchloric acid method (Erickson *et al.*, 1949)

(i) Remove the paraffiin and pass the slides through a gradual alcohol series to water.

(ii) Before treatment of perchloric acid, coat the sections with 0.5% celloidin solution in alcohol ether (1 :1).

(iii) Leave the sections (slide) in the 1 N perchloric acid for 12-24 h in a refrigerator at 4°C. This steps remove the RNA.

(iv) Wash well in running tap water.

(v) Place the sections in 0.5 perchloric acid at 70°C for 20 to 40 min. This step removes the DNA.

(vi) Immerse the section in a 1% sodium carbonate solution for 5 min and then wash in running tap water.

(vii) Stain using the above procedure.

Proteins High in Basic Amino Acids (Histones)

Histones are high in basic amino acids and they have isoelectric points, which are more acidic than other proteins. At the isoelectric point, dye binding by proteins is minimal. When the tissue is placed at pH 8.0, most of the proteins are above or near their isolectric point. So, by placing the tissue at pH 8.0 in an acid dye such as Fast Green, the only protein having groups available to bind the dye are the histones. When the conditions are standardized, the colour developed shows to the amount of histones present.

Fast Green Test (Alfert and Geschwind, 1953)

Materials

Celloidin, TCA, fast green (FCF), sodium nitrite, acetic acid, acetic anhydride, pyridine, alcohol and ether.

Procedure

(i) Use chemically fixed tissue (tissue fixed for 3-6 h in 10% neutral formalin, washed overnight in running water) dehydrated and embedded in paraffin.

(ii) Section at 8-10 μm and mount on slides with Haupt's adhesive.

(iii) Deparaffinize and bring the slides through a gradual alcohol series to water. Coat the sections with 0.5% celloidin solution in alcohol-ether (1:1) and rinse in distilled water.

(iv) Place the slides for 15 min in 5% trichloroacetic acid in a boiling water bath, or hydrolyse for 90 min in 5% TCA in the oven at 60°C to extract the nucleic acids.

(v) Place in 70% ethyl alcohol and make three changes to remove the traces of TCA followed by distilled water.

(vi) Stain at room temperature for 30 min in 0.1% aqueous solution of fast green (FCF), at pH 8.0-8.1 with minimum NaOH. Do not use a buffer in making the stain solution.

(vii) Rinse the sections in distilled water for 5 min.

(viii) Place the slides directly in 95% ethyl alcohol, dehydrate, clear in xylol and mount in D.P.X.

The basic proteins (histones) will appear green.

(3) Total Proteins [Mercuric Bromophenol Blue (Mazia *et al.*, 1953)]

The proteins, constitute all living cells. Maximum histochemical analysis of plant tissue have been made for proteins than of any other cellular constitutent, because proteins content express the amount of protoplasm present in the cell,as a measure of cell growth. The method of Mazia *et al.* (1953) is based on the formation of complex between the bromophenol dye and the proteins. The protein and the dye form an insoluble complex that will appear blue.

Materials

Mercuric chloride, bromophenol blue, acetic acid, sodium nitrite, acetic anhydride, pyridine and alcohol.

Procedure

(i) Use chemically fixed tissue (F.A.A. fixed tissue), dehydrate and embed in paraffin.

(ii) Section at 8-10 μm and mount on slides with Haupt's adhesive.

(iii) Deparaffinize and pass the slides through a gradual alcohol series to water.

(iv) Stain at room temperature for 15 min in 0.1% mercuric bromophenol blue in ethanol.

(v) Rinse in 0.5% acetic acid for 20 min and transfer to distilled water.

(vi) Dehydrate rapidly in 100% ethanol, clear in xylol and mount in D.P.X.

The proteins will appear blue.

Control

For control run the following blocking procedures for histones and total proteins between steps iii and iv.

(i) **Deamination** : Place the section in mixture (20 ml of 605 sodium nitrite + 60 ml of 15 acetic acid) at room temperature for 1-24 h.

(ii) **Acetylation** : Place the sections in 10% solution of acetic anhydride in pyridine at room temperature for 2-20 h.

Precautions

(i) Use the specific fixative for particular metabolite.

(ii) For every new tissue, test for the optimum time for oxidation and hydrolysis is necessary.

(iii) Stain Schiff's reagent should be clear and colourless before use.

REFERENCES

Alfert, M. and Geschwind, I.I. 1953. A selective staining method for the basic proteins of cell nuclei. *National Academy of Science* (USA) 39 : 991-998.

Dixon, M. and Webb, E.C. 1979. *Enzymes.* Academic Press, New York. pp. 7-22.

Ekes, M. 1971. The use of diaminobenzidines (DAB) for histochemical demonstration of cytochrome oxidase activity in unfixed plant tissues. *Histochemie* 27 : 103-108.

Erickson, R.O., Sax, K.B. and Ogur, M. 1949. Perchloric acid in the cytochemistry of pentose nucleic acid. *Science* 110 : 472-473.

Goyal, S.C. and Pillai, A. 1985. Morpho-histological studies of shoot apex in papaver. *Phytomorphology* 35 : 257-263.

Hall, J.L., Al-Azzawi, M.J. and Fielding J.L. 1977. Microscopic cytochemistry in enzyme localization and development. In : *Regulation of Enzyme Synthesis and Activity in Higher Plants* (Ed. H. Smith). pp. 329-363. Academic Press, London.

Hattersley, P.W., Watson, L. and Osmond, C.B. 1977. *In situ* immunofluorescent labeling of ribulose 1, 5-bisphosphate carboxylase in leaves of C_3 and C_4 plants. *Australian Journal of Plant Physiology* 4 : 523-539.

Hoyer, L.C. and Bucana, C. 1982. Principles of immunoelectron microscopy. In : *Antibody as a Tool* (Eds. J.J. Marchalonis and G.W. Warr.) pp. 233-271. Willey, Chicester, USA.

Hotchikiss, R.D. 1948. A microchemical reaction resulting in the staining of polysaccharide structure in fixed tissue preparations. *Archieves of Biochemistry* 16 : 131-141.

Jense, W.A. 1962. *Botanical Histochmistry.* W.H. Freeman & Co. San Francisco, London, 402 p.

Lessler, M.A. 1953. The nature and specificity of the Feulgen nuclear reaction. *International Review of Cytology* 2 : 231-247.

Longley, J.B. 1952. Effectiveness of chief variatns in the periodic Schiff and Feulgen nuclear technique. *Stain Technology* 27 : 161-169.

Malik, C.P. and Singh, M.B. 1980. *Plant Enzymology and Histo-Enzymology-A Text Manual.* Kalyani Publishers, New Delhi, 431 p.

Mazia, D., Brewser, P.A. and Alfert, M. 1953. The cytological staining and measurements of proteins with mercuric bromophenol blue. *Biological Bulletin* 104 : 57-67.

Mc Manus, J.F.A. 1948. Histological and histochemical uses of perchloric acid. *Stain Technology* 23 : 99-108.

Seligman, A.M., Karnosky, M.J., Wasserrug, H.L. and Hanker, J.S. 1968. Non droplet ultrastructural demonstration of cytochrome oxidase activity with a polymerizing osmophilic reagent, diaminobenzidine (DAB). *Journal of Cell Biology* 38 : 1-14.

Tepper, H.B. and Gifford, E.M. Jr. 1962. Detection of ribonucleic acid (RNA) with pyronin stain. *Stain Technology* 37 : 52-57.

CHAPTER - 21

METHODS OF ROOT STUDIES

INTRODUCTION

Observations and measurement of plant root system requires laborious and time consuming effort. Numerous techniques and modifications have been developed to speed the process and reduce labour requirements. There can be no ideal method for measuring all questions. Often it will be advisable, sometimes also necessary, to use two methods simultaneously.

CLASSIFICATION OF DIFFERENT METHODS

A unique classification of existing root study methods over systematic basis is not possible because several methods, different in principles have certain similar features. Therefore, the grouping of methods is more or less on personal decision.

In the simplest way a classification into old and new, or direct and indirect or field and container methods can be made. Bohm (1979) grouped these methods as given below :

(1) Excavation methods

(2) Monolith methods

(3) Auger methods

(4) Profile wall methods

(5) Glass wall methods

(6) Indirect methods

(7) Container methods

(8) Other methods

1. EXCAVATION METHODS

To expose the complete root system of a plant, one must remove the surrounding soil, which is normally accomplished by careful excavation of the soil surrounding the individual roots. Depending upon the objective of the study, this procedure may be accompanied by simultaneously drawing or photographing the position and relationship of the various roots exposed. The excavation method also referred to as *Skeleton method,* is the oldest method used in ecological research.

PROCEDURE

(1) *Selection of the plant :* The plant/plants should be representative one. After choosing the plant, its height, development stage and special morphological features should be determined and recorded. Before starting excavation, the top parts of the plant must be removed or secured and the root crown firmly secured to maintain in its original position.

(2) *Digging the trench :* A trench is dug with spade and shovel at distance far enough from the plant stem to ensure that no laterally growing roots are destroyed, yet close enough to minimise the amount of soil that must be excavated. For grasses or herbs 20-80 cm will generally be appropriate, depending on the species and growing stage.

(3) *Width and depth :* The trench should be at least 1 m wide to facilitate working. Its depth should be extended 20-30 cm below the deepest roots. The width and depth of the trench varies not only with the plant species but also with soil type. In very sandy soils, the trench should be wide in order to minimise the danger of the walls caving in on to the worker.

(4) *Initial Trench:* The initial trench should be dug with nearly vertical walls. If the maximum rooting depth is known or expected to be shallow, the trench should be dug in one step. For deep growing plants, it can be advisable to dig the trench in steps, so that digging alternates with excavating the roots.

(5) *Excavating the root system :* The next step is to expose the root system. One should begin by carefully removing the soil from the plant side of the trench, starting with the surface soil, and gradually working downward and into the face of the exposed profile. The soil must be removed particle by particle to avoid root destruction.

(6) *Removing soil particles:* The careful removal of soil particles from around the root system requires a wide variety of small hand tools like small metal forks, screw drivers, forceps, spatulas, small dental picks and sharp pointed needles of different sizes. For finer root systems such as grasses, fine and delicate picks, sharp-ened needles and camel hair brushes should be used.

(7) Clayey soils when dry, should be moistened with water overnight to facilitate excavation. But soil should not be too wet, because finer roots may be broken off and lost during the process of remvoing the soil.

(8) *Drawings and photographs*: Simultaneously with the excavating procedure, careful measurement and drawings should be made from the root system. The reduction scale used depends on the original size of the root system. For annual field crops, Weaver (1926) recommended a reduction of two-thirds which permits drawing even the finest roots. Where the root system exceeds 1 m in depth, greater reduction will be necessary. Different root diameters can be indicated through the use of different shades.

In many cases the number of trenched roots in a given soil layer may be too large for all of them to be drawn on the graph paper at the selected scale. It is then recommended that the major roots be drawn on the original scale and a supplementary drawing made over a larger scale. The original root drawnings can be transferred or covered, then photographically reduced or enlarged to the desired size (Bohm, 1979).

Advantages

(a) The excavation method provides a clear picture of the entire root system of a plant as it exists naturally. The length, size, shape, colour, distribution and other characteristics of the individual roots making up the root system may be studied directly.

(b) Quite useful in evaluating tracer techniques.

(c) Only useful method in soil having cracks and biological channels.

(d) The excavated root systems can be converted into quantitative data (Atkinson and Gwiyneth, 1976). The information on the drawing can also be quantified.

(e) On very stony or in dry sandy soils this will often be the only effective means which can be used to study the root systems.

Disadvantages

(a) Requires large amount of physical labour and is very time consuming.

(b) More than one specimen is needed to be sure that the root system information obtained is representative of the plant species growing on the given site.

2. MONOLITH METHODS

These methods require the taking of soil monoliths and separating the soil from the roots by washing. This can be done directly in the field, but generally the soil samples are transported to a special washing place. These methods are frequently applicable for investigation, where the aim is to make a quantitative determination of the roots, but in many cases this is supplemented with visual pictures.

Small monoliths can be taken with simple tools, such as spades, but with increasing size of the monoliths, power driven mechanised machines must be used. In most cases trenches must be made before starting the sampling procedure. Monolith methods are mainly used by researchers in agriculture.

SIMPLE SPADE METHOD

To obtain a rough estimation about the roots in the upper soil horizons, small monoliths about 20 cm in square can be taken with a spade. Generally, it is more common to put the monoliths dug out by a spade in small boxes with a sieve bottom, and after soaking, to wash the root systems free with a spray. Then the root systems can be classified visually by size.

Spade methods, with additional visual estimations of the exposed root systems, are relatively quick and can be used for solving some specific questions as preliminary studies.

Most commonly used monolith methods are the following :

Square Monoliths

It consists of digging a trench about 1 m long to the maximum rooting depth, and then taking away monoliths from the side-wall layer by layer. The height of the monolith sample depends on how accurately the root distribution in a soil profile is to be determined. Taking samples from 10 cm layers is common from a profile wall of 1 m length, about five or more subsamples can be taken.

The size of such soil monoliths varies from small soil prisms of about 10 x 10 x 10 cm to large blocks of 1 ton in weight and more, depending on plant species and the objective of the studies. Generally, the volume of a common monolith varies between 1000 and 5000 cm^3.

Before starting the removal of the monoliths from the soil, the profile wall must be flattened, using a Plumb line. All loose soil should be removed from the bottom of the trench. Then the size of the monoliths to be taken is marked on the profile wall with a needle.

To extract the soil monoliths, broad knives and metal sheets sharpened at one side which can be beaten into the soil with a hammer are used. The complete soil sample is put into a pail and the roots are separated by washing.

If only roots in the upper soil layers are to be studied, the technique of preparing free-standing monoliths without digging trenches can be used. After the monoliths have been prepared with a spade, they are cut at the bottom and transferred on sieve boxes to a washing place. This technique has been widely used for investigating the total root mass in the plough layer of the soil.

Needle Board Methods

Probably the most common method which combines pictorial presentation with quantitative determination of the root system of the plants is the needle board, nail board or pin board method. By this method, a soil monolith with a representative sample of the root system is taken by means of a special wooden board. Needles or nails positioned in this board keep the roots nearly in their natural position while the soil is removed by soaking and washing. The root systems can be investigated and photographed as a whole or can be sectioned for more quantitative determinations.

Construction and Preparation of the Needle Boards

The dimension of the boards and the length of their needles can vary according to the size of the root system to be sampled. To take a small soil monolith, a board size of 50 x 50 cm with needles 5 cm long is common for larger samples size up to 60 x 100 cm with needles up to 20 cm can be used. Still larger dimensions are possible, but they will be governed by the mechanical equipment and man power available to handle such large samples.

These needle boards can be constructed from two similar sized pieces of plywood of 2 cm and 1 cm thickness. Holes are drilled into the thicker board in vertical and horizontal rows 5 cm apart. Then knitting needles or stainless steel wires or even bicycle spokes are cut to length, sharpened at the cut ends and bent into U-shapes. These bent needles are pushed through the holes of the 2 cm board and backed up by the thinner piece of plywood. Brass screws can hold the two boards together and keep the needles in place when the board is driven into the soil (Schuurman and Goedewaagen, 1971). Instead of *U*-bent wires of needles, nails or other pins can also be driven directly into the board.

The 5 cm distance between the needles on the boards has been proved to be the most suitable. If the needles are placed closer together, the washing procedure can become diffisult in case of hard soil layers. If the needles are spaced wider apart, too many roots move from their natural position and the root picture will be distorted.

Needle board is painted black to provide a constructing background for photographing the exposed root systems. Instead of painting black, polythene sheet can be stretched over the board and pressed between the needles until it is against the board. This has the advantage that after washing the root system as a whole can be lifted easily from the board.

A further improvement is to cover the black painted board with a nylon net of about 5 mm mesh or to insert a hardware cloth among the needles which can be lifted during the washing procedure to allow water and soil to pass underneath. In this manner, the abrasion of roots by soil particles can be minimised. To prevent rusting, the needles should also be painted.

Excavating the Monoliths

First a trench is dug at least 1 m long and wide. If the plants to be excavated are growing in rows, the trench can be dug parallel or transversely to the rows. The distance from the face of the trench to the stem of the plant to be sampled should be determined in such a way that their crowns come to lie in the middle of the soil monolith to be held by the needles on the board after excavation. The face of the trench can be cut back to the desired distance from the plants, and then the wall is smoothened to an exact vertical plane.

After preparation of the profile wall, the needle board is placed against this vertical wall. The top row of the needles should be almost at the ground level. Then the board is pressed into the soil by means of a jack or a hammer to the fullest extent of the needles. The enclosed soil monolith will be supported by the jack or by wooden planks.

The soil from the profile face right and left of the board and the bottom is then sliced way with a spade and knives Cutting the bottoms free can be done with a steel plate driven into the wall. Such a bottom plate also supports the monolith, which is important in sandy and loose soils. To minimise the danger of soil clods breaking off during the excavation procedure, a wooden or metal frame can be pushed over the sides of the monothis.

At last the inner face of the monolith is cut free. This is done either by working inwards with a spade and knives starting from the two sides, or by driving a steel sheet into the ground from the soil surface. Such a steel sheet can also be driven into the soil before the trench is dug, which is especially desirable in very loose soils.

The cutting free of the inner face of the monolith should not be done directly at the tips of the needles. It is advisable to prepare the monolith a little thicker because part of the soil can break away during the cutting procedure. Finally the needle board with the soil monolith is laid down flat, after which the excess soil from the inner face is cut away.

Washing Procedure

If the monolith consists of sandy or loamy soil, it is directly submerged in a tank of water until the soil is saturated. Generally a soaking overnight is enough. The frame which encases the monolith should not be removed, to avoid breaking off of parts of the monolith. Instead of the frame, however, a nylon net can be placed over the whole monolith, which also prevents the roots at the edtes of the monolith from being severed during soaking and the following washing procedure.

The needle board with the soaked monolith is then raised so that the water level of the tank is always 2-3 cm below the surface of the soil monolith and tilled slightly to allow soil and water to flow away. With low water pressure of a ard sprinkler, starting from the bottom of the board and working upwards to the plant stems, the soil is washed away. The strength and volume of water stream must be controlled by a hand-operated valve. Too much pressure is to be avoided. Instead of using a hand sprinkler, a rotary sprinkler with three or four arms or an oscillating sprinkler can be used.

Washing out soil monoliths with high clay content is extremely difficult, and often only possible with a considerable loss of roots. To minimise the loss of roots, Schuurman and Goedewaagen (1971) recommended drying the entire monolith gently at 100°C, soaking it in a solution of sodium pyrophosphate and then washing it as described above. Often this process needs repetition.

Advantages

(a) Simultaneously qualitative and quantitative studies of the roots are possible.

(b) The cleaning procedure of the washed root is reduced considerably in comparison with other methods.

Disadvantages

(a) Many of the dead and old broken roots are washed away and mainly the live roots remain on the board. Thus, it seldom contains the entire root system. Roots running towards the front and back of the monolith are cut off.

(b) High labour requirement.

(c) Does not work in stony soils and may be problematic in very sandy soils especially when dry.

3. AUGER METHODS

Of all sampling techniques, the auger methods are the most suitable for taking volumetric soil root samples. The methods involve taking samples by hand augers or by mechanical sampling machines, and separating the roots from the soil by washing. Except for special purposes no attempt is made to keept the roots in their natural positions. The size of samples generally is smaller than those from the monolith methods.

Sampling Procedures

The simplest way of taking soil cores is by means of a hand auger. From the many types of augers, the model developed by Goedewaagen and described in details by Schuurman and Goedewaagen (1971) consists of a cylindrical tube 15 cm high with an inside diameter of 7 cm. Above the tube a hollow shaft about 100 cm long is fixed so it can be used for sampling up to a depth of 100 cm. On the outside of the tube and also on the shaft there are marks at 10 cm intervals. At the end of the shaft, auger is provided with a T-handle which makes it possible to rotate the auger while driving it into the soil and to pull it out again. The cutting edge of the cylindrical tube is serrated. The hollow shaft contains a rod with a disc at the bottom which works as a plunger inside the cylindrical tube to force the soil cores out of the auger.

For taking soil root samples, the auger is pressed into the soil twisting it at the same time, until the first 10 cm mark is reached, rotated several times and then pulled out. The easiest way to force out the core from the tube is to turn the auger up side down with one foot. Sampling is continued by replacing the auger in the bore hole again and the whole procedure is repeated down to the next mark. When drilling in clay soils, the work can be facilitated if the auger is dipped into a pail (bucket) of water before every drilling operation. If continuing samplings are required on the same site then the holes are filled with soil immediatelhy after each sampling.

The diameter of the core sampler is a point of important consideration. As recommended by Schuurman and Goedewaagen (1971) the most common diameter of soil cores for root studies with hand auger should be about 7 cm. Sampling with lesser diameter auger requires large number of replications.

Number of Replications

For most purposes, samples from three to six bore holes are commonly practiced. In row crops at least half should be taken in the rows and the other half between rows. It can be recommended that with 7 cm augers, samples should be taken from at least five bore holes. The number of samples taken from each experimental unit is the primary factor governing the variability.

4. PROFILE WALL METHOD OR TRENCH PROFILE METHOD

In this method, plant roots are excavated and only a part of the whole root system is exposed. Bohm (1976) described the method in a modified and simplified form, which is very commonly in use. The method has been known mostly by the term trench profile method.

PROCEDURE

(a) Digging the Trench

The position of the trench depends on the plant species to be studied. In broadcast crop, the position of the trench is of no importance. On plots with plants growing in rows, the trench is generally dug transversely to the rows, because in this case the root distribution is between the rows and therefore, the variation in the whole soil profile, can be inspected much better.

The length, width and depth of the trench will mainly be influenced by the development stage of the plants to be studied and the aim of the research. Generally, the working face should be 1 m in width and has to be dug about 1.20-1.50 m. The depth should be 10-15 cm below the expected rooting depth and minimum should be at least 1 m even if only the upper soil layer contain roots. After digging the trench with spade and shovel, the working face of the profile is smoothed with a sharp blade. Roots exposed from this prepared profile wall are removed with a pair of scissors.

(b) Preparing the Profile Wall

A nearly 5 cm layer of soil is then washed down with a fine sprayer (around three atmospheric pressure). Where hard soil layers exist in the profile, the wall is soaked before spraying. The washing procedure may be aided by using small brush without disturbing the exposed roots.

(c) Mapping and Counting Procedure

Immediately after spraying, a wooden or metal frame (interior 100 x 100 cm) with 5 x 5 cm grid made up of copper or white nylon threads is laid against the profile wall. The legnth of roots exposed by washing is then estimated by counting the number of 5 mm length roots in each 5 x 5 cm grid area. In this procedure, roots which are 10, 20 or 30 mm long are counted two, four or six times.

During the counting procedure, the profile is always kept moist so that the roots would not dry out. To avoid direct sunshine, the working profile facing north should be prepared. Counting is done up to the maximum rooting depth.

The classical way of determining the exposed roots by mapping them in their natural position on cross-section paper (mm paper) with a scale of 1 : 10 or 1 : 5. At first, the soil depths are marked on the map, horizons are lettered and their boundaries are indicated by lines. Large stones or special soil features such as cracks and earthworn channels, can also be mapped.

DETERMINATION OF ROOT LENGTH

Bohm (1976) tried to estimate root length directly at the profile wall with the assumption that the root distribution in the soil is nearly uniform. If 5 mm of soil has been washed away from a 100 x 100 cm^2 profile wall, and if estimated root length, a total count of 1000 root units in the whole profile corresponds to a root length of 1000 m/m^3, it is equal to 10,000 km root length/ha.

Calculation

Counted root = 5 mm = 0.005 m

e.g., 1,000 counts = 1,000 x 0.005 = 5 m/0.005 m^3 if the frame size is 100 x 100 cm^2

$$\text{Therefore, in } 1\ m^3 = \frac{5}{0.005} m = 1000\ m,\ i.e.\ 1000\ m/m^3$$

But accuracy of this estimated length is about one-half of that measured with root counting machine.

The main reason for the underestimation seems to be that the fine roots which are still wet during the counting procedure often adhere closely to the roots so that they can not be distinguished at the profile wall. But the difference is not there with relative densities.

Foil Method

A very good improvement was proposed by Reijmerink (1973) who placed a transparent plexiglass plate (5 mm thick) which was covered with a sheet of clear plastic foil (0.1 mm thick) in front of the profile wall. Then with a special pencil were traced the exposed roots as well as the limits of the soil and structural horizons on the plastic foil.

One drawbeck of the foil method is that the operator's direct contact with the exposed roots at the profile wall is lost which may lead to some errors. The advantage is that natural pictures of the real root distribution in a soil profile can be obtained.

One reason of underestimation of counting may be that with the counting procedure, the investigator tends to count less root length units than are visible at the profile wall, especially if many roots are in one grid area. For this reason the greater deviation is found in the A_p-horizon. Secondly, an added disadvantage is that the counted roots are not makred so that investigator can not make a quick chekc if he is right or wrong.

Trench-profile Method for Studies of Nodule Distribution on Roots (Grubinger *et al.*, 1982)

The trench-profile method of root system examination can also be employed for the studies of nodule distribution on roots with relatively greater precision. For this a smooth vertical surface of the base of plant to be examined is prepared. Using a sprayer filled with water, a hand-held wand, approximately 5 cm of the vertical face of the profile is washed away. Then a wooden frame, preferably 100 cm x 60 cm, and sectioned into squares 10 cm on a side, with white string is entered below the plant. Nodules are counted and a few, selected at random, split open to determine their internal colour. Red nodules are assumed to be actively fixing nitrogen.

5. GLASS WALL METHOD (RHIZOTRON OR MINIRHIZOTRON)

Investigating plant roots in the field frequently involves the removal of one or more soil cores, and inherently destructive process.

Observing plant roots in rhizotron is a valuable technique, but initial construction costs are high, and root growth is restricted in at least one direction. The method is good and probably the only method for studying phenologic changes of roots and their decay.

A pit 2 m x 3.5 m x 2 m is dug and a glass wall is constructed over a concrete floor. The space between the glass and the pit is filled with representative soil. Then the soil is wetted and allowed to stand for a week to permit settling before planting. Seeds are sown near the glass wall. Plant weight and root depth are measured periodically and the total root length visible at the glass soil interface is determined at intervals using a map mileage reader.

MINI-RHIZOTRON METHOD

The method of Waddington (1971) who inserted glass tubes into boxes that contained soil and observed plant root development at the soil glass interface. Bohm (1974) adapted the technique for plant root studies in field soils. He marked 5 x 5 cm grids on eight acrylic plastic tubes that were 185 cm long and had external diameter of 7.5 cm. He used a bucket auger to bore 170 cm deep holes vertically, four holes in the row and four midway between 100 cm wide rows. Refilled the free space between the tube-wall and the soil-wall with seived and dried from the same horizon. The projecting tubes were covered with aluminium foil.

Roots growing at the soil tube inter-face were lighted by four (6V) bulbs positioned around a concave mirror fastened to steel rod. Roots were observed by keeping down the tube and observing their reflected images in the mirror. Rooting densities are obtained from the number of roots that crossed horizontal transects at 5 cm depth increments (Taylor *et al.,* 1970). The number of root intersection as a function of depth are observed at 7 to 10 day intervals. It was found advisable to put the tube into the soil at an angle to the vertical to minimise the problem of roots growing along its length.

This method enables estimates of root penetration and rate and size of root growth to be made from time to time without disturbing the root system (Taylor and Klepper, 1973, 1975).

The technique is not free of problems, in many positions of the probe, a bright spot of light was reflected from the light guide into the image and prevented good photographs from being obtained through the equipment. In addition the fibre nature of the image transmission system limited resolution to roots of an apparent diameter greater than 0.2 mm. However, the method should be useful for investigating root systems in disturbed soil such as is used in green house, growth cabinet, and field lysimeter experiments, and possibly also in some ecological work where investigations extending over several growing seasons will minimise the effects of soil disturbance which occur when the transparent tube is installed.

6. INDIRECT METHODS

High input of time and labour required to obtain information about roots in the soil by direct observations or sampling methods has led to the development of indirect study methods.

Principle : Determining changes in water or nutrient in different soil layers by successive sampling occassions. These changes provide information on the root distribution in the soil profile.

Applicable : For ecological investigations, especially if the activity and not the absolute amount of roots in a soil profile is the primary research aim.

Limitations : Success depends on several assumptions, if they can not be met it may lead to erroneous results.

SOIL WATER DEPLETION METHOD

Assumptions : That correlation exists between soil water depletion rate and the quality of roots present in soil.

Method : 2.0 cm diameter soil tube auger is used to sample from different depths normally 0-15, 15-30, 30-60, 60-90, 90-120, 120-150 cm and so on up to the maximum rooting depth.

Soil samples are dried to a constant weight at 105°C; soil with high organic matter content should be dried at 50°C in a vacuum oven.

It is an elegant method to measure directly how much water the plants have absorbed since previous sampling occasion. It assumes that during the period of two soil water measurements there is nearly no transfer of water from one profile to another, and there is no loss of soil water except by transpiration. Water loss by evaporation is not a serious problem if ground surface is covered by vegetation.

Limitations

(1) Used in homogeneous soil profile where discontinuity with grooved horizons or soil structure can not affect water movement.

(2) Period of water depletion commences with a uniform distribution of available water in the soil profile which is approximately equal to field capacity.

(3) It is difficult to obtain accurate results if rainfall occurs between two sampling occassions.

Direct study methods are preferred. Measuring water depletion should be used as a complementary method to obtain information on water uptake by roots in different soil layers.

NEUTRON METHOD

The number of neutrons attenuated in soil is proportional to the volumetric water content. Since count rates are not linearly related to volume of water, calibration is necessary.

Limitation : High content of organic matter, boron, iron or chlorine can falsify the results.

RADIOACTIVE TRACER METHODS

Radio tracer techniques, involving the placement of a readily detectable tracer in the soil at marked depths in the soil, have been used in a number of investigations of plant root distribution and activity (Fox and Lipps, 1964).

The following radioactive material can be used for the purpose.

(1) ^{32}P : half-life = 14.3 days, it is a β-emitter; suitability is due to the following reasons :
   (a) Mobile enough in plants to become rather uniformly distributed throughout the root system in a shorter time.
   (b) Relatively inexpensive.
   (c) Its easy detection in the above ground parts.

(2) ^{86}Rb : half-life = 18.7 days; suitability is due to the following reasons :
   (a) Strong γ-emitter and hence no need of ashing.
   (b) Decays fast enough to minimise health hazard when used in field.

(3) ^{131}I : Costly

(4) ^{42}K and ^{45}Ca : Problems with measurements.

Principle

Root activity : A solution containing suitable isotope is injected at various distances and depths from the base of plant into soil methods. As complementary to other measurements, the radiotracer methods will certainly keep their place in root ecology (Bohm, 1979).

ESTIMATION OF ROOT PARAMETERS FROM Ca ABSORPTION VALUES IN SHOOT

In the technique suggested by Bar-Yosef and Kafkafi (1970, 1972) and Sacchan and Sharma (1980), the underlying principle takes advantages of the fact that since the amount of Ca uptake by the plant is relatively small (Russel, 1961), in a moist soil the mass flow transports several times more Ca to the root surface than can be absorbed by the plant. Thus the Ca concentration around the roots is always such that is satisfies the maixmum absorption capacity of the roots. Therefore, the amount of Ca absorbed is related to the total amount of root length or surface area that has developed and fully participated in the uptake process. In order to apply the suggested technique, requirements that should be met are (i) Ca absorption per unit root is uniform at least during the major active grwth period, (ii) Ca absorption is not affected by fluctuation in temperature and humidity encountered by the plants when grown in the field conditions. To verify the above points one must know the Ca concentration that satisfied the maximum absorption capacity of roots.

The relationships between the Ca content in the shoot and the root parameters are established by pot culture experiment and tested in regression equation for estimation of wheat parameters from Ca absorption values in shoot (Table 1). These estimations are termed as equivalent root parameter.

Table 1. Regression of Ca uptake (Y) on various root parameters (X) of wheat (Sacchan and Sharma, 1981).

Root Parameters	Regression equation	`r'
Fresh weight	Y = 1.5368 + 1.8035 X	0.982
Dry weight	Y = 3.3438 + 39.4195 X	0.991
Length	Y = 1.4622 + 0.2677 X	0.983
Surface area	Y = 1.4627 + 0.0581 X	0.983

ROOT WASHING METHODS

Roots can be separated from soil by many ways. The common washing methods are as follows :

Floatation Methods

With this washing technique, the soil sample is soaked in a pail of water until thoroughly dispersed. Then the entire content of the pail is poured into a larger metal can equipped with an overflow spout. A hose with a nozzle is placed a few centimetres above the bottom of the can so that the soil root water mixture can circulate and overflow on to a sieve placed directly under the spout. The upward circulation of the suspension can be regulated by changing the position of the nozzle. The suspended soil particles which come through the overflow spout run through the sieve, while the floated roots remain on the sieve. The overflow procedure is continued by occasionally moving the hose until no more roots float out and the running water is closed.

The floating method needs two to five times more time than the common technique of washing roots by hand, but it is less destructive for the root systems. As the separation of the roots from the soil is done under water, the root damage from strong water pressure is reduced to a minimum. The density of the dispersed soil root suspension can be increased up to 1.5 g/cm^3 by adding pure $CaCl_2$. Roots and organic debris will float to the surface of the container and can be skinned off with wire seives.

Root Washing Assembly for Soil Root Cores

A single low-cost root washing assembly was developed and described by Sharma *et al.* (1981). The assembly can be fabricated with the materials available easily in a research laboratory.

The apparatus consists of a polyethylene carboy (inner diameter 30 cm) of 20 litre capacity. The top portion of the carboy is cut off at 28 cm height from the bottom. Through a hole made in the centre of the bottom portion of carboy, a PVC pipe of 4 cm inner diameter is fixed protruding it. After wraping the thread at appropriate place around PVC pipe, hot water is poured to check any leakage along the sides of the pipe through the hole.

The horizontally slanting holes at about 45° against the wall are made on the two sides of the carboy at 2 cm height from the bottom. Through these holes, jet type polyethylene tubes tappered on one side and 8 cm longer are fixed and sealed to deliver water into the carboy. The slanting holes and the jet type tubes allow the delivery of water into the carboy with violent whirling action. These jet-type tubes are connected to a water tap through suitable tubings. Alternatively, the water to the carboy could also be delivered by clamping the delivery tube, leading from the tap, to the wall of the carboy and its open end. The assembly is rested on a raised platform so as to allow the PVC pipe protruding outside the carboy to delier soil water root suspension to the set of sieves kept below it. In order to allow the soil to pass through the sieves, the sieve set is kept in such a way that about 4 to 5 cm space is left between the platform and the lower sieve. The set consist of two sieves. Keep one over the the other, upper one with 0.7 mm aperture and the lower one with 0.4 mm openings.

PROCEDURE

Soil root cores of 20 x 20 x 15 cm (6,000 cm^3 volume) are transferred from polyethylene bags to the assembly without any chemical pre-treatment and washed thoroughly. Some clods present in the sample are broken by the fingers, when in the process of being washed. The sample is stirred two to three times by hand during the process of washing. The last stirring should be done when the water becomes clear to help a very few roots settle in the bottom (along with the heavy particles) to float in the water and come down to the sieves. The water supply to the assembly is then cut off. The water which remained in the carboy to the level of PVC pipe is drained off by tilting the assembly and allowing this water to pass through the same set of sieves. The heavy soil particles and gravel settled in the bottom of the carboy are then washed off.

The whole process takes about 16 to 25 min depending on the clay content of the sample. The roots collected on sieves are transfetrred to polyethylene sheet laid on a table top by inverting and gently tapping the sieves over it. The roots on the sieve net are removed with the help of tweezers. In the mean time, a second sample is transferred to the assembly with another set of sieves.

For smaller cores, the assembly can be fabricated by using a smaller circular container.

Cleaning Roots from Debris

After the roots have been separated from the soil, they must be cleaned, because in the washing procedure a separation of live roots from dead ones or from other organic matter is not possible. The aim of the cleaning procedure is mostly to separate the live roots from dead ones and from all other organic debris. The only practical way of cleaning- the washed root samples are put in flat black dishes filled with water and the separation is done with the aid of tweezers. When there are considerable plant residues or dead roots in a sample, the separation can be facilitated by stirring the roots in the dish. Since dead plant material is lighter than living roots, it floats to the water surface quicker and can be collected with a sieve. However, this pre-separation seldom is completed. Often live roots are also floated within the debris, and must then be picked out by tweezers. Generally, the final cleaning process is done in tap water. Cleaning is done by distilled water only when nutrient content is to be determined.

MEASUREMENT OF ROOT PARAMETERS

The parameters commonly used to express root growth and distribution are number, weight, surface area, volume, diameter, length and the number of root tips. The manner in which the root data are best expressed should be considered before starting an experiment. In some methods, e.g., by the use of the profile wall methods, there is little choice, but in many other study methods several approaches can be used. Not only the research aim, but also the time and labour needed for the determination of the root parameter which has been choosen must be considered.

Root Number

Counting root number is the common procedure when working with the core break method, profile wall methods and glass wall methods. Root numbers counted in the different soil layers generally give a good impression of the rooting density in a soil profile. In studies where the entire undamaged root system of a single plant can be extracted, the number of main and lateral roots is counted to provide an estimate of their total length. Although root number is not an ideal parameter, as long and short roots are regarded and counted as equal units, high correlations with other root parameters are possible.

Root Weight

Root weight is the most commonly used parameter for studies of root growth in response to environment. Generally the washed roots are dried and then their weight is determined.

To determine dry weight, the washed and cleaned roots are dried in an oven at 75°C for about 10 to 20 hours depending on the amount of the roots. The drying can also be done at 60 to 75°C which will take longer time.

Even after thorough washing and cleaning, soil particles can still adhere to the roots and cause errors. So generally it is recommended to place the dried roots into a muffle furnace at about 650°C. The cooled ash is finally treated with hydrochloric acid, filtered by decanting, dried at 105°C, and weighed. By treating the ash residue with hydrochloric acid it is assumed that the original matter in the roots is dissolved. The weight of the ash residue therefore, represents the weight of unsolved soil particles in the root sample. The difference between the root dry weight, including the soil particles adhering at the roots, and the weight of the ash residue can be termed the ash-free organic root matter. The final treatment of the ash residue with hydrochloric acid is not certainly necessary in every case. In soils with free lime, an acid treatment should be avoided because the dissolved lime is then included in the root weight. So the term ash-free organic root matter should be used in both cases whether ash residue is treated with acid or not.

Advantages

Root weight is a good parameter for characterising the total mass of roots in a soil. It can be regarded as a fundamental measure of photosynthate storage in a plant. The known shoot-root relations are based mainly on shoot and root weight.

Disadvantages

By expressing root data as root weight, the amount of fine roots represents only a small fraction of the total weight, although these fine roots can be the most active part of the root system. Therefore, a high total root weight in a soil layer may not be identical with a zone of high water and nutrient uptake.

Root Surface Area

Surface area is one of the best parameters when experiments on nutrient or water uptake are made. Following methods are commonly used.

(1) Direct method

The most direct method is to determine the average diameter of a large number of individual roots, and to measure the total root length per sample. From this data surface area is calculated (Adepeu and Akapa, 1977; Evans, 1977).

Atkinson and Gwiyneth (1976) calculated surface area separately for the three root diameter classes (e.g. 0.5, 0.5 to 2.0 mm and > 2 mm) as :

$$\text{Area (cm}^2) = 2\ [\text{volume (cc)} \times \pi \times \text{length cm}]^{1/2}$$

where root volume was estimated by displacement of water.

$$\text{length (cm)} = \text{weight (mg)} \times 0.890$$

or measured directly on mm graph paper for a number of samples.

Root area can also be calculated by determining root diameter and root volume.

(2) Photoelectric Measurement

Rhizometer (same as leaf area meter) : It consists of a light source in which roots are placed, photocell for measuring light reduction due to roots and a galvanometer for measuring decrease in output from photocell. After adjusting the device with pieces of black paper of known area, a calibration curve is made. The projected area is then converted into total root surface by multiplying by π, assuming that measured roots are circular in cross-section.

Working with this technique, it is important that (i) roots should not overlap, (ii) if roots are transparent, it is stained in an opaque dye.

(3) Absorption Methods

In principle a methylene blue solution (0.02 - 5 mg/litre) is made and the concentration is recorded by a colorimeter. The washed root samples are then dipped into the dye solution where it remains for a standard time with gentle swirling. The absorbed amount of methylene blue per root sample is calculated from the difference between initial and final dye concentration.

By this method active and inactive root surfaces can be distinguished. After the root sample has been immersed in the methylene blue solution, it is transferred to a solution of calcium chloride, in which the methylene blue from the dead inactive root surface is displaced by calcium ions. In this way the active roots in a sample can be determined to an approximate result about the activity of the roots for water and nutrient uptake.

Root Volume

(1) **Calculation from other parameters :** Root volume can be calculated by measuring the average root diameter and the root length.

(2) **Displacement technique :** The water displacement technique is to be preferred. The measurement is done in a special container with an overflow spout. This container is filled with water until it overflows from the spout. Then fresh washed roots which have been carefully dried with a soft cloth are inserted and the overflow water volume is measured in a graduated cylinder.

Limitations

(1) Root volume measurements from species with large amounts of small fibrous roots are not adequate. So the volume data alone are of limited value for most research problems in root ecology.

(2) Data from volumetric measurements in general may be used only to supplement other parameters.

Root Diameter

Root diameter is used for calculating root surface area/root volume. It also gives an idea about pore size of soil and potential of root penetration. Small roots are measured with the aid of a microscope filled with an occular micrometer. To measure large roots, a hand lens, a micrometer screw/calliper graduated to a tenth of a millimetre can be used. If individual roots differ in diameter they are measured at regular intervals throughout their length. Before starting the measurements, the roots are placed for some hours in water because many roots dry irregularly.

Limitation

It would be incorrect to identify root diameter with any particular kind of root function, even fibrous roots which correlate to root surface. Many roots of this category can be lignified and subscrised and therefore, they can not be considered as absorbing parts.

Root Length

(1) DIRECT MEASUREMENT

(A) Wet roots are placed in a flat glass dish containing a small amount of water. Graph paper is placed under the dish. Roots are straightened by foreceps so that they do not overlap and held in position by a glass plate. The length of the given roots or root segments are then estimated to the nearest millimetre by eye inspection or by the use of a magnfying glass.

(B) Branches are cut, joined end to end by gum, placed on a graph paper and length is measured instead of arranging them randomly in a dish.

Limitations

(a) Tedious and time consuming.

(b) Recommended only for estimating the length of single roots, not as a routine method.

(2) INTERSECTION METHOD

Instead of tedious direct measurement, root length can be calculated more rapidly by counting the intersections between roots and a regular pattern of lines. Head (1966) counted the total number of intersections between the roots and the vertical and horizontal lines of a grid on the glass observation windows and found a linear relation between the number of intersections and the actual root length.

Newmann (1966) derived formula R = $\frac{\pi NA}{2H}$ for estimating the total length of roots in an extracted sample. Root length (R) was measured by counting the number of intercepts (N) of roots in a regular area (A) with oriented lines of total length (H). In principle, the longer the root, the more intercepts it made with the randomly arranged lines. Thus, when a sample of roots is randomly dispersed over an area 32 x 54 cm^2 (A) within which are total number of root intersections (N) made with 100 x 4 cm line segments (H), the total root length R, within A, is given by

$$R = \frac{(\pi \times N \times (32 \times 54))}{2 \times (100 \times 4)}$$

Today root length can be determined without any expensive equipment, in a shallow dish made of transparent plastic or glass. A convenient dish size is 30 x 40 cm. A grid is placed under the bottom of the dish. The wet roots are then poured into the dish with some water, and they are positioned randomly over the grid with forceps or needless so that they do not overlap. If necessary, long branched roots are cut into smaller pieces. Finally, counts are made of the intersections of the roots with the vertical and horizontal grid lines. The use of hand tally counter facilitates the counting procedure.

Grid Size: Grid size depends on the amount of roots to be measured. For small root samples with total length below 1m, a 1 cm grid; for larger samples to about 5 m, a 2 cm grid, and for total root length up to 15 m, a 5 cm grid is recommended.

It has been reported that the number of intersections to be counted in one root sample should not exceed 400, due to possible fatigue of the operator and should not be below 50 because then the accuracy of the results decreases.

Tennant (1975) simplified the Newman's equation. For a grid of indeterminate dimensions the intersection counts can be converted to centimetre measurements using the equations :

$$\text{Root length (R)} = \frac{11}{14} \times number\ of\ intersections\ (N) \times grid\ unit.$$

As proposed by Tennant (1975) the 11/14 of the equation can be combined with the grid unit, to give a length conversion factor. The factors for 1, 2 and 5 cm grid square are 0.786, 1.57 and 3.93 respectively. The conversion factors can also be derived from the diagrams first proposed by Head (1966) which shows the linear relationship between actual root length and the number of intersections. The modified formula therefore, tool the easily managed form :

Root length (R) = number of intercepts (N) x length conversion factor.

Reicosky *et al.* (1970) while comparing three methods for estimating root length showed that the direct count method is the most precise, and more estimates using the inch counter (opisometer) and line intercept methods are required to achieve the same precision.

Estimates by the inch counter method tend to be slightly smaller and those by the intercept method tend to be slightly larger than those by the direct count method. The latter difference is probably due to the slight error in lifting and resetting the unidirectional inch counter on the root image. This error is more for cut roots because the inch counter is lifted more times from the table top. The tendency for the line intercept method to overestimate the root length is probably due to underdispersion of roots found at the edges of the glass tray. Newman's method for estmating root length holds strictly only when the straight lines are placed randomly in the field of view. Despite the variations, the intercept method appears to be as reliable as the direct count method.

REFERENCES

Adepeu, J.A. and Akapa, L.K. 1977. Root growth and nutrient uptake characteristics of some cowpea varieties. *Agronomy Journal* 69: 940-943.

Atkinson, D.N. and Gwiyneth, A.C. 1976. The effect of tree spacing on the apple root system. *Horticulture Research* 16 : 89-105.

Bar-Yousef, B. and Kafkafi, U. 1970. A new approach to evaluating pre root area of field crops. In R.M. Samish (ed.) *Recent Advances in Plant Nutrition*. Vol. I. 281-291. Gorton and Breach Pub., New York.

Bar-Yousef, B. and Kafkafi, U. 1972. Rates of growth and nutritional uptake of irrigated corn as affected by N and P fertilizers. *Soil Science Society of America Proceedings* 36: 931-936.

Bohm, W. 1976. *In situ* estimations of root length at natural soil profiles. *Journal of Agricultural Sciences* 87: 365-368.

Bohm, W. 1979. *Methods of Studying Root Systems. Ecological Studies* 33: Springer-Verlag. New York, 188 p.

Evans, P.S. 1977. Comparative root morphology of some pasture grasses and clovers. *Newzealand Journal of Agricultural Research* 20: 331-335.

Fox, R.L. and Lipps, R.C. 1964. A comparison of stable strontium and P^{32} as tracer for estimating alfalfa root activity. *Plant and Soil* 20: 337-350.

Grubinger, V., Zobel, R. and Cortes, P. 1982. Nodules distribution on roots of field grown soybeans in sub-surface horizons. *Crop Science* 22: 153-155.

Head, G.C. 1966. Estimating seasonal changes in the quality of while unsubscrized root on fruit trees. *Journal of Horticultural Sciences* 41: 197-206.

Newmann, J. 1966. A method of estimating the total length of root in a sample. *Journal of Applied Ecology* 3: 139-145.

Reicosky, D.C., Millington R.J. and Peters, P.B. 1970. A comparison of three methods for estimating root length. *Agronomy Journal* 62: 451-453.

Reijmerink, A. 1973. Micro-structure soil strnegth and root development of asparagus on loamy sands in the Netherlands. *Netherland Journal of Agricultural Sciences* 21: 24-43.

Russell, E.N. 1961. *Soil Conditions and Plant Growth.* Longman Green and Company, London.

Sacchan, R.S. and Sharma, R.B. 1980. Calcium absorption as an inter of root parameters of cucumber (*Cucumas sativus* L.). *Plant and Soil* 36: 117-125.

Sacchan, R.S.and Sharma, R.B. 1981. Estimation of wheat root parameters from calcium absorption values in shoot. *Indian Journal of Ecology* 8: 167-170.

Schuurman, J.J. and Goedewaagen, M.A.J. 1971. *Methods for the Examination of Root Systems and Roots.* Centre for Agri. Publication and Documentation. Wageningeon, Netherlands.

Sharma, R.B., Sacchan, R.S. and Ghiedyal, B.P. 1981. A simple technique for root washing from soil cores collected from field. *Journal of Indian Society of Soil Science* 29: 283-284.

Taylor, H.M. and Klepper, B. 1973. Rooting density and water extractors patterns for corn (*Zea mays* L.). *Agronomy Journal* 65: 965-968.

Taylor, H.M. and Klepper, B. 1975. Water uptake by cotton roots systems: an examination of assumptions in the single root model. *Soil Science* 120: 57-67.

Taylor, H.M., Huck, M.G., Klepper, B. and Lund Z.F. 1970. Measurement of soil grown roots in rhizotrons. *Agronomy Journal* 62: 807-809.

Tennant, D. 1975. A test of modified line intersect method of estimating root length. *Journal of Ecology* 63: 995-1001.

Waddington, J. 1971. Observation of plant roots *in situ*. *Canadian Journal of Botany* 49: 1850-1852.

Weaver, J.F. 1926. *Root Development of Field Crops.* McGraw Hill, New York, 291 p.

Chapter - 22

SYMBIOTIC NITROGEN FIXATION RELATED ESTIMATIONS IN PLANTS

INTRODUCTION

By far the most important process by which atmospheric N_2 is converted to forms utilizable by plants is performed by certain kinds of plant organisms and is therefore, known as biological N-fixation. Chemically, biological N-fixation consists of the reduction of N_2 to ammonia :

$$N = N \xrightarrow[Nitrogenase]{+6e^- + 6H^+} 2NH_3$$

Protons and electrons for the endergonic process are generated in metabolic reactions, and catalysis is performed by an enzyme system called nitrogenase.

Table 1. Classification of N-fixing Organisms.

Type(s) of organism	Example(s)
I. Free-living bacteria	
A. Heterotrophs.	
1. Aerobic	*Azotobacter vinelandii*
2. Anaerobic	*Clostridium pasterurianum*
B. Autotrophs.	*Rhodospirillum rubrum*
II. Blue-green algae.	
A. Free-living	*Anabaena* spp., *Nostoc* spp.
B. Symbiotic	In various lichens and other symbiotic associations, e.g. the fern *Azolla* and an *Anabaena* endophyte
III. Bacteria in root nodules of angiosperms	
A. Legume hosts	Soybean (*Glycine max)* + *Rhizobium japonicum*
B. Non-legume hosts.	Alder (*Alnus rubrum*), *Ceanothus velutinus,* and *Myrica gale* + unidentified microbial symbionts

Many kinds of plants perform this process, either indenpendently or in symbiotic association with other kinds of plants, as the following classification shows (Table 1).

It is estimated that a vigorous crop of nodulated legumes can fix 200 or more pounds of N per acre per year.

Symbiotic N-fixation in nodules of legumes is truly a symbiotic function, since neither the legume host nor the *Rhizobium* symbiont can fix nitrogen apart from the other. Of course the legume can live and function very well without symbiotic association with *Rhizobium,* obtaining its N by absorbing chiefly nitrate from the soil. The same species of *Rhizobium* found in nodules also live very well as free living saprophytes in the soil;, however, unlike certain other free-living species of bacteria (e.g. *Clostridium* and *Azotobacter* spp.), free living *Rhizobia* do not fix nitrogen. Although N-fixation in nodules of legumes strictly depends upon the symbiotic association of the two species, in fact it is the transformed *Rhizobium* ``bacteroids" which perform the actual biochemistry of reducing N_2 to ammonia.

A highly significant advance was made independently in several laboratories in about 1966, when it was discovered that the nitrogen-fixing enzyme system, nitrogenase, from *Azotobacter vinelandii, Clostridium pasteurianum* and soybean nodules (*Glycine max* + *Rhizobium japonicum*) is not exclusively specific for N_2 but is capable of catalyzing reduction of several other compounds as well, including reduction of acetylene to ethylene. This discovery has greatly expedited subsequent research on analysis of N-fixation. Formerly it was necessary to assay N-fixation using the heavy isotope ^{15}N, the analysis of which required mass spectrometry, or some other relatively insensitive, time-consuming procedure. Now N-fixation can be easily assayed at low cost in comparatively little time by measuring the production of ethylene from nodulated plants nodules and bacteroids isolated from nodules and cell-free extracts.

Assay of Nitrogenase Activity in Free-living Bacteria and Nodules

Principle

All the nitrogen fixing micro-organisms exhibit a unique enzyme system "nitrogenase" either in free-living state or in association with a plant host. Nitrogenase catalyses the reduction of a number of substrates, in addition to nitrogen, e.g., N_2O, N_3^-, CN^-, NC^-, C_2H_2. The ability of nitrogenase to reduce acetylene has been used as a method for its assay (Dilworth, 1966). The reactions may be designated as follows :

$$N_2 + 6H^+ + 6e^- \text{ — } 2NH_3$$

$$C_2H_2 + 2H + 2e^- \text{ — } C_2H_4$$

The technique involves incubation of nitrogenase containing system (e.g. nodules, free-living bacteria) in a known volume of acetylene and after some time ethylene produced is measured by a gas-chromatograph, using flame ionisation detector (Stewart *et al.*, 1967, Hardy *et al.*, 1968).

Procedure

(A) Free-living Bacteria : Free living bacteria like *Klebsiella, Azotobacter* express nitrogenase on a simple medium having a carbon source but a few organisms like *Rhizobium* require stringent conditions of growth for the expression of acetylene reduction activity. Several media recipes for induction of nitrogenase as indicated by acetylene reduction in free-living rhizobia are now available. The composition of one, the most commonly used media CS7 (Pagan *et al.*, 1975) is as follows :

Sodium succinate	02 mM	KI	6 um
L-arabinose	2.5 mM	$CuSO_4 . 5H_2O$	0.8 uM
Myoinositol	5.6 mM	$Na_2MO_4 . 2\ H_2O$	0.4 uM
Glutamine	2 mM	$CoCl_2.6H_2O$	0.4 uM
$MgSO_4, 7H_2O$	0.14 mM	$FeSO_4 . 7H_2O$	54 uM
KCl	0.9 mM	Na_2 . EDTA	54 uM
$CaCl_2.2H_2O$	7 mM	Thiamine-HCl	15 uM
KH_2PO_4	2.2 mM	Nicotimic acid	41 uM
$MnSO_4 . 4\ H_2O$	58 um	Pyridoxine-HCl	2.4 uM
H_3BO_3	82 um	Agar (Difco)	1% (w/v)
$ZnSO_4 . 7H_2O$	3.5 um	pH	5.9

The following procedure is followed for the acetylene reduction assay of nitrogenase :

(1) Prepare slants by taking 5 ml CS7 media in 15 ml corning glass tubes. Keep at least 10 replications.

(2) Inoculate slants and grow rhizobia for 8 days at 28 ± 1°C.

(3) Replace cotton plugs of slants with leak proof rubber stoppers and replace 10% volume of test tubes by acetylene using air tight syringe. Acetylene is prepared by using calcium carbide.

(4) Incubate tubes at 28±1°C for 24 h.

(5) Inject 1 ml gas from test-tubes into GLC for ethylene assay using erature Flame Ionization Detector (FID). GLC is fitted with a Porapak-R column, injection port temperature 90°C and column temperature 60°C.

(6) Inject 1 ml of standard ethylene into GLC.

(7) Record the GLC value on chart for experiment and the standard ethylene.

(8) Estimate total protein of each test tube by collecting cells in 1 N NaOH. Keep for 10 min in boiling water, neutralise NaOH and estimate proteins by Lowry's method.

(9) Calculate acetylene recduction activity using the following formula.

mM acetylene reduced / mg protein / h

$$= \frac{Ce \ x \ U \ x \ As \ x \ Vt}{22.4 \times 10^{6} \times T \times P \times Ue \times Ae}$$

Where Ce : Conc of standard ethylene in vpm.

Us : GLC chart units for 1 ml standard ethylene.

As : atternation used for recording 1 ml of standard ethylene.

Vt : Volume of test tube in ml.

T : Time of incubation in hrs.

P : protein content of slant in mg.

Ue : GLC charge units for 1 ml sample gas.

Ae : atternation used for recording 1 ml of sample gas.

(B) Nodules

Acetylene reduction can be measured eiher in detached nodules or in intact root system. The following procedure is followed :

1. Legume plants having sufficient nodules are uprooted and gently shaken to remove soil. Washing with water is not advised as the water forms an acetylene impermeable film around nodules. Weigh the nodules for detached nodule assay.
2. The intact root system or immediately detached nodules are transferred to the assay flask. 10 per cent air is replaced with equal volume of acetylene.
3. Incubate the flasks at 28°C for 30 min. Amount of acetylene reduced is assayed using GLC.

4. Acetylene reduction activity is expressed as C_2H_2 reduced/h/mg fresh wt of nodules of as per plant.

Comments

Acetylene is a dangerous gas, forming exposive mixture with air, it should therefore be handled carefully. The gas mxiture for incubation may have to be suitably manipulated as nitrogenase is sensitive to oxygen. Many time vessels in which assay is conducted leak and therefore, results in low gas peaks. The gas peaks could also be low if there is a blockage in the needle of the syringe. Any drop in the pressure of the carrier gas, due to a leaking injection serum or declining cylinder supply, or exhaustion of H_2 or air cylinders lowers both C_2H_2 and C_2H_4 thus stoppers are removed from the assay vessels immediately after the assay is completed and allowed to rest for 2-4 weeks before reuse. The volume of the incubation vellel should be adequate to minimise changes in the conditions during assay.

A major problem in the use of the acetylene reduction assay to estimate absolute rates of N_2 fixation is the calibration of the assay. Theoretically, a C_2H_2 to N_2 ratio of 3 : 1 can be justified, however, in practice, rates from 5 : 1 upto 25 : 1 have been recorded (Hardy *et al.*, 1973).

ACC Oxidase

The activity, of ethylene forming enzyme (EFE) is measured by the method given by Fearn and LaRue (1991).

Reagent

(1) 20 mM ACC

Procedure

The roots and nodules are separated from the shoot and shaken gently to remove sand. Then 1.0 g of roots and nodules are incubated in 15 cc reaction vials separately containing 0.5 ml of 20 mM ACC. The vials are then sealed with Teflon/silicon discs and kept in dark for 3 h at 25°C. Then 1.0 ml of gaseous sample is taken from each vial and analysed for ethylene on the Gas Chromatograph (Nucon, 5700).

The activity of ACC oxidase is expressed as n moles C_2H_4 g^{-1} DW h^{-1}.

1-Aminocyclopropane-1-Carboxylic Acid (ACC) Content

Free ACC content in the roots and nodules is assayed following the method of Miller and Pengelly (1984).

Reagents

(i) 5% (w/v) sulfosalicylic acid (ph 1.8)

(ii) 50 mM $HgCl_2$

(iii) 2.6% (w/v) NaOCl in 5 N NaOH

Procedure

One g of roots and nodules is ground separately in 2 ml of 5% 5-sulfosalicylic acid with a mortar and pestle and the extract is centrifuged at 30,000 g for 30 min at 4°C. Then 1.0 ml of supernatant and 0.5 ml of 50 mM $HgCl_2$ are added to 3.0 ml of 5% 5-sulfosalicylic acid in 15 ml reaction vials (pierce). The vials are then sealed with Teflon/silicon discs and 1.0 ml of 2.6% NaOCl in 5 N NaOH is injected into the vials. The vials are then vortexed for 5 seconds and incubated on ice for 2 h. Then 1.0 ml of gaseous sample is taken from each vial and assayed for ethyelene production on the Gas Chromatograph (Nucon, 5700). The ACC content is measured in terms of n moles C_2H_4 g^{-1} DW h^{-1} using standard ethylene (110 vpm) for comparison purpose.

Acetylene Reduction Assay (ARA)

Nitrogenase activity of nodules is determined by acetylene reduction assay (Hardy *et al.*, 1968).

The nodulated roots are separated from the shoots and shaken gently to remove sand. These ndoulated roots are placed in 50 cc assay vials which are immediately closed with airtight rubber stoppers. 2.5 ml of air is taken out from each flask and an equal volume of acetylene is injected. These are incubated for 30 minutes. Ethylene produced is estimated by injecting 1.0 ml of gas sample from the assay flasks into a steel column (2 m x 2 mm) fitted with Porapak-R (Water Associates Inc.) fitted in a Gas Chromatograph (Nucon 5700 Hydrocarbon Analyser) using a flame ionization detector (FID). Nitrogen at a flow rate of 25 ml min^{-1} is used as a carrier gas and hydrogen is used as fuel gas. Standard ethylene (110 ppm) is used for quantification of data.

Activity is expressed as μ mol C_2H_4 produced g^{-1} nodules DW h^{-1} (specific activity). The formula used is

$$\frac{A \times Cu \times V}{t \times g}$$

where,

A = Attenuation factor

Cu = Chart unit

V = Volume of flask

t = time (sec.)

g = dry weight of nodules

Ethylene Evolution

It is measured by the method of Fearn and LaRue (1991).

Procedure

The roots and nodules are placed separately in 50 ml reaction vials containing 0.1 ml of sterile water. The vials are then sealed with teflon/silicon discs and kept in dark for 4 h at 25°C. Then 1 ml of gas sample is taken from each vial and assayed for ethylene production on Gas Chromatograph (Nucon 5700). Standard ethylene (110 vpm) is used for comparison purpose.

Amount of ethylene evolved is calculated as n moles C_2H_4 g^{-1} DW h^{-1}.

ENZYME ASSAY IN NODULES

Extraction

500 mg of roots and 500 mg of nodules are washed in chilled distilled water, homogenized separately with a chilled pestle and mortar in 6 ml extraction buffer (0.1 M phosphate buffer, pH 7.0). The extract is then centrifuged at 10,000 rpm at 4°C for 20 minutes. The supernatant is then used for the estimation of following enzymes.

Catalase (EC 1.11.1.6)

Catalase activity is estimated by the UV method of Aebi (1983).

Reagents

(i) 10 mM H_2O_2

(ii) 30 mM potassium phosphate buffer (pH 7.0)

Procedure

600 µl of enzyme extract is taken and to this added 0.1 ml of 10 mM H_2O_2 and 2 ml of 30 mM potassium phosphate buffer. The enzyme sample is added immediately at the time of taking the absorbance and incubated for 3 min. The change in absorbance is recorded at 240 nm at an interval of 10 sec. for one min. The enzyme activity is expressed

as μ moles of H_2O_2 decomposed mg^{-1} protein by using H_2O_2 extinction coefficient 3.6 μM^{-1} cm^{-1}.

Peroxidase (EC 1.11.1.7)

Reagents

(i) Phosphate buffer (0.1 M, pH 7.0)

(ii) Guaiacol solution 20 mM

(iii) H_2O_2 solution 12.3 mM

Procedure

3 ml buffer solution, 0.05 ml guaiacol solution, 0.1 ml enzyme extract and 0.03 ml H_2O_2 solution are taken in a cuvette and placed in spectrophotometer. Wait until the absorbance has increased by 0.05. Time required in minutes (Δt) to increase the absorbance by 0.1 is recorded.

The enzyme activity is expressed as units mg^{-1} protein min^{-1}. 0.01 change in absorbance = 1 unit.

Glutathione Reductase (EC 1.11.1.9)

Glutathione reductase (GR) activity is estimated by the method of Goldberg and Spooner (1983) with little modifications.

Reagents

(i) 100 mM potassium phosphate buffer (pH 7.0)

(ii) 0.015 M EDTA (15 μM)

(iii) 9.6 μM NADH

(iv) 0.065 μM oxidised glutathione (GSSG)

Procedure

0.1 ml enzyme extract is taken and diluted to 1 ml. Of this, 0.3 ml is taken. To this 1.7 ml of phosphate buffer, 0.1 ml of EDTA and 0.1 ml of oxidised GSSG are added. 0.025 ml of NADH is added after 5 minutes and mixed thoroughly. The reaction is monitored and readings are taken at 340 nm after every 10 second. Under the assay conditions change in 0.1 absorbance corresponds to 1 unit and enzyme activity is expressed as units mg^{-1} protein min^{-1}.

Ascorbate Peroxidase (EC 1.11.1.11)

Ascorbate peroxidase (ASC-POX) activity is measured by the method of Nakano and Asada (1981).

Reagents

(i) 50 mM potassium phosphate buffer (pH 7.0)

(ii) 0.5 mM ascorbic acid

(iii) 1 mM H_2O_2

Procedure

0.1 ml enzyme extract is taken and diluted to 2.0 ml. Of this, 0.3 ml is taken and 1.9 ml of potassium phosphate buffer containing 0.5 mM ascorbic acid is added. The peroxidase reaction is started by addition of 0.1 ml of H_2O_2 and the absorbence decrease is recorded at 5 sec. intervals after this addition, at 260 nm. Under the assay conditions a decrease in 0.01 absorbance corresponds to 3.6 n mole ascorbic acid oxidised.

Superoxide Dismutase (EC 1.15.1.1)

The superoxide dismutase (SOD) activity is estimated by the method of Giannopolitis and Ries (1977) with little modifications.

Reagents

(i) 60 μM riboflavin

(ii) 17 mM methionine

(iii) 0.5 mM nitroblue tetrazolium (NBT)

(iv) 1.5 M Na_2CO_3

(v) 3 mM EDTA

Procedure

The reaction mixture contains 0, 0.2 ml, 0.3 ml, 0.5 ml and 1 ml of enzyme extract in separate sets with addition 1 ml of each of methionine. NBT, EDTA and Na_2CO_3 and the total volume of 5 ml is made with buffer in each set, adjusting the pH at 10.2. Then 1 ml of riboflavin is added to each set in the last. The tubes are shaken and placed 30 cm from light bank consisting of two 10-15 W fluorescent lamps. The reaction is allowed to run for 15 minutes and then stopped by switching off the light. The tubes are immediately covered with a black cloth. Then the absorbance is recorded at 560 nm.

A non-irradiated reaction mixture which did not develop a colour, serve as the control. However, in the presence of SOD, the reaction is inhibited and the amount of inhibition is used to quantitate the enzyme.

Log A_{560} is plotted as a function of volume of enzyme extract used in reaction mixture. From the resultant graph, volume of enzyme extract corresponding to 50 per cent inhibition of the photochemical reaction is obtained and considered as one enzyme unit.

The enzyme activity is expressed as unit mg^{-1} protein min^{-1} (0.01 change in absorbance = 1 unit).

Glutathione Transferase (EC 2.5.1.18)

Glutathione transferase activity is estimated by thiolysis method (Habig and Jakoby, 1981).

Reagents

(i) p-nitrophenylacetate (0.2 mM) in 0.1 M potassium phosphate buffer (pH 7.0).

(ii) 50 mM GSH solution.

Procedure

0.3 ml of enzyme extract is taken and 1 ml of phosphate buffer containing p-nitrophenyl acetate is added. Then 0.1 ml of GSH solution is added at the time of taking absorbance. Change in OD is recorded at 400 nm at time interval of 15 sec for 5 min. The extinction coefficient for p-nitrophenolate at pH 7.0 is taken as 8.79 mM^{-1} cm^{-1}. The enzyme activity is expressed as μ moles p-nitrophenol mg^{-1} protein min^{-1}.

Enzymatic Protein

The enzymatic protein is determined by Folin-cio-Calteau reagent (Lowry *et al.*, 1951).

The enzymatic extract prepared for enzyme assays is used for estimation of enzymatic protein.

Reagents

(i) 2% aqueous Na_2CO_3 (in 0.1 N NaOH).

(ii) 0.5% $CuSO_4$. 5 H_2O in 1% potassium sodium tartarate.

(iii) Alkaline $CuSO_4$ solution : Mix 50 ml of Na_2CO_3 solution with 1 ml of $CuSO_4.5H_2O$ solution.

(iv) Folin-cio-calteau reagent : 1 N.

Procedure

To 0.1 ml of appropriately diluted protein extract 5 ml of alkaline $CuSO_4$ solution is added. Mix well and allow to stand for 10 min. Then 0.5 m of Folin reagent is added and mixed vigorously on cyclo mixture. The reaction mixture is kept for 30 minutes and then the OD is taken at 500 nm. Standard curve is prepared using graded concentrations of Bovine serum albumin (BSA).

Enzyme protein is expressed as mg g^{-1} DW.

Hydrogen Peroxide

H_2O_2 content of the roots and nodules is determined by a modified Patterson *et al.* (1984) method.

Reagents

(i) 5% trichloroacetic acid (TCA).

(ii) 100 mM potassium phosphate buffer (pH 8.4)

(iii) Colorimetric reagent : It is prepared by mixing 0.6 mM M_4 - (2 pyridyzalo) resorscinol and potassium titanium oxalate in 1 : 1 (v/v) ratio. The mixture is kept on ice until use.

Extraction

500 mg of roots and nodules are homogenized separately with 0.2 g of activated charcoal and 10 ml of 5% TCA. The homogenate is filtered through Whatman No. 1 filter paper and centrifuged at 8000 x g for 10 min. The supernatant is filtered through three layers of Whatman No. 1 filter paper.

Procedure

200 μl of the extract is brought to 4 ml with potassium phosphate buffer. The pH is adjusted to 8.4 with concentrated ammonia solution. Then 2 ml of colorimetric reagent is added to the reaction mixture. The absorbance is then read at 508 nm. H_2O_2 is determined from the different in absorbance between sample and blank. The H_2O_2 content is calculated using its molar extinction coefficient of 3.6 x 10^{-4} moles and the H_2O_2 content is expressed as moles g^{-1} DW.

Lipid Peroxidation

The level of lipid peroxidation is measured in terms of malondialdehyde (MDA) present in the tissues (roots and nodules). MDA is a product of lipid peroxidation and is measured by thiobarbituric acid (TBA) reaction with some modifications of the method of Heath and Packer (1968).

Reagents

(i) 0.1% Trichloroacetic acid (TCA).

(ii) 20% TCA containing 0.5% thiobarbituric acid (TBA).

Extraction

500 g of roots and nodules are homogenized separately with 5 ml of 0.1% TCA. The homogenate is centrifuged at 8000 rpm for 15 min. The supernatant is then directly used for the assay.

Procedure

One mol of the supernatant is taken and precipitated by 4 ml of 20% TC containing TBA. The mixture is heated in a water bath shaker at 95°C for 3 min. and quickly cooled in an ice-bath. After centrifugation at 8000 rpm for 10 min the absorbance of the reaction mixture is read at 532 nm and the value for non specific absorption at 600 nm is subtracted. The concentration of MDA is calculated using its extinction coefficient of 155 mM^{-1} cm^{-1}.

Leghemoglobin (Lb) Content

Immediately after uprooting the plants, the nodules are detached, washed and weighed. Their leghemoglobin is determined by the method of Hartree (1955) with some modifications. This method is based on the conversion of haematin in pyridine hemochromogen.

Reagents

(i) Diluent buffer : 0.1 M potassium phosphate buffer (pH 7.0).

(ii) Alkaline pyridine reagent : Dissolve 0.8 g of NaOH in 50 ml water and cool. Add 33.8 ml pyridine, dissolve and dilute to 100 ml with water. This produces 4.2 M pyridine in 0.2 M NaOH.

(iii) Sodium dithionite ($Na_2S_2O_4$) : Grind finely and store in small stoppered tubes in desiccator.

Extraction

The whole plant nodules are homogenised in 6 ml of chilled 0.1 M phosphate buffer (pH 7.0) in a pestle and mortar placed in an ice-bath.

The extract is centrifuged at 5000 rpm for 30 min at 4°C in refrigerated centrifuge.

Procedure

To 2.0 ml aliquot of supernatant, 2.0 ml of alkaline pyridine reagent is added. 50 mg of sodium dithionite is added to this alkaline mixture. After half an hour, the mixture is centrifuged at 5,000 rpm for 20 min. and the absorbance of supernatant is taken at 555 nm. A standard curve is prepared using graded concentrations of hemin dissolved in 0.1 N NaOH.

ESTIMATION OF UREIDES AND UREIDE FORMING ENZYMES

Leguminous plants have adopted different strategies for assimilating and transporting fixed nitrogen. The amides are predominant form of organic nitrogen in xylem exudate of many nitrogen fixing plants including lupin, pea, lyrica and guar. The ureides, allantoin and allantoic acid are the principal forms of organic nitrogen exported from nodules of soybean, cowpea, pigeonpea and other tropical legumes (Herridge *et al.*, 1978; Schubert, 1981; Luthra *et al.,* 1981).

Estimation of Ureides by Acid and Alkaline Hydrolysis

This method involves the estimation of glyoxylate. Glyoxylate is converted to its phenyl hydrazone, which is oxidised in strong acid with ferricyanide to produce an intensely coloured formazan. Although, specific sensitive and accurate, this method involves numerous reagent additions, both water and ice baths, and large volumes of concentrated HCl. Analyzing many samples in a selection of breeding program would be too expensive and laborious using this procedure.

This method is based on Trijbel and Vogel (1966). To 1 ml of diluted extract, 1 ml of 0.5 N NaOH and 2 drops of freshly prepared phenylhydrazine hydrochloride solution (100 mg in 30 ml water) is added. The mixture is placed in a boiling water bath for 10 min. After cooling, it is subjected to acid hydrolysis using 1 ml of 0.65 N HCl. Subsequently, 1 ml of phenyl-hydrazine solution is added and the mixture is placed in boiling water bath for 4 min. then cooled in ice bath. After 15 min, 4 ml of concentrated HCl is added followed by 1 ml of freshly prepared potassium ferrcyanide solution (100 mg in 30 ml water). The tubes are shaken and after 5 min, the absorbance is read at 525 nm. This is designated as A1.

Another 1 ml of extract is directly subjected to acid hydrolysis by adding 0.15 N HCl and processed further as mentioned above. The absorbance obtained is referred to as AII.

The concentration of allantic acid is calculated using a known concentration of allantonic acid. Absorbance A_I measures the total concentration of allantoin and allantonic acid. Allantoin concentration is computed from the difference of A_I and A_{II}.

Alternative Method

This method is based on Patterson *et al.* (1982). Plant extracts are treated with cation exchange resin. The principal advantages of this method are low cost, simple reagents, few reagnet additions, reaction at room temperature and greater sample output per day. The procedure described is useful in the routine estimation of ureides.

Procedure

Materials

Dowes HCR - a strongly acidic cation - exchange resin (nuclear ulfonic acid) obtained from Sigma Company. It is washed on sintered glass filter with 4 vol water, 2 ml 0.5 M NaOH, 2 ml 1 M HCl and enough water until the washing are neutral. The resin is drained and suspended in % ethanol, well drained with vacuum filteration on the glass filter, and stored in a sealed bottle until use.

Potassium phthalate, 2 M is adjusted to pH 4.0. Commercial household bleach is mixed with water (3 : 7, v/v) just before use in the oxidation step. Chramotography grade phenol saturated with water is mixed with methanol (1 : 1 v/v) and stored in a dark bottle. Just before use it is mixed (1 : 4 v/v) with 20% NaOH.

Procedure

Approximately 2 g (w/w) of resin is added to the vial containing the plant extract, and the vials are shaken on a reciprocating shaker for 30 min. The solutions are decanted from the resin and are stored for later analysis.

A sample containing 50-1000 nmol ureide is placed in a 18 x 15 mm test tube and diluted to 1 ml with water. One ml of phthalate buffer and 0.5 ml of oxidant are added and mixed. After 5 min at room temperature, 2.0 ml phenol reagent is added and mixed. After 10 min, 5.5 ml of water is added and mixed and the absorbance determined at 625 nm.

ENZYMES OF UREIDE METABOLISM IN NODULES

Preparation of Nodule Extract

The medium used for preparing extract will depend on which enzymes are being assayed. All the steps in the preparation of enzyme extracts are carried out at 0-4°C. Nodules are directly picked into homogenising medium, quickly blotted dry on filter paper to remove excess of liquid, weighed and then crushed in 1 volume (w/v) of 0.1 M Phosphosphate buffer (pH 7.8) containing 0.4 M sucrose, 10 mM DTT, 10 mM KCl, 1 mM $MgCl_2$, 10 mM EDTA and 1% polyvinylpolypy olidone (PVP) using a pestle and mortar.

For IMP dehydrogenase, the homogenizing buffer used is 50 mM Hepes-KOH (pH 7.5), containing 25 mM KCl, 5 mM $MgCl_2$, 1% (w/v) PVP 15% (w/v) glycerol, 10 mM 2-mercaptoethanol, 5mM DTT, 2mM GSH and 0.25 M sucrose.

For 5' nucletidase and allantoinase, the buffer used is 0.1 M tris-HCl (pH 7.5). The brei is filtered through two layers of cheese cloth previously wetted with homogenising medium, and then centrifuged at 10,000 xg for 30 min. The plant cytosol fraction can be assayed directly or preferably, after passing through sephadex G-25 equilibr-ated with a buffer suitable for use in subsequent enzyme assays.

IMP dehydrogenase (EC 1.2.1.14)

IMP dehydrogenase is the first enzyme involved in the metabolic pathway leading to the formation of ureides. The assay procedure described is based on that of Shelp and Atkins (1983).

$$\text{IMP} + \text{NAD}^+ \xrightarrow{\text{IMP DH}} \text{XMP} + \text{NADH} + \text{H}^+$$

Assay mixture (2.0 ml) consists of 100 mM tris-HCl (pH 7.5) 2 mM DTT, 10 mM $MgCl_2$, 100 mM KCl, 4 mM NAD^+, 2 mM allopunnol and enzyme extract. After equilibration at 25°C, the reaction is started by the addition of 2.7 mM IMP, and the increase in absorbance at 340 nm is monitored. IMP dehydrogenase activity is expressed as (u mol NADH formed) min^{-1} (mg protein)$^{-1}$. E 340 (NADH = 6.22 x 10^3 1 mol^{-1} cm^{-1}.

5'-Nucleotidase (EC 3.1.13.5)

The first step in the purine catabolism involves the hydrolysis of nucleotides to nucleosides. This reaction is carried cut by 5' nucleoti-dase. The assay procedure described here is based on that of Christensen and Johnsen (1983) with minor modifications.

$$\text{XMP} \xrightarrow{\text{Nucleotidase}} \text{Xanthosine + Pi}$$

Assay mixture (1.5 ml) consists of 100 mM tris-maleate, buffer (pH 5.3), 12 mM $MgCl_2$, 1 mg BSA and ezyme extract. After equilibration at 37°C, the reaction is started by the addition of 3 mM XMP. After 15 min, the rection is terminated by the addition of 1 ml of 10% TCA. The protein precipitate is removed by centrifugation and the inorganic phosphate in the supernatant is estimated by the method of Ozbun *et al.* (1973). Blanks are prepared by adding TCA before the addition of enzyme extract.

Estimation of Inorganic Phosphate

Inorganic phosphate is estimated by the method of Ozbun *et al.* (1973). To the supernatant obtained, add 3.0 ml of molybdate reagent. (1NH_2SO_4 containing 1% ammonium molybdate) and the contents are mixed thoroughly. The absorbance is then read at 700 nm against a suitable blank.

Nucleosidase (EC 3.2.2.1)

Nucleosidase hydrolyses the purine nucleosidase xanthosine, inosine and adenosine to the corresponding purine bases and ribose. The assay procedure is based on that of Triplett *et al.* (1982) by coupling with the activity of xanthine dehydrogenase present in the enzyme extract.

$$\text{Inosine} \xrightarrow{\text{Nucleosidase}} \text{Hypoxanthine + Ribose}$$

$$\text{Hypoxanthine + NAD} \xrightarrow{\text{XDH}} \text{Uric acid + NADH} + H^+$$

Assay mixture (1.5 ml) consists of 3.2 mM NAD+ and enzyme extract. After equilibration at 25°C, the reaction in started by the addition of 0.33 mM inosine, and the increase in absorbance at 340 nm is monitored. All the reagents listed above are dissolved in 100 mM phosphate buffer (pH 7.8). Nucleosidase activity is expressed as (μ mol NADH formed) min^{-1} (mg protein)$^{-1}$, E 340 (NADH) = 6.22 x 10^3 1 mol^{-1} cm^{-1}

Xanthine Dehydrogenase (EC 1.2.1.37)

Xanthine dehydrogeanse catalyses the conversion of hypoxanthine /xanthine to uric acid. The assay procedure described here is based on that of Triplett *et al.* (1982).

$$\text{Xanthine} + NAD^+ \xrightarrow{\text{XDH}} \text{Uric acid + NADH} + H^+$$

Assay mixtures (1.5 ml) consists of 3.2 mM NAD^+ and enzyme extract. After equilibration at 25°C the reaction is started by the addition of 0.33 mM xanthine and the increase in absorbance at 340 nm is monitored. All the reagents listed above are dissolved in 100 mM phosphate buffer (pH 7.8). Xanthine dehydrogenase activity is expressed as (μ mol NADH formed) min^{-1} (mg protein)$^{-1}$

$$E\ 340\ (NADH) = 6.22 \times 10^3\ 1\ mol^{-1}\ cm^{-1}$$

Uricase (EC 1.7.3.3)

Uricase or urate oxidase catalyses the oxidation of urate formed from xanthine to allantoin. The assay procedure described here is based on that of Tajima and Yamamato (1975).

$$\text{Uric acid} + 2H_2O + B_2 \xrightarrow{\text{Uricase}} \text{Allantoin} + H_2O_2 + CO_2$$

Assay mixture (2 ml) consists of 100 mM tricine KOH buffer (pH 9.0) and enzyme extract. After equilibration at 25°C, the reaction is started by the addition of 0.03 M uric acid, and the decrease in absorption at 292 nm is monitored. Uricase activity is expressed as (u mol uric acid degraded) min^{-1} (mg protein)$^{-1}$.

Allantoinase (EC 3.5.2.5)

Allantoinase catalyses the hydrolysis of allantoin to allantoic acid. The asay procedure described here is based on that of Tajima and Yamamato (1975).

$$\text{Allantoin} + H_2O \xrightarrow{\text{Allantoinase}} \text{Allantoic acid.}$$

Assay mixtures (2.0 ml) consists of 50 mM tris-HCl (pH 7.7), 3.5 mg allantoin. After equilibration at 37°C, the reaction is started by the addition of enzyme extract. After 15 min, the reaction is terminated by adding 1 ml of 0.15 N HCl. Allantoic acid formed is measured by acid hydrolysis. Allantoinase activity is expressed as (μ mol allantoin degraded min^{-1} (mg protein)$^{-1}$.

REFERENCES

Aebi, H.E. 1983. *Catalase. In: Methods of Enzymatic Analysis* (Ed. H.U. Bergemeyer) Vol. III. pp. 278-282. Weinhiem Germany, Verlag Chemic.

Bergerson, F.J. 1980. *Methods for Evaluating Biological Nitrogen Fixation.* John Wiley and Sons, New York.

Christensen, I.M.I.E. and Jochimsen, B.U. 1983. Enzymes of ureide synthesis in pea and soybean. *Plant Physiology* 72 : 56-59.

Dilworth, M.T. 1966. Acetylene reduction by nitrogen fixing preparation from *Clostridium pasteurianum. Biochemica and Biophysica Acta* 127 : 285-294.

Fearn, J.C. and LaRue, T.A. 1991. Ethylene inhibitors restore nodulation of sym. 5 mutants of *Pisum sativum* L. cv. Sparkle. *Plant Physiology* 96: 239-249.

Giannopolits, C.N. and Ries, S.K. 1977. Superoxide dismutase I. occurence in higher plants. *Plant Physiology* 59: 309-314.

Goldberg, D.M. and Spooner, R.D. 1983. Glutathione reductase. In: *Methods of Enzymatic Analysis.* Vol. II (Ed. H.U. Bergmeyer) pp. 258-265.

Habig, W.H. and Jakoby, W.B. 1981. Assay for differentiation of glutatione-S-transferases. In: *Methods of Enzymology* (Ed. Jakoby, W.B.) Academic Press, New York, pp. 398-405.

Hardy, R.W.F., Burns, R.C. and Holsten, R.D. 1973. Applications of the acetylene - ethylene assay for measurement of nitrogen fixation. *Soil Biology and Biochemistry* 5 : 47-81.

Hardy, R.W.F., Holston, R.D., Jackson, E.K. and Burns, R.C. 1968. The acetylene ethylene assay for N_2 fixation. Laboratory and field evaluation. *Plant Physiology* 43: 1185-1207.

Hartree, E.F. 1955. Heamation compounds. In: *Modern Methods of Plant Analysis* (Eds. Peach K. and Tracy, M.V.). Springer-Verlag, Berlin, pp. 197-211.

Heath, R.L. and Packer, L. 1968. Photoperoxidation in isolated chloroplasts. I-Kinetics and stiochemistry of fatty acid peroxidiation. *Archives of Biochemistry and Biophysics* 125: 189-198.

Herridge, D.F., Atkins, C.A., Pate, J.S. and Rainbird R.M. 1978. Allantoin and allantoic acid in nitrogen economy of cowpea (*Vignas unquiculata* L. Walp.). *Plant Physiology* 62 : 495-498.

Lowry, O.H., Rosenbrough, N.J., Farr, A.L. and Randall, R.J. 1951. Protein measurement with Folin-phenol reagent. *Journal of Biological Chemistry* 193: 265-275.

Luthra, Y.P., Sheoran, I.S. and Singh, R. 1981. Pigeonpea (*Cajanus cajan* L.) A ureide producing legume. *Current Science* 50 : 270-271.

Miller, A.R. and Pengelley, W.L. 1984. Ethylene production by shoot forming and unorganized cram-ball tumor tissues of *Nicotiana* and *Lycopersicon* cultivated *in vitro. Planta* 161: 418-424.

Nakano, Y. and Asada, K. 1981. Hydrogen peroxide is scavenged by ascorbate-specific peroxidase in spinach chloroplasts. *Plant and Cell Physiology* 22: 867-880.

Ozbun, J.L., Hawker, J.S., Greenberg, E., Lommel, C. and Preiss, J. 1973. Starch synthetase, Phosphorylase, ADP glucose pyro-phosphorylase and UDP glucose pyrophosphorylase in developing maize kernels. *Plant Physiology* 51 : 1-5.

Pagan, J.D., Child, J.J., Scowcroft, W.R. and Gibson, A.H. 1975. Nitrogen fixation by *Rhizobium* cultured on a defined media. *Nature* 256 : 406-407.

Patterson, B.D., Machae, E.A. and Ferguson, J.G. 1984. Estimation of hydrogen peroxide in plant extracts using tiatnium. *Analytical Biochemistry* 139: 487-492.

Patterson, T.G., Glenister, R. and LaRue, T.A. 1982. Simple estimate of ureides in soybean tissue. *Analytical Biochemistry* 119 : 90-95.

Schubert, K.R. 1981. Enzymes of purine biosynthesis and catabolism in *Glycine max*. I. Comparison of activities with N_2 fixation and composition of xylem exudates during development. *Plant Physiology* 68 : 1115-1122.

Shelp, B.J. and Atkins, C.A. 1983. Role of inosine monophosphatoxidoreductase in the formation of ureides in nitrogen fixing nodules of cowpea (*Vigna unguiculata* L. Walp). *Plant Physiology* 72: 1019-1034.

Stewart, W.D.P., Fitzgerald, G.P. and Burris, R.H. 1967. *In situ* studies on N_2 fixation using the acetylene reduction technique. *Proceedings of National Academy of Sciences,* USA. 58 : 2071-2078.

Tajima, S. and Yamamato, Y. 1975. Enzymes of purine catabolism in soybean plants. *Plant and Cell Physiology* 16 : 271-282.

Trijbels, F. and Vogels, G.D. 1966. Degradation of allantoin by *Pseudomonas acidovorans. Biochemica et Biophysica Acta* 113 : 292-301.

Triplett, E.W., Blevins, D.G. and Randall, D.D. 1982. Purification and properties of soybean nodule xanthine dehydrogenase, *Archives of Biochemistry and Biophysics* 219 : 39-46.

CHAPTER 23

ASSAY OF GROWTH HORMONES

INTRODUCTION

One of the important growth-controlling systems in plants is provided by the so-called plant growth substances or plant hormones. The term hormone is, of course used by animal physiologists too, and indeed the concept of such chemical messengers in the body was first developed by them in the early part of this century. Animal hormones differ from plant hormones in a number of ways, of which perhaps the most important is that they are produced in specific organs or glands and often have very highly specific effects. Plant hormones, on the other hand, though they may be produced in fairly restricted regions of plants, are manufactured by unspecialized cells and frequently have many different effects upon the plant, depending upon other circumstances.

A plant growth substance (or plant hormone) is an organic substance which is produced within a plant and which will at low concentrations promote, inhibit or qualitatively modify growth, usually at a site other than its place of origin. Its effect does not depend upon its calorific value or its content of essential elements.

Basically there are five different group of plant hormones which are outlined briefly here.

(1) Auxins : These are substances chemically related to indole acetic acid (IAA) which itself appears the principal auxin of many plants. They are characterized by their ability to promote growth in certain biological tests involving the use of excised parts of plants freed as far as possible from their own endogenous auxins.

(2) Gibberellins: These are substances chemically related to gibberellic acid (usually abbreviated to GA_3), which is a metabolic product of the fungus *Gibberella fujikuroi* and can be obtained from the liquid medium in which the fungus has been cultured. The gibberellin molecule is based on a gibbane skeleton.

(3) Cytokinins : These are substances which are derivatives of the purine adenine, which is well known as one of the nitrogenous bases

in the molecules of the nucleic acids DNA and RNA. They are characterized by their ability to interact with IAA to promote cell division in cultures of plant cells grown on artificial media, and especially by their property of affecting the patterns of differentiation which occur in such cultures. They have many other properties but this one is critical. An example of a naturally occurring cytokinin is zeatin, which has been obtained from maize grains.

(4) Inhibitors : These are many substances present in plant cells, which generally inhibit plant processes. Such substances most able to be considered as inhibitors in the hormone sense are those similar to structure and properties to abscisic acid (ABA) which is characterized by its ability to inhibit many growth phenomena in plants, but perhaps especially by its association with bud dormancy in woody plants and with the abscission of leaves in the cotton plant.

(5) Ethylene : This simple substance has long known to affect the growth of plants especially in triggering the climateric rise in respiration in fruits. Ethylene is a gas that forms through the *yang cycle* from the breakdown of methionine which is in all cells.

ASSAY OF AUXINS

Extraction of IAA from Culture Filtrates

Fungi and bacteria produce IAA in culture medium containing L-tryptophan which is the precursor of IAA (Mahadevan, 1984). Usually, L-tryptophan (0.005 M) is added to the culture medium, inoculated with the microorganism and incubated in the dark for a suitable period (10-15 days for fungi; 2-3 days for bacteria). The auxin is extracted and identified by either biological assays or chemical tests.

Materials

Filtrates from 10-day-old cultures of *F. oxysporum f. vasinfectum* or *P. oryzae* grown on Czapek's medium supplemented with 0.005 M L-tryptophan (or 0.01 M DL-tryptophan).

Separatory funnel

Centrifuge

pH meter

Rotary evaporator or water bath with hot plate

Reagents

HCl, 1 N, 6 N

Sodium bicarbonate, 5%

Ethyl ether (peroxide free)

Methanol (analytical grade)

Method

Centrifuge 100 ml of the culture filtrate at 2,000 g for 30 min to remove mycelial fregments and spores. Acidify the supernatant to pH 3 with 1 N HCl employing a pH meter and extract with 100 ml of ethyl ether in a spearatory funnel. Repeat the extraction with ether three more times, using 100 ml for each extraction. Complete extraction is obtained if ether with the filtrate is kept at 4°C for at least 1 hr between each solvent change. Pool the ether extracts and reduce the volume to about 25 ml either by flash evaporation or in a rotary evaporator at 35-40°C.

Partition the ether fraction with 25 ml of sodium bicarbonate. Repeat it three more times, each time with 25 ml of bicarbonate. Acidify the bicarbonate fraction to pH 3 with 6 N HCl using a pH meter and extract with 50 ml of ethyl ether. Repeat the extraction with 3 x 50 ml of ether. Evaporate the ether in vaccum and dissolve the residue in 2 ml of methanol. Store in a vial at -5 to -10°C.

Extraction of IAA from Plant Tissues

The isolation of auxin from plants still poses some problems. Plant extracts are usually tested by bioassays and caution should be exercised since sugars, amino acids, vitamins, inorganic salts and phenols exhibits some auxin-like activity (Sridhar, 1967; Chandramohan *et al.*, 1968).

Materials

Plant seedlings

Blender or pestle and mortar

Cheese cloth

(Other requirements as above)

Reagents

Absolute alcohol (methanol or ethyl acetate can be used)

(Other requirements as above)

Method

Freeze the tissue (200-250 g) immediately after collection. Cut into pieces of 1-2 cm and homogenize with pre-chilled absolute alcohol (alternatively methanol or ethyl acetate can be used) in a blender.

Filter through cheese cloth and repeat the extraction with the residue. Reduce the volume of the pooled filtrate on a rotary evaporator to about 50 ml with reduced pressure at 40-50°C. Adjust the pH of the aqueous residue to 3 with 1 N HCl and extract three to four times with equal volumes of peroxide-free ether over a period of an hour using a separatory funnel. Combine the ether extracts and partition with 4 x 100 ml of $NaHCO_2$ solution to separate acidic and neutral auxins. The ether layer contains the neutral auxins while the bicarbonate fraction contains acidic substances. Acidify the alkaline fraction to pH 3 with N HCl and extract with 4 x 100 ml of peroxide-free ether. Pool the ether extracts which contains acidic auxins. Evaporate both the ether extracts containing neutral and acidic auxins seperately on a rotary evaporator to dryness. Dissolve the residue in minimum quantity of methanol. Use this extract for chromatography. Store the extract at -5 to -10°C.

Alternatively, IAA from the frozen tissue samples can be extracted directly with ether. Grind tissues to a fine paste in a mortar with pestle. Extract the homogenate with 200 ml of peroxide-free ethyl ether by shaking in a seperatory funnel and allowing the layers to separate at 4°C. Repeat the extraction with 3 x 100 ml ether. It is advisable to prolong the time of extraction. Reduce the volume of ether to represent 2 ml extract for 5 g tissue (40 ml) by evaporation at 40-50°C either on a water bath or an a rotary evaporator with reduced pressure. This fraction is shaken with 40 ml of $NaHCO_3$ in a separatory funnel. Proceed as above to separate neutral and acidic substances.

Estimation of Auxins

Presence of auxins in the extracts is either chemically or biologically detected and the auxin level is measured.

Chemical Detection of IAA

Separate the extract containing auxins by chromatography: paper, thin-layer or column; the first two methods are widely used.

Materials

Auxin extract

Micropipettes

Glass sprayer

Chromatography set up for paper or TLC kit

Hot air oven

Reagents

IAA reference solution, 5 x 10^{-4} M

NH_4OH

Isopropanol

Detecting reagents

Salkowski reagent : Mix 2 ml of 0.05 M $FeCl_3$ and 100 ml of 5% $HClO_4$. Use fresh reagent prepared from stock solutions. Examine the chromatogram after 30 min of spraying. Appearance of pink spots is characteristic of indoles.

Ehrlich's reagent : 2% p-Dimethyl aminobenzaldehyde in 2 N HCl in 80% ethanol. Spray the chromatogram dry in an oven at 100°C for about 8-10 min. Examine the development of gray colour which is typical of indole.

Method

Spot 100 μl of the extract on Whatman No. 1 paper and develop descendingly in isopropanol-ammonia-water (80 : 10 : 10 v/v), IAA separation, especially from tryptophan is very sharp in the solvent. Run reference samples of IAA. Developing time is about 12 to 14 h.

IAA is readily separated on thin-layer plates coated with silica gel G (250 μ thickness). Developing solvents used for paper chromatography are usually used for TLC. The sensitivity is high compared with paper chromatography.

Spray with either Salkowski or Ehrlich's reagent to detect IAA. Ehrlich's reagent is more sensitive than Salkowski.

Quantitative Measurement of IAA

Colorimetric Method

Gordon and Paleg (1957) standardized a method to measure IAA. This method permits the detection of IAA in crude extract. The methanol extract from the culture filtrates of microorganisms can be directly estimated for IAA as there are no interfering substances. However, plant extracts need purification.

Materials

IAA extract from culture filtrate of microorganisms or chromatograms of plant extract

Pipettes, 1 ml, 2 ml

Test tubes

Colorimeter

Reagent

Salper-reagent : Mix 1 ml of 0.5 M $FeCl_3$ in 50 ml of 35% $HClO_4$; prepare frssh.

Method

Remove the spot containing IAA from the chromatogram corresponding to the authentic IAA and elute in 1 or 2 ml of methanol or ethanol (elution time, 30 min).

To 1 ml of extract containing IAA add 2 ml of Salper reagent. Add the reagent dropwise but rapidly with continuous agitation. Incubate the samples in the dark (35 min for diethyl ether dissolved IAA; 60 min for methanol sample). The pink colour is stable. Measure its absorbance at 565 nm against a solvent-reagent blank. If the intensity is deep, dilute the reaction mixture with the solvent. From a standard curve drawn from known concentrations of IAA, find out the quantity of IAA in the extract.

Gas-liquid Chromatographic Estimation

Materials

Healthy and blast-infected plant leaves

Blender

Cheese cloth

Whatman No. 41 filter paper

Rotary evaporator

Separatory funnell

Activated charcoal

Gas-liquid chromatograph equipped with flame ionization dectector and OV 101 column.

Reagents

Acetone

Ethyl acetate

$NaHCO_3$, 5%

H_2SO_4

Petroleum ether (40-60°C)

Nitrosomethylurea

KOH, 40%

Method

Cut 25 g leaves into 2 to 3 mm sections. Homogenize with 75 ml acetone and filter. Concentrate the acetone extract under reduced pressure in a rotatry evaporatory at 60°C to 20 ml. Add sufficient quantity of ethyl acetate to the concentrated extract. Adjust the pH of the solution to 8.4 by adding $NaHCO_3$. This will extract IAA in aqueous layer. Readjust the pH to 2.5 with H_2SO_4. Extract the aqueous phase thrice with equal volume of ether. Dehydrate the ether solution with Na_2SO_4 and evaporate the solvent. Dissolve the residue in acetone and pass through activated charcoal. Evaporate the acetone. Metylate IAA with diazomethane in ether. Diazomethane can be prepared in dist 40-60°C petroleum ether using nitrosomethylurea in 40% KOH solution. Analyse IAA using a gas-liquid chromatograph equipped with N-P flame ionization detector and OV 101 column in a temperature range of 190 to 230°C using fenthion as an internal standard.

Bioassays for IAA

There are several biological assays to measure auxin activity; some of them require special rooms to grow seedlings while a few tests are relatively easy to perform. Avena straight growth test, pea curvature test and rice root inhibition assay are easy to conduct in the laboratory and are described (Sridhar *et al.,* 1968; Vyas and Jain, 1977).

Avena Straight Growth Test

Cell extension of Avena coleoptiles is stimulated when floated on solutions containing auxin. This test is sensitive to inhibitors that retard growth as well as to auxins that stimulate elongation.

Materials

Oat seeds, locally available cultivar

Wooden trays to grow seedlings

Sharp razor

Petri dishes (4- cm)

Pipettes, 5 ml

Dark room or a dark chamber

Photographic enlarger set

Ruler

Extract having auxin

Reagent

Phosphate citrate buffer, pH 5.5 (0.01 M) containing 1-2% sucrose

Method

Sow the oat seeds in vermiculite or sand in wooden trays; avoid waterlogging and place them in a dark room. Use 3-4 day-old seedlings having coleoptiles of 1.4 to 1.6 cm. Reject the coleoptile tip with the surrounding tissue (3 mm); cut 10 mm length sections and float them in a 4-cm Petri dish containing a suitable aliquot (5-10 ml) of the test solution.

Chromatogram strip can be readily assayed for auxin activity by cutting the paper into small pieces, approximately of 0.05 Rf units. Cut each piece into smaller pieces and immerse in 5 ml of the buffer in Petri dish. The strip is allowed to elute overnight in the buffer and the resulting solution is assayed without removing the paper strips.

Perform the test at 25°C at room temperature and under dark. Incubate the dishes for 24-48 h and usually measurable response is obtained in presence of auxin after 18 h. Measure the section lengths with a photographic enlarger set at a magnification of 4. Measure the images of the sections with a ruler to the nearest mm. Alternatively, the sections can be photographed and measured. The linear growth of the segment is proportional to the log conc. of auxin in the test solution. One mm of coleoptile growth is achieved by 2.2×10^{-1} mg IAA. Calculate IAA conc. from a dosage response curve. Interpret the data statistically.

Comment

Chromatogram strips can be eluted in dist water. However, better results are obtained with 1-2% sucrose and/or 0.01 M phosphate citrate buffer, pH 5.5; but control growth is increased proportionately. Work in reduced light.

Rice Root Inhibition Assay

Materials

Rice seeds, any cultivar

Agar

Petri dishes

Glass vials, 10 ml

Filter paper
Ruler
Incubator at 25°C
Water bath with heater
Extract having auxin

Method

Dilute the test solution in test tubes and add bacteriological grade agar (final conc 3%). Warm the solution in a water bath to melt the agar. Distribute 5 ml of this medium when warm into 10-ml glass vials and autoclave at 120°C for 15 min. Similarly, maintain suitable controls containing only aqueous agar and vials containing known conc of IAA (10^{-5} to 10^{-1} mg/ml).

Surface sterilize the rice seeds and soak in the sterile dist water for 24 h at 25 ± 1°C in the dark in an incubator. Spread the seeds on moist filter paper in Petri dishes. Keep for another 24 h in the dark. Transfer the germinated seeds of uniform development to the sterile agar in the vials containing either the auxin extract or IAA. Allow the seedlings to grow for 48 h in the dark at 25 ± 1°C in an incubator. The primary roots measuring about 1 mm would have emerged by this time. The mean root length after 48 h incubation in expressed as per cent of the control. Alternatively, it can be expressed as per cent inhibition over the control. Prepare a standard curve by plotting the logarithms of conc of IAA (ug/l) with either percent growth compared to control or per cent inhibition over control. The amount of IAA in ug/g fresh weight of the tissue can be calculated.

Indoleacetic Acid Oxidase of Plant Tissues

The accumulation or destruction of auxin is associated with IAA oxidase, an enzyme that destroys the auxin. Inhibition of IAA oxidase leads to the accumulation of IAA and *vice versa*. Disease development is usually associated with alterations in oxidative enzymes, particularly peroxidase, phenol oxidase, ascorbic acid oxidase and catalase. However, changes in IAA oxidase during infection have not received much attention. Such studies should help us to correlate between changes in auxin level and enzyme activity.

Preparation of Enzyme Extract

Materials

Cotton or rice seedlings (15 to 20 days old)

Blender or mortar and pestle

Centrifuge

Dialyis tubing

Reagent

Mcllvaine's buffer, pH 4.8

Method

Wash the seedlings in tap water; cut into 1 to 2 cm pieces and freeze in dry ice or at -10°C. Grind the plant tissues while frozen to a fine paste. Suspend 10 g of the paste in 50 ml of Mcllvaine buffer; stir the mixtur vigorously for 30 min at 4°C. Dialyze the supernatant in dialyzing membrane against several volumes of dist water for 24 h at 4°C, changing the water at least twice. Use the dialyzed extract to assay IAA oxidase.

Measurement of IAA Oxidase

IAA oxidase activity in assayed by determining the residual IAA in the reaction mixture quantitatively using Salper reagent which reacts with IAA and produces a pink colour. The intensity of the coloured product is measured in a colorimeter (Purushothman *et al.,* 1971).

Materials

Enzyme extract

Pipettes, 1 ml, 5 ml

Erlenmeyer flasks, 100 ml

Water bath at 30 + 1°C

Colorimeter

Reagents

IAA solution, 10^{-3} M

2, 4-Dinitrophenol, 10^{-9} M

$MnCl_3$, 5 x 10^{-1} M

Mcllvaine's buffer, pH 4.8

Salper reagent : Add 1 ml of 0.5 M $FeCl_3$ in 50 ml of 35% perchloric acid; use fresh reagent

Method

To 3 ml of IAA kept in 100 ml Erlenmeyer flasks, add 0.5 ml of 2, 4-dinitrophenol, 0.5 ml of $MnCl_3$ 6 ml of Mcllvane's buffer and 2 ml of the enzyme extract. Incubate the mixture at 30°C in a water bath. At intervals of 15 min, add 3 ml Salper reagent to each flask. Read the absorbance of the solution at 535 nm in a colorimeter.

Maintain black reagent and samples withdrawn at zero time as controls.

Comment

Conduct the assay in reduced light.

GIBBERELLINS

Gibberellins are cyclic diterpenes. They contain the skeleton enantiomer of gibberellin (ent-gibberelline) for its advantage of using a uniform numbering system similar to other cylic diterpenes. So far, more than 100 gibberellins have been identified. Like auxin, they are important in growth and development of plants. One well documented instance is the bakkanae (foolish seedling) disease of rice caused by *Gibberella fujikuroi* (*Fusarium moniliforme*). The fungus produces many types of gibberellins in culture media as well as in the inoculated rice plants. The symptoms in rice plants develop rapidly when gibberellic acid conc is high. Gibberellins as a group act as a toxin to bring pathologial effects in rice plants. However, the exaggerated growth response of infected plants is not uniform in all infected plants. Both enhanced and decreased levels of gibberellins due to disease have been observed in host-parasite interactions (Mahadevan, 1984). Gibberllins are important in seed germination, affecting enzyme production that mobilizes food production used for growth of new cells. This is done by modulating chromosomal transcription.

Extraction of Gibberellins from Culture Filtrates of *G. fujikuroi*

The acidic metabolites containing gibberellins present in the culture filtrates can be extracted with ethyl acetate and the concentrated extract may be used for bioassays.

Materials

G. fujikuroi grown for 10-15 days in Czapek's medium.

Separatory funnel

Pipettes, 10 ml

Rotary evaporator or glass distillation assembly with hot water bath pH meter

Reagents

HCl, 0.5 N

Saturated $NaHCO_3$

Na_2SO_4

Ethyl acetate

Tween 20, 0.5%

Method

Mix 250 ml of saturated $NaHCO_3$ solution with 100 ml of culture filtrate kept in a separatory funnel. Partition it twice with 300 ml of ethyl acetate and discard the organic phase. Acidify the aqueous layer to pH 2.5 in a pH meter using HCl. Add an equal volume of ethyl acetate and shake vigorously for 5 min in a separatory funnel. Separate the ethyl acetate fraction; re-extract the aqueous layer two more times with 300 ml of the solvent for each extraction. Pool all the ethyl acetate fractions. Dry the organic extract over anhydrous Na_2SO_4. Evaporate the ethyl acetate extract to dryness under vacuum on a rotary evaporator at 40-45°C or distill off the solvent on a hot water bath. Dissolve the residue in 5 ml of dist water containing Tween 20. Use this extract for bioassays. For chromatography, dissolve the residue in 1-2 ml of ethanol or acetone.

Comment

Use only distilled solvents.

Extraction of Gibberellins from Plant Tissues

Gibberellins occur in plants in bound form. The existence of glucosyl-GA and GA-protein complexes has been reported. Binding influences the mechanism involved in the hormonal action especially under stress conditions. The method to extract free and bound forms of gibberellins from plant tissues is described.

Extraction of Free Gibberellin like Substances

Materials

Cotton seedlings, 4-week old infected by *F. oxysporum* f. sp. *vasinfectum*

Blender

Beakers, 500 ml

Pipettes, 5 ml, 10 ml

Buchner funnel

Separatory funnel

Whatman No. 41 filter paper

Rotary evaporator

Water bath at 60°C

pH meter

Reagents

Methanol absolute, 80%

Ethyl acetate

HCl. 1.6 and 3.2 M

$NaHCO_3$, 0.48 M

Polyvinylpyrrolidone (PVP)

Method

Homogenize 250 g of shoot portions in blender at 4°C with 80% pre-chilled methanol (10 ml/g fresh tissue). Transfer the homogenate to a beaker and allow it to stand overnight at 4°C. Filter through Buchner funnel using Whatman No. 41 filter paper. Re-extract the residue twice with ethanol in cold. Wash the residue after the final extraction with methanol until the residue is colourless. Combine the methnaol extracts and evaporate at 40°C in a rotary evaporator. Freeze the water phase of the extract; thaw it to room temperature and filter it through Whatman No. 41 filter paper. This would remove chlorophyll and precipiate proteins. Stir the aqueous phase of the extract with PVP, acidify it with 3.2 M HCl to pH 2.5 and extract thrice with equal volumes of ethyl acetate using separatory funnel.

Save the aqueous phase to extract bound gibberellins.

Combine ethyl acetate extracts and partition it with 0.48 M $NaHCO_3$. Adjust the pH of the bicarbonate phase to 2.5 with 1.6 M HCl and partition two or three times again with ethyl acetate. Combine the ethyl acetate fractions and evaporate to dryness in a rotary evaporator at 40°C. Dissolve the residue in minimum quantity of methanol. The methanol extract contains `free' acidic fractions of gibberellin-like substances.

Extraction of Bound Gibberellin like Substances

Materials

Extract containing bound gibberellins

pH Meter

Water bath at 60°C

Rotary evaporator

Pipettes, 5 ml

Reagents

Ethyl acetate

Methanol

NaOH, 3.75 M

HCl, 3.2 M

Method

Adjust the pH of the aqueous phase after extraction with ethyl acetate to 11 with 3.75 M NaOH and hydrolyse it in a water bath maintained at 60°C for 1 h. This releases the bound gibberellins. After hydrolysis, cool the content to 25°C and adjust the pH to 2.5 with HCl and partition it three times with equal volumes of ethyl acetate. Combine the ethyl acetate fraction and evaporate it in a rotary evaporator to dryness. Dissolve the residue in methanol. The extract contains the `bound' fraction of gibberellin-like substances.

Estimation of Gibberellins

Presence of gibberellins in the extracts can be detected and quantified either by chemical or biological tests. The biological assays are specific.

Chemical Detection of Gibberellins

The extracts can be separated either on paper chromatoram or on TLC plates. Paper chromatography is widely used, with isopropanol ammonia water solvent system as it leaves no toxic residues in eluants. Presence of gibberellins in the chromatoplates can be detected either by acid spray or examination under UV light (Murakami, 1959).

Materials

Gibberellin extract

Paper chromatography unit or TLC kit

Micropipettes

Glass sprayer

UV light

Hot air oven/hot plate

Reagents

Solvent for paper chromatography

Isopropanol-ammonium hydroxide-water (10 : 1 : 1, v/v)

Solvents for TLC

Benzene-n-butanol-acetic acid (30 : 15 : 5, v/v)

di-Isopropyl ether-acetic acid (95 : 5, v/v)

Isopropanol-water (4 : 1, v/v)

Spray reagents

Ethanol-conc. sulphuric acid (95 : 5, v/v)

Water-conc. sulphuric acid (30 : 70, v/v)

Method

Develop the chromatogram in a solvent system and dry at room temperature. Spray with either ethanol conc. H_2SO_4. Heat the chromatograms at 120°C for 10 min either in a hot air oven or on a hot plate. Remove the plates, cool and examine under UV light (275 mm) for fluorescent spots.

With ethanol-sulphuric acid spray, gibberellins, A_1, A_3, A_5, A_6 and A_3 which possess 7-hydroxyl group would appear as blue fluorescent spot under UV light, while gibberellins A_2, A_6, A_7 and A_9 which have no 7-hydroxyl group would appear as purpose spots. When sprayed with water-conc. sulphuric acid and the chromatograms are viewed under UV light without heating, gibberellins A_1, A_5, A_6 and A_8 would appear as blue green fluorescent spots : A_2, A_6 and A_9 as blue-purple spots and A_5 and A_7 as light yellow and brown fluorescent spots.

Comment

Bromocresol green (pH 4.0-5.6) may be used as a spray reagent to detect acid substances.

Quantification of Gibberellins

Spectrophotometric Method

The amount of GA in an extract can be estimated in a spectrophotometer by the method of Holdbrook *et al.* (1961). It is based on the conversion of gibberellenic acid followed by the measurement of its absorption at 254 nm.

Materials

Gibberellin extract

Test tubes

Pipettes, 2 ml, 10 ml

Centrifuge

Incubator at 20°C

Spectrophotometer

Reagents

HCl, 5%, 30%

Standard GA solution

Zinc acetate solution : Dissolve 21.9 g of zinc acetate in 80 ml dist water, add 1 ml of glacial acetic acid and make up the volume to 100 ml with dist water.

Potassium ferrocyanide solution : Dissolve 10.6 g of AR grade potassium ferrocyanide in 100 ml of water.

Method

Pipette 15 ml of the culture filtrate or the extract containing GA to the test tube and add 2 ml of zinc acetate to it. After 2 min, add 2 ml of potassium ferrocyanide. Centrifuge at 2,000 g for 15 min. To 5 ml of the supernatant, add 5 ml of 30% HCl in a test tube and incubate the mixture at 20°C for 75 min. Treat the blank simple with 5% HCl. Measure the absorbance of the sample and the blank at 254 nm in a spectrophotometer. Calculate the amount of GA present in the extract from a standard curve prepared with known GA, and express the results as GA, equivalent.

Colorimeteric Method

For routine quantitative measurement of gibberellins present in purified preparations, colorimetric method developed by Graham and

Handerson (1961) based on the reduction of phosphomolybdic acid to molybdenum blue can be used.

Materials

Gibberellin extract

Volumetric flask, 25 ml

Hot water bath

Ice water bath

Colorimeter

Reagent

Phosphomolybdic acid reagent : Dissolve 35 g of molybdic acid (MoO_3, 85%) and 5 g of sodium tungstate in 200 ml of 10% NaOH solution kept in a beaker, and add 200 ml of dist water. Boil the contents for 40 min to remove any ammonia present in molybdic acid. Cool the mixture to room temperature and dilute to 350 ml with dist water. Add 125 ml of conc (85%) phosphoric acid to the mixture and make up the volume to 500 ml dist water.

Method

Pipette 1 ml of the gibberellin extract into 25 ml volumetric flask and add 15 ml of phosphomolybdic acid reagent to it. Mix the contents of the flask thoroughly and place the flask in boiling water bath. After 1 h, remove the flask and immerse them in an ice water bath. After cooling to room temperature, make up the volume to 25 ml with dist water. Measure the absorbance at 780 nm in a colorimeter using dist water blank. Maintain reagent blank and subtract the absorbance of the reagent blank from the values obtained with samples.

Comment

The method is sensitive for concentrations of GA ranging from 5 to 50 μg/ml and for gibberellin from 10 to 50 μg/ml. Since amino acids and sugars would respond to the test, the method should be restricted to the determination of pure solutions of GA and gibberellins. If analysis of mixed extracts is contemplated, previous separation and purification are necessary. Prepare standard curve for each batch of reagents. If sodium tungstate is omitted from the reagent, measure the absorbance at 660 nm.

Bioassay of Gibberellins

Bioassays sensitive for gibberellins have been developed. The gibberellin-like activity may be assayed directly with partially purified extracts. While they are chromatographically separated for detailed information, it is usual practice to bioassay all the Rf fractions of the chromatogram, since more than one GA with different Rf values may be cut into 10 equal sections accroding to Rf eluted in methanol at 4°C overnight. It is necessary to concentrate the eluate to minimum volume before proceeding to the bioassays. Some of the specific and commonly used bioassays are described.

Rice Seedling Test

Stimulation of plant growth is a prominent effect of gibberellins and is used to develop bioassays. A simple bioassay based on this principle using rice seedlings is easy to perform.

Materials

Gibberellin extract

Standard GA_3 solutions of 0.01, 0.1, 1 and 10 ppm

Rice seeds

4-cm Petri dishes

Pipettes, 2 ml, 5 ml

Enamel tray

Ruler

Reagent

Hoagland's nutrient solution

Method

Soak the seeds for 2-3 days in water at room temperatures. Spread sterile cotton for 2-3 mm thickness at the bottom of 4 cm Petri dishes and add 3 ml of gibberellin extract to it. Air dry the cotton and place 12 uniformly germinated rice seeds with 0.5 mm long coleoptiles in each petri dish. Add 5 ml of dist water or half strength Hoagland's nutrient solution and cover them with their lids. Arrange the Petri dishes on a tray and place them in light for 3 days. After 2 days, remove the lids of the Petri dishes and add 2-3 ml of dist water. Place them in a moist chamber under light. After 5-7 days, reject the longest and shortest plants and measure the shoot lengths of the other ten plants from the mesocotyl to the second leaf sheath. From a dosage response curve calculate the GA content. A linear relation exists between log conc of gibberellic acid and tissue response.

Barley Endosperm Test

Gibberellins trigger the release of a-amylase from the aluerone layer of embryoless seed pieces of cereals such as barley, wheat or rice into the starchy endosperm where starch is hydrolyzed to reducing sugars. Either the enzyme induced the production of reducing sugars as a result of the enzyme action may be determined and correlated with gibberellin level. The method of Nicholls and Paleg (1963) which measures the amount of reducing sugars formed from barley endosperms is described. This assay is sensitive between 10^{-5} and 10^{-2} ug GA_3/ml.

Materials

Gibberellin extract (free from solvent)

GA_3 standard solutions

Barley seeds

Glass stoppered vials, 5 ml

Pipettes, 0.1 ml, 1 ml

Incubator at 30°C

Razor

Reagents

Sodium hypochlorite, 1%

Streptomycin solution (10 ug/ml)

Method

Sterilize the seeds by soaking in hypochlorite solution for 20 min and rinse in several changes of sterile dist water. Cut the seeds transversely 3 mm with a razor from the distal end and discard the embryo. Pipette 0.05 ml of streptomycin and either 0.5 ml of standard GA_3 solution or gibberellin extract into glass vials. Add 0.45 ml of dist water to make up the volume to 1 ml. Weight the embryoless seed pieces in groups of four and place them in the vials containing the test solution. Stopper the vials and incubate them at 30°C in an incubator for 48 h. Draw out the samples of the ambient medium and estimate the reducing sugars released by α-amylase from starch. Maintain control with dist water instead of GA extract. Calculate the amount of reducing sugars released per gram dry wt of barley endosperm and find out the GA content in the extract from a dosage response curve.

Comment

Traces of solvent or residues, helminthosporol and heliminthorsporic acid interfere with the stimulation of α-amylase.

Release of a-Amylase from Seeds

The barley endosperm assay has been modified by Jones and Varner (1967) using barley seeds to avoid interference by the endogenous reducing sugars present in the extracts and the residues of some of the organic solvents used. In this test, the enzyme activity is determined and it is insensitive to solvent residues.

Materials

Barley seeds, any variety

Gibberellin extract

GA_3 standards

Pipettes, 1 ml, 5 ml

Seitz filter/Millipore filter

Glass mortar and pestle

Whatman No. 1 filter paper

Incubator at 30°C

Water bath at 40°C

Colorimeter

Reagents

Sodium hypochlorite solution

Calcium chloride, 0.2%

Sodium acetate buffer, pH 4.8 containing 0.255 (w/v)

Starch (potato starch)

Acetic acid 0.5 M

Iodine solution 0.03 N

Method

Cut the dehulled barley seeds transversely with a razor. Sterile the apical halves without embryo with sodium hypochlorite solution for 10-20 min and rinse in several changes of sterile dist water. Weigh them in groups of five. Sterilize the test solution (gibberellin extract/ GA standard) by filtering through Seitz filter or by steaming for 15 min twice at 24 h interval in an autoclave. Pipette 1 ml of the test solution into test tubes and transfer the weighed embryoless seed halves under aseptic condition to the test solution. Incubate the seeds for 4 days at 30°C. Discard the test solution and rinse them with 5 ml of 0.25 calcium chloride in a glass mortar. Filter the homogenate through whatman No. 41 filter paper (alternatively, it can be centrifuged at 2,000 g for 10-15 min). Pipette 1 ml of the filtrate to test tubes containing 4 ml of sodium acetate buffer, pH 4.8 containing 0.25% starch and incubate the reaction mixture at 40°C in a constant temperature bath for 30 min. After incubation, pipette 0.5 ml of the reaction mixture into another test tube containing 5 ml of 0.5 M acetic acid to stop the reaction. Add 5 ml of 0.03 N iodine and measure the absorbance at 700 nm in a colorimeter. Maintain zero hour control. Calculate α-amylase activity using the formula :

$$\frac{C-E}{C} \times 100(\%)$$

where, C is the critical density of the initial starch solution used and E is the optical density of starch solution at the end of the reaction.

Express the enzyme activity per unit wt of seeds. Use a dosage response curve prepared with GA_3 to calculate GA in the test solution.

Comment

In double logarithmic plot, the increase in enzyme activity is proportional to GA_3 concentration. This method is sensitive in the range from 10-5 to 10-2 μg GA_3/ml. The hydrolysis of starch in this test is solely due to α-amylase, since the activity is stable on heat treatment in presence of calcium chloride.

CYTOKININS

Substances that induce cell division in plant tissues in presence of auxins are called cytokinins. The group includes both naturally occurring and synthetic compounds. Naturally occurring cytokinins represent another important group of plant hormones associated with growth and development of a plant. During recent years, plant scientists have investigated cytokinins in relation to symptomatology and encouraging results have been obtained particularly with rusts, fasciation disease and crown gall development in diseased plants

(Mahadevan, 1984). Although both pathogenic and non-pathogenic microorganisms produce cytokinins in culture, the enhanced cytokinin levels observed in diseased plant tissues might be primarily due to the deranged host metabolism. Cytokinins counter the apical dominance induced by auxins. They in conjunction with ethylene promote abscission of leaves, flower parts and fruits. They also help delay senescence of the aging of tissues and affect internodal length and leaf growth.

Extraction of Cytokinins from Culture Filtrates of Microorganisms

Microorganisms produce cytokinins in culture media, Besides, cytokinin has been extracted from both bacteria and fungi.

Materials

H. turcicum grown on Richard's medium for 2 weeks.

Buchner funnel

Whatman No. 1 filter paper

Pipettes, 5 ml, 10 ml

Separatory funnel

Rotary evaporator

Dowex 50 column

pH meter

Reagents

HCl, 0.1 N

NH_4OH, 0.1 N

Diethyl ether

n-Butanol

Method

Filter the fungal growth using a Buchner funnel and Whatman No. 1 filter paper, Discard the mycelium. Acidify the filtrate to pH 3 with 0.1 N HCl. Extract it with diethyl ether thice using separatory funnel to remove the auxins. Adjust the pH of the aqueous phase to 7.8 with NH_4OH and extract thrice with equal volumes of n-butanol. Combine the fractions and evaporate under reduced pressure at 40°C using a rotary evporator. Dissolve the residue in 10 ml of 0.1 N HCl and pass through a column of Dowex 50 Elute the cytokinins with 200 ml of 0.1 N NH_4OH. Evaporate the eluate to dryness, dissolve the residue in minimum quantity of dist water and use for bioassay.

Extraction of Cytokinins from Plant Tissues

Extraction of cytokinins from the plant tissues poses certain problems due to the presence of interfering substances which are difficult to separate and due to losses occurring during purification. The method used by Sridhar *et al.* (1978) is described. Further purification of the extracts by column chromatography and paper or thin-layer chromatography before bioassaying the extracts may increase the reproductibility of results (Sridhar *et al.,* 1979).

Materials

Rice seedlings, 15-20 day old.

Blender

Separatory funnel

Centrifuge

Rotary evaporator

pH meter

Reagents

HCl, 1 N

NaOH, 1 N

Mcllvaine's buffer, pH 5.6

Ethanol

Ether

n-Butanol

Method

Homogenize 25 g fresh tissues in 25 ml of buffer, in a blender. Add 150 ml of ethanol and after 6-8 h, centrifuge at 2,000 g for 20 min. Decant the supernatan and evaporate it to 10 ml under reduced pressure on a rotary evaporator at 40°C. Adjust the pH of the solution to 2.9 with 1 N HCl. Shake the extract in a separatory funnel thrice with 10 ml of ethyl ether to remove chlorophylls, auxins and gibberellins. Discard the ether layer. Adjust the pH of the aqueous solution to 7.8 with 1 N NaOH. Extract with 5 ml portions of n-butanol six times in a separatory funnel. Pool the butanol layers and evaporate to dryness under reduced pressure at 40°C. Dissolve the residue in 2 ml of buffer and use it for chromatography.

Chromatographic Separation of Extract

Materials

Cytokinin extract

Paper or TLC kit

UV light (254 nm)

Reagents

Solvents for paper chromatography

Iso-propanol-25% - ammonium hydroxide - water (10 : 1 : 1, v/v)

Water saturated sec-butanol

n-Butanol - conc ammonium hydroxide - water (3 : 1 : 1, v/v)

Ethyl acetate-formic acid - water (60 : 5 : 35, v/v, upper phase)

Solvent for TLC

n-Butanol-1 - N ammonium hydroxide - water (7:1:2, v/v, upper phase)

Method

Separate the cytokinins extract using either paper chromatography or TLC using any of the solvents. View the dried chromatograms under UV light (254 nm) and mark the fluorescing spots. Divide the chromatograms into 10 equal strips. Use the strips for bioassay directly or elute in small volume of n-butanol. Evaporate the eluate and suspend the residue in either dist water or in the buffer and use it for bioassays.

Estimation of Cytokinins

The quantity of cytokinin present in the extract is measured by using bioassays. Gas chromatography may be employed only with highly purified extracts. Cytokinins may be assayed using leaf senecence, chlorophyll production, cotyledon enlargement or callus growth. Callus bioassay is the most specific and dependable, but takes more time and needs a well equipped tissue culture laboratory. The commonly used bioassay are described (Mandahar and Garg, 1976).

Chlorophyll Retention Test

The test is based on the observation that cytokinins delay senescence of leaves. Since the introduction of chlorophyll retention test, screening an extract for cytokinin activity is relatively simplified. This test is more reliable than root inhibition. The method is sensitive between 0.1 and 10 mg Kinetin/l.

Materials

Cytokinin extract

Kinetin standards (0.1 to 10 mg/l)

Rice seedlings 4-cm Petri dishes

Pipettes, 1 ml, 5 ml

Test tubes

Whatman No. 41 filter paper

Razor or cork borer

Enamel tray

Hot water bath

Aluminium foil

Colorimeter

Reagent

Ethanol, 80%

Method

Collect the first or second leaves from the plants and float them in Petri dishes containing dist water in dimlight for 24 h. Select leaves with uniform light green colour and cut out 3 cm apical sections of the leaves or prepare 1 cm leaf discs with cork borer. Float them in groups of five with their adaxial side up (the lower side of the blade containing the solution or the filter paper) on 5 ml of test solutions kept in 4 cm Petri dishes or place them on filter paper kept in the bottom of Petri dishes which were previously moistened with 0.5 ml of either the test solution of water (control). Arrange the dishes in trays lined with moist filter paper. Cover them with aluminium foil and Acubate for 48 h at room temperature. Pool the leaf bits from each dish and transfer them to test tubes containing 5 to 6 ml of 80% ethanol. Extract chlorophyll from the tissues by boiling in a hot water bath. Cool the tubes in cold water, filter and dilute the content to 10 ml with 80° ethanol. Measure the absorbance of the solution at 665 nm (chlorophyll a). Plot the absorbance of the standards against the log conc. of Kinetin and calculate the amount of cytokinins present in the extract with reference to Kinetin from the standard curve.

Radish Cotyledon Test

Cytokinins induce the growth of radish cotyledons which is primarly due to cell expansion rather than cell division (Letham, 1971). This test is sensitive to Kinetin (10 μg to 25 mg/l) and is insensitive to IAA. Although GA shows weak activity in this test, it does not interfere with the response to cytokinin.

Materials

Cytokinin extract

Kinetin standards (10 μg to 25 mg/l)

Radish seedlings (2-3 day old) grown in dark

10-cm Petri dishes

Filter paper

Pipettes, 10 ml

Balance

Method

Select uniform sized cotyledons. Float the preweighed cotyledons in groups of 5 in 7 ml of either water (control) or test solutions in 10 cm Petri dishes or place them on filter paper moistened with test solution. Incubate the leaves under fluorescent light for 3 days. After incubation, remove the leaves, blot them dry and determine their fresh weight or measure the leaf area. Correspondingly, run a standard with known concentrations of Kinetin and draw a dosage response curve. Calculate the conc of cytokinins present in the extract from a standard curve.

ABSCISIC ACID

Abscisic acid (ABA), a sesquiterpenoid, is considered as the only naturally occurring grown inhibitor. It was referred to as *dormin* and abscicin II. ABA is present in almost all plants. It has pronounced effects on dormancy, leaf abscission, leaf senecescence, stomatal opening, transpiration, flowering and synthesis of certain enzymes at the cellular level and antagonises GA action. ABA has been implicated in disease syndrome involving stunting, wilting and defoliation of diseased plants. Endogenous ABA increased in diseased plants and palnts under water and salt stress (Mohanty *et al.*, 1979). In plants under water stress, ABA plays a role in closing the stomata by modulating the potassium and sodium uptake within the guard cells, which lose turgidity, closing the stomata. The effect of ABA on plant parasites has not been critically investigated.

Extraction of ABA from Fungal Culture

A few fungi produce ABA in culture media (Mahadevan, 1984). Production of ABA by *Diplocarpon rosse* has been claimed (Wani *et al.*, 1980).

Materials

D. oryzae grown on PDA

Erlenmeyer flasks, 250 ml

Separatory funnel

Blender

Cheese cloth

Refrigerated centrifuge

Water bath at 35°C

pH Meter

Reagents

Dialyzed potato-dextrose-broth

1 N HCl

Aqueous $NaHCO_3$, 5%

Anhydrous sodium sulphate

Petroleum ether

Method

Grow the fungus in 100 ml of dialyzed potato-dextrose broth dispensed in 250 ml Erlenmeyer flasks under stationary condition for 5 to 10 days at 25°C. Homogenize the culture fluid and the mycelium in a blender and filter the homogenate through two layers of cheese cloth. Centrifuge the filtrate at 12,000 g for 20 min in cold. Concentrate the supernatant by flash evaporation at 35°C to one fourth of its original volume. Adjust the pH of the concentrated fluid to 2.8 and extract thrice with petroleum ether. Pool the ether fractions and extract with small volumes of sodium bicarbonate. Adjust the pH of the aquesou fraction to 2.8 and extract it thrice with petroleum ether. Shake the petroleum ether with anhydrous sodium sulphate and evaporate the extract with compressed air. Dissolve the residue in petroleum ether and use for chromatorgraphy and bioassays.

Comment

The extract may be separated using TLC in (benzene ethyl acetate-acetic acid (50 : 5 : 2, v/v) solvent system and ABA, located by its fluorescence under short wave UV light

Extraction of ABA from Plant Tissues

ABA accumulates in plant tissues in response to environmental stresses and virus and bacterial infections. The method employed by Mohanty *et al.* (1979) to extract ABA from plant tissues is described.

Materials

Rice leaves infected by tungro virus/plant leaves of stressed plants

Blender

Buchner funnel

Separatory funnel

Glass vials, 5 ml

Whatman No. 41 filter paper

Rotary evaporator

pH Meter

Reagents

Methanol, 70%

1 N $NaHCO_3$

1 N HCl

Method

Homogenize fresh leaves in methanol using a blender. FIlter the homogenate through Buchner funnel using Whatman No. 41 filter paper and wash the residue with excess methanol Pool the filtrate and washings and reduce its volume to 15% of the original volume at 40°C. Adjust the pH of the concentrated extract to 8.5 with 1 N $NaHCO_3$. Partition the alkaline extract with petroleum ether thrice and discard the petroleum ether fraction. Adjust the pH of the aqueous fraction to 2.7 with 1 N HCl. Partition the acidic extract twice with ethyl acetate. Pool the ethyl acetate fractions and evaporate to dryness at 40°C in a rotary evaporator. Dissolve the residue in minimum volume of methanol and use if for chromatographic separation of ABA and for subsequent bioassays.

Comment

ABA present in the extract may be separated using TLC in isopropanol-ammonia-water (10 : 1 : 1, v/v) and ABA located by viewing the chromatograms under short wave UV light.

Chromatographic Separation of ABA

Materials

Extract containing ABA

Paper or TLC kit

UV light (short wave)

Reagents

Solvents for paper chromatography

Isopropanol-ammonia-water (80 : 0 : 1 : 19, 9, v/v)

Isopropanol-ammonia-water (90 : 1 : 9, v/v)

Isopropanol-ammonia-water (10 : 1 : 1, v/v)

Solvents for TLC

Toleune - ethyl acetate - acetic acid (15 : 3 : 2, v/v)

n-Propanol-n-butanol-ammonia - water (6 : 2 : 2 : 1, v/v)

Chloroform - benzene-acetic acid - water (100 : 100 : 1 : 5, v/v)

Ethyl acetate - hexane (4 : 1, v/v)

Method

Streak 50 ul of the extract (equivalent to 100 to 200 mg fresh weight of the original sample) onto chromatograph paper or TLC plate (if available, use Whatman No. MM paper or silica gel HF_{254} containin; fluorescent dye). Separate the ABA by viewing the chromatogram under short wave UV light. Do not expose to UV the ABA separated from the extracts on the chromatogram, if it is to be used for bioassay, for it causes photolysis to trans ABA. Elute the area corresponding to authentic ABA in minimum quantity of redistilled ethanol or methanol or buffer and use for bioassays.

Estimation of ABA

Sophisticated methods are available for precise measurement of ABA. The most specific method involves a combination of gas chromatography and mass spectrometry. ABA may be estimated using biological tests. Preferably use as many bioassays as possible to detect ABA-like activity.

Rice Seedling Growth Inhibition Test

The rice seedling growth inhibition is used to measure ABA-activity in extracts from various plants (Mohanty *et al.,* 1979).

Materials

ABA extract

ABA standard solutions

Rice seeds

Petri dishes, 10 cm dia

Pipettes, 1 ml, 10 ml

Cotton wool (absorbent)

Fluorescent tubes (40 watts)

Ruler

Reagent

Hoagland's solution

Method

Soak the surface sterilized seeds in water for 2 or 3 days at room temperature. Select seedlings with uniform sized roots and coleoptiles and transfer 12 of them to 10 cm petri dish containing cotton wool and irrigate with 10 ml of half-strength Hoagland solution and 0.5 ml of ABA extract. Maintain controls with dist water and standards with known concentrations of ABA, incubate the seedlings 15 to 20 cm below a light source consisting of two fluorescent tubes. Provide light for 16 h per day. Add dist water to the coton pad daily to compensate loss of water by evaporation. After 2 to 3 days, remove the Petri dish lids and continue incubation. After 6 days of growth measures the length of the second leaf sheath of 10 plants for each treatment omitting two plants of extreme leaf sheath length. Express the results as length of second leaf sheath in mm and compare with standards.

Inhibition of α-amylase Synthesis in Barley Endosperm

ABA inhibits the synthesis of α-amylase in the aleurone layers which is triggered by gibberellins. A bioassay has been developed to determine ABA activity. Mohanty *et al.* (1979) used this test to detect ABA-like activity in extracts of diseased plants.

Materials

ABA extract

ABA standard solutions

Barley seeds

Glass-lidded vials, 5 ml

Pipettes, 1 ml

Incubator at 28 + 2°C

Razor

Reagents

Sodium hypochloride, 1%

Streptomycin sulphate solution (1 mg/ml)

GA solution, 10^{-3} M

Method

Sterilize the seeds by soaking in hypochlorite solution for 20 min and rinse in several changes of sterile dist water. Cut the seeds transversely with a razor 3 mm from the distal end and discard the piece with embryo. Pipette 1 ml of GA solution, 0.5 ml of streptomycin sulphate, 0.2 ml of ABA extract and 0.8 ml of dist water to make up the volume to 2.5 ml into 5 ml glass lidded vials. Weigh the embryoless barley seed halves in groups of four and transfer them to the vials. Maintain controls with dist water and standards with known conc of ABA. Incubate the vials for 32 h at 28 + 2°C in the dark. Estimate the reducing sugars released into the ambient solution by the action of α-amylase on the endosperm starch, employing either Nelson's or DNS method. Express the results as amount of reducing sugars released per mg dry weight of the endosperm.

Comment

This test may be performed with rice or wheat seeds. Instead of assaying the reducing sugars released, the enzyme activity may be measured.

Chlorophyll Loss Test

Materials

ABA extract

ABA standards

Rice seedlings, 10-day-old

Method

Collect the third leaves from rice seedlings and float them in Petri dishes containing dist water in dim light for 24 h. Select leaves with uniform green colour and cut out 3 cm apical sections. Use the excised leaf pieces to perform the bioassay as described earlier Maintain dist water control and standards with known concentrations of ABA. After incubation in the dark for 24 h. extract and estimate the chlorophyll content of leaf sections. Express the results as absorbance at 665 nm and compare with standards.

Gas-liquid Chromatographic Assay

The method developed by Quarrie (1978) for the assay of ABA using ethyl abscisate as an internal standard is described.

Materials

Fresh rice leaf

Sonicator

TLC set

UV light

Gas-liquid chromatograph equipped with electron capture detector

Nitrogen cylinder

Reagents

Acetone

Ethyl acetate

ABA standard

Nitrosoethylurea

Petroleum ether 40-60°C

KOH 40%

Cyclohexane

Method

Cut leaves (100 to 300 mg) into 2 to 3 mm pieces. Extract them in the dark in a cold room by ultrasonic distintegation in acetone-water (9 : 1, v/v) employing a solvent volume to fresh weight ratio of 10 : 1 for 5 min at setting 3 (50 W). Allow the extract to settle in a refrigerator for 24 to 48 h. Use the supernatant for TLC.

Separate 300 ul of extract on TLC plates (5 x 5 cm) precoated with silica gel GF_{254} using ethyl acetate as the solvent. Similarly, develop a separate plate with the extract containing added ABA and locate the position of ABA band using UV light.

Remove a 4 mm band of silica gel containing ABA from each plate and elute in 2 x 175 μl of water-saturated dist ethyl acetate using suction. Prepare ethyl abscisate from ABA and diasoethane which can be obtained in 40-60°C. Petroleum ether solution from nitrosoethylurea and 40% KOH solution. Add 50 (2.92 ng) or 100 μl (5.84 ng) of 2 x 10^{-7} M of EtABA in cyclohexane to the elute as an internal standard. Esterify the ABA with diazomethane which can be prepared in dist 40-60°C petroleum ether using nitrosomethylurea and 40% KOH solution. Add 50 μl of a weak diasomethane solution pale yellow) to the ABA solution. Leave the mixture in the dark until the reaction is complete (in about 5 min). Remove the solvents by evaporation under a streak of oxygen-free nitrogen in a fume hood. Dissolve the residue in cyclohexane.

Inject 100 μl and 2 μl of this solution into a gas-liquid chromatograph equipped with electron capture detector. Use the ratios of the areas for MeABA and EtABA to calculate the ABA content (ng/g fresh weight of leaves).

ETHYLENE

Ethylene is the simplest growth regulator responsible for the regulation of a large number of responses in plants. It is produced by many diseased and stressed plants and has been implicated in the development of disease syndromes. In a few cases, the pathogens produce ethylene *in vivo*. Therefore, ethylene production in diseased tissue is primarily considered to be a response of the host to the pathogen (Mahadevan, 1984). Ethylene is produced at a faster rate in rapidly growing and dividing cells, especially in darkness. New growth and newly germinated seedlings produce more ethylene than can escape the plant which leads to elevated amounts of ethylene inhibiting leaf expansion (Hyponastic response). Ethylene affects fruit ripening due to *climacteric* event.

Detection of Ethylene Production *in vitro*

Ethylene is produced by both bacteria and fungi. It is essential to know ethylene production by parasites in assessing its participation in disease development. The method developed by Chalutz and DeVay (1969) is easy to use.

Materials

- An isolate of *Penicillium* sp. grown on PDA
- Petri dishes (modified) with a glass tube of 2 cm long welded to a 13 mm opening in its cover
- Rubber stopper
- Whatman No. 1 filter paper
- Hypodermic syringe
- Pipettes, 1 ml
- Incubator at 24-25°C

Medium

PDA

Method

Place a sterile Whatman No. 1 filter paper disc (5 mm dia) in the centre of the modified Petri dish containing 20 ml of PDA medium. Plug the opening of the welded glass tube in the cover with cotton and seal it with a rubber stopper. Inoculate the paper discs with 0.1 ml of spore suspension (*ca* 10^6 endo conidia/ml) of *Penicillium* sp.

Maintain control dishes with uninoculated medium and dishes containing medium inoculated with an autoclaved spore suspension. Incubate the dishes at 24°C. Keep the opening of the glass tube covered with the rubber stopper for a period of 12 h. Withdraw gas samples by means of a hypodermic syringe from the atmosphere above the culture at the end of 12 h and estimate ethylene conc using the bioassay. At the end of the period, keep the rubber stopper opened for about 1 h to equilibrate CO_3 and O_2 conc inside the dishes with that in the atmosphere.

Comment

Alternatively, grow the fungus on agar medium in the modified Petri dish and measure ethylene. Here it is possible to record the growth rate of the fungus by measuring colony diameter.

Detection of Ethylene Production *in vivo*

Chalutz and De Vay (1969) developed a technique to detect ethylene production *in vivo*. This model may be used to study ethylene production during the initial phases of host parasite interaction.

Materials

Uniform sized whole roots of carrot

Penicillium sp. grown on PDA

Cork borer

Pipette, 1 ml

Tygon tubing (5 cm long)

Rubber stopper

Hypodermic syringe

Modified Petri dishes

Reagents

Detergent solution

Alcohol, 70%

Method

Wash the roots with detergent, followed by water and surface sterilize with a alcohol. Make a 7 cm long hole perpendicular to the cut surface with a sterile cork borer. Inoculate the hole in the root with a spore suspension (*ca* 100 spores). Maintain suitable controls. Seal the hole with 5 cm long sterile tygon tubing by inserting it 1 cm into the hose (choose the tube in such a way that the outside dia of the uncompressed tubing is slightly large than the hose made in the roots so that it will be gas tight when fitted in the plant material). Plug the open end with cotton and rubber stopper. Withdraw gas samples at appropriate intervals to detect ethylene.

Comment

Cut uniform carrot discs of 1 cm thickness. Immerse them partially in the spore suspension for about 30 sec. Place them on moist filter paper discs with inoculated surface up, in the modified culture dishes. Maintain suitable controls. Withdraw gas samples at suitable intervals to detect ethylene.

Measurement of Ethylene

Ethylene is measured by gas-liquid chromatograph, colorimetry or by bioassay techniques. The gas-liquid chromatographic technique and biological tests are relatively sensitive and easy to conduct. The colorimetric method is less sensitive.

Gas liquid Chromatographic Estimation

Materials

Healthy and rust infected groundnut leaves

Sealed serum flasks, 40 ml

Gas-liquid chromatograph equipped with activated alumina column and flame - ionization detector

Method

Place leaf discs (25 mm dia) or segments in 40 ml sealed serum flasms (use 14-ml test tubes sealed with a rubber serum cap while using leaf segments of grass leaves) containing 1 ml of water. Incubate them at 20°C in a fluorescent light (10 W/m^2). At predetermined interval, withdraw 1 ml gas samples by means of hypodermic syringe and inject into a gas liquid chromatograph equipped with activated alumina column and a flame-ionization detector. Perform at least three independent incubations and calculate the amount of ethylene produced based on peak area (or peak height). Use reagent grade ethylene standard.

Colorimetric Estimation

Ethylene trapped in aqueous mercuric chloride can be released and oxidized to formaldehyde to measure its conc in a colorimeter (LaRue and Kurz, 1973). The probable first products of oxidation of ethylene are ethylene glycol and some glyoxal. These are oxidized to formaldehyde, formic acid CO_3. The amount of formaldehyde formed is determined with Nash reagent which is based on the Hantzsch reaction between acetyl acetone ammonia and formaldehyde, giving 3, 5-diacetyl-1, 4-dihydrolutidine. The reaction is specific for formaldehyde. Although this method is economical, safe and simple, it is not sensitive and rapid.

Materials

Ethylene sample

Erlenmeyer flask, 10 ml sealed with rubber serum bottle cap

Pipettes, 1 ml, 2 ml

Syringe

Rotary shaker

Colorimeter

Reagents

Sodium metaperiodate 0.05 M : Dissolve 16.5 g of $NaIO_4$ in 1 L of water. Store in dark colour bottole. Reagent keeps for 4 weeks.

Potassium permangnate, 0.005 M : Dissolve 0.79 g of $KMnO_4$, in 1 l water and store in dark in well stoppered bottle. A slight precipitate usually forms after 1 or 2 days. If more precipitate occurs, use fresh reagent.

Sodium arsenite, 4 M: Dissolve 52 g of $NaASO_4$ in 100 ml water

Nash reagent : Dissolve 150 g of ammonium acetate in water and add 3 ml of acetic acid and 2 ml of dist. acetyl acetone (2,5-pentanedione) and dilute the mixture to 1 l with water.

Method

Place 1.5 ml of permanganate solution in 10 ml Erlenmeyer flask and seal it with a rubber serum bottle cap. Transfer 1 to 5 ml of gas sample, having up to 1 u mole of C_2H_4 (from a chamber in which the plants are incubated or from reaction vials used to released trapped ethylene using a syringe). Replace some air in the flask with a syringe prior to injection of the gas sample. Agitate the flask on a rotary shaker at 500 g for 90 min at room temperature. Add 0.25 ml of sodium arsenite and mix well to destroy the excess permanganate. Add 1 ml Nash reagent and after 60 min, measure the absorbance of the solution at 412 nm in a colorimeter. Calculate ethylene conc from a standard curve prepared with known amounts of gas.

Comment

This method is sensitive to measure ethylene concentrations ranging from 0.1 to 1 μ mole per sample. The low levels of ethylene produced by plants cannot be detected by the procedure. The permanganate periodate reagent would yield formaldehyde not only from ethylene but from compounds with a terminal olefin group. Periodate would yield aldehydes from many naturally occurring compounds including carbohydrates. However, such compounds are not expected to appear in the gas phase during the release of trapped ethylene.

REFERENCES

Auxins

Chandramohan, D., Sridhar, R. and Mahadevan, A. 1968. IAA synthesis from tryptamine by *Fusarium oxysporum* f. *vasinfectum*. *Indian Journal of Experimental Biology* 6 : 261-262.

Gordon, S.A. and Paleg, L.G. 1957. Qunatitative measurement of indole acetic acid. *Physiologia Plantarum* 10 : 37-48.

Mahadevan, A. 1984. *Growth Regulators, Microorganisms and Diseased Plants.* Oxford and IBH, New Delhi.

Purushothaman, D., Kandawamy, D. and Ragunathan, V. 1971. Indole acetic acid synthesis and virulence of different races of *Pyricularia oryzae. Current Science* 40 : 113-114.

Sridhar, R. 1967. Production of indole acetic acid by *Pyricularia oryzae. Indian Phytopathology* 20 : 379-380.

Sridhar, R., Chandramohan, D. and Mahadevan, A. 1968. Role of tryptamine in indole acetic acid synthesis by *Fusarium oxysporium* f. *vasinfectum* and *Pyricularia oryzae. Indian Journal of Experimental Biology* 6: 59-60.

Vyas, K.M. and Jain, S.K. 1977. Production of auxins by microorganisms. *Hindustan Antibiotics Bulletin* 16 : 20-21.

Gibberellins

Graham, H.D. and Henderson, J.H.M. 1961. Reaction of gibberellic acid and gibberellins with Follin-Wu phosphomolybdic acid reagent and its use for quantitative assay. *Plant Physiology* 36 : 405-408.

Holbrook, A., Edge, W.J.W. and Bailey, F. 1961. Spectrophotometric method for determination of gibberellic acid, in *Gibberellin*, pp. 159-167; ACS Washington, DC.

Jones, R.L. and Varner, J.E. 1967. The bioassay of gibberellins. *Planta* 72 : 155-161.

Mahadevan, A. 1984. *Growth Regulators, Microorganisms and Diseased Plants.* Oxford & IBH, New Delhi.

Murakami, Y. 1959. A paper chromatographic survey of gibberellins and auxins in immature seeds of leguminous plants, *Botanical Magazine* Tokyo 72 : 36-43.

Nicholls, P.B. and Paleg, L.G. 1963. A barley endosperm bioassay for gibberellins. *Nature* 199 : 823-824.

Cytokinins

Letham, D.S. 1971. Regulator of cell division in plant tissues. XII. A cytokinin bioassay using excised radish cotyledons. *Physiologia Plantarum* 25 : 391-396.

Mandahar, C.L. and Garg, T.D. 1976. Cytokinin activity of powdery mildew infected leaves of *Abelmoschus esculentus. Phytopathologische Zeitschrift. Z.* 85 : 86-89.

Sridhar, R., Mohanty, S.R. and Anjaneyulu, A. 1978. Physiology of rice tungro virus disease : Increased cytokinin activity in tungro infected rice cultivars. *Physiologia Plantarum* 43 : 363-366.

Sridhar, R., Mohanty, S.R. and Anjaneyulu, A. 1979. The role of abscisci acid and cytokinin-like substances in rice tungro virus diseases. In : A. Mahadevan (ed.) *Physiology of Host Pathogen Interaction.* pp. 301-308. Today & Tomorrow, New Delhi.

Abscisic Acid (ABA)

Khangura, R.K. and Sokhi, S.S. 1995. Hormonal make up of stageheads of *Brassica* infected by *Aebugo candida. Indian Phytopathology* 48: 32-34.

Mahadevan, A. 1984. *Growth Regulators, Microorganisms and Diseased Plants.* Oxford & IBH, New Delhi

Mohanty, S.K., Anjaneyued, A. and Sridhar, R. 1979. Physiology of rice tungro disease: Involvement of abscisic acid like substances in susceptible host-virus introductions. *Physiologia Plantarum* 45: 132-136.

Quarrie, S.A. 1978. A rapid and sensitive assay for abscisic acid using ethyl abscisate as an internal standard. *Biochemistry* 87: 143-156.

Wani, S.P., Raj, P.V. and Patil, R.B. 1980. Abscsic acid production by *Diplocarpon rosae* and its role in the disease production. *Current Science* 49: 238-239.

Ethylene

Chalutz, E. and De Vay, J.E. 1969. Production of ethylene *in vitro* and *in vivo* by *Ceratocystis fimbriata* in relation to diseases development, *Phytopathology* 59: 750-755.

LaRue, T.A. and Kurz, W.G.W. 1973. Estimation of nitrogenase using a colorimetric determination for ethylene. *Plant Physiology* 51: 1074-1075.

APPENDIX-1

UNITS AND SYMBOLS

1. SI (International System of Units) base units and symbols)

Quantity	*Unit*	*Symbol*	*Quantity*	*Unit*	*Symbol*
Amount of substance (N)	mole	mol	Luminous intensity	candela	cd
			Mass (m)	kilogram	kg
Electric current (I)	ampere	A	Temperature (T)	kelvin	K
Length (l)	meter	m	Time (t)	second	s

2. Important derived SI units

	Unit	*Symbol*	*Equivalent in SI units*
Electric charge, quantity of electricity	coulomb	C	$1\ C = 1\ A\ s = 1\ J\ V^{-1}$
Electric potential, potential difference	volt	V	$1\ V = 1\ J\ A^{-1}\ s^{-1} = 1\ W\ A^{-1}$
Electric resistance	ohm	Ω	$1\ \Omega = m^2\ kg\ s^{-3}\ A^{-2}$
Energy, work, quantity of heat	joule	J	$1\ J = 1\ W\ s = 1\ kg\ m^2\ s^{-2}$
	electron volt	eV	$1\ eV = 1.602 \times 10^{-19}\ J$
Force	newton	N	$1\ N = 1\ kg\ m\ s^{-2}$
Frequency	hertz	Hz	s^{-1}
Light flow, luminous flux	lumen	lm	1 lm = 1 cd sr (sr = steradian)
Light flux, illuminance	lux	lx	$1\ lx = 1\ lm\ m^{-2}$
Power, radiant flux	watt	W	$1\ W = 1\ kg\ m^2\ s^{-3}$
Pressure, stress	pascal	Pa	$1\ Pa = 1\ N\ m^{-2} = 1\ kg\ m^{-1}\ s^{-2}$
Radioactive disintegrations	becquerel	Bq	$1\ Bq = 1\ s^{-1}$

3. Units for photochemically active radiation

Energy fluence	$J\ m^{-2}$	Light quantity - photometric	lm s

Energy flow	J s^{-1}	Particle (photon) flow	mol s^{-1}
Energy fluence rate	J m^{-2} s^{-1} = W m^{-2}	Particle (photon) fluence	mol m^{-2}
Energy quantity	J	Particle (photon) fluence rate (flux)	mol m^{-2} s^{-1}
Fluence - photometric	lm m^{-2} s		

4. Units for transport phenomena

Flow (I)	mol s^{-1} or m^3 s^{-1}
Flux (J)	mol m^{-2} s^{-1} or m^3 m^{-2} s^{-1}

5. Other processes (for example, chemical reactions, growth) are characterized by :

Catalytic activity (in enzyme reactions)	mol s^{-1} = kat
Intensity	mol s^{-1}, m s^{-1} etc.
Velocity (phenomena of movement)	m s^{-1}

6. Frequently used physiological units (not contained in but accepted under the SI)

Absorbance (A)	log l_o l (l_o - incident quantum flux; l transmitted quantum flux)
Absorptance	(l_o - l) l_o

The expression ``absorption" is commonly used as an inclusive term for absorbance and absorptance.

	Unit	*Symbol*	*Equivalent in SI units*
Concentration (c)			
Molarity	mole per liter of solution	M	mol l^{-1}, mol dm^{-1}
Molality	mole per kg of solvent		mol kg^{-1}
Osmolality	mole of osmotically effective particles per kg of solvent (water)		osmol kg^{-1}
Molecular mass (``molecular weight")	gram per mole		10^{-3} kg mol^{-1}
Particle mass	dalton	Da	
Pressure (P)	bar	bar	
Sedimentation constant	svedberg	S	
Temperature (T)	degree Celsius	°C	
	minute	min	
	hour	h	

	day	d	
Time (t)			
Volume, liquid (V)	liter	L. l	l l = 10^{-3} m^3

7. Conversion factors for units which have been widely used but are no longer allowed under SI

Unit	*Symbol*	*Conversion to SI units*	*Unit*	*Symbol*	*Conversion to SI units*
angstrom	A	1 A = 10^{-10} m = 10^{-1} nm	einstein	E	1 E = 1 mol photons
atmos-phere	at	1 at (760 mm Hg) = 1.013 x 10^5 Pa = 1.013 bar	erg	Erg	1 erg = 10^{-7} J (1 erg cm^{-2} s^{-1} = 10^{-3} W m^{-2})
calorie	cal	1 cal = 4.1868 J	rad	rd	1 rd = 0.01 J kg^{-1}
curie	Ci	1 Ci = 3.77 x 10^{10} Bq	roentgen	R, r	1 R = 2.58 x 10^{-4} C kg^{-1}
			torr	torr	$1 torr = \frac{101125}{760} Pa$

8. Multiples of units (prefixes)

peta-(P)	10^{15}	hecto-(h)	10^2	micro-(m)	10^{-6}
tera-(T)	10^{12}	deka-(da)	10^1	nano-(n)	10^{-9}
giga-(G)	10^9	deci-(d)	10^{-1}	pico-(p)	10^{-12}
mega-(M)	10^6	centi-(c)	10^{-2}	femto-(f)	10^{-15}
kilo-(k)	10^3	milli-(m)	10^{-3}		

9. Some fundamental constants

Avogradro's number	N = 6.022 x 10^{23} mol^{-1}	Gas constant	R = k N = 8.314 J mol^{-1} K^{-1}
Boltzmann's constant	k = R N^{-1} = 1.381 x 10^{-23} J K^{-1}	Gravitational acceleration (sea level, 45° latitude)	g = 9.806 m s^{-2}
Elementary electric charge	e = F N^{-1} = 1.602 x 10^{-19} C (= A s)	Planck's constant	h = 6.626 x 10^{-34} J s
Faraday constant	F = e N = 9.649 x 10^4 C mol^{-1} (= A s mol^{-1} = J V^{-1} mol^{-1})	Velocity of light	e = 2.998 10^8 m s^{-1}

SOME USEFUL PARAMETERS, SYMBOLS AND THEIR UNITS

Symbol	Parameter	Units
A	CO_2 assimilation rate	(μmol m^{-1} s^{-1})
c	Concentration of CO_2 in air	(μmol mol^{-1})
C	Crop growth rate	(g m^2 d^1)
D	Dry weight of dead vegetation	(g m^{-2})
D	Leaf area duration	(d)
d	Loss or disappearance of plant material	(g m^{-2} d^{-1})
E	Unit leaf rate	(g m^{-2} d^{-1})
E	Rate of transpiration or evapotranspiration	(μmol m^{-1} s^{-1})
E	Extinction coefficient	–
ε	Photosynthetic efficiency of solar energy conversion	–
e	Partial pressure of water vapour	(kPa)
F	Leaf area ratio	–
g	Conductance	(s m^{-1} or μmol m^{-2} s^{-1})
g_s	Stomatal conductance	
I	Irradiance, radiant energy flux	(W m^{-2})
I_A	Solar radiation absorbed by the canopy	(W m^{-2})
I_b	Solar radiation at the base of the canopy	(W m^{-2})
I_h	Solar radiation incident in the horizontal	(W m^{-2})
I_p	Irradiance in the range 400-700 nm	(W m^{-2})
L	Leaf area index (LAI)	–
P_R	Gross primary production	(g m^{-2} d^{-1})
P_n	Net primary production	(g m^{-2} d^{-1})
Q	Quantum flux of photons	(μmol m^{-1} s^{-1})
R	Sum of respiration losses	(g m^{-2} d^{-1})

Rϕ	Relative water content	–
R	Relative growth rate	**(g g^{-2} d^{-1})**
s	Photosynthetic surface area of leaf	**(m^2)**
T	Temperature	**(°C or K)**
T_l	Leaf temperature	**(°C)**
T'	Web-bulb temperature	**(°C)**
W	Dry weight of biomass (live vegetation)	**(g m^{-2})**
ψ	Water Potential	**(MPa)**

APPENDIX -2

CONVERSION FACTORS FOR SI AND NON-SI UNITS

To convert Column 1 into Column 2 multiply by:	Column 1 SI Unit	Column 2 non-SI Unit	To convert Column 2, into Column 1, multiply by:
		Length	
0.621	kilometer, km (10^3 m)	mile, mi	1.609
1.094	meter, m	yard, yd	0.914
3.28	meter, m	foot, ft	0.304
1.0	micrometer, μm (10^{-6} m)	micron	1.0
3.94×10^{-2}	micrometer, μm (10^{-3} m)	inch, in	25.4
10	nanometer, nm (10^{-9} m)	Angstrom, A	0.1
		Area	
2.47	hectare, ha	acre	0.405
247	square kilometer, km^2 $(10^3 m)^2$	acre	4.05×10^{-3}
0.386	square kilometer, km^2 $(10^3 m)^2$	square mile, mi^2	2.590
2.47×10^{-4}	square meter, m^2	acre	4.05×10^3
10.76	square meter, m^2	square foot, ft^2	9.29×10^{-2}
1.55×10^{-3}	square millimeter, mm^2 (10^{-3} m^2)	square inch, in^2	645
		Volume	
9.73×10^{-3}	cubic meter, m^3	acre-inch	102.8
35.3	cubic meter, m^3	cubic foot, ft^3	2.83×10^{-2}
6.10×10^4	cubic meter, m^3	cubic inch, in^3	1.64×10^{-5}
3.53×10^{-2}	liter, L (10^{-3} m^3)	cubic foot, ft^3	28.3
0.265	liter, L (10^{-3} m^3)	gallon	3.78
33.78	liter, L (10^{-3} m^3)	ounce (fluid), oz	2.96×10^{-2}
2.11	liter, L (10^{-3} m^3)	pint (fluid), pt	0.473
		Mass	
2.20×10^{-3}	gram, g (10^{-3} kg)	pound, lb	454
3.52×10^{-2}	gram, g (10^{-3} kg)	ounce (avdp), oz	28.4
2.205	kilogram, kg	pound, lb	0.454
0.01	kilogram, kg	quintal (metric), q	100

1.10×10^{-3}	kilogram, kg	ton (2000 lb), ton	907
1.102	megram, Mg (tonne)	ton (U.S.), ton	0.907
1.102	tonne, t ton (U.S.) ton	0.907	
	Yield and Rate		
0.893	Kilogram per hectare, kg ha^{-1}	pound per acre, lb $acre^{-1}$	1.12
0.107	liter per hectare, L ha^{-1}	gallon per acre	9.35
893	tonnes per hectare, t ha^{-1}	pound per acre, lb $acre^{-1}$	1.12×10^{-3}
893	megagram per hectare, Mg ha^{-1}	pound per acre, lb $acre^{-1}$	1.12×10^{-3}
0.446	megagram per hectare, Mg ha^{-1}	ton (2000 lb) per acre, ton $acre^{-1}$	2.24
	Pressure		
9.90	megapasacal, MPa (10^6 Pa)	atmosphere	0.101
10	megapasacal, MPa (10^6 Pa)	bar	0.1
1.00	megagram per cubic meter, Mg m^{-3}	gram per cubic centimeter, g cm^{-3}	1.00
2.09×10^{-2}	Pascal, Pa	pound per square foot, lb ft^{-2}	47.9
1.45×10^{-4}	Pascal, Pa	pound per square inch, lb in^{-2}	6.90×10^3
	Temperature		
1.00 (K-273)	Kelvin, K	Celsius, °C	1.00 (°C + 273)
(9/5°C) + 32	Celsius. °C	Fahrenheit, °F	5/9 (°F-32)
	Electrical Conductivity, Electricity, and Magnesium		
10	siemen per meter, S m^{-1}	millimho per centimeter, mmho cm^{-1}	0.1
10^4	tesla, T	Gauss, G	10^{-4}
	Water Measurement		
9.73×10^{-3}	cubic meter, m^3	acre-inches, acre-in	102.8
9.81×10^{-3}	cubic meter per hour, m^3 h^{-1}	cubic feet per second, ft^3 s^{-1}	101.9
4.40	cubic meter per hour, m^3 h^{-1}	U.S. gallons per minute, gal min^{-1}	0.227
8.11	hectare meter, ha-m	acre-feet, acre-ft	0.123
97.28	hectare meters, ha-m	acre-inches, acre-in	1.03×10^{-2}

8.1×10^{-2}	hectare - centimeters, ha-cm	acre-feet, acre-ft	12.33
	Concentrations		
1	centimole per kilogram, cmol kg^{-1} (ion exchange capacity)	milliequivalents per 100 grams, meq 100 g^{-1}	1
0.1	gram per kilogram, g kg^{-1}	percent, %	10
1	milligram per kilogram, mg kg^{-1}	parts per million, ppm	1
	Plant Nutrient Conversion		
2.29	P	P_2O_5	0.437
1.20	K	K_2O	0.830
1.39	Ca	CaO	0.715
1.66	Mg	MgO	0.602

APPENDIX - 3

SOLUTION CONCENTRATIONS

System Name	Abbreviation	Definition
Molar	M	gram-molecular weight (mole of solute) per liter of solution.
Molal	M	gram-molecular weight (mole of solute) per kilogram of solvent.
Formal	F	gram formula weight of solute per liter of solution
Normal	N	gram-equivalent weight of solute per liter of solution.
Weight per volume, percent	w/v, %	number of grams of solute x 100 per volume of solvent (mL).
Volume percent	Vol % or v/v %	Volume of solute x 100 per volume of solution.
Weight percent	wt % or w/w %	Weight of solute x 100 per weight of solution.
Parts per million	ppm	milligras of solute or milligrams per liter of solution kilogram.
Parts per billon	ppb	micrograms of solute or micrograms per liter (kilogram) of solution.

APPENDIX - 4

SOME USEFUL RELATIONSHIPS

---

1 g = 1000 mg = 1,000,000 μg

1 μg = 0.001 mg = 0.000001 g

1 L = 1000 ml

1 mL = 0.001 l

ppm = μg./ml (solid per liquid)

ppm = mg/l (solid per liquid)

ppm = μg/g (solid per liquid)

ppm = μL/l (liquid per liquid)

ppm x 2 = 1 lbs/A

ppm x 10^{-4} = %

1% = 1 gm/100 ml

1% = 10,000 ppm

APPENDIX - 5

CONCENTRATION NORMALITY AND AMOUNT OF CONCENTRATED ACIDS AND BASES TO MAKE 1-L of 1 N SOLUTION

Acid or Base	Chemical Properties				Solution Needed[1] (ml)
	Specific Gravity	Percent by Weight	Grams per Liter	Approximate Normality (N)	
Acetic acid	1.05	99.0	1042.0	17.45	58
Ammonium hydroxide	0.90	28.3	255.0 (NH_3)	15.0	67
Hydrochloric acid	1.19	38.0	451.6	12.4	81
Hydrofluoric acid	1.16	50.0	577.5	28.8	35
Nitric acid	1.42	72.0	1024.0	16.2	62
Phosphoric acid	1.69	85.0	1436.0	44.0	23
Perchloric acid	1.66	70.0	1165.0	11.6	86
Sodium hydroxide	1.53	50.0	762.7	19.0	53
Sulfuric acid	1.84	96.0	1742.0	35.5	28

[1]To make up 1-L of 1 N.

APPENDIX - 6

SOIL pH LEVELS AND ASSOCIATED CONDITIONS

Soil pH	Indications	Associated Conditions
<5.5	Soil is deficient in Ca and/or Mg, and should be lined	Poor crop growth due to low cation exchange capacity and possible Al^{3+} toxicity. Expect P deficiency.
5.5-6.5	Soil is lime-free, should be closely monitored	Satisfactory for most crops
6.5-7.5	Ideal range for crop.	Soil cation exchange capacity is production near 100% base saturation
7.5-8.4	Free lime ($CaCO_3$) exists in soil	Usually excellent filtration and percolation of water due to high Ca content of clays. Both P and micronutrients are less available.
>8.4	Invariably indicates sodic soil	Poor physical conditions. Infiltration and percolation of soil water is slow. Possible root deterioration and organic matter dissolution.

APPENDIX - 7

SUGGESTED PLANT TISSUE SAMPLING PROCEDURES FOR SELECTED DRYLAND CROPS[1]

Growth Stage	Plant Part to Sample	Plants Sampled
	Wheat and Barley	
Seeding stage (<30 cm tall)	All the aboveground portion	50-100
Before head emergence	Flag leaf	25-50
	Corn	
Seedling stage (<30 cm tall)	Al the aboveground shoot.	20-30
Prior to tasselling	The entire leaf fully developed below the whorl	15
From tasselling to silking	The entire leaf at the ear node (or immediately above or below it)	15-25
	Sorghum	
Prior to or at heading	Second or 3rd leaf from top of plant	15-25
	Soybean or other Beans	
Seedling stage (<30 cm tall)	All the aboveground portion	20-30
Prior to or during	Two or three fully developed	20-30
initial flowering	Leaves at the top of the plant	
	Peanut	
Maximum tillering	Recently matured leaflets	25
	Alfalfa, Clover and other Legumes	
Prior to or at 1/10th bloom stage	Mature leaf blades taken about One-third of the way down the plant	40-50
	Food Legumes including Chickpea and Lentil	
Vegetative growth stage	Whole shoots	40-50
Bloom initiation	Recently matured leaf	50-200

[1]When specific guidelines are unknown, the general rule of the thumb is to sample upper mature leaves at flower initiation.

APPENDIX - 8

GENERALIZED INTERPRETATION OF CEREAL TISSUE ANALYSIS DATA

Nutrient	Nutrient Concentration in Dry Tissue			
	Deficient	Low	Sufficient	High
Nitrogen				
(winter grains)	<1.25	1.25-1.74	1.75-3.00	>3.00
(spring grains)	<1.50	1.50-1.99	2.00-3.00	>3.00
Phosphorus	<0.15	0.15-0.19	0.20-0.50	>0.50
Potassium	0.125	1.25-1.49	1.50-3.00	>3.00
Calcium				
(wheat, oats)		<0.20	0.20-0.50	>0.50
(barley)		<0.30	0.30-1.20	>1.20
Magnesium		<0.15	0.15-0.50	>0.50
Sulfur		<0.15	0.15-0.40	>0.40
		ppm		
Manganese	<5	5-24	25-100	>100
Zinc		<15	15-70	>70
Copper		<5	5-25	>25

APPENDIX - 9

OSMOTIC POTENTIAL

Calculation of approximate osmotic potential (Ψs) from the molal concentration of a solution can be found from Van't Hoff equation :

$$\Psi s = \text{mi Rt}$$

Where,

Ψs = Osmotic potential.

m = molality of the solution (moles of solutes/ 100 g H_2O).

i = a constant that accounts for ionization of the solute and deviations from perfect solutions.

R = gas constant 0.00831 liter MPa $mol^{-1.}$

T = absolute temperature (K) = degree C + 273.

Thus, if m, i and T are known Ψs can be calculated. For non-ionised dilute solutions of glucose, mannitol etc. i = 1.0. For 1 m solution of glucose at 30°C (room temperature). Ψs would be :

$$\psi_s = \left(1.0 \frac{mol}{liter\ H_2O}\right)(1.0)(0.0083)\frac{liter\ MPa}{mol\ K}(303\ K)$$

$$= -2.518 \text{ MPA (at 30°C or 303 K)}$$

APPENDIX - 10

WATER POTENTIAL OF NaCl SOLUTION AT TEMPERATURE BETWEEN 0-40°C (LANG, 1967)

Temperature Molalty	0°C	5°C	10°C	15°C	20°C	25°C	30°C	35°C	40°C
0.05	–2.14	–2.18	–2.22	–2.26	–2.30	–2.34	–2.38	–2.42	–2.45
0.1	–4.23	–4.31	–4.39	–4.47	–4.54	–4.62	–4.70	–4.77	–4.85
0.2	–8.92	–8.52	–8.68	–8.84	–9.00	–9.15	–9.30	–9.46	–9.61
0.3	–12.72	–12.72	–12.97	–13.21	–13.44	–13.68	–13.91	–14.15	–14.37
0.4	–16.58	–16.58	–17.27	–17.59	–17.91	–18.23	–18.55	–18.86	–19.17
0.5	–20.70	–21.15	–21.58	–22.00	–22.41	–22.81	–23.22	–23.62	–24.02
0.6	–24.84	–25.39	–25.93	–25.44	–26.94	–27.44	–27.94	–28.43	–28.91
0.7	–29.01	–29.67	–30.30	–30.91	–31.51	–32.10	–32.70	–33.28	–33.85
0.8	–33.20	–33.98	–34.72	–35.43	–36.12	–36.82	–37.91	–38.18	–38.85
0.9	–37.43	–38.32	–39.17	–39.98	–40.79	–41.58	–42.27	–43.14	–43.90
1.0	–41.69	–42.70	–43.66	–44.59	–45.50	–46.40	–47.19	–48.15	–49.01
1,1	–45.96	–47.13	–48.20	–49.24	–50.26	–51.27	–52.26	–53.22	–58.35
1.2	–50.32	–51.60	–52.78	–53.94	–55.07	–56.20	–57.30	–58.35	–59.40
1.3	–54.70	–56.11	–57.42	–58.69	–59.94	–61.19	–62.39	–63.54	–64.71
1.4	–59.12	–60-68	–62.10	–63.50	–64.87	–66.23	–67.54	–68.80	–70.00
1.5	–63.59	–65.29	–66.84	–68.37	–69.86	–71.34	–72.76	–74.11	–75.48
1.6	–68.11	–69.96	–71.63	–73.30	–74.91	–76.52	–78.05	–79.50	–81.00
1.7	–72.60	–74.60	–76.40	–78.20	–80.00	–81.70	–83.30	–84.90	–86.50
1.8	–77.30	–79.40	–81.30	–88.40	–85.20	–87.00	–88.80	–90.40	–92.10
1.9	–81.90	–84.30	–86.30	–88.40	–90.40	–92.40	–94.30	–96.00	–97.00
2.0	–86.70	–89.20	–91.30	–93.60	–95.70	–97.80	–99.80	–101.60	–103.50

WATER POTENTIAL (BARS)

APPENDIX - 11

PHYSICAL CONSTANTS AND CONVERSION FACTORS

Physical Constants	Conversion Factors
Atomic mass unit	1 amu = 1.6605×10^{-24} g
Avogadro's number	N = 6.0221×10^{23} mol^{-1}
Gas constant	R = 8.2056×10^{-2} L atm K^{-1} mol^{-1}
Ideal gas molar voume	V_m = 22.414 L mol^{-1}
Masses of fundamental particles :	
Electron (e^-)	M_e = 0.0005486 amu = 9.1096×10^{-26} g
Proton (p^-)	M_p = 1.007277 amu = 1.67261×10^{-24} g
Neutron (n)	M_n = 1.008665 amu = 1.67492×10^{-24} g
Speed of light (in vacuum)	C = 2.997925×10^8 ms^{-1}
Length	1 kilometer (km) = 0.621 miles = 10^3 meters (m)
	1 meter (m) = 39.4 inches (in.)
	= 10^2 centimeters (cm)
	= 10^3 millimeters (mm)
	= 10^6 micrometers (μm)
	= 10^9 nanometers (nm)
	1 inch (in.) = 2.54 centimeters (cm)
	1 Angstrom (A) = 10^{-8} centimeters (cm)
Volume	1 cubic meter (m^3) = 10^3 liters (l) = 264.2 gallons (gall)
	1 liter (l) = 10^3 milliliter (ml)
	= 10^3 cubic cenmtimeters (cm^3)
	= 1.06 quarts (qt)
	1-millimeter (ml) = 1 cubic centimeter (cm^3)
	1 gallon (gal) = 4 quarts (qt)
	1 quart (qt) = 2 pints (pt) = 946 millimeters (ml)
Mass	1 kilogram (kg) = 10^3 grams (g) = 2.20 pounds (lb)
	1 pound (lb) = 16 ounces (oz) = 454 grams (g)
	1 ounce (oz) = 28.4 gram (g)
Energy	1 kilocalorie (kcal) = 10^3 calories (cal) = 4.18 kilojoules
	1 calorie (cal) = 4.18 joules (J)

	1 British thermal unit (BTU) = 1055 joules (J)
Pressure	1 atmosphere (atm) = 760 millimeters of mercury (mml) = 760 Torr = 101.3 kilopascals (kl)
Temperature	$^{o}C = \dfrac{(^{o}F - 32)\times 5}{9}$
	$^{o}F = \dfrac{^{o}C \times 9}{5} + 32$
	$K = {^{o}C} + 273$

APPENDIX - 12

PREPARATION OF PLANT GROWTH HORMONE SOLUTION

Various PGR are commercially available in solid form, except ethylene gas, ethrel and cycocel (both liquid). The concentration of PGR can be expressed as parts per million (ppm), molarity (μM or mM) or in terms of net amount applied (n mol/ng or μmol/μg). All these expressions are interconvertible.

ppm or μg/ml

Part per million, mg of PGR present in one litre of solution. Weight of one litre of solution = 1000 g or 1000,000 mg (106 mg). To it if we have added 10 mg of PGR. The solution contains 10 ppm of PGR i.e. 10 mg of PGR in 106 mg of water. Similarly, if 10 mg of nutrients are supplied to one kg of soil. Concentration of the nutrient will be 10 ppm. Such application on land area basis will require 10 kg ha^{-1} of a nutrient (assuming soil weight ha^{-1} is 106 kg). These solutions can easily be diluted by using C1V1 = C2V2.

Where,

C1 and C2 are concentration of stock and diluted solutions, respectively.

V1 is volume of stock solution required to dilute to V2 volume.

Example: There is stock solution of 250 ppm of PGR (C1). You want to prepare 200 ml (V2) of 100 ppm (C2). How much volume (V1) should be used to prepare the dilute solution?

C1 = 250   C2 = 100

V1 =   V2 = 200

$$VI = \frac{C2 \times V2}{C1} = \frac{100 \times 200}{250} = 80\text{ ml}$$

You are required to dilute 80 ml of stock solution to 200 ml with water. If more than one PGR are to be used i.e. 3 compounds. Then prepare stock solution 3 times more concentrated. For individual PGR solution, dilute each stock solution 3 times. However, for interaction purpose mix any two stock solution and increase volume by two-third or mix the same volume of each stock solution. It will be clear from following example:

You have 300 ppm stock solution of IAA, GA3 and BAP. Prepare 100 ppm solution of individual PGR and varius possible interactions.

S.N.	Vol. (ml) of 300 ppm of			Final volume with water	Concentration of diluted sol.
	IAA	GA3	BAP		
	100	—	—	300	100 ppm IAA
	—	100	—	300	100 ppm GA3
	—	—	100	300	100 ppm BAP
	100	100	—	300	100 ppm IAA+ 100 ppm BAP
	100	—	100	300	100 ppm IAA+ 100 ppm BAP
	—	100	100	300	100 ppm GA3+ 100 ppm BAP

Molar solution

For molar solution gram MW of a substance is dissolved and volume is made to one litre. This will be one molar or 1 M. If mg MW of a compounds is used then it will be one m molar or mM or 10^{-3}M. It will have μ mol. ml^{-1} amount. Dilution or interaction solution can be prepared as described for ppm solution.

Net amount

For calculating the net amount of PGR used, we should know the volume and concentration, Net amount = Volume of solution x concentration (ppm or molarity).

Example: You have sprayed 100 ml of 250 ppm or 1 mM PGR solution. Calculate how much amount of PGR have been used.

250 ppm = 250 μg m1-1 or mM = 1 μ mol ml-1

Net amount of PGR - 100 x 250 μg ml-1 = 25000 μg or 2.5 mg or

100 x 1 μmol ml-1 = 100 μ mol

If PGR is not 100 per cent pure and its ingredient is less, say 40 per cent. Then amount required is to be divided by 0.4. It will be the actual amount of PGR to weigh. In case of liquid formulations.

Weight of a litre of PGR solution = 1000 x sp. density = x gl^{-1}

$$\text{Actual amount of PGR present in 1 ml} = \frac{X}{1000} = g\ ml^{-1}$$

Solution is 100 per cent. In case it is less concentrated i.e. 60 per cent. Then actual amount in

$$1\,\text{ml} = \frac{X}{1000} \times \frac{1000}{\text{Per cent}} = \frac{X}{6.0}\,\text{g ml}^{-1}$$

Preparation of stock solution

Stock solution (10-100 time more concentrated) can be easily prepared by weighing the required amount of PGR and dissolve in minimum volume of organic solvent and slowly add distilled water with continuous shaking.

IAA, GAs, ABA can be solubilised in ethanol (cold or slight warming). Cytokinis are to be dissolved either in 0.1 M HCl or NaOH. Before making final volume check pH and adjust, if required. In case of NAA, stock solution gives precipitate when diluted with water. Add alternately small amount of 0.1 M NaOH solution and dilute with water with shaking. Check pH before making the final volume of the solution, adjust it if required.

Preparation of PGR linolin/petroleum jelly paste

When PGR is to be applied locally then solution is prepared in an inert, semisolid material. For this purpose required amount of PGR is weighed and dissolved in known vlume or weight of molten linolin or petroleums jelly (on warming these are liquified), allow it to cool down so that it becomes semisolid. Apply a known weight of the preparation with a toothpick to stem apex or leaf axil or distal and or proximal end of explant.

SUBJECT INDEX